Examples
of the Design of
Reinforced
Concrete
Buildings
to BS8110

BOOKS ON STRUCTURAL CONCRETE FROM E & FN SPON

For details of these and other titles, contact the Promotion Department, E & FN Spon, 2-6 Boundary Row, London SE1 8HN, Tel: 071-865 0066

Examples of the Design of Reinforced Concrete Buildings to BS8110

FOURTH EDITION

Charles E. Reynolds,
BSc (Eng), CEng, FICE

and

James C. Steedman,
BA, CEng, MICE, MIStructE

E & FN SPON
An Imprint of Chapman & Hall
London · Glasgow · Weinheim · New York · Tokyo · Melbourne · Madras

Published by E & FN Spon, an imprint of Chapman & Hall,
2-6 Boundary Row, London SE1 8HN, UK

Chapman & Hall, 2-6 Boundary Row, London SE1 8HN, UK

Blackie Academic & Professional, Wester Cleddens Road, Bishopbriggs, Glasgow G64 2NZ, UK

Chapman & Hall GmbH, Pappelallee 3, 69469 Weinheim, Germany

Chapman & Hall USA., 115 Fifth Avenue, New York, NY 10003, USA

Chapman & Hall Japan, ITP-Japan, Kyowa Building, 3F, 2-2-1 Hirakawacho, Chiyoda-ku, Tokyo 102, Japan

Chapman & Hall Australia, 102 Dodds Street, South Melbourne, Victoria 3205, Australia

Chapman & Hall India, R. Seshadri, 32 Second Main Road, CIT East, Madras 600 035, India

First edition 1952
Fourth edition 1992
Reprinted 1995 (twice)

© 1992 E & FN Spon

Typeset in 10/12½pt Times by EJS Chemical Composition, Midsomer Norton, Bath, Avon
Printed in Great Britain by TJ Press (Padstow) Ltd, Padstow, Cornwall

ISBN   0 419 16990 3 (HB)   0 442 31417 5 (USA/HB)
        0 419 17000 6 (PB)   0 442 31422 9 (USA/PB)

British Library Cataloguing in Publication Data
Reynolds, Charles E.
    Examples of the design of reinforced concrete buildings to
    BS8110.-4th ed.
    I. Title   II. Steedman, James C.
    693
ISBN 0-419-16990-3
ISBN 0-419-17000-6 (pbk)

Library of Congress Cataloging-in-Publication Data available

∞ Printed on acid-free paper, manufactured in accordance with ANSI/NISO Z39.48-1992 and ANSI/NISO Z39.48-1984 (Permanence of Paper)

Contents

Preface

Like its predecessor, this edition of what is often colloquially known as the *Examples* book has been delayed repeatedly during the past few years. As with its sister publication, the *Reinforced Concrete Designer's Handbook*, since it last appeared it has passed through the ownership of two publishers before coming to rest in the safe hands of Spon.

Once again, sincere thanks must go to two groups of people. Firstly, I am grateful to the many editorial and production staff at E. & F. N. Spon, who are equally involved with myself in the production but whose names do not appear in the 'credits'. And secondly, I would like to thank Freda Reynolds and the other members of her family for their continued encouragement and support. All of us hope that this edition will prove as useful to those designers for whom it is intended as did its predecessors.

The present edition follows the same plan as earlier editions. Part One describes the various British Standard and Code of Practice requirements relating to the design of various parts of reinforced concrete buildings. Part Two consists of drawings and calculations for a reasonably typical six-storey framed building. Most of the dimensions of this structure are the same as of that designed to meet the requirements of CP110 in the previous edition, and the building is as close as possible to the previous imperial design. This affords an interesting and sometimes enlightening, although of course seldom typical, comparison between the designs that result from the use of BS8110 and those based on its predecessors, although it is becoming clear that more radical changes in the examples provided will be needed in the future.

Because of the differing nature of the two books, a far larger proportion of the present work has to be rewritten each time a new edition is prepared than is the case with subsequent editions of the *Reinforced Concrete Designer's Handbook*. Nevertheless, I have taken care to retain from the previous editions all Charles Reynolds's ideas and helpful advice where these are still relevant, while updating and supplementing this information where necessary.

Although it would be gratifying to discover otherwise, experience with both this and the *Reinforced Concrete Designer's Handbook* has shown that it is a practical impossibility entirely to eliminate errors in books of this nature, however hard one tries. I would therefore like to take this opportunity to apologize for such mistakes and to thank those readers who take the trouble to write and point them out. It is such people who help to make this book and its companion, the *Handbook*, the useful reference works that they are.

J.C.S.
Upper Beeding

The authors

Charles Edward Reynolds was born in 1900 and educated at Tiffin Boys School, Kingston upon Thames, and Battersea Polytechnic. After some years with Sir William Arroll, BRC and Simon Carves, he joined Leslie Turner and Partners, and later C. W. Glover and Partners. He was for some years Technical Editor of Concrete Publications Ltd and later became its Managing Editor, combining this post with private practice. In addition to the *Reinforced Concrete Designer's Handbook*, of which well over 150 000 copies have been sold since it first appeared in 1932, Charles Reynolds was the author of numerous other books, papers and articles concerning concrete and allied subjects. Among his various appointments, he served on the council of the Junior Institution of Engineers and was the Honorary Editor of its journal at his death on Christmas Day 1971.

The current author of *Examples of the Design of Reinforced Concrete Buildings to BS8110*, James Cyril Steedman, was educated at Varndean Grammar School and was first employed by British Rail, whom he joined in 1950 at the age of 16. In 1956 he commenced working for GKN Reinforcements Ltd and later moved to Malcolm Glover and Partners. His association with Charles Reynolds commenced when, following the publication of numerous articles in the magazine *Concrete and Constructional Engineering*, he took up an appointment as Technical Editor of Concrete Publications Ltd in 1961, a post he held for seven years. Since that time he has been engaged in private practice, combining work for the Publications Division of the Cement and Concrete Association with his own writing and other activities. In 1981 he established Jacys Computing Services, an organization specializing in the development of microcomputer software for reinforced concrete design, and much of his time since then has been devoted to this project. He is also the joint author, with Charles Reynolds, of the *Reinforced Concrete Designer's Handbook*.

vii

Introduction to fourth edition

The data, calculations and designs in this edition conform to the recommendations of the current British Standard codes of practice for reinforced concrete, and in particular to Parts 1 to 3 of BS8110:1985 'Structural use of concrete', BS6399 'Design loading for building. Part 1: Code of practice for dead and imposed loads: 1984' and CP3 'Functional requirements of buildings. Chapter V: Loading. Part 2: Wind loads: 1972'. These codes have been prepared by the British Standards Institution and the writers thank the Institution for permission to refer in detail to their contents.

In Part One the recommendations of these codes and other supplementary data relating to the design of buildings are considered in the same sequence as in design calculations: namely, loads, bending moments, design strengths, resistance to ultimate limit-state, resistance to shearing and torsional forces, bond and serviceability limit-states. Structural parts are then considered in the order in which design proceeds: namely, slabs, beams, columns and load-bearing walls, stairs, basements and foundations. The overall stability of the structure is then considered. The final chapter in this part is devoted to a consideration of the fire resistance of reinforced concrete in general and in relation to the building in Part Two in particular.

The application of the recommendations of the codes is illustrated by designing the principal parts of a reinforced concrete building, many of the calculations and drawings for which are given in Part Two. Although the codes are here applied to a specific structure, the data and comments in Part One and the design procedure in Part Two are of general application. The writers' interpretation of the intention of the recommendations is given where there may be ambiguity; no doubt usage will in time eliminate uncertainties, and precedents will be established for matters now left to the discretion of the designer.

The building in Part Two has been planned to incorporate as many as possible of the matters dealt with by the codes. Although the structural design complies with the codes, the general planning of the building may not necessarily comply with the bylaws of all local authorities or with other mandatory regulations. Alternative designs are given for some structural parts of buildings: namely, floors of slab-and-beam construction, flat slabs, hollow-tile slabs and precast slabs, columns with and without taking into account the effects of wind, and so on. Alternative designs of column bases are also indicated, as well as some designs for other simple types of foundations not necessarily related to the building in Part Two.

Note that current UK practice does not favour the use of inclined bars to resist shear or to provide top steel over supports, bars with hooked or bobbed ends, and so on. However, the use of these techniques is still discussed in detail and illustrated in the example forming Part Two of this book, and elsewhere. This is done because the adoption of such procedures still remains valid in situations where different economic circumstances prevail.

Readers who knew the original editions of this book will observe that this edition includes many more charts, graphs and similar design aids than its predecessors. Although many of the aids contained in the current edition of the *Reinforced Concrete Designer's Handbook* were specially devised to facilitate rapid design according to BS8110, for various reasons it was not possible to include all the design aids that were desirable. Therefore, in preparing the present book, the opportunity has been taken to incorporate as many of these as possible. For example, more than twenty charts are provided to simplify the determination of deflections by the rigorous analytical procedure set out in Part 2 of BS8110.

It should be emphasized that the information and material provided on the following pages are intended to supplement and not to supplant that given in companion publications dealing with the same subject. In addition to the *Reinforced Concrete Designer's Handbook* by C. E. Reynolds and J. C. Steedman, these publications include the *Handbook to British Standard BS8110:1985* by R. E. Rowe *et al.*, and *Reinforced Concrete Design to BS8110 – Simply Explained* by A. H. Allen. For brevity these books, which are mentioned frequently on later pages, are there referred to as *RCDH*, *Code Handbook* and *Allen* respectively.

Where, to solve a particular problem, it has been possible to devise alternative but equally valid graphical aids to those that are given in the *Reinforced Concrete Designer's Handbook*, this has been done. As with methods of structural analysis, often one method will appeal more to one designer and another more to the next. However, in the few cases where one form of chart or method appears distinctly superior to its rivals it is included here, even if this has meant reproducing a limited amount of material in a near-identical form to that in the *Reinforced Concrete Designer's Handbook*. Thus, while access to the above-mentioned publications (particularly the *Reinforced Concrete Designer's Handbook*) is desirable, it is certainly not essential, as the present book is self-contained. Reference to a copy of Part 1 of BS8110 itself is, however, important. (Note that specific table references to the *Reinforced Concrete Designer's Handbook* refer to the tenth edition; these numbers may not necessarily remain unchanged in subsequent editions.)

BS8110 permits two different bases to be employed when designing reinforced concrete sections at the ultimate limit-state. Of these two rigorous methods, design charts corresponding to that requiring the use of a so-called parabolic-rectangular concrete stress-block form Part 3 of BS8110. The other rigorous method involves the employment of a uniform rectangular concrete stress-block, and some design formulae based on this assumption are provided in Part 1 of the Code. The basis of both methods and the derivation of these formulae are discussed in Chapters 5 and 14, and the design of slab, beam and column sections in the calculations in Part II and elsewhere in this book is undertaken using design charts based on the uniform rectangular concrete stress-block. Owing to limitations on space, only a single design chart is provided for the design of beam and slab sections, but it is hoped to publish a comprehensive series of charts separately soon.

In accordance with the recommendations of BS3921 'Clay bricks and blocks', the format (i.e. the so-called standard designated size) of a metric brick (including the joint) is taken as 225 mm by 112.5 mm by 75 mm. Where dimensions are controlled by brick widths, this has unfortunately led to the need to introduce such cumbersome dimensions as 5.3375 m, for example, but it was thought that rounding such values to the nearest millimetre might obscure their derivation.

Notation

The notation employed in this book is based on that used in BS8110. This in turn takes as its basis the internationally agreed procedure for preparing notations produced by the European Concrete Committee (CEB) and the American Concrete Institute, which was approved at the 14th biennial meeting of the CEB in 1971 and was outlined in Appendix F of CP 110. In the following list, terms specifically defined and used in BS8110 are indicated in bold type. Only the principal symbols are listed here; all others are defined where they appear.

A	Area
A_c	Area of concrete
A_s	Area of tension reinforcement
A_s'	Area of compression reinforcement
A_{s1}'	Area of reinforcement near more highly compressed face
A_{s2}	Area of reinforcement near less highly compressed face
A_{sc}	Total area of longitudinal reinforcement (in columns)
A_{sl}	Area of longitudinal reinforcement provided for torsion
$A_{s\,req}$	Area of tension reinforcement required
A_{sv}	Cross-sectional area of two legs of link reinforcement
a	Dimension (as defined); deflection
a_b	Distance between bars
a_{cr}	Distance between point at which crack width is evaluated and face of nearest longitudinal bar
a_u	Deflection of column at ultimate limit-state
a'	Distance between compression face and point at which crack width is evaluated
b	Width of section; dimension (as defined)
b_e	Effective breadth of strip of flat slab transferring moment to column
b_t	Breadth of section at level of tension reinforcement
b_w	Breadth of web or rib of member
C	Torsional constant
C_f	Force coefficient when evaluating wind loading
C_{pe}	External pressure coefficient when evaluating wind loading

C_{pi}	Internal force coefficient when evaluating wind loading
c	Cover to reinforcement; column width
c_{min}	Minimum cover to reinforcement
d	Effective depth to tension reinforcement
d'	Depth to compression reinforcement
d_{min}	Minimum effective depth that may be provided
E_c	Short-term modulus of elasticity of concrete
E_n	Nominal earth load
E_s	Modulus of elasticity of steel
e	Eccentricity
e_a	Additional eccentricity due to deflection of column
e_x	Resultant eccentricity of load at right angles to plane of wall
e_{x1}	Resultant eccentricity calculated at top of wall
e_{x2}	Resultant eccentricity calculated at bottom of wall
F	Total design ultimate load
FEM	Fixed-end moment
F_s	Force due to ultimate load in bar or group of bars
f_{bu}	Anchorage-bond stress due to ultimate load
f_c	Actual compressive stress in concrete (deflection analysis)
f_{ct}	Tensile stress in concrete at centroid of tension steel (deflection analysis)
f_{cu}	Characteristic cube strength of concrete
f_s	Service stress in reinforcement
f_y	Characteristic strength of reinforcement
f_{yd}	Maximum design stress in reinforcement
f_{yd1}	Actual design stress in compression reinforcement
f_{yd2}	Actual design stress in tension reinforcement
f_{yl}	Characteristic strength of longitudinal torsional reinforcement
f_{yv}	Characteristic strength of shearing reinforcement
G	Shear modulus
G_k	Characteristic dead load
g	Distributed dead load
g_k	Characteristic dead load per unit area

h	Overall depth or diameter of section	M_2	Larger initial end moment in column due to ultimate load
h_c	Diameter of column head in flat-slab design	N	Ultimate axial load
h_f	Thickness of flange	N_{bal}	Ultimate axial load giving rise to balanced condition in column
I	Second moment of area		
I_e	Transformed second moment of area of cracked section (in concrete units)	N_{uz}	Ultimate resistance of section to pure axial load
I_g	Transformed second moment of area of uncracked section (in concrete units)	n	Total distributed ultimate load per unit area ($= 1.4g_k + 1.6q_k$)
K	A constant; stiffness of member	n_0	Number of storeys
K_u	Link-reinforcement factor	Q_k	Characteristic imposed load
k	A constant	q	Distributed imposed load
k_1, k_2, k_3	Factors determining shape of parabolic-rectangular concrete stress-block	q_k	Characteristic imposed load per unit area
l	Span	r	Radius; internal radius of bend of bar
l_e	Effective span or effective height of member	$1/r_{cs}$	Curvature due to shrinkage
		$1/r_{ip}$	Instantaneous curvature due to permanent load
l_{ex}	Effective height for bending about major axis	$1/r_{it}$	Instantaneous curvature due to total load
l_{ey}	Effective height for bending about major axis	$1/r_{tp}$	Long-term curvature due to permanent load
l_0	Clear height of column between end restraints	$1/r_{tt}$	Long-term curvature due to total load
l_x	Length of shorter side of rectangular slab	s_b	Spacing of bars
l_y	Length of longer side of rectangular slab	s_v	Spacing of links
l_1	Length of flat-slab panel in direction of span measured between column centres	S_1, S_2, S_3	Non-dimensional factors for evaluating wind loading
l_2	Width of flat-slab panel measured between column centres	T	Torsional moment due to ultimate loads
		u	Effective length of shear perimeter
M	Bending moment due to ultimate loads	u_{crit}	Length of critical perimeter
M_{ds}	Design bending moments in flat slabs	u_s	Effective perimeter of reinforcing bar
M_i	Maximum initial moment in column due to ultimate load	V	Design shearing force due to ultimate loads; basic wind speed
M_{ix}	Initial moment about major axis of slender column due to ultimate load	V_b	Total shearing resistance provided by inclined bars
M_{iy}	Initial moment about minor axis of slender column due to ultimate load	V_s	Characteristic wind speed
M_{sx}, M_{sy}	Bending moments at midspan on strips of unit width and of spans l_x and l_y respectively	v	Shearing stress on section due to ultimate loads
		v_c	Ultimate shearing resistance per unit area provided by concrete alone
M_t	Total moment on column due to ultimate load	v_{max}	Limiting ultimate shearing resistance per unit area when shearing reinforcement is provided
M_{tx}	Total moment about major axis of slender column due to ultimate load		
M_{ty}	Total moment about minor axis of slender column due to ultimate load	v_t	Shearing stress due to torsion
		$v_{t\,min}$	Ultimate torsional resistance per unit area provided by concrete alone
M_u	Design ultimate moment of resistance of section	v_{tu}	Limiting ultimate combined resistance (i.e. shear torsion) per unit area when torsional reinforcement is provided
M_{ux}	Maximum moment capacity of short column under action of ultimate load N and bending about major axis only		
		W_k	Characteristic wind load
M_{uy}	Maximum moment capacity of short column under action of ultimate load N and bending about minor axis only	w_k	Characteristic wind load per unit area
		x	Depth to neutral axis
		x_1	Lesser dimension of a link
M_x, M_y	Moments about major and minor axes of short column due to ultimate load	y_1	Greater dimension of a link
		z	Lever-arm
M_1	Smaller initial end moment in column due to ultimate load	$\alpha, \beta, \zeta, \psi$	Factors or coefficients
		α_e	Modular ratio (for serviceability calculations)
		γ_f	Partial safety factor for loads

Notation

γ_m	Partial safety factor for materials	ϱ	Proportion of tension reinforcement ($= A_s/bd$)
ϵ_{cs}	Free shrinkage strain in concrete		
ε_h	Average surface strain at tension face (crack-width analysis)	ϱ'	Proportion of compression reinforcement ($= A_s'/bd$)
ε_m	Adjusted surface strain (crack-width analysis)	ϱ_1	Proportion of total reinforcement in terms of gross section ($= A_s/bd$ or A_{sc}/bh)
ε_{mh}	Adjusted average surface strain at tension face (crack-width analysis)	ϕ	Bar size; continuity factor for precise moment distribution; creep coefficient
ϵ_1	Strain at level considered (crack-width analysis)	θ	Angle

Part One

Design of Reinforced Concrete Buildings

Chapter 1
Introduction to limit-state theory

More than twenty years have elapsed since the appearance of the preliminary version of CP110, the first Code of Practice for concrete wholly based on limit-state principles, caused shock waves to pass through the structural engineering profession. Yet for the generation of designers who have come into the profession since 1969 it is probably difficult to comprehend what all the fuss was about, since the principle on which CP110 was conceived, the so-called limit-state method of design, is in many respects simply a reworking and extension of principles that had already been embodied in such codes for more than thirty years. As early as 1924 George Manning (ref. 1) had suggested that, because of the discrepancies between the behaviour predicted by elastic analysis and that occurring in practice, the only logical theory to employ for reinforced concrete design was one based on the conditions existing in an actual structure when it had just reached its ultimate load. In 1934 the Code of Practice published by the Department of Scientific and Industrial Research (DSIR) permitted axially loaded columns to be designed by summing the individual resistances of the concrete and the reinforcement (i.e. ignoring the differences between the strains in the two adjacent materials). However, to conform to the basis adopted in the rest of the document, suitable factors of safety were incorporated solely by specifying low permissible material stresses, rather than working on actual ultimate loads and strengths.

A further step in this direction came with the appearance of the 1957 version of CP114 (the predecessor to CP110), where the concept of load-factor design was specifically stated and now extended to beams, slabs and eccentrically loaded columns. However, as before, design by elastic-strain (i.e. modular-ratio) principles was permitted in the same Code. Therefore it was thought necessary, to avoid any possibility of confusion arising due to the use of working (i.e. service) loads and strengths, and ultimate loads and strengths in the same document, to modify the load-factor method in such a way that the calculations were undertaken in terms of working

loads and stresses. Unfortunately such an approach has led to some confusion in the minds of those using the Code as to exactly what their calculations were predicting.

The implementation of the limit-state design method presented in CP110 avoids such confusion. In addition it extends the logic of load-factor design, by permitting the relative uncertainty by which each individual type of load and material strength can be assessed to be considered individually, instead of needing to adopt a single global factor of safety to cater for all the possible uncertainties. As pointed out in the introduction to BS8110, an immediately apparent advantage of such a procedure occurs when a critical situation is brought about by a combination of loads, such that one load is at its maximum while the other is at a minimum. This happens, for example, where vertical load on a frame is combined with lateral wind forces. In such a case the greatest likelihood of overturning is when the least vertical load is combined with the greatest wind force. However, the use of a single global loading factor causes both loads to be increased.

Since CP110 was first published in 1972, there has been a gradual acceptance by the majority of engineers of the principles embodied in this document. When BS8110 appeared in 1985, it contained no basic changes in principle from its predecessor, although many minor modifications were introduced and it was considerably rewritten. BS8110 states that the redrafting and alterations were made in the light of experience of the practical convenience in using CP110, and that they were also undertaken to meet the criticisms of engineers preferring the form of CP114. Although going some way to achieve this aim, the rewording and rearrangement have sometimes introduced confusion as to whether a particular change is merely cosmetic or whether it indicates a definite change of policy.

Fortunately, there are two publications that help to resolve some of these doubts. As with CP110, the authors of the Code have produced a *Handbook to British Standard BS8110:1985*, which explains in detail the basis of many Code requirements; on later pages this is referred to as the *Code Handbook* for brevity. In addition Arthur Allen, who has for many years lectured on Cement and Concrete Association design courses dealing with CP110 and BS8110 and who has had long and detailed discussions with the BS8110 authors, has produced an invaluable book entitled *Reinforced Concrete Design to BS8110 – Simply Explained*, to which the present author is greatly indebted; it is referred to here as *Allen*.

Two other publications should be mentioned. In October 1985 a joint committee formed by the Institutions of Civil and Structural Engineers published the *Manual for the Design of Reinforced Concrete Building Structures*, which deals with those aspects of BS8110 of chief interest to reinforced concrete designers and detailers. The advice contained in this document, which generally but not always corresponds to the Code requirements, is presented concisely in a different form from that in BS8110 (and one clearly favoured by many designers); elsewhere in this book it is referred to as the *Joint Institutions Design Manual*. The *Standard Method of Detailing Structural Concrete* (ref. 2) is the product of another joint committee, this time of the Institution of Structural Engineers and the Concrete Society, and the drawings for the example that forms Part Two of this book have been prepared in accordance with the proposals put forward in this important publication.

In accordance with the current policy of the British Standards Institution, the present document BS8110:1985 is designated as a British Standard whereas its predecessor CP110 was a BS Code of Practice; however, this is not intended to indicate any change of status. The formal subtitles of Parts 1 and 2 of BS8110 are 'Code of practice for design and construction' and Code of practice for special circumstances', respectively.

Although many engineers have accepted the limit-state design philosophy, a vociferous body known as the Campaign for Practical Codes of Practice (CPCP) has fought long and hard to retain a revised version of CP114 as an alternative document based on permissible-stress principles and having similar status to BS8110. This group instigated a referendum of members of the Institution of Structural Engineers in 1987 and successfully obtained support for its proposal that the Institution should produce a draft document for public comment on its value as a code of practice. The final document (*Recommendations for the Permissible-Stress Design of Reinforced Concrete Building Structures*: ref. 3) has now been published.

At the same time, Part 1 of Eurocode 2 (EC2) for concrete structures has been made available in draft form to enable member states of the CEB (European Concrete Committee) to familiarize themselves with its requirements. Like BS8110, this is firmly rooted in limit-state principles. The current aim is to publish the document as an ENV (or Euronom) during 1991. Such a document would then have an equivalent status to a BSI Draft for Development, with validity for three years and the possibility of an extension for a further two. By such

time (the mid 1990s) it will have been reassessed and revised for fully operational use as a possible replacement for BS8110. For further information, see ref. 4.

1.1 LIMIT-STATE DESIGN

The limit-state concept is the rational outcome of a rethinking of the fundamental purpose of structural design, which is to produce structures that are safe, serviceable and economic (ref. 5). When used correctly the method gives a clearer idea than previous design procedures of the actual factors of safety employed, and enables these to be adjusted to cater for the degrees of uncertainty involved in the analysis and the seriousness of any resulting failure, taking account of variations that may occur in the loadings and material strengths, and inadequacies in the analytical methods and qualities of construction. The aim is to produce a structure that will not become unfit for its intended purpose during its planned lifetime.

A structure will become unfit for use, of course, if part or all of it collapses, but it will also become unfit if it deflects too much, if large cracks form, or if vibration is so great that discomfort or alarm is caused to the occupants or the operation of machinery is interfered with. Similar factors that affect a structure are fatigue, lack of durability and so on. Such conditions which cause unfitness are classified as *limit-states*; the limit-state at which collapse occurs is known as the ultimate limit-state. The three principal considerations that together constitute the limit-states of serviceability are the prevention of excessive deflection, the prevention of excessive cracking and the prevention of excessive vibration. Special types of structure may also be subject to additional limit-states.

As well as the foregoing limit-states, other phenomena may require consideration during structural design. For example, for a structure such as a machine foundation that is subjected to cyclic loading, the effects of fatigue on the materials may require consideration. Although the Code prescriptions are designed to meet normal durability requirements, it may also be necessary to take additional measures to combat exceptional exposure conditions such as those encountered when substances that are injurious to concrete are to be stored. In such cases reference should be made to specialist literature. A further consideration is that of fire resistance; details of the Code requirements in this respect are set out in Chapter 19. The robustness and stability of the entire structure must also be considered: see Chapter 18.

1.2 CHARACTERISTIC LOADS AND STRENGTHS

Limit-state design is carried out in terms of characteristic loads and characteristic strengths of materials. In theory, a characteristic load is obtained by adding to the mean load the product of the standard deviation from the mean and a factor K. The value of K is chosen to ensure that the probability of the characteristic load actually being exceeded is remote. In practice, however, it is not yet possible to specify dead and imposed loads in such statistical terms, and therefore BS8110 states that such loads should be taken as the dead load and imposed load, respectively, defined in Part 1 of BS6399. In the case of wind loads, however, the values given in Part 2 of CP3, Chapter V, incorporate a multiplying factor S_3 which takes account of the probability of the basic wind speed being exceeded during the specified life of the structure and ensures that the resulting loads are characteristic values.

The characteristic strength of each principal material is similarly theoretically found by subtracting from the mean strength of the material the product of the standard deviation and a factor K_1. The value of K_1 presently adopted is 1.64, which ensures that not more than one test result in twenty will fall below the characteristic value. For concrete, the method of specification is designed to achieve the correct characteristic strength; for reinforcement, the characteristic strength is taken as the minimum yield-point stress specified in the appropriate British Standard. Further information is given in Chapter 4.

Although the characteristic values adopted for loading take account of anticipated variations, they do not allow for loads that differ significantly from those assumed, for employing inadequate analytical methods or imprecise calculations during design, or for errors made during construction, such as incorrectly positioning bars or making minor errors in the sizes and spacing of members. To cater for these variations from the idealized design model, partial safety factors are employed to give so-called *design loads*. The requisite design load is obtained by multiplying the characteristic load by the appropriate partial safety factor for load γ_f. The actual partial safety factors prescribed in BS8110 depend on the particular limit-state being considered, and so take some account of the seriousness of this limit-state being attained or exceeded.

In a similar manner, since the quality of the materials actually used will probably differ from those tested and their performance may deteriorate during their lifetimes, partial safety factors are also employed to convert the characteristic strengths of the materials to design strengths. Here, of course, the design strength is obtained by dividing the characteristic strength by the appropriate partial safety factor for materials γ_m. Furthermore, since different values of γ_m can be employed for concrete and steel, it may be arranged that the mode of collapse that would theoretically result if the structure were loaded to failure is an acceptable one. This occurs, for example, when a beam fails due to yielding of the tension reinforcement, since such an action is preceded by excessive deflection, giving ample warning of imminent collapse, rather than by sudden explosive crushing of the concrete in compression. Again, BS8110 suggests the adoption of different values of γ_m for different limit-states. It should be noted, however, that many design expressions, such as the formulæ for the design of sections given in clause 3.4.4.4 of Part 1 of BS8110 itself,

incorporate the correct values of γ_m for the appropriate limit-state concerned, so that calculations are carried out using the characteristic strengths for materials f_y and f_{cu}.

The values of γ_f and γ_m specified in the Code and the resulting design loads and strengths are set out in *Data Sheet 1*. At present the overall (i.e. so-called 'global') safety factor can be obtained by simply multiplying γ_f by γ_m. The values adopted for γ_f are so chosen that the resulting global safety factor corresponds to the possibility of failure occurring being acceptably low; however, since different values of γ_m are adopted for steel and for concrete, the actual overall safety factor determines the critical condition. In future it may be considered advisable to introduce additional individual partial safety factors relating to the nature of the structure (i.e. whether imminent failure would be indicated), the behaviour of the material (i.e. whether brittle or ductile) and the economic or social consequences of collapse. In addition, the partial safety factor for concrete could be varied depending on the standard of workmanship adopted when mixing or placing the material. A principal advantage of the limit-state method adopted in BS8110 is that it enables such developments to be introduced in the future without the need for any fundamental changes in the basic design procedure.

The critical ultimate forces and moments required for designing the members forming the structure are calculated as described in Chapter 3. The normal design procedure is then first to design each member to withstand the bending moments and shearing forces at the ultimate limit-state using the formulæ developed in Chapters 5 and 6. When this has been done, each section should be checked to confirm that it also satisfies the serviceability requirements of BS8110, particularly the prevention of excessive cracking and deflection. As explained in detail in Chapter 8, to avoid much unnecessary repetitive calculation, sets of rules have been developed and are presented in the Code. If these rules are complied with, the general serviceability requirements specified in BS8110 are automatically observed. However, the designer has the option of not complying with these simplified rules if he so wishes, provided that he can show by calculation that the general requirements regarding serviceability set out in the Code are still met. As regards deflection, the simplified rules consist of establishing limiting span/effective-depth ratios which are then modified according to the section shape and resistance provided. Cracking is controlled by such rules by limiting the spacing of the tension steel.

It is also necessary to ensure that the reinforcing bars are bonded sufficiently tightly to the surrounding concrete for the applied forces to be transferred between the two materials without slipping occurring. The requirements of the Code in this respect are discussed in Chapter 7. For certain members it may also be necessary to consider the effects of torsion (i.e. the rotation of the member about its longitudinal axis), and the appropriate requirements of BS8110 are dealt with in Chapter 6.

Chapter 2
Loads

As explained in the previous chapter, although ideally the dead and imposed loads employed in limit-state design should be expressed in statistical terms, this is not yet possible. Therefore the characteristic values of dead and imposed loading should at present be taken as the values of dead and imposed load recommended in BS6399, Part 1. In the case of wind loading, the pressures given in Part 2 of CP3, Chapter V, are already expressed as characteristic values.

2.1 IMPOSED LOADS ON BEAMS AND SLABS

Part 1 of BS6399:1984 specifies ten types of occupancy upon which the imposed loadings that must be considered are based. There are three types of residential property, namely self-contained dwellings, buildings such as boarding or guest houses, and hotels or motels. The remaining types are institutional or educational premises, public assembly buildings, offices, retail premises, industrial buildings, warehouses and similar stores and structures supporting vehicles. For each type of occupancy the Standard specifies a load per unit area, together with an alternative minimum concentrated load. The principal loadings recommended are set out in *Data Sheet 2*, where they are presented in a different arrangement to that adopted in the Code. For certain storage areas BS6399 specifies a uniform load per metre of height, on some occasions together with a minimum total uniform load per unit area; the most important of these cases are listed near the foot of *Data Sheet 2*. The alternative imposed concentrated load is assumed to be applied over an area 300 mm by 300 mm. This concentrated load need not be considered where the floor slab is capable of effectively distributing the load laterally, as would be the case with a solid reinforced concrete floor. Where no concentrated load is specified (marked 'nil' in the second column on the data sheet), the uniform load is considered adequate for design purposes.

If one span of a beam of a floor that is not used for storage purposes supports at least 40 m² of floor at one level, the intensity of the imposed load may be reduced by 5% for each 40 m² of floor supported. The greatest reduction permitted is 25%; i.e. no further reduction is permitted if the area of the floor supported by one span exceeds 240 m². Thus if q_k is the imposed load per unit area acting on an area A, the reduced total imposed load Q_k that must be considered is given by the expression $Q_k = (1.05 - A/800)q_k A$, where A must be between 40 and 240. If A exceeds 240 m², the corresponding expression becomes $0.75q_k A$. These reduced loads should only be adopted if the designer is assured that it is unlikely that the entire floor will be fully loaded.

Beams spaced at not more than 900 mm apart may be designed as floor slabs.

The intensities of the ordinary imposed loads are minimum values and should be increased if the specified load seems likely to be exceeded. The normal imposed loads include the ordinary effects of impact and acceleration, but not extraordinary loads.

The weight of heavy equipment should be allowed for in the design of floors carrying machinery and the like. Heavy computing or data-processing equipment, for example, should be considered independently of the recommended imposed load, and provision should be made in the design for moving the equipment into position. Large safes are similar items requiring special consideration. In the case of moving loads, the effects of vibration, impact, acceleration and deceleration must be taken into account. In cases such as the supports of lifts, cranes and similar items, the static loads should be increased by the amounts specified in BS6399 and BS2655, or by the makers, to allow for these effects.

2.2 IMPOSED LOADS ON GARAGE FLOORS

The floors of garages are classified in Part 1 of BS6399 as those for parking passenger vehicles and light cars not exceeding 2500 kg in gross weight, and those for parking heavier vehicles and which act as repair workshops for vehicles of all kinds. The weights of large commercial vehicles are such that suspended floors for garages catering for them would have to be designed to withstand loads comparable with those specified for highway bridges in order to allow for the possibility of the garage being occupied by loaded vehicles. It is therefore generally advisable for the floors of garages for such heavy vehicles to be laid directly on the ground where possible.

For vehicles not exceeding 2500 kg in weight, the uniform loading specified in BS6399 is 2.5 kN/m^2. The alternative concentrated load prescribed is 9 kN on a 300 mm square, but since the reinforced concrete garage-floor slab would be capable of effectively distributing the load laterally, this need not normally be considered. These same loadings also apply to ramps and driveways leading to such garages.

For heavier vehicles it may be necessary to determine the actual maximum loading that may be applied due to the wheel loads, but in no circumstances may a load of less than 5 kN/m^2 be considered. The alternative concentrated load specified is again that resulting from the worst possible combination of wheel loads. The gross weight of most cars does not exceed 2500 kg, but the fully laden weight of a few of the largest private cars may be slightly greater. Thus if a garage is to provide parking for all normal types of private and commercial vehicle it must be designed to withstand the greater loading prescribed in BS6399. In a multistorey garage it is theoretically possible to reserve the lower floor or floors for the heaviest vehicles and to use the upper ones to store light cars only, the floors then being designed accordingly. In assessing the greatest wheel load caused by vehicles where the floors are designed to support actual vehicle loads, it is necessary to consider the possibility of vehicles being stored while fully laden. Private cars are unlikely to impose a wheel load exceeding 7.5 kN. The maximum wheel load of a vehicle having a gross weight not exceeding 4 tonnes (which was the upper limit of the class 150 vehicles in a predecessor to BS6399, and is a reasonable value for a 'normal' commercial vehicle) is likely to be about 12.5 kN, and this load is used in the example in section 12.1.1 and is discussed at the end of this chapter.

Although not stated in the Code, a concentrated wheel load may presumably be spread over a certain area to allow for 45° dispersion through the slab. Thus, assuming a contact area of not less than 100 mm by 100 mm say, a concentrated wheel load of F kN would be spread over a width of $2h + 100$ mm, where h is the slab thickness in millimetres. To assist in designing garage floors, *Data Sheet 3* enables the equivalent uniformly distributed load per unit area of slab to be determined, in order to calculate the bending moment due to a concentrated load F/b per unit width on a span l. The curves given are for freely supported slabs, for slabs fixed at both ends and for slabs fixed at one end and freely supported at the other. Intermediate conditions of continuity can be interpolated.

The chart is used as follows. Assume a slab thickness h, and divide F by $2h + 100$ in order to obtain the load per unit width. Also $\alpha = (2h + 100)/l$. Read off the appropriate value (or values) of K corresponding to the value of α and the fixity. Then the equivalent uniform load per unit width is given by $n = KF/bl$ and, by

substituting these values of n in the formulæ given against the curves on *Data Sheet 3*, the required moments can be obtained. For example, with a 150 mm fully fixed slab spanning 2.5 m and supporting a wheel load of 12.5 kN, the equivalent uniform load for the span moment, since $b = 0.4$ m and $\alpha = 0.4/2.5 = 0.16$, is $2.54 \times 12.5/(0.4 \times 2.5) = 31.8$ kN/m^2, and the equivalent uniform load for the support moment is $1.49 \times 12.5/(0.4 \times 2.5) = 18.6$ kN/m^2. The resulting ultimate span and support moments are thus $1.6 \times 31.8 \times 2.5^2/24 = 13.23$ kN m and $1.6 \times 18.6 \times 2.5^2/12 = 15.52$ kN m per metre width respectively.

The curves in *Data Sheet 3* apply to one wheel load in such a position on a span that the maximum bending moment is produced. One wheel load usually applies to small spans, as only on large spans is it possible to arrange the wheels of vehicles so that two or more loads act on one span in such a way that the bending moment exceeds that due to a single wheel. Also for large spans the minimum uniform imposed load of 5 kN/m^2 is often the critical load for a slab. Similarly, a beam is only likely to be subjected to one wheel load, as assumed in *Data Sheet 3*, if the span is small. Beams having larger spans therefore require special consideration, as described in Chapter 12. Concentrated loads on one-way slabs are discussed again in section 9.4 and two-way slabs in section 10.7.

2.3 IMPOSED LOADS ON STAIRS, LANDINGS, CORRIDORS ETC.

2.3.1 Stairs and landings

For stairs and landings in self-contained dwellings, BS6399 specifies a uniform load of 1.5 kN/m^2, with an alternative concentrated load of 1.4 kN. For flats etc. the specified loadings are 3 kN/m^2 and 4.5 kN respectively. For all other types of floor (although BS6399 does not specify values for stairs and landings in storage structures) a uniform load of 4 kN/m^2 or an alternative concentrated load of 4.5 kN should be considered. Note that in certain types of stair design the alternative concentrated load is the critical factor.

2.3.2 Corridors, balconies etc.

The corridor loadings specified in BS6399 for self-contained dwellings and flats are identical to those specified for stairs and landings. For hotels, retail premises and buildings storing vehicles, the prescribed loadings are 4 kN/m^2 (uniform) and 4.5 kN (concentrated). For all other structures, loadings of 5 kN/m^2 or 4.5 kN must be considered. For industrial structures, BS6399 differentiates between corridors which may be used by wheeled vehicles such as trolleys, where a 5 kN/m^2 uniform load or an alternative concentrated load of 4.5 kN should be considered, and other situations, where a 4 kN/m^2 uniform load is applicable, although the alternative concentrated load remains the same.

All balconies should be designed for the same loadings as the floor areas to which they give access, but as an alternative a concentrated load of 1.5 kN per metre run along the edge is specified.

2.3.3 Footpaths, terraces and plazas

Where no positive obstruction to vehicular traffic is provided, footpaths, terraces and plazas leading from ground-floor level should be designed for a uniform load of 5 kN/m^2 or an alternative concentrated load of 9 kN. Where access is definitely restricted to pedestrians only, these loads may be reduced to 4 kN/m^2 (uniform) and 4.5 kN (concentrated).

2.3.4 Balustrades and parapets

The balustrades and parapets of 'light-access' stairs and gangways not more than 600 mm wide should be designed to resist a horizontal load of 220 N per metre run acting at a height of 1100 mm above the foot of the balustrade or parapet, irrespective of the height of the handrail or coping level. For balustrades to other stairs in residential buildings and for balconies, ramps, landings and floors serving only an individual dwelling, the corresponding load should be 360 N per metre, while for structures of all other types except those designed for public assembly this load is increased to 740 N per metre. For stairs etc. provided for individual dwellings, BS6399 states that alternative horizontal loadings of 0.5 kN/m^2 distributed uniformly or a concentrated load of 0.25 kN acting on the infilling must be considered. For other stairs in residential buildings (except light-access stairs etc. as specified above) and stairs, landings etc. in all non-public assembly buildings, the alternative horizontal loadings specified are 1 kN/m^2 and 0.5 kN respectively. For structures designed for public assembly, BS6399 specifies horizontal loads on parapets and balustrades of 3 kN per metre at handrail or coping level, with alternative horizontal loads on the infill of 1.5 kN/m^2 (uniform) and 1.5 kN (concentrated). Where fixed seating is provided in balconies and stands to within 530 mm of the barrier concerned, BS6399 relaxes these values to 1.5 kN/m, 1.5 kN/m^2 and 1.5 kN respectively. Where the parapets are to footways or pavements adjoining access roads or similar areas, values of only 1 kN/m, 1 kN/m^2 and 1 kN need be considered.

2.4 IMPOSED LOADS ON ROOFS AND PARAPETS

Roofs are classified in BS6399 as flat and sloping. The imposed loads discussed in this section are those due to snow, access etc.; wind loads are considered in section 2.6.

For both flat roofs and roofs sloping at less than 45°, in order to cater for loads that may occur during maintenance, all roof coverings (other than glazing) must be able to support a load of 900 kN on a square 125 mm by 125 mm in plan.

2.4.1 Flat roofs

A roof is considered to be flat if the slope does not exceed 10°, i.e. if it is not steeper than about 1 in 5.7. The ordinary imposed load, excluding wind, on slabs and beams to be considered is 1.5 kN/m^2 of plan area supported, if general access to the roof is provided in addition to access for cleaning and repairing; the alternative concentrated load is 1.8 kN. If no general access is provided, the imposed load may be reduced to 0.75 kN/m^2 and the equivalent concentrated load to 0.9 kN.

2.4.2 Sloping roofs

For a roof inclined at more than 10° but not more than 30°, the minimum imposed load excluding wind should be 0.75 kN/m^2 of plan area. If the slope exceeds 75° no imposed load need be considered, but for a slope of between 30° and 75° (i.e. about 1 in 1.73 and 1 in 0.27) the imposed load may be obtained by linear interpolation between 0.75 kN/m^2 and zero.

When designing the slabs and possibly the beams also, it is necessary to know the load acting at right angles to the slope. The effect of the wind is specified in Part 2 of CP 3, Chapter V, in this manner, but the imposed loads are specified in Part 1 of BS6399 per unit area of plan. One of the curves on *Data Sheet 4* gives the imposed load on a unit area of sloping slab. The self-weight of the slab should also be converted into a force per unit area at right angles to the slab; and the dead-load factors, by which the weights of a unit area of slab and finishes should be multiplied to give this force, are also given in *Data Sheet 4*. The total load per unit area of slab acting at right angles to the slab is therefore the converted imposed load, the wind pressure and the converted dead load.

2.5 IMPOSED LOADS ON COLUMNS, PIERS, WALLS AND FOUNDATIONS

The imposed loads on columns and similar members are the same as the ordinary uniform loads for the floors that they support. However, for columns, piers, walls and other similar supporting members, and their foundations, which support several floors that are not garages, warehouses or floors used for storage or filing purposes, the imposed loads may be reduced in accordance with *Data Sheet 4*, the greatest reduction being 50%. The load on a column is generally the product of the area of the floor A supported by the column in question and the total imposed load (q_k per unit area) for the type of floor concerned. The reduction factors K in *Data Sheet 4* enable the total imposed load (in kN) on a column supporting two or more floors, if A is the same for each floor, to be calculated from Kq_kA. A smaller reduction is recommended for floors of factories and workshops designed for imposed loads of 5 kN/m² or more; the recommendation is that the reductions for non-storage floors can be adopted but the reduced load should not be less than that which occurs assuming all floors to be loaded at 5 kN/m²; see *Data Sheet 4*.

Since it is generally easier to calculate the load on a column commencing from the top of a building, the reduction of the load on the floors varies as each storey is considered. To enable the total load to be calculated as on *Calculation Sheets 17*, the difference between successive factors K denotes the equivalent increment of imposed total load at each storey.

The reduced load that may be assumed if a single span of beams supports more than 40 m² of floor (see section 2.1) may be used in the design of columns and similar supporting members in place of the reductions described above if the resulting reduction proves to be greater.

2.6 WIND FORCES ON BUILDINGS

The characteristic wind pressure per unit area w_k depends on various factors including the locality, degree of exposure and height of the structure concerned. In Part 2 of CP 3, Chapter V, two principal methods of determining wind forces are described. The general procedure is first to determine the characteristic wind speed V_s. This is done by multiplying V, the basic wind speed which depends only on the locality and may be read from the map on *Data Sheet 5*, by three non-dimensional factors S_1, S_2 and S_3.

Factor S_1 relates to the topography of the site. In the majority of instances it should be taken as 1.0, although a value of 1.1 is recommended on exposed hills or in narrowing valleys, and the factor may be reduced to a minimum of 0.9 for an enclosed valley. Factor S_3 results from the statistical concept, and depends on the anticipated life of the structure and the probability of V being exceeded during that period. For practical use a value of unity is recommended; this corresponds to a probability of 0.63 that V will be exceeded once in 50 years. Factor S_2 relates the terrain, the plan size (or overall height) of the area considered and the height above ground to the top of the section involved. Appropriate values of S_2 can be read from the curves on *Data Sheet 6*. At heights below the general level of obstructions (i.e. to the left of the chain lines) the figures given should be treated with caution because of the likelihood that higher speeds may result from eddies due to the wind funnelling between the buildings.

When V_s has been determined, it is then converted into the corresponding characteristic wind pressure w_k by employing the expression w_k (in N/m²) $= 0.613V_s^2$, where V_s is in metres per second. This conversion can conveniently be done by reading from the scale on *Data Sheet 6*. The next step is to determine the appropriate external pressure coefficient C_{pe} for a building of the given shape; the required coefficients for rectangular buildings with flat roofs are set out in *Data Sheet 5*.

The total wind force on a given area is then obtained by multiplying together the characteristic wind pressure, the external pressure coefficient and the area concerned. Thus to find the total wind force F acting on part of a rectangular building presenting an area A frontal to the wind, the appropriate expression is $F = w_kA(C_{pe1} - C_{pe2})$, where C_{pe1} and C_{pe2} are the external pressure coefficients for the windward and leeward faces respectively. Normally, however, except when stability is being considered, the designer wishes to know the forces acting on a particular face of a structure in order to design the structural members: then $F = w_kAC_{pe}$. Note from *Data Sheet 5* that the maximum suction to which a surface is subjected occurs when the wind is blowing parallel to the face in question.

In the case of cladding, the total force F on an element of area A is given by $F = w_kA(C_{pe} - C_{pi})$, where C_{pi} is the appropriate internal pressure coefficient. Typical values of C_{pi} are given on *Data Sheet 5*; for additional information on selecting suitable values, reference should be made to the Code itself.

The foregoing method may also be used to obtain the total wind force acting on a building by dividing it into component areas, determining the force on each area in turn, and then vectorially summing the results. An alternative method of obtaining the total wind force (in order, for example, to investigate the stability of the building) is to use the force coefficients C_f also provided in CP 3, Chapter V:Part 2. If the characteristic wind pressure w_k is found as described above, the total force $F = w_kAC_f$, where A is the frontal area presented by the building and C_f is the appropriate force coefficient. Some values of C_f for rectangular buildings are given in *Data Sheet 5*; values of C_f for structures of many other shapes are included in the Code itself, and some are reproduced in *RCDH*. The use of a value of w_k appropriate to the top of the building in this calculation corresponds to the assumption of a constant pressure over the entire height with the total force acting at a centroid of one-half of this height, thus overestimating the total force and also the overturning moment. A more accurate calculation may be made by dividing the height into a series of convenient lengths (usually corresponding to the storey heights) and employing values of S_2 corresponding to the height of the top of each length (see section 14.10).

The frame of the building should be designed to resist the wind pressures that act on the faces of the structure and on the roof. In the previous version of CP 3, Chapter V, it was considered unnecessary to so design such a frame if the building, the height of which did not exceed twice the width, was sufficiently stiffened by walls or by walls and floors. This concession has been eliminated from the current version of CP 3, however. In the building considered in Part Two it is assumed that sufficient longitudinal resistance to wind is provided for the whole building by the partition walls and staircase shafts.

Loads

2.6.1 Roofs

According to Part 2 of CP 3, Chapter V, the wind pressure coefficients for a roof depend on the part of the roof concerned, the angle of the wind and the slope of the roof. The Code gives general coefficients for the windward and leeward areas of roof, together with local coefficients for designing individual cladding components. Examination of these coefficients indicates that the resulting general wind pressures are negative (i.e. suctions occur) with flat roofs and on the leeward sides or halves of pitched roofs, and also on the windward sides or halves of pitched roofs sloping at not more than about 30°. All of the local coefficients given are also negative. As it is most unlikely that, on a roof sloping at less than 30°, no wind pressure ever occurs under any circumstances, it might seem that the present single values should be replaced by alternative maximum values of pressure and suction for roofs of any inclination.

If allowance must be made for wind pressure, the effect is to cause bending moments and shearing forces on the columns, and additional bending moments on the beams in line with the direction of the wind. A simple method of calculating the bending moments and shearing forces at any floor level, that complies with the requirements of BS8110, is to first calculate the total horizontal wind force on one bay above the level of the floor being considered. This gives the total horizontal shearing force on a single row of columns at the level considered, and should be so divided between the columns that each external column resists one-half of the shearing force resisted by an internal column. The bending moment on a column is then one-half of the product of the storey height and this shearing force, and the additional bending moment on the floor beam is the sum of the bending moments on the columns above and below the floor being considered. This method and similar methods of taking lateral forces into account when designing columns are discussed in more detail in section 14.10.

Part 2 of CP 3, Chapter V, gives much information on the determination of wind forces on tall structures of various shapes, including sheeted towers and unclad structures of a similar nature (but not chimneys, for which a BSI Draft for Development is in preparation). However, such structures are outside the scope of this book.

2.7 DEAD LOADS

The primary dead or permanent load of a reinforced concrete structure is the weight of the structural members, of finishes on walls, floors, ceilings, stairs and elsewhere, of brickwork, masonry, steelwork, partitions, fixed tanks and other permanent construction supported by the structural members. BS6399 refers the designer to BS648 for the weights of building materials. This standard gives the weight of plain broken-brick concrete as 19.6 kN/m^3, plain ballast concrete as 22.6 kN/m^3, and

concrete containing about 2% of reinforcement as between 23.1 kN/m^3 and 24.7 kN/m^3. The nominal weight of reinforced concrete generally is given in BS648 as 23.6 kN/m^3; this value is employed in the examples in the following chapters. However, a more convenient and frequently used value is 24 kN/m^3, and this is the value adopted in the calculations in Part Two. For high percentages of reinforcement the weight may increase to as much as 25.6 kN/m^3 according to BS648.

2.7.1 Partitions

According to BS8110 the loads on floors due to partitions should be included in the dead load. Where the weights and positions of the partitions are known, the floor should be designed to support them. If the positions are not known, an addition should be made to the dead load of the floor to allow for the partitions, the addition being, according to BS6399, Part 1, a uniform load on each square metre of floor of one-third of the weight of a 1 m length of finished partition, but in the case of offices, of not less than 1 kN/m^2. This latter condition is normally not restrictive unless only timber or similar light partitions are to be used, as brick or clinker-block partitions demand a greater allowance. For example, for a partition 3 m high and plastered on both faces, an allowance of 2.5 kN/m^2 would be required if the partition were of 115 mm brickwork, and 1.2 kN/m^2 if of 75 mm clinker blocks.

2.8 DESIGN LOADS FOR THE BUILDING IN PART TWO

It is now possible to consider how the foregoing rules apply to the building designed in Part Two (*Drawings 1 and 2*).

2.8.1 Roofs

The slabs and beams of the main roof, to which there is access, should be designed for an imposed load of 1.5 kN/m^2. The small flat roof slabs over the stair and lift wells and the tank room may be designed for 0.75 kN/m^2 as there is no general access, but it is advisable to consider that people may walk on such flat roofs and there is little to be gained by designing them for less than 1.5 kN/m^2.

2.8.2 Upper floors

The offices on the second, third and fourth floors are intended for general use and could be designed for an imposed load of 2.5 kN/m^2 with an additional load of 1 kN/m^2 for lightweight partitions. However, it is not advisable to restrict storage and filing rooms to any particular part of an office floor, and it would therefore be preferable to design these floors for an imposed load of 5 kN/m^2 with an additional load of 1 kN/m^2 for lightweight partitions, as is done on *Calculation Sheets 1*.

Partitions other than lightweight partitions must be over the beams. In the design of the flat-slab (i.e. beamless) floor in Chapter 11 there are no beams, so an extra allowance must be made for heavy partitions, say an additional 1 kN/m^2 as allowed for on *Calculation Sheets 1*. The showrooms on the first floor may be considered as shop floors, and therefore designed for an imposed load of 4 kN/m^2. The residential flats on the top floor may be designed to carry 1.5 kN/m^2. Since it might be required to convert the showroom floor and residential floor into offices in the future it is, however, proposed to design both of these floors for the same load as the office floors. The flat roof or terrace on the same level as the residential floor could be designed to carry 1.5 kN/m^2, but it is not worth while making any reduction for such a small area.

The canopy cantilevering from the first floor should be considered as a flat roof to which there is access and designed for an imposed load of not less than 1.5 kN/m^2. The parapet of the terrace on the level of the top floor should be designed for a horizontal load of 740 N per metre acting at a height of 1.1 m.

Each of the ordinary main and secondary beams supports less than 40 m^2 of floor and therefore no reduction in the imposed load is permissible.

2.8.3 Ground floor

Part of the ground floor is to be used as a garage for vehicles not exceeding 4 tonnes in weight. The greatest load from a rear wheel of such a vehicle may be about 12.5 kN. Since each span of the slab is continuous over both supports, the conditions at midspan are intermediate between a freely supported slab and a slab fixed at both supports. Therefore for a concentrated load of 12.5 kN on a 150 mm slab, the equivalent uniformly distributed load for the bending moment at midspan is, from *Data Sheet 3*, about 27.4 kN/m^2. The conditions at the supports are slightly less rigid than fixity at both supports, and the equivalent uniform load for the bending moment at the support is therefore about 22.0 kN/m^2. These design loads are therefore in excess of the minimum uniform load of 5 kN/m^2 specified for such floors, and the slabs are designed for the greater loads as described in section 12.1.1.

The uniform load equivalent to a wheel load of 12.5 kN on a length of 0.4 m of a fully continuous secondary beam having a span of 6 m is, by a consideration similar to that for slabs, about 4.94 kN per metre for the bending moment at midspan and about 3.71 kN per metre for the bending moment at the support. The minimum imposed load of 5 kN/m^2 results in a load of 2.5 × 5.0 = 12.5 kN per metre of secondary beam, and each secondary beam should be designed for a load of not less than this amount. It is necessary to consider, as is done in section 12.1.2, whether two or more wheels acting on a beam produce an ultimate bending moment greater than does a uniform load of 12.5 kN per metre. The load on the main beams is considered in the same way (section 12.1.3).

The front part of the ground floor is occupied by shops, and the beams and slabs for this area could be designed for an imposed load of 4 kN/m^2 as in the alternative comparative designs in sections 12.2.1 to 12.2.3. It is nevertheless worth while considering designing the entire ground floor for the garage load so that it may be used without restriction in the same way as all the upper floors and would be capable of carrying a greater load than is likely at the time of the first occupancy. Such measures enhance the value of a property without adding much to the initial cost.

2.8.4 Stairs and landings

Since the front stairs above the ground floor serve offices, they should be designed for an imposed load of 4 kN/m^2. The back stairs to the basement should be designed for 5 kN/m^2 as they would probably be used for delivering goods. The design of the front stairs is dealt with in section 16.1.

2.8.5 Columns

The columns are designed to support the same imposed loads as the roof and floors that they carry, but the reductions given on *Data Sheet 4* are applicable because the building is not intended for storage purposes. The application of the loading reductions on *Data Sheet 4* is illustrated on *Calculation Sheets 17 to 20*.

2.8.6 Wind

The height of the building above ground level in *Drawing 1* is about 23 m to the top of the parapet. If the degree of exposure is assumed to be condition 4 in the Code and the building is to be located in Plymouth, taking S_1 and S_3 as unity gives values of $V = 44$ and thus, at the top of the structure, $V_s = 44 × 0.85 = 37.4$ m/s for the cladding and $44 × 0.8 = 35.2$ m/s for the face of the building as a whole, since the total value of C_{pe} for the longitudinal faces of the building (with $h/a = 1.5$ and $b/a = 1.87$) is $0.7 + 0.3 = 1.0$; the resulting values of w_k are then 858 N/m^2 and 760 N/m^2 respectively. If it is assumed for simplicity that these loads apply over the entire building, the total wind force on any storey can then be obtained by multiplying the surface area of the storey concerned by this value of w_k.

Some reduction in these forces can be achieved by determining the appropriate value of S_2, and hence w_k, at each individual floor level (see section 14.10). However, inspection of the calculations involving the use of these wind forces (i.e. those on *Calculation Sheets 20*) shows that in the present example the additional refinement obtained by summing the forces at individual floor levels is not justified, since the wind forces obtained by the approximate method are insufficient to affect the amounts of reinforcement required. Nevertheless, in more-critical cases there are clear advantages in using the more detailed method.

It can be assumed that the floors act as horizontal beams transferring the lateral force from the wind on the longitudinal faces of the building to end shear-walls of reinforced or plain concrete. If instead, the end walls are constructed as reinforced concrete frames with infilling brick panels, the resulting columns and beams must be designed to resist the wind loading. Both of these possibilities are considered in the appropriate sections of the book (see section 14.10). The forces acting longitudinally (i.e. at right angles to the end walls) must be resisted by the walls surrounding the stairs and the partitions. All exterior wall panels, whether of reinforced concrete or some other material, should be designed to resist an inward or outward pressure of about $1.2 \times 0.858 = 1.03$ kN/m^2.

The wind pressure on the parapet is 0.76 kN/m^2. If the parapet is about 1.2 m high, the alternative load of 0.74 kN acting at the top, as previously described, produces the greater bending moment.

The flat roofs are assumed to be subjected to a maximum suction of 0.76 kN/m^2 on the windward half and 0.46 kN/m^2 on the leeward half, since the corresponding coefficients of C_{pe} given in the Code are -1.0 and -0.6 respectively. These are the general wind loads on the roof, but on strips along the edge of the roof that are 0.15 times the width (or length) of the building a suction of 1.52 kN/m^2 should be considered, since $C_{pe} = -2.0$ for local effects in these areas. However, the self-weight of the roof slab acting downwards is far greater than these upward-acting forces, so that they can be ignored when designing the roof slab.

The projections above the general roof level are the lift-motor room and the tank room. The height of the roof of the motor room above the ground is about 25.8 m, the corresponding basic wind pressure for which only slightly exceeds 0.86 kN/m^2, but the weight of the slab far exceeds the equivalent upward pressure.

Chapter 3
Bending moments on structural members

When the characteristic loads for which a building is to be designed have been established (as described in the previous chapter), the next design stage is to determine the appropriate bending moments induced in the slabs and beams comprising the floors and roof. In a reinforced concrete building the beams and slabs are often formed monolithically with each other and with the supports; i.e. they are designed as continuous over the supports. In some cases the beams are also designed to act monolithically with the supporting columns, the structural frame then being analysed as a whole. Nowadays, however, it is quite common for precast reinforced or prestressed concrete units to be utilized in the construction of multistorey buildings, particularly to form the slab areas; alternatively a proprietary flooring system may be employed.

The bending moments on simple beams and slabs that are freely supported on two supports or cantilevered can be calculated from simple statics. The bending moments on continuous slabs, beams and frames are statically indeterminate and their determination is discussed in this chapter. Slabs spanning in two directions are considered in Chapter 10, flat slabs in Chapter 11 and other types of floor in Chapter 12. Bending moments due to concentrated loads on solid slabs are discussed in sections 9.4 and 10.7.

3.1 CONTINUOUS BEAMS AND SLABS

BS8110 specifies, in effect, three different methods of calculating the bending moments on beams and slabs that span in one direction and are continuous over several supports. If there are three or more nearly equal spans (for beams the Code states that the maximum permissible difference is 15% of the greatest span) supporting predominantly uniform loads, and in the case of beams the imposed load does not exceed the dead load, simple approximate expressions utilizing coefficients given in BS8110 may be used. The corresponding Code

coefficients for slabs only apply where the area of the bay comprising the slab considered exceeds 30 m^2, and where the imposed load does not exceed the dead load by more than 25% and is not greater than 5 kN/m^2 (excluding partitions). Alternatively, for both beams and slabs, a theoretical analysis may be made assuming that the members are free to rotate about their supports (i.e. the frequently adopted assumption of knife-edge support). Otherwise, the members may be considered to be part of a monolithic frame and analysed as such. Each of these methods is now discussed in turn.

When calculating the maximum bending moments in the spans and at the supports, BS8110 states that the spans should be loaded as shown in *Figure 3.1a*. The arrangement of the imposed load prescribed to obtain the support bending moments does not give the theoretical maximum moments at the supports, to obtain which it is necessary to arrange the imposed loads as indicated in *Figure 3.1b*. The difference between the maximum negative bending moments resulting from the two sequences of loading is sometimes significant; for example, for a theoretically infinite number of spans and a uniform load, the loading specified in BS8110 gives bending moments at the internal supports that are 74% of the maximum values that result when the critical loading condition illustrated in *Figure 3.1b* is applied.

As shown in *Figure 3.1a*, BS8110 requires the analysis to be made considering a maximum dead load of $1.4G_k$ and a minimum dead load of $1.0G_K$, these loads being so arranged as to induce maximum moments. This requirement may be dealt with conveniently by considering instead a total 'imposed load' of $0.4G_k + 1.6Q_k$ and a dead load of $1.0G_k$, as shown in *Figure 3.1c* for the loading arrangement giving the maximum support moment.

For the purpose of calculating bending moments, BS8110 defines the effective span of a continuous member as the distance between the centres of the supports or from the face of a cantilever to the centre of its support.

13

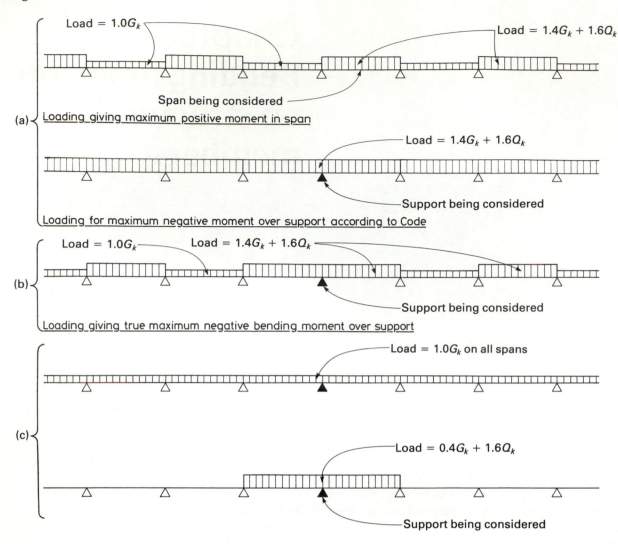

Figure 3.1 Loading arrangements for maximum moments.

For freely supported members, however, either the distance between the centres of the supports or the clear distance between the supports plus the effective depth, whichever is the lesser, may be adopted.

3.2 EQUAL SPANS: APPROXIMATE METHOD

For slabs spanning in one direction only, and for beams continuous over three or more approximately equal spans and carrying substantially uniform loads, where the ratio of characteristic imposed load Q_k to characteristic dead load G_k is not greater than 1 (for beams) and 1.25 (for slabs), and where for slabs the span exceeds 30/(breadth of building), and the imposed load is not greater than $5 \, kN/m^2$ (excluding partitions), BS8110 gives the following approximate coefficients for the calculation of the ultimate bending moments at the critical sections:

	For beams	For slabs
At end support	0	0
In end span: near midspan	+0.09	+0.086
At penultimate support	−0.11	−0.086
In interior spans: near midspan	+0.07	+0.063
At internal supports	−0.08	−0.063

To obtain the ultimate bending moments, the above coefficients must be multiplied by the product of the intensity of the total ultimate load (i.e. $1.4G_k + 1.6Q_k$) and the span. Consecutive spans may be considered to be equal if the difference in length does not exceed 15% of the longer span. These coefficients appear to be based on the assumptions that the characteristic dead load and characteristic imposed load are equal and that the member is freely supported on the outer supports. Since this latter condition is rarely obtained in the case of solid slabs, and since the torsional resistance of the supporting beams is generally effective in restraining a slab from acting freely as if it were supported on knife-edge supports, it is questionable whether it is necessary to be so precise in the estimation of the bending moments on solid slabs as is indicated by the Code coefficients. Perhaps the well-known coefficients of one-tenth for the end spans and the penultimate supports and one-twelfth for all other spans and supports, which have been widely adopted for many years, are sufficiently realistic for such slabs.

A comparison of the ultimate bending moments calculated on the assumption of equal spans when the inequality is less than 15%, and by a more accurate method, is given in section 9.5.1 when considering the secondary beams in the building in Part Two.

3.3 THEORETICAL ANALYSIS: KNIFE-EDGE SUPPORTS

If the spans are unequal, if the loads are not distributed uniformly, if the limiting ratio of Q_k/G_k is exceeded, or if two or more of these conditions apply, the ultimate bending moments can be calculated by means of one of the exact theoretical methods, such as the theorem of three moments, slope deflection, least work etc. Such a procedure is often complex, even when permissible simplifying assumptions are made; in such cases iterative methods such as the well-known Hardy Cross moment-distribution method and its variants are simpler and sufficiently accurate. The application of one of these variants of moment distribution is described later in this chapter.

When a strict theoretical analysis is employed, BS8110 permits any section to be designed for a resistance moment that is not less than 70% of the bending moment obtained at that point from an elastic bending-moment analysis taking account of all appropriate combinations of dead and imposed load, provided that two further conditions are met. These are that the internal and external forces are in equilibrium, and that the maximum depth to the neutral axis assumed when designing the concrete section at that point is related to the percentage of moment redistribution that has been adopted at the section concerned. The latter condition results in a maximum neutral-axis depth of $0.6d$ (where d is the effective depth to the tension steel) if no redistribution is employed, and $0.3d$ when the maximum 30% redistribution is adopted.

The requirements relating to moment redistribution in the Code have been framed to ensure that, since the positions of the points of contraflexure are normally altered by the redistribution process, sufficient reinforcement is provided at these locations (which would otherwise theoretically only require nominal reinforcement, of course) to limit the formation of cracks due to the moments that arise at these points from service loads.

An important point to note is that BS8110 places no corresponding limit on the maximum percentage *increase* in moment that may be adopted.

Moment redistribution is discussed in some detail in the *Code Handbook*. The justification for redistributing bending moments in this way is that, as the actual moments occurring in a member reach the values which the critical sections can withstand, so-called 'plastic hinges' form which permit rotation to occur at these points without any increase in the moment-carrying capacity. Thus further increases in the load on the member are resisted by increases in the moments elsewhere, leading to the formation of further plastic hinges. Finally the system fails when the last plastic hinge

to form renders the system unstable. The resulting collapse-moment diagram can be produced by evaluating an elastic-moment diagram using ultimate loads and then redistributing the moments as permitted by BS8110. For further details, reference should be made to the *Code Handbook*.

The application of moment redistribution to beams and slabs that are continuous over a number of equal spans is discussed in detail below, but the redistribution procedure is, of course, equally applicable to any continuous system. The major practical benefit of redistribution is that it enables the congestion of reinforcement that would otherwise occur at the supports, i.e. at the intersections of the beams and columns in an ordinary building, to be reduced. It is also convenient, especially in solid slabs, to have the same bending moment at the support as at midspan; and, although absolute uniformity of maximum ultimate bending moments can clearly not be obtained simultaneously for all types of load and at all the critical sections, advantage can be taken of using the permitted redistribution to keep the bending moments within allowable limits, with a view to reducing the inequality between peak moments. In flanged beams, however, it is generally an advantage to reduce the support moments as much as possible, since the area of concrete in compression in the rib at the support is so much less than that in the flange at midspan.

As explained above, the maximum percentage redistribution that may be undertaken is related to the adopted ratio of x/d by the expression $x/d \leq (\beta_b - 0.4)$, where β_b is the ratio of the bending moment after redistribution to that before redistribution, at the section considered. Thus if large redistributions of moment are contemplated it must be remembered that there will be a corresponding restriction on the maximum value of x/d that may be employed; if the ratio of d'/d is high, this may limit f_{yd1} to less than its maximum possible value. However, it must be remembered that the limit of 70% applies to *each particular combination* of load separately. Consequently, with imposed loads both the maximum span and maximum support moments may normally be reduced as they usually arise as a result of different loading conditions. Thus the overall adjustment is normally much less than the 30% limit, especially when taking the moments due to dead load into account as well. For example, considering the three-span beam carrying central concentrated loads which is examined in detail in the next section, if $g = q$ and making the full 30% reduction at the supports:

maximum moment in end span before adjustment

$$= 0.175 \times 1.0gl^2 + 0.213(1.6 + 0.4)ql^2 = 0.600gl^2$$

maximum moment in end span after adjustment

$$= 0.198 \times 1.0gl^2 + 0.198(1.6 + 0.4)ql^2 = 0.594gl^2$$

Thus the percentage adjustment made near midspan is only $(0.600 - 0.594) \times 100/0.6 = 1\%$ and the corresponding maximum value of x/d is 0.59. However, the support moment has been reduced by the full 30% and so the maximum permissible ratio of x/d here is only 0.3.

3.4 EQUAL SPANS AND UNIFORM MOMENTS OF INERTIA

For members that are continuous over two or more equal spans, which have a uniform second moment of area throughout all spans and are freely supported at end supports, formulæ giving the critical bending moments at each support and near the middle of each span are given on *Data Sheets 7* and *8*, for uniform loads and for the loading transferred from two-way slabs according to BS8110 respectively. However, for the common case of such beams and slabs also supporting equal loads, it is worth while also tabulating the limiting bending moment coefficients. On *Data Sheet 9* such coefficients are given for two, three and a theoretically infinite number of spans with a uniform load extending over the entire span, a uniform load extending over the central 75% of the span, and a central concentrated load. Both the condition of all spans loaded (e.g. as in the case of dead load) and the various conditions of incidental (e.g. imposed) load producing the greatest bending moments are considered. If the coefficients are calculated by a so-called 'exact' method (as is the case with those on *Data Sheet 9*), the maximum values may be reduced to not less than 70% of the original values as described below. *Data Sheet 9* also give the coefficients for the positive bending moments at the supports and the negative bending moments on the spans which result from some conditions of imposed load; these enable the relevant bending-moment diagrams to be sketched and thus the reinforcing bar stopping-off points to be estimated.

The method of calculating the basic and adjusted coefficients, such as those tabulated on *Data Sheet 9*, is illustrated by the example shown in *Figure 3.2* of a beam that is continuous over three spans and is loaded with dead and imposed loads concentrated at midspan only. Although this is not a very practical example, as in real life there would always be some uniform load due to the self-weight of the beam, it does illustrate particularly clearly the numerical procedures and adjustments involved. The theoretical bending moments are calculated for all spans loaded (i.e. dead load) in *Figure 3.2a*, and for each of the three cases of imposed load that produce maximum bending moments according to BS8110: i.e. in *Figure 3.2b* for the middle of the central span (positive), in *Figure 3.2c* for the middle of an end span (positive), and in *Figure 3.2d* for a support (negative). (As explained earlier, for simplicity BS8110 permits the assumption to be made that the maximum negative moments at the supports occur when all the spans are loaded. In reality, the true maximum negative moment at any support occurs when the two spans adjoining the support are loaded, together with all alternate spans, and the true maximum positive moment at any support occurs when the two adjoining

spans and all other alternate spans are unloaded, all remaining spans being loaded. The Code simplification means that for a continuous system of n spans, to determine the maximum support moments only a single loading condition needs to be considered rather than $2(n-1)$ conditions.)

For each case in *Figures 3.2a–d* the theoretical bending-moment diagram is adjusted as follows. For the diagram of maximum negative support moments, the theoretical negative bending moments at the supports are reduced by 30% and the corresponding positive bending moments in the spans are increased accordingly. Then for the respective diagrams of maximum positive bending moments in the spans, the theoretical positive bending moments are reduced by 30% provided that the corresponding negative moments at the supports are not, as a result, increased to values greater than those obtained by making a 30% reduction in the maximum values of negative moment. If this would occur, the percentage reduction of positive moment made is limited to that which makes the corresponding increased negative support moments equal to the reduced negative support moments corresponding to the loading that produces the maximum negative moments. For example, it is only possible to decrease the span moments in *Figure 3.2c* by 7% without increasing the support-moment coefficient to more than -0.105, which is the value obtained by reducing the basic maximum support-moment coefficient of -0.150 by 30%. Similarly, the maximum reduction in the span moments that can be made in *Figure 3.2d* is 17%.

Figure 3.2e shows the resulting envelope of maximum bending moments due to imposed load only, both before and after a reduction of 30% has been applied to the support moments. Two points resulting from such an adjustment are worth noting. Firstly, the maximum negative moments throughout the spans are increased considerably. These moments are normally of little practical importance, however, since they act in opposition to the moments resulting from dead loading, but they may become more important with high ratios of imposed load to dead load. Secondly, the lines defining the areas of the bending-moment envelope after redistribution in the vicinity of the points of contraflexure need to be adjusted to meet the requirement in BS8110 that at least 70% of the elastic-moment values must be considered at all points. The adjusted lines, marked X on *Figures 3.2a–d*, influence the areas shaded on *Figures 3.2e* (and *3.2f*). When designing reinforcement for the lengths within a distance of about one-quarter of the span from any support, remember that the maximum bending moment and shear force generally result from a combination of only partial load on the span in question with full loading on others, a condition not considered in the Code.

In preparing the foregoing tabulated coefficients it has been considered of prime importance to reduce the support moments by as much as possible. Although this is true in normal cases of frame construction for the reasons

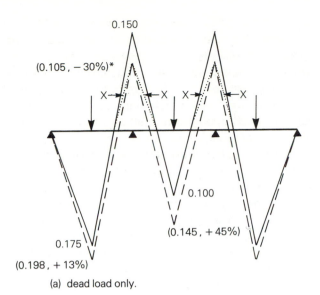

(a) dead load only.

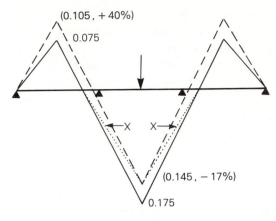

(b) maximum negative moment in end spans, maximum positive moment in central span.

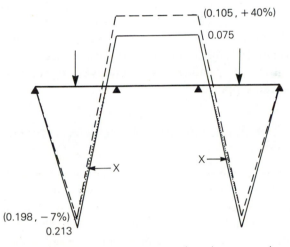

(c) maximum positive moment in end spans, maximum negative moment in central span.

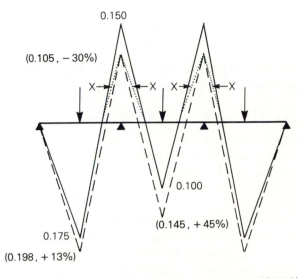

(d) maximum negative moment at support (as BS8110).

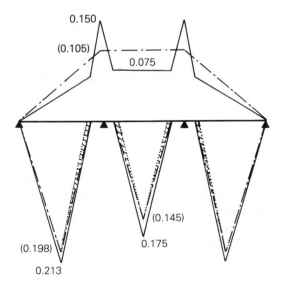

(e) with maximum reduction of negative moments.

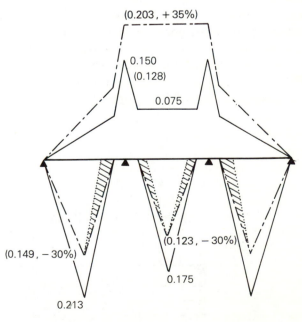

(f) with maximum reduction of positive moments.

Figure 3.2 (a)–(d) Maximum bending-moment diagrams before redistribution (full lines) and after (broken lines); (e)–(f) combined imposed-load envelopes before redistribution (full lines) and after (chain lines).

* Figures within brackets correspond to redistributed moments and percentage adjustments made to adjoining values.

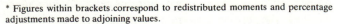

explained in section 3.3, it is sometimes advantageous to reduce the span moments to the fullest extent (perhaps if upstand beams are employed) or to partially equalize the span and support moments. *Figure 3.2f* illustrates the bending-moment envelope obtained when all the span moments are reduced to 70% of their original values and the support moments are increased accordingly.

The position and magnitude of the maximum positive bending moments are readily calculated for beams carrying concentrated loads. For a beam such as LR of span l in *Figure 3.3*, carrying a uniformly distributed load of intensity n and subjected to theoretical negative support moments M_{LR} at L and M_{RL} at R, the position and magnitude of the maximum positive bending moment are given by

$$X = \tfrac{1}{2}l + \frac{M_{LR} - M_{RL}}{nl}$$

$$M_{max} = \frac{n}{2}\left(\frac{M_{LR} - M_{RL}}{nl} + \frac{l}{2}\right)^2 - M_{LR}$$

If the negative bending moments at the supports are each increased by $x\%$, the modified formulae are

$$X' = \tfrac{1}{2}l + \frac{1 + 0.01x}{nl}(M_{LR} - M_{RL})$$

$$M'_{max} = \frac{n}{2}\left[\frac{1 + 0.01x}{nl}(M_{LR} - M_{RL}) + \frac{l}{2}\right]^2$$
$$- (1 + 0.01x)M_{LR}$$

The coefficients for the maximum positive bending moments tabulated in *Data Sheet 9* have been calculated from these expressions by taking $l = 1$ and $n = 1$.

Unless the support moments M_{LR} and M_{RL} are greatly dissimilar, it is generally sufficiently accurate to determine the maximum positive moment by subtracting the mean of the two support moments from the free moment.

In calculating the foregoing bending-moment coefficients it is assumed that the beams are freely supported at the outer supports. In the cases of beams framing into columns and of slabs which are monolithic with large supporting beams, however, negative bending moments occur at the end supports. These bending moments must be resisted and allowance should be made for their effect on the bending moments on adjoining spans. *Data Sheet 14* indicates the effect of a unit bending moment applied at one end, or at both ends simultaneously, of a series of continuous equal spans. The corresponding bending-moment diagrams should be superimposed upon the normal diagrams resulting from the case of free end support. It is seen from the coefficients in *Data Sheet 14* that a bending moment which is applied at an end support affects to any extensive degree only the bending moment at the penultimate support; other than in exceptional conditions, the effect on other supports and spans (beyond the first and second) may be ignored. Since the effect is to reduce the negative bending moment at the penultimate support, the bending moment applied at the end must not be overcalculated when making this reduction. The magnitude of the bending moment applied at the end may be calculated from the formulæ given in BS8110 for bending on exterior columns (see section 14.2) or from considering the beam as being a member of a monolithic frame (see section 3.8), or it may be produced by a cantilever extending beyond the end support.

3.5 FORMULÆ FOR UNIFORM LOADS AND EQUAL SPANS

The use of tabulated coefficients is most convenient when the loads are concentrated and the resulting maximum moments occur beneath the loads. With distributed loads, however, it is more difficult to combine the maximum values that occur due to dead and imposed loads (since they occur in slightly different positions) and to sketch the resulting overall envelope of bending moments. In such cases it may be simpler to calculate the critical total moments produced by the combined dead and imposed loads from bending-moment formulæ. The expressions given in *Data Sheets 7 and 8* enable the moments at the supports and near midspan for beam and slab systems that are continuous over two, three or a theoretically infinite number of spans to be determined when loaded uniformly. Separate expressions are given for the moment at each point due to each arrangement of loading; these enable the bending-moment envelope to be sketched to a fair degree of accuracy if required. For many purposes, however, only a knowledge of the maximum values is necessary, and the appropriate formulæ for this purpose are enclosed in boxes on *Data Sheets 7 and 8*. If g is the characteristic dead load and $n = g + q$, the resulting moments are the normal elastic values. However, if g is taken as $1.0g$ and $n = 1.4g + 1.6q$, the resulting moments are the critical ultimate values required for design to BS8110.

The use of formulæ catering directly for the combined moments due to dead and imposed loads, such as those given on *Data Sheets 7 and 8*, is less convenient if moment redistribution is to take place. For this reason, *Data Sheet 9* gives the critical moment coefficients for various

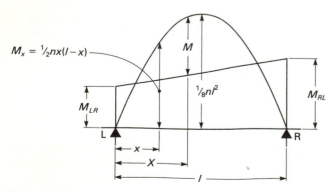

$M_x = \tfrac{1}{2}nx(l - x)$

M

$\tfrac{1}{8}nl^2$

M_{RL}

M_{LR}

L   R

x

X

l

Figure 3.3

continuous-beam systems supporting uniform or central concentration loads for three conditions, namely: (a) with no moment redistribution; (b) with 10% redistribution; and (c) with a maximum redistribution of 30%. When using *Data Sheet 9* it is necessary to determine the moments due to dead loads of $1.0G_k$ and 'imposed' loads of $0.4G_k$ separately (taking into account the degree of redistribution required), and then to sum the resulting values.

3.6 ANALYSIS OF CONTINUOUS SYSTEMS

If the beams or slabs are continuous over several unequal spans, if the moment of inertia differs from span to span, or if the applied loading is not one of the three arrangements considered above, the system must be analysed by means of one of the so-called exact methods. These methods may be divided into two basic groups. The first category, which includes such methods as the theorem of three moments, slope deflection, least work etc., consists basically of setting up and solving a series of simultaneous equations to obtain exact results. Such equations are tedious to solve when a solution has to be obtained by hand methods, but are ideal when there is convenient access to a computer. If, however, the values obtained from a preliminary design require adjustment, the analysis must usually be repeated in full; this recycling process is continued until the resulting design is satisfactory, and it may therefore be impracticable if long delays occur between submitting the values for computation and receiving the results. In such cases it may well be advantageous to prepare an approximate design by trial and adjustment, using one of the methods described below, and only to employ the computer to check the design finally chosen and to calculate the exact bending moments for which the sections must be designed.

The other category of analytical methods consists of various iterative processes including relaxation and moment distribution. These methods involve a cyclic procedure whereby the final solution is approached in stages, each successive adjustment bringing the interim results progressively closer. If a computer is not available, these methods have two basic advantages over those involving simultaneous equations. The first is that the actual computational procedure is usually extremely simple. For example, in the well-known method of moment distribution the moments at each end of a span are first calculated on the assumption that the ends of each individual span are fully fixed. The ends of the members meeting at each particular intersection in turn are then assumed to be released, and the resulting out-of-balance 'fixed-end' moments at that joint are distributed between them in direct proportion to their stiffnesses. The next step is to 'carry over' a proportion (one-half in the case of prismatic members) of the distributed moments to the opposite ends of the members. These carried-over moments are then again distributed between the

intersecting members at each individual joint, a further carry-over operation takes place, and so on. As the differences between the moments at the ends of the members meeting at a particular intersection become less and less, so the values of unbalanced moments to be distributed and then to be carried over become progressively smaller; thus a summation of the total moment at the end of a particular member more nearly approaches its true value. In theory, the cyclic procedure should be repeated until the moments being distributed and carried over are negligible. However, in practice two, or at the most three, complete cycles are usually sufficient with prismatic members to obtain results that are within a few per cent of their exact values, an accuracy that is quite sufficient in view of the uncertainties made in the basic assumptions regarding the stiffnesses of the members etc.

The second basic advantage of iterative methods is that it is often fairly clear, even after only one distribution cycle, whether or not the final values will be acceptable. If they are not, the analysis need not be continued further, thus saving much unnecessary work.

3.7 PRECISE MOMENT DISTRIBUTION

The original Hardy Cross method of moment distribution is too well known to warrant a detailed description here, since there are already many books dealing specifically with the subject. For the generation of reinforced concrete designers that came to structural analysis after the method was first conceived in the early 1930s and before the widespread use of computers, it became the most popular method of analysing series of continuous beams, and its continued popularity has led to the introduction of numerous developments and extensions of the original concept.

One such variant is a hybrid, combining features of both of the foregoing types of solution. Known most commonly as precise moment distribution, although it has also been referred to as the coefficient-of-restraint method, it is closely related to the method of fixed points and the degree-of-fixity method. The analytical procedure is extremely similar to and only slightly less simple than normal moment distribution, but the distribution and carry-over factors are so adjusted that an exact solution is obtained after only a single distribution in each direction. The method thus retains the advantage, when using hand computation, of eliminating the need to decide when to terminate the successive approximation procedure. Since it is perhaps less widely known than should be the case, its use to analyse series of continuous spans formed of prismatic members is now described. Owing to considerations of space, details of the theoretical basis of the method are kept to a minimum, since these are available elsewhere (refs 6–8). The few formulæ that are required are easy to memorize and the use of graphs or nomograms is not necessary, although the analysis may be undertaken even more quickly if they are employed. Alternatively, programming the formulæ for a

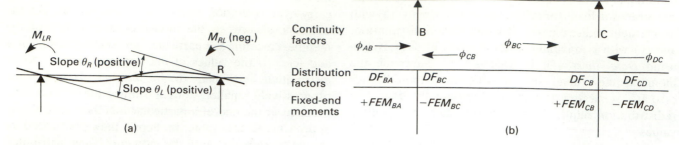

Figure 3.4

computer or calculator is simple and advantageous if such a device is available. The method may be extended to analyse frames and systems that include non-prismatic members (ref. 8).

For an unloaded prismatic member (*Figure 3.4a*) the slope-deflection equations are

$$M_{LR} = EK_{LR}(4\theta_L + 2\theta_R)$$

and

$$M_{RL} = EK_{LR}(2\theta_L + 4\theta_R) \qquad (3.1)$$

Thus if $M_{RL}/M_{LR} = \phi_{RL}$,

$$\phi_{RL} = \frac{EK_{LR}(2\theta_L + 4\theta_R)}{EK_{LR}(4\theta_L + 2\theta_R)}$$

from which

$$\theta_L = \frac{\theta_R(4 - 2\phi_{RL})}{4\phi_{RL} - 2} \quad \text{and} \quad \theta_R = \frac{\theta_L(2 - 4\phi_{RL})}{2\phi_{RL} - 4}$$

Substituting in equations (3.1),

$$M_{LR} = EK_{LR}\left[4\theta_L + \frac{2\theta_L(2 - 4\phi_{RL})}{2\phi_{RL} - 4} \right] = \frac{6EK_{LR}\theta_L}{2 - \phi_{RL}}$$

Since the restraint factor R_{LR} at support L is defined as M_{LR}/θ_L,

$$R_{LR} = \frac{6EK_{LR}}{2 - \phi_{RL}}$$

Similarly,

$$M_{RL} = \frac{6EK_{LR}\theta_R}{2 - (1/\phi_{RL})}$$

and

$$R_{RL} = \frac{6EK_{LR}}{2 - (1/\phi_{RL})} = \frac{6EK_{LR}\phi_{RL}}{2\phi_{RL} - 1}$$

Known as the *continuity factor*, ϕ represents the proportion of the bending moment at one support which is transferred to the next support, and is thus analogous to the carry-over factor of conventional moment distribution. Successive values of ϕ in a continuous system are interrelated, the first being determined by the known or assumed end conditions. Thus if the end is free, no moment is transferred to the penultimate support and therefore $\phi = 0$; if the end is fixed, $\phi = 0.5$. Intermediate fixity conditions may be assumed if desired.

The restraint factors R may be thought of as modified stiffness factors, and are consequently analogous to the true stiffness factors of conventional moment distribution. Since the sum of the moments in all the members meeting at a support is zero, for successive spans l_n and l_{n+1}, by definition $-R'_n = R_{n+1}$, so that

$$-\frac{6K_n}{2 - \phi_n} = \frac{6K_{n+1}}{2 - (1/\phi_{n+1})}$$

Thus

$$\phi_{n+1} = \frac{1}{2 + \dfrac{K_{n+1}}{K_n}(2 - \phi_n)} \qquad (3.2)$$

Equation (3.2) enables the continuity factors for successive spans to be determined once the first is known. The left-hand nomogram on *Data Sheet 10* gives successive values of ϕ_{n+1} corresponding to given values of ϕ_n and the appropriate ratio of K_{n+1}/K_n.

The continuity factors are determined throughout the system in each direction. Then at a given support B between adjoining members AB and BC, say, the *distribution factor* DF_{BA} (*Figure 3.4b*) is

$$DF_{BA} = \frac{R_{BA}}{R_{BA} + R_{BC}}$$

$$= 1 \bigg/ \left[1 + \frac{K_{BC}(2 - \phi_{AB})}{K_{AB}(2 - \phi_{CB})} \right] = \frac{1 - 2\phi_{BA}}{1 - \phi_{AB}\phi_{BA}}$$

Thus in the notation employed above,

$$DF_{LR} = \frac{1 - 2\phi_{LR}}{1 - \phi_{RL}\phi_{LR}} \qquad (3.3)$$

Equation (3.3) enables the distribution factors at each side of each support to be determined from the continuity factors already established. The right-hand nomogram on *Data Sheet 10* facilitates the calculation of the distribution factors. Since the distribution factor relating to one side of any support is determined solely from the continuity factors relating to that span, it is recommended that the distribution factors on each side of any support be determined independently in this way. Then, since the sum of the distribution factors at any support must equal unity, an additional check is provided that both nomograms (or the corresponding formulæ) have been used correctly.

Note that the values of the distribution factors are *not* the same as would be obtained if conventional moment distribution were used.

3.7.1 Design procedure

Perhaps the most convenient method of dealing with a design problem when using precise moment distribution is as follows:

1. Except in the case where the moment of inertia throughout all the spans is uniform, determine the moment of inertia of the gross concrete section for each span (see *Data Sheet 49*) and hence calculate the stiffness $K(=I/l)$ for each span.
2. Calculate the fixed-end moments for the dead and imposed loads separately in exactly the same way as is done in the case of normal moment distribution (i.e. the fixed-end moments correspond to the support moments when the loaded member is assumed fully fixed at both supports) using *Data Sheet 48*. Remember, however, that for an analysis in accordance with the requirements of BS8110 it is most convenient to consider a dead load of $1.0G_k$ and an 'imposed' load of $1.6Q_k + 0.4G_k$.
3. Knowing the stiffness and the continuity factors corresponding to the known or assumed end conditions, determine the remaining continuity factors in each direction by using expression (3.2) or the appropriate nomogram on *Data Sheet 10*.
4. Knowing the continuity factors, calculate the distribution factors at each side of each support by using expression (3.3) or the appropriate nomogram.
5. Distribute the moments due to the dead load as shown in *Figure 3.5*. For example, since the distribution factor at the free end A is 1, the balancing moment at this point is $+45\,kN\,m$. The proportion of this balancing moment that is transferred to B is now obtained by multiplying by the continuity factor ϕ_{BA}; thus $45.0 \times 0.215 = +9.7\,kN\,m$. The moment required to balance the difference between the fixed-end moments on spans AB and BC of $45.0 - 101.3 = -56.3\,kN\,m$ is now divided between spans AB and BC

in proportion to their distribution factors; i.e. that on span BC is $56.3 \times 0.431 = +24.2\,kN\,m$. This is now added algebraically to -9.7, to give $+14.5\,kN\,m$ which is multiplied by ϕ_{CB} to give the proportion of the moment transferred to support C. Thus $-3.4\,kN\,m$ is added to the value of $+34.1\,kN\,m$, obtained by multiplying the moment of $78.7\,kN\,m$ required to balance the difference between the fixed-end values of 180.0 and 101.3 by the distribution factor DF_{CD} of 0.433, which gives $+30.7\,kN\,m$. Then by multiplying by the value of ϕ_{DC}, i.e. 0.333, the moment of $10.2\,kN\,m$ transferred to support D is obtained, and so on. When the right-hand end of the system is reached, the process is repeated in reverse: the proportion of the balancing moment at each support obtained by multiplying the out-of-balance moment by the left-hand distribution factor at that support is added algebraically to the previously carried-over moment. This sum is then multiplied by the appropriate continuity factor to obtain the moment transferred to the subsequent support, and so on. One point that should be noted in particular is that, to obtain the moments carried over from left to right, the appropriate moment sums are multiplied by the corresponding continuity factors obtained by working *from right to left* and vice versa.

6. Distribute the moments due to the imposed load on each separate span in turn. The process is similar to that described for dead load in step 5, but of course only two support moments need to be considered for each loaded span. The procedure necessary for a typical span is shown in *Figure 3.6*.
7. Sum those support moments giving the critical conditions. For example, the maximum support moments that require consideration according to BS8110 occur when all spans carry both dead and imposed load. Similarly, the support moments corresponding to the maximum moment near midspan occur when the dead load covers all the spans while the imposed load occurs on the span under consideration and the corresponding alternate spans. Under such conditions the moments near midspan of the spans not carrying imposed loads attain their minimum values.

Relative stiffnesses		$\frac{1}{6}$		$\frac{1}{9}$		$\frac{1}{12}$		$\frac{1}{8}$	
Continuity factors $\phi\rightarrow$			0		0.300		0.305		0.220
$\leftarrow\phi$	0.215		0.237		0.333		0.500		
Distribution factors	1.000	0.569	0.431	0.567	0.433	0.371	0.629		0
Fixed-end moments	-45.0	$+45.0$	-101.3	$+101.3$	-180.0	$+180.0$	-80.0		$+80.0$
Distribution and carry-over procedure	$+45.0 \rightarrow$	$+9.7$ $[+32.1$	$-9.7]$ $+24.2]$	$+3.4$ $[+44.6$	$-3.4]$ $+34.1]$	$+10.2$ $[-37.1$	$-10.2]$ $-62.9]$	$\rightarrow -36.6$	
	$0 \leftarrow$	$[-16.8$	$+16.8 \leftarrow$	$[+11.3$	$-11.3 \leftarrow$	0	$0 \leftarrow$		0
Summations	0	$+70.0$	-70.0	$+160.6$	-160.6	$+153.1$	-153.1		$+43.4$

Figure 3.5

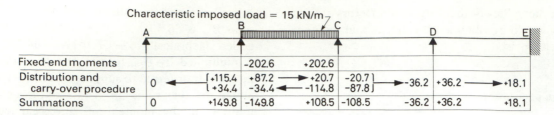

	A		B		C		D		E
Fixed-end moments			−202.6		+202.6				
Distribution and carry-over procedure	0	← {+115.4 / +34.4}	+87.2 → / −34.4 ←	+20.7 / −114.8	{−20.7 / −87.8}	→ −36.2	+36.2 →		+18.1
Summations	0		+149.8	−149.8	+108.5	−108.5	−36.2	+36.2	+18.1

Figure 3.6

An alternative, and sometimes simpler, analytical procedure is first to consider the effect of a nominal support moment of say 100 units at each individual support in turn, and then to adjust the resulting values in proportion to the actual unbalanced fixed-end moments due to the dead and imposed loads.

Yet another possibility with uniform loads only is to consider two loading conditions, (a) when loads occur on alternate spans 1,3,5 etc. and (b) when spans 2,4,6 etc. are loaded, and to calculate the fixed-end moments for a load of unity. Then combining the results given by (a) and (b) and multiplying by the uniform dead load gives the moments due to dead load only. Adding these to the values obtained by multiplying results (a) separately, results (b) separately, and then both combined, by the uniform imposed load produces the support moments corresponding to the conditions for the two sets of maximum span moments and for the maximum support moments, respectively.

The sign convention adopted here is the same as that usually employed with conventional moment distribution. Thus with normal (i.e. downward-acting) loads a negative fixed-end moment is assumed at the left-hand end of the span and a positive fixed-end moment occurs at the right-hand end; the distributed balancing moments at the supports are so arranged that the resulting sum of the moments at both sides of each support is zero.

An alternative, and perhaps preferable, convention is to consider normal fixed-end moments as negative throughout and to so arrange the signs when distributing the out-of-balance support moments that the resulting sum of the moments at each support is still zero. Thus, considering *Figure 3.5*, at B, of the difference between the fixed-end moments of -45.0 kN m from AB and -101.3 kN m from BC, i.e. 56.3 kN m, $0.569 \times 56.3 = 32.0$ kN m is 'added' to -45.0 to give -77.0 kN m, and $0.431 \times 56.3 = 24.3$ is 'subtracted' from -101.3 to give -77.0 kN m. If this sign convention is adopted, it is necessary to reverse the sign when carrying over moments from one support to the next.

When the loading on a span is symmetrically arranged, the resulting fixed-end moments are equal, of course. In such a case a further simplification may be introduced into the analysis which reduces the entire operation for each loaded span to a single distribution. The procedure involves the introduction of a further set of factors which may be designated symmetrical *residual-moment factors* (RMFs), each factor being the proportion by which the fixed-end moment at that point must be multiplied in order to obtain the final support moment. The manner in which the symmetrical residual-moment factors are derived from the distribution and continuity factors is shown in *Figure 3.7a*, from which it is seen that, for example, the symmetrical residual-moment factor

(a)

	Support L			Support R	
Continuity factors		$\leftarrow \phi_{RL}$		$\phi_{LR} \rightarrow$	
Fixed-end moments		-1		$+1$	
Distribution	$+\dfrac{(2\phi_{LR}-\phi_{LR}\phi_{RL})}{1-\phi_{LR}\phi_{RL}}$	$+\dfrac{(1-2\phi_{LR})}{1-\phi_{LR}\phi_{RL}}$	$-\dfrac{(1-2\phi_{RL})}{1-\phi_{LR}\phi_{RL}}$	$-\dfrac{(2\phi_{RL}-\phi_{LR}\phi_{RL})}{1-\phi_{LR}\phi_{RL}}$	
Carry-over	$+\dfrac{\phi_{LR}(1-2\phi_{RL})}{1-\phi_{LR}\phi_{RL}}$	$-\dfrac{\phi_{LR}(1-\phi_{RL})}{1-\phi_{LR}\phi_{RL}}$	$+\dfrac{\phi_{RL}(1-2\phi_{LR})}{1-\phi_{LR}\phi_{RL}}$	$-\dfrac{\phi_{RL}(1-2\phi_{LR})}{1-\phi_{LR}\phi_{RL}}$	
Summation	$+\dfrac{3\phi_{LR}(1-\phi_{RL})}{1-\phi_{LR}\phi_{RL}}$	$-\dfrac{3\phi_{LR}(1-\phi_{RL})}{1-\phi_{LR}\phi_{RL}}$	$+\dfrac{3\phi_{RL}(1-\phi_{LR})}{1-\phi_{LR}\phi_{RL}}$	$\dfrac{3\phi_{RL}(1-\phi_{LR})}{1-\phi_{LR}\phi_{RL}}$	

(b)

	Support L			Support R	
Continuity factors		$\leftarrow \phi_{RL}$		$\phi_{LR} \rightarrow$	
Fixed-end moments		$+1$		$+1$	
Distribution	$-\dfrac{(2\phi_{LR}-\phi_{LR}\phi_{RL})}{1-\phi_{LR}\phi_{RL}}$	$+\dfrac{(1-2\phi_{LR})}{1-\phi_{LR}\phi_{RL}}$	$+\dfrac{(1-2\phi_{RL})}{1-\phi_{LR}\phi_{RL}}$	$-\dfrac{(2\phi_{RL}-\phi_{LR}\phi_{RL})}{1-\phi_{LR}\phi_{RL}}$	
Carry-over	$+\dfrac{\phi_{LR}(1-2\phi_{RL})}{1-\phi_{LR}\phi_{RL}}$	$-\dfrac{\phi_{LR}(1-2\phi_{RL})}{1-\phi_{LR}\phi_{RL}}$	$\dfrac{\phi_{RL}(1-2\phi_{LR})}{1-\phi_{LR}\phi_{RL}}$	$+\dfrac{\phi_{RL}(1-2\phi_{LR})}{1-\phi_{LR}\phi_{RL}}$	
Summation	$-\dfrac{\phi_{LR}(1+\phi_{RL})}{1-\phi_{LR}\phi_{RL}}$	$+\dfrac{\phi_{LR}(1+\phi_{RL})}{1-\phi_{LR}\phi_{RL}}$	$+\dfrac{\phi_{RL}(1+\phi_{LR})}{1-\phi_{LR}\phi_{RL}}$	$\dfrac{\phi_{RL}(1+\phi_{LR})}{1-\phi_{LR}\phi_{RL}}$	

Figure 3.7 Derivation of (a) symmetrical (b) antisymmetrical residual-moment factors.

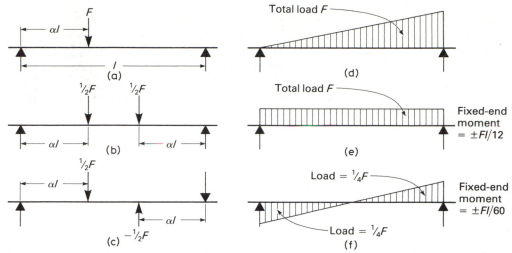

Figure 3.8 Method of representing unsymmetrical loads by combination of symmetrical and antisymmetrical loading.

RMF_{LR} for span LR at support L is

$$RMF_{LR} = \frac{3\phi_{LR}(1 - \phi_{RL})}{1 - \phi_{LR}\phi_{RL}} \quad (3.4)$$

Symmetrical residual-moment factors can be read from the left-hand nomogram on *Data Sheet 11*. The procedure for distributing moments is now exactly the same as that described above, apart from the fact that symmetrical residual-moment factors are used instead of distribution factors and that no moments are carried over within the span under consideration.

In cases where a combination of symmetrically and unsymmetrically loaded spans occurs, the foregoing methods may be combined, using distribution factors at the supports of the unsymmetrically loaded spans and symmetrical residual-moment factors elsewhere.

This device may be extended to cover unsymmetrically arranged loads by introducing the concept of antisymmetrical loading, where the antisymmetrical load produces fixed-end moments that are equal in value but of opposite sign at the supports of the loaded span. As shown in *Figure 3.7b*, the residual end moments due to such a loading system can be determined directly from the continuity factors, and thus read from the right-hand nomogram on *Data Sheet 11*.

Now any unsymmetrical load, for example the concentrated load in *Figure 3.8a*, can be represented by the sum of the symmetrical loading shown in *Figure 3.8b* and the antisymmetrical loading in *Figure 3.8c*. Similarly, the triangular load in *Figure 3.8d* can be represented by the combination shown in *Figures 3.8e* and *3.8f*. Thus, multiplying the fixed-end moments due to the resulting

symmetrical and antisymmetrical loads by the respective symmetrical and antisymmetrical residual-moment factors and summing the resulting values will give the final support moments due to the original load.

The necessary fixed-end moments can normally be found in either of two ways. The first method is to calculate the values FEM_{LR} and FEM_{RL} for the unsymmetrical load in the normal manner. Then the required fixed-end moments for the symmetrical load are $\pm\frac{1}{2}(FEM_{LR} + FEM_{RL})$ and those for the antisymmetrical load are $-\frac{1}{2}(FEM_{LR} - FEM_{RL})$. Alternatively, for individual concentrated loads the values may be found by using the coefficients read from the scales in the centre of *Data Sheet 11*. For each load F, the fixed-end moments for symmetrical loading are then $\pm k_s Fl$ and for antisymmetrical loading $+k_a Fl$. Systems of antisymmetrical loads can be considered by determining and summing the values due to the individual loads.

If the span carries both symmetrically and unsymmetrically disposed loads the foregoing procedure still applies, the fixed-end moments due to the symmetrical loading being multiplied by the symmetrical RMFs only, of course. However, if the loading is as complex as this, it is probably simpler to use the basic precise moment distribution procedure, making use of distribution factors. This is especially true where systems of loaded spans are considered. Residual-moment factors come into their own when considering the effects of the loading on an individual span and are particularly useful when all the loading is symmetrical.

Figure 3.9 illustrates the use of symmetrical residual-moment factors to solve the same problem as *Figure 3.5*.

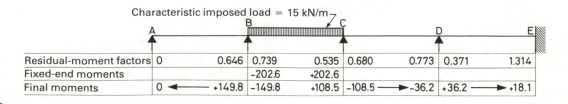

	A		B		C		D		E
Residual-moment factors	0	0.646	0.739	0.535	0.680	0.773	0.371		1.314
Fixed-end moments			−202.6	+202.6					
Final moments	0 ←	+149.8	−149.8	+108.5	−108.5 →	−36.2	+36.2 →		+18.1

Figure 3.9

23

Bending moments on structural members

	A	B		C		D		E
Dead load		+70.0	−70.0	+160.6	−160.6	+153.1	−153.1	+43.4
Imposed load on AB	0	+58.1	−58.1	−13.8	+13.8	+4.6	−4.6	−2.3
on BC	0	+149.8	−149.8	+108.5	−108.5	−36.2	+36.2	+18.1
on CD	0	−73.4	+73.4	+244.7	−244.7	+278.4	−278.4	−139.2
on DE	0	+5.4	−5.4	−18.1	+18.1	+59.4	−59.4	+210.3
For max. support moments	0	+209.9	−209.9	+481.9	−481.9	+459.3	−459.3	+130.3
For max. mts in AB and CD	0	+54.7	−54.7	+391.5	−391.5	+436.1	−436.1	−98.1
For max. mts in BC and DE	0	+225.2	−225.2	+251.0	−251.0	+176.3	−176.3	+271.8

Figure 3.10

The fixed-end moment of 202.6 kN m at C is multiplied by the symmetrical residual-moment factor RMF_{CB} of 0.535 at C to give 108.5 kN m, and by multiplying this by $\phi_{DC} = 0.333$ the moment of 36.2 kN m transferred to D is obtained. Further multiplication by $\phi_{ED} = 0.500$ gives the moment transferred to E. At B, the appropriate symmetrical residual-moment factor is 0.739, so that the final moment at B is $202.6 \times 0.739 = 149.8$ kN m.

Symmetrical imposed loads on the other spans can be dealt with in the same way, and *Figure 3.10* gives the final support moments due to an imposed load on each span in turn. Then the maximum support moment that must be considered at B, for example, occurs when all spans are loaded with imposed load and is thus $+58.1 + 149.8 - 73.4 + 5.4 = 139.9$ kN m plus the moment due to the dead load. Similarly, the maximum moments in spans BC and DE occur when these spans are loaded with dead and imposed loads while the other spans carry dead load only. Thus the support moments due to the imposed loads to be considered for span BC, for example, are $149.8 + 5.4 = 155.2$ kN m and $108.4 - 18.1 = 90.3$ kN m (plus the dead-load moments).

If the moments are to be redistributed as permitted by BS8110 the first step, assuming that the beams are of normal rectangular, or flanged sections, is to reduce the maximum support moments by 30% (i.e. to 70% of their calculated values), provided that the resulting limitation on the design stress in the compression reinforcement (see section 5.2.1) due to the related restricted value of x/d is not too severe. These reduced maximum support moments then represent maximum values to which the support moments, corresponding to the loading condition that gives the maximum moments in the span, may be increased. The span moment corresponding to these support moments should then be calculated.

For instance, in *Figure 3.10* the support moments at B and C due to dead load plus imposed load throughout are 209.9 kN m and 481.9 kN m respectively, which, when reduced by 30%, become 146.9 and 337.3 kN m respectively. With a total load of $(1.4 \times 15) + (1.6 \times 15) = 45$ kN m on the 9 m span, the corresponding positive bending moment is

$$\frac{n}{2}\left(\frac{M_{BC} - M_{CB}}{nl} + \frac{l}{2}\right)^2 - M_{BC}$$

$$= \frac{45}{2}\left(\frac{146.9 - 337.3}{45 \times 9} + \frac{9}{2}\right)^2 - 146.9 = 218.5 \text{ kN m}$$

The maximum positive moment in span BC which results from the normal critical loading occurs when the support moments at B and C are 225.2 and 250.9 kN m respectively. Then the maximum positive bending moment in the span is

$$\frac{45}{2}\left(\frac{225.2 - 250.9}{45 \times 9} + \frac{9}{2}\right)^2 - 225.2 = 217.7 \text{ kN m}$$

Thus to reduce the maximum support moments by the full 30% permitted by BS8110 it is only necessary to *increase* the corresponding maximum moment considered near the midspan of BC from 217.7 kN m to 218.5 kN m; i.e. by less than 1%.

3.8 CONTINUOUS BEAMS AS MEMBERS OF A FRAME

The calculation of the bending moments on continuous beams that are considered as parts of a monolithic frame is necessary for irregular structures or for beams that are loaded irregularly. In ordinary rectangular buildings it is seldom necessary to consider the interaction of the columns and beams more accurately than by applying the formulæ given in sections 14.1.1 and 14.1.2.

3.8.1 Recommended Code method

A far less approximate method that is suitable for complex beams and is applicable when considering the effects of vertical loading only is described below. Since the effects of lateral deformation (or 'sway' as it is commonly known) are ignored, the method is not strictly accurate if large out-of-balance loads are present, and it cannot be used in isolation if lateral loading, such as the effects of wind, occurs. According to BS8110, however, the method may be used in all cases where the lateral stability of the entire structure does not depend on the lateral stability of the frame in question but is provided by some other means, such as the use of shear-walls etc. Furthermore, where the frame does provide lateral stability for the whole structure, the analysis due to vertical loading alone may be combined with a separate simplified analysis due to the effects of wind loading only, to obtain the moments that must be considered when designing the frame, as described in more detail in section 14.10.

The simplified method of analysis outlined in BS8110 (clause 3.2.1.2.1) involves the division of the complete structural frame into a series of sub-frames, each of which consists of the system of beams at one level together with the columns above and below that level, these columns being assumed fully fixed at the ends furthest from the beams considered (unless it is clearly more reasonable to assume that the end is pinned). Each sub-frame is analysed elastically under the following arrangements of loading. To obtain the maximum moments in the spans, alternate spans carry a total load of $1.4G_k + 1.6Q_k$ while the remaining spans support a minimum dead load of $1.0G_k$ only. For the maximum moment at any support, all the spans carry $1.4G_k + 1.6Q_k$.

The elastic moments in the simplified sub-frame may be determined by any of the methods normally used to analyse statically indeterminate structures, and thus once again two general types of solution are possible. Since more members are involved, the resulting analysis is somewhat more complex than in the case of a simple run of continuous beams and, unless computer methods are available to solve the necessary simultaneous equations, the use of an iterative method is even more advantageous when analysing a frame. Normal moment distribution is quite straightforward, and the variant known as precise moment distribution, discussed above, can also be employed as now described.

The procedure necessary to analyse a sub-frame using precise moment distribution is only slightly more involved than when analysing a system of continuous beams. If more than two members meet at a support it can be shown by slope-deflection methods that the continuity factors ϕ for successive spans are given by the expression

$$\phi_{n+1} = \frac{1}{2 + \dfrac{K_{n+1}}{\Sigma[K_N/(2-\phi_n)]}} \qquad (3.2a)$$

where ϕ_{n+1} is the continuity factor, K_{n+1} is the stiffness of the span being considered, and $\Sigma[K_n/(2-\phi_n)]$ is the sum of the values of $K_n/(2-\phi_n)$ of all the remaining members meeting at that joint. Since the far ends of all the columns are assumed to be fully fixed, $\phi_n = 0.5$ and thus $K_n/(2-\phi_n) = \frac{2}{3}K_n$ for each column. Thus if K_u and K_l are the stiffnesses of the upper and lower columns respectively, $\Sigma[K_n/(2-\phi_n)] = \frac{2}{3}(K_u + K_l) + K_n/(2-\phi_n)$, where ϕ_n and K_n are the continuity factor and stiffness of the previous span. Starting at the end beam in a sub-frame, successive values of ϕ are determined for the beams in each direction as with a continuous-beam system but using expression (3.2a); note that the corresponding nomogram *cannot* be used.

When the appropriate continuity factors have been found, the distribution factors for the beams at each side of each joint are again given by formula (3.3) or the corresponding nomogram on *Data Sheet 10*. However, in the case of a frame, the sum of the distribution factors obtained at a support will not equal unity because of the interaction of the columns. At a support B, say, the sum of the distribution factors for the upper and lower columns is equal to $1 - DF_{BA} - DF_{BC}$, where DF_{BA} and DF_{BC} are the distribution factors for the beams on each side of the support. The distribution factor relating to each individual column is then obtained by dividing the total column distribution factor in proportion to the stiffnesses of the columns (i.e. if the stiffness of the lower column is twice that of the upper, the distribution factor will be twice as great).

Since there is now no automatic check when calculating the distribution factors, as is the case when analysing continuous-beam systems, especial care is essential when calculating the continuity and distribution factors.

When the continuity and distribution factors and fixed-end moments have been determined, in the case of a frame the mathematical distribution procedure differs slightly from that described earlier. After distributing the unbalanced moment at the first support, the amount to be carried over is determined as before by multiplying by the appropriate right-to-left continuity factor. The moment carried over to the far end of the span is now divided between the remaining members meeting at that point in proportion to their values of $K_{LR}/(2-\phi_{RL})$, i.e. their 'relative restraint factors' (which correspond to their restraint factors divided by a constant value of $6E$). At each joint it is therefore worth while calculating $K_{LR}/(2-\phi_{RL})$ for each member. The amount of carried-over moment transferred to the next beam is thus obtained by multiplying it by the ratio of the relative restraint factor for that beam to the sum of the factors for the beam and the columns. The out-of-balance fixed-end moment at this joint is now distributed between all the members in proportion to their distribution factors, and the algebraic sum of the moments on the next span is carried over to the succeeding support as before.

Thus in *Figure 3.11*, which corresponds to the beam system considered previously but now forms part of a frame, the columns are 4 m long and have a value of I of one-tenth of the beams. The continuity factors are found as follows:

$$\phi_{AB} = \frac{1}{2 + \dfrac{K_{AB}}{\frac{2}{3}(K_{Au} + K_{Al})}} = \frac{1}{2 + \dfrac{\frac{1}{6}}{\frac{2}{3}(\frac{1}{40} + \frac{1}{40})}} = 0.143$$

$$\phi_{BC} = \frac{1}{2 + \dfrac{K_{BC}}{\left(\dfrac{K_{AB}}{2-\phi_{AB}}\right) + \frac{2}{3}(K_{Bu} + K_{Bl})}}$$

$$= \frac{1}{2 + \dfrac{\frac{1}{6}}{\left(\dfrac{\frac{1}{9}}{2-0.143}\right) + \frac{2}{3}(\frac{1}{40} + \frac{1}{40})}} = 0.344$$

25

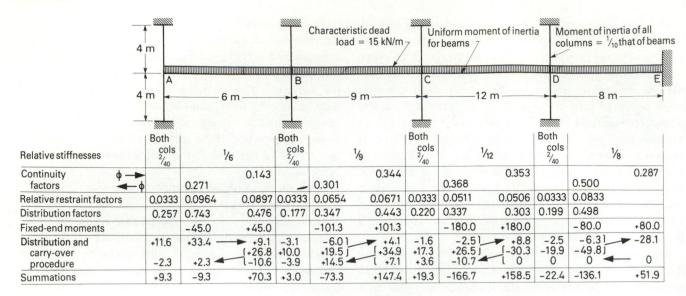

Relative stiffnesses	Both cols $\frac{2}{40}$	$\frac{1}{6}$		Both cols $\frac{2}{40}$	$\frac{1}{9}$		Both cols $\frac{2}{40}$	$\frac{1}{12}$		Both cols $\frac{2}{40}$	$\frac{1}{8}$	
Continuity factors $\phi \rightarrow$ $\leftarrow \phi$		0.143 0.271			0.344 0.301			0.353 0.368			0.287 0.500	
Relative restraint factors	0.0333	0.0964	0.0897	0.0333	0.0654	0.0671	0.0333	0.0511	0.0506	0.0333	0.0833	
Distribution factors	0.257	0.743	0.476	0.177	0.347	0.443	0.220	0.337	0.303	0.199	0.498	
Fixed-end moments		−45.0	+45.0		−101.3	+101.3		−180.0	+180.0		−80.0	+80.0
Distribution and carry-over procedure	+11.6 −2.3	+33.4 → +2.3 ←	+9.1 −3.1 {+26.8 +10.0 {−10.6 −3.9		−6.0} +4.1 −1.6 +19.5} {+34.9 +17.3 +14.5 { +7.1 +3.6		−2.5} +8.8 −2.5 +26.5} {−30.3 −19.9 −10.7← { 0 0		−6.3} →−28.1 −49.8} 0 ←	0
Summations	+9.3	−9.3	+70.3 +3.0		−73.3 +147.4 +19.3		−166.7 +158.5 −22.4		−136.1	+51.9

Figure 3.11

and so on. Similarly, $\phi_{ED} = 0.5$ since E is fully fixed. Then

$$\phi_{BC} = \cfrac{1}{2 + \cfrac{K_{CD}}{\left(\cfrac{K_{DE}}{2 - \phi_{ED}}\right) + \frac{2}{3}(K_{Du} + K_{Dl})}}$$

$$= \cfrac{1}{2 + \cfrac{\frac{1}{12}}{\left(\cfrac{\frac{1}{8}}{2 - 0.5}\right) + \frac{2}{3}(\frac{1}{40} + \frac{1}{40})}} = 0.368$$

$$\phi_{CB} = \cfrac{1}{2 + \cfrac{\frac{1}{9}}{\left(\cfrac{\frac{1}{12}}{2 - 0.368}\right) + \frac{2}{3}(\frac{1}{40} + \frac{1}{40})}} = 0.301$$

and so on.

Now since $\phi_{CB} = 0.301$, $\phi_{BC} = 0.344$ and $K_{BC}/6E = 1/9$, the relative restraint factors $R'_{BC} = 1/[9(2 - 0.301)] = 0.0654$ and $R'_{CB} = 1/[9(2 - 0.344)] = 0.0671$. The remaining relative restraint factors may be found in a similar manner.

Then from equation (3.3) or the right-hand nomogram on *Data Sheet 10*, the distribution factors are found as

$$DF_{BC} = \frac{1 - 1\phi_{BC}}{1 - \phi_{BC}\phi_{CB}} = \frac{1 - (2 \times 0.344)}{1 - (0.344 \times 0.301)} = 0.347$$

$$DF_{BA} = \frac{1 - 1\phi_{BA}}{1 - \phi_{BA}\phi_{AB}} = \frac{1 - (2 \times 0.271)}{1 - (0.271 \times 0.143)} = 0.476$$

Consequently $DF_{Bu} + DF_{Bl} = 1 - 0.347 - 0.476 = 0.177$, so that $DF_{Bu} - DF_{Bl} = 0.0885$. The remaining distribution factors are determined in the same way.

The fixed-end moments for the beams are identical to those determined in *Figure 3.5*. Then, considering dead loads only for the present, the distribution procedure is as

follows. Of the unbalanced moment of +45 kN m at A, $0.743 \times 45.0 = 33.4$ kN m is resisted by AB. The moment carried over to joint B is obtained by multiplying this value by ϕ_{BA}, i.e. $33.4 \times 0.271 = 9.1$ kN m. This must be balanced by an equal and opposite moment of −9.1 kN m and, of this amount, $-9.1 \times 0.0654/(0.0654 + 0.0333) = -6.0$ kN m is resisted by BC and the rest (i.e. $9.1 - 6.0 = 3.1$ kN m) by the two columns at B. The out-of-balance fixed-end moment at B is $45.0 - 101.3 = -56.3$ kN m, which requires balancing moments of $56.3 \times DF_{BA} = 56.3 \times 0.476 = 26.8$ kN m in BA, $56.3 \times 0.347 = 19.5$ kN m in BC and 10.0 kN m in the columns. The algebraic sum of the moments in BC at B is thus $19.5 - 6.0 = 13.5$ kN m, so that the resulting moment carried over to C is $13.5 \times \phi_{CB} = 4.1$ kN m. This moment is distributed between CD and the columns at C and added algebraically to the values required to balance the out-of-balance fixed-end moments at C, and so on. The same procedure is then repeated starting from the right-hand end of the span; the moment balancing the carried-over moment is divided between the subsequent beam and the columns in proportion to their relative restraint factors, and the moment to be carried over in the next span is determined by summing algebraically the appropriate value obtained in this operation and that already obtained during the distribution of the unbalanced fixed-end moments. The complete distribution and carry-over procedure is tabulated in *Figure 3.11*.

If the loading on any span is symmetrical (i.e. if the fixed-end moments are the same at both supports), symmetrical residual-moment factors obtained by using expression (3.4) or the left-hand nomogram on *Data Sheet 11* may be employed, as described in section 3.7.1. (The concept of antisymmetrical residual-moment factors may also be employed for unsymmetrical loads.) The residual moment at the end of the loaded span must now, however, be balanced by moments that are distributed

	A		B			C			D			E
Residual-moment factors		0.325	0.725		0.806	0.662		0.770	0.822		0.502	1.249
Fixed-end moments due to 'imposed' load		−90.0	+90.0		−202.5	+202.5		−360.0	+360.0		−160.0	+160.0
Dead load	+9.3	−9.3	+70.3	+3.0	−73.3	+147.4	+19.3	−166.7	+158.5	−22.4	−136.1	+51.9
Imposed load on AB	+29.3	−29.3	+65.3	−22.0	−43.3 →	−13.0	+5.1	+7.9 →	+2.9	−0.8	−2.1 →	−1.0
on BC	−17.0	+17.0 ←	+119.0	+44.1	−163.1	+134.0	−52.9	−81.1 →	−29.9	+8.6	+21.3 →	+10.7
on CD	+6.6	−6.6 ←	−46.5	−17.3	+63.8 ←	+185.2	+92.0	−277.2	+295.8	−84.5	−211.3 →	−105.6
on DE	−0.4	+0.4 ←	+2.9	+1.0	−3.9 ←	−11.4	−5.7	+17.1 ←	+48.4	+31.9	−80.3	+199.9
For max. support moments	+27.8	−27.8	+211.0	+8.8	−219.8	+442.2	+57.8	−500.0	+475.7	−67.2	−408.5	+155.9
For max. mts in AB and CD	+45.2	−45.2	+89.1	−36.3	−52.8	+319.6	+116.4	−436.0	+457.2	−107.7	−349.5	−54.7
For max. mts in BC and DE	−8.1	+8.1	+192.2	+48.1	−240.3	+270.0	−39.3	−230.7	+177.0	+18.1	−195.1	+262.5

Figure 3.12

between the remaining members meeting at that joint in proportion to their relative restraint factors. Thus in *Figure 3.12* the fixed-end moments at B and C on span BC due to an 'imposed' load of $0.4G_k + 1.6Q_k$ are 202.5 kN m. Since the symmetrical residual-moment factor at C is 0.662, the final moment in BC at C is $202.5 \times 0.662 = 134.0$ kN m. Then, of the moment of −134.0 kN m required to balance this value, $-134.0 \times 0.0511/(0.0511 + 0.0333) = -81.1$ kN m is transferred to CD and $-134.0 \times 0.0333/0.0844 = -52.9$ kN m is transferred to the columns. The moment carried over to D is thus $-81.1 \times \phi_{DC} = -81.1 \times 0.368 = -29.9$ kN m and, to balance this, moments of $29.9 \times 0.0833/(0.0833 + 0.0333) = 21.3$ kN m and $29.9 \times 0.0333/0.1167 = 8.6$ kN m are set up in DE and in the columns at D respectively. The moment carried over to E is thus $\frac{1}{2} \times 21.3 = 10.7$ kN m. Similarly the residual moment at B is $-202.5 \times 0.806 = -163.1$ kN m, since the symmetrical residual-moment factor at B for span BC is 0.806. Thus balancing moments of $163.1 \times 0.0897/(0.0897 + 0.0333) = 119.0$ kN m and $163.1 \times 0.0333/0.1230 = 44.1$ kN m occur in AB and the columns at B respectively. The moment transferred to joint A is thus $119.0 \times \phi_{AB} = 119.0 \times 0.143 = 17.0$ kN m.

As in the case of continuous-beam systems, it is possible to combine the use of symmetrical residual-moment factors and distribution factors, using the former for those spans that support symmetrical loading and the latter elsewhere.

The moments in each member at each joint due to a minimum dead load of $1.0G_k$ on all spans and to an 'imposed' load of $0.4G_k + 1.6Q_k$ on each individual span in turn are tabulated in *Figure 3.12*. The summations giving the maximum support moments and the support moments corresponding to the maximum span moments are also tabulated. The maximum positive and negative moments transferred to the columns occur under the same loading conditions that give the maximum span moments. The maximum moments in the span can now be determined as described earlier. For example, the maximum moment in span BC is about $(1.5 \times 303.8) - \frac{1}{2}(240.3 + 270.0)$ or 200 kN m.

Comparing the maximum values obtained in *Figure 3.12* with those resulting earlier from the continuous-beam analysis, it is seen that, except for the previously

'free' support, the maximum moments in the beams at the supports are very slightly greater (a maximum of about 5% at the internal supports) in the case of a frame. The maximum moments in the spans, however, are somewhat less (about 8.5% in the two inner spans) in the frame.

3.8.2 Simplified sub-frame

An alternative analytical method that is also permitted by BS8110 when considering the effects of vertical loading only, is to determine the moments in each beam in a floor in turn by considering a simplified sub-frame that consists of the beam in question with its adjoining beams and columns assumed fixed at their further ends (unless it is clearly more reasonable to assume a pinned end), as shown in *Figure 3.13*. The stiffnesses of the beams adjoining that for which the analysis is being undertaken should be taken as one-half of their calculated values.

The adoption of this simplified sub-frame may at first sight appear not to save a great deal of calculation since, to analyse a system comprising x bays, x separate three-bay sub-frames must be investigated, each under the action of three different arrangements of loading. However, the advantage of the system chosen is that the final moments can be expressed explicitly by relatively simple formulæ, thus eliminating the need for complex structural analysis. The derivations of such formulæ by slope-deflection methods, and their use to determine typical span and support moments in the frame considered in *Figures 3.11* and *3.12*, are now given.

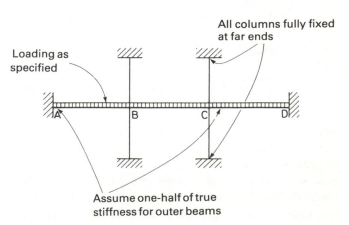

Figure 3.13 Simplified three-bay sub-frame.

Bending moments on structural members

As before, for a prismatic member the slope-deflection equations are

$$M_{BC} = K_{BC}(\theta_{BC} + \tfrac{1}{2}\theta_{CB}) \quad \text{and} \quad M_{CB} = K_{BC}(\tfrac{1}{2}\theta_{BC} + \theta_{CB})$$

where the stiffness K of any member is equal to $4EI/l$. For the frame being considered, it can be shown that

$$\theta_{BC} = \frac{\begin{array}{c}\tfrac{1}{2}K_{BC}(FEM_{CB} - FEM_{CD}) \\ + \Sigma K_C(FEM_{BC} - FEM_{BA})\end{array}}{\Sigma K_B \Sigma K_C - \tfrac{1}{4}K_{BC}^2}$$

$$\theta_{CB} = \frac{\begin{array}{c}-\tfrac{1}{2}K_{BC}(FEM_{BC} - FEM_{BA}) \\ -\Sigma K_B(FEM_{CB} - FEM_{CD})\end{array}}{\Sigma K_B \Sigma K_C - \tfrac{1}{4}K_{BC}^2}$$

where $\Sigma K_B = K_{BC} + K_{BA} + K_{Bu} + K_{Bl}$ and $\Sigma K_C = K_{BC} + K_{CD} + K_{Cu} + K_{Cl}$, and FEM_{BC} is the fixed-end bending moment at B on span BC, and so on. Now if the distribution factor

$$D_{BC} = \frac{K_{BC}}{K_{BC} + K_{BA} + K_{Bu} + K_{Bl}} = \frac{K_{BC}}{\Sigma K_B}$$

and so on, it is possible to eliminate θ_{BC} and θ_{CB} and to express the final support moments at B and C on span BC in terms of the distribution factors and fixed-end moments only. Note that, to denote the distribution factors, a symbol (D_{XX}) is employed here which is slightly different from that (DF_{XX}) used in precise moment distribution. This is because the present distribution factors represent directly the ratio of the stiffness of the member concerned to the sum of the stiffnesses of all the members meeting at that joint, and are thus the same as the distribution factors used in normal moment distribution. They therefore differ from the modified distribution factors employed in precise moment distribution, which are derived from the restraint factors.

Considering now an arbitrary span LR, the final support moments M_{LR} and M_{RL} are thus

$$M_{LR} = FEM_{LR} + \frac{D_{LR}}{4 - D_{LR}D_{RL}}$$

$$\times \left[2D_{RL}\left(\frac{1}{D_{LR}} - 1\right)F_R' - (4 - D_{RL})F_L' \right] \quad (3.5)$$

$$M_{RL} = FEM_{RL} + \frac{D_{RL}}{4 - D_{LR}D_{RL}}$$

$$\times \left[2D_{LR}\left(\frac{1}{D_{RL}} - 1\right)F_L' - (4 - D_{LR})F_R' \right] \quad (3.6)$$

where F_L' and F_R' are respectively the out-of-balance fixed-end moments at L and R for the particular loading condition considered. These out-of-balance moments are assumed positive if, under normal (i.e. downward-acting) loads, the fixed-end moment at the side of LR that is being considered exceeds the fixed-end moment on the other side of that support.

Writing

$$\alpha = \frac{(4 - D_{RL})D_{LR}}{4 - D_{LR}D_{RL}} \quad \text{and} \quad \beta = \frac{2D_{RL}(1 - D_{LR})}{4 - D_{LR}D_{RL}}$$

enables equation (3.5) to be expressed as

$$M_{LR} = FEM_{LR} - \alpha F_L' + \beta F_R' \quad (3.5a)$$

Similarly, transposing D_{RL} and D_{LR} in equation (3.6) to obtain α' and β',

$$M_{RL} = FEM_{RL} - \alpha' F_R' + \beta' F_L' \quad (3.6a)$$

The chart on *Data Sheet 12* gives the values of α and β for given distribution factors D_{LR} and D_{RL}. By transposing the distribution factors, the appropriate values of α' and β' may then be read from the same chart, and by substituting values of F_L' and F_R' corresponding to the appropriate critical loading conditions in formulæ (3.5a) and (3.6a), the final support moments M_{LR} and M_{RL} may be obtained.

It is also necessary to determine the support moments in the adjoining columns and outer beams. All these members are assumed to be fully fixed at their further ends and, for such a member as AB say, the slope-deflection equation is

$$M_{BA} = K_{AB}\theta_{BA}$$

where $K_{AB} = 4EI_{AB}/l_{AB}$. Then, since the rotation of all the members meeting at a joint is equal, $\theta_{BA} = \theta_{BC}$ and thus

$$M_{BA} = \frac{K_{AB}(\tfrac{1}{2}K_{BC}F_C' + \Sigma K_C F_B')}{\Sigma K_B \Sigma K_C - \tfrac{1}{4}K_{BC}^2}$$

Since $K_{AB}/\Sigma K_B = D_{BA}$, $K_{BC}/\Sigma K_B = D_{BC}$ and $K_{BC}/\Sigma K_C = D_{CB}$, this may be written as

$$M_{BA} = \frac{D_{BA}(2D_{CB}F_C' + 4F_B')}{4 - D_{BC}D_{CB}}$$

Thus for any such beam or column meeting at L, the far end of which is fully fixed, employing the previously adopted notation and sign convention, the final moment M in the member at the joint is

$$M = FEM + \text{distribution factor for member concerned}$$

$$\times \frac{2D_{RL}F_R' + 4F_L'}{4 - D_{LR}D_{RL}} \quad (3.7)$$

where FEM is the fixed-end moment (if any) in the member being considered at the joint concerned. The simplicity of formula (3.7) makes the use of a chart unnecessary.

To obtain the maximum moment at each support, all the spans forming the sub-frame should carry the total load (i.e. $1.4G_k + 1.6Q_k$). For the critical moment near midspan on the central span the total load should be applied to this span only, while the outer spans support the minimum dead load. This latter arrangement of loading also produces the maximum moments in the columns. These moments may result when either of the beams adjoining a particular column forms the middle span of the sub-frame, but normally the maximum column moments occur when the central span of the sub-frame is the longer of the two spans adjoining the column, and this is the criterion specified in BS8110.

The foregoing formulae and chart are equally applicable when the 'central' span of the sub-frame corresponds to the end span of the actual structure. This is illustrated in the following example.

3.8.3 Sub-frame method: worked example

Determine the maximum moments at support B and in span BC of the system considered in *Figures 3.11* and *3.12* using the simplified sub-frame method.

Consider first span BC as the central span of the three-bay sub-frame. The required distribution factors are as follows:

$$D_{BC} = \frac{K_{BC}}{^1/_2 K_{AB} + K_{BC} + K_{Bu} + K_{Bl}}$$

$$= \frac{^1/_9}{(^1/_2 \times {}^1/_6) + {}^1/_9 + {}^1/_{40} + {}^1/_{40}} = \frac{5}{11} = 0.455$$

$$D_{CB} = \frac{K_{CB}}{^1/_2 K_{CD} + K_{BC} + K_{Cu} + K_{Cl}}$$

$$= \frac{^1/_9}{(^1/_2 \times {}^1/_{12}) + {}^1/_9 + {}^1/_{40} + {}^1/_{40}} = \frac{40}{73} = 0.548$$

Thus from *Data Sheet 12*, $\alpha = 0.418$ and $\beta = 0.159$, and transposing D_{BC} and D_{CB}, $\alpha' = 0.518$ and $\beta' = 0.110$.

The fixed-end moments are as follows:

Span AB: for total load
$$FEM_{AB} = 45 \times 6^2/12 \ = 135 \text{ kN m}$$
for dead load only
$$FEM_{AB} = 15 \times 6^2/12 \ = 45 \text{ kN m}$$
Span BC: for total load
$$FEM_{BC} = 45 \times 9^2/12 \ = 303.8 \text{ kN m}$$
for dead load only
$$FEM_{BC} = 15 \times 9^2/12 \ = 101.3 \text{ kN m}$$
Span CD: for total load
$$FEM_{CD} = 45 \times 12^2/12 = 540 \text{ kN m}$$
for dead load only
$$FEM_{CD} = 15 \times 12^2/12 = 180 \text{ kN m}$$

Now for the maximum moment at support B, all three spans must carry dead + imposed load. Thus $F'_B = 303.8 - 135 = 168.8 \text{ kN m}$ and $F'_C = 303.8 - 540 = -236.2 \text{ kN m}$. Then

$$M_{BC} = FEM_{BC} - \alpha F'_B + \beta F'_C$$
$$= 303.8 - (0.418 \times 168.8)$$
$$+ (0.159 \times -236.2) = 195.7 \text{ kN m}$$

and since

$$D_{BA} = \frac{^1/_2 K_{AB}}{^1/_2 K_{AB} + K_{BC} + K_{Bu} + K_{Bl}}$$

$$= \frac{^1/_{12}}{^1/_{12} + {}^1/_9 + {}^1/_{40} + {}^1/_{40}} = \frac{15}{44} = 0.341$$

then

$$M_{BA} = 135.0 + 0.341$$
$$\times \left(\frac{2 \times 0.548 \times -236.2 + 4 \times 168.8}{4 - 0.455 \times 0.548} \right)$$
$$= 172.9 \text{ kN m}$$

For the maximum moment in span BC, this span carries total load while AB and CD carry dead load only. Thus now $F'_B = 303.8 - 45 = 258.8 \text{ kN m}$ and $F'_C = 303.8 - 180 = 123.8 \text{ kN m}$. Then

$$M_{BC} = FEM_{BC} - \alpha F'_B + \beta F'_C$$
$$= 303.8 - (0.418 \times 258.8) + (0.159 \times 123.8)$$
$$= 215.2 \text{ kN m}$$

$$M_{CB} = FEM_{CB} - \alpha' F'_C + \beta' F'_B$$
$$= 303.8 - (0.518 \times 123.8) + (0.110 \times 258.3)$$
$$= 268.0 \text{ kN m}$$

Consequently the maximum moment in span BC is approximately

$$(1.5 \times 303.8) - {}^1/_2 (215.2 + 268.0) = 214 \text{ kN m}$$

To ensure that the maximum value of the support moment at B has been determined, it is now necessary to consider the sub-frame formed by spans AB and BC only, the former being taken as the 'central' span. Now the fixed-end moments remain as before, and the distribution factors are

$$D_{AB} = \frac{K_{AB}}{K_{Au} + K_{Al} + K_{AB}}$$

$$= \frac{^1/_6}{^1/_{40} + {}^1/_{40} + {}^1/_6} = \frac{10}{13} = 0.769$$

$$D_{BA} = \frac{K_{AB}}{K_{AB} + {}^1/_2 K_{BC} + K_{Bu} + K_{Bl}}$$

$$= \frac{^1/_6}{^1/_6 + {}^1/_9 \times {}^1/_2 + {}^1/_{40} + {}^1/_{40}} = \frac{30}{49} = 0.612$$

giving, from *Data Sheet 12*, $\alpha = 0.738$, $\beta = 0.080$, $\alpha' = 0.560$ and $\beta' = 0.169$. For the maximum support moment at B, both AB and BC must be fully loaded. Thus $F'_A = 135.0 - 0 = 135.0 \text{ kN m}$, and $F'_B = 135.0 - 303.8 = -168.8 \text{ kN m}$, so that

$$M_{BA} = FEM_{BA} - \alpha' F'_B + \beta' F'_A$$
$$= 135.0 - (-168.8 \times 0.560) + (0.169 \times 135.0)$$
$$= 252.4 \text{ kN m}$$

and since

$$D_{BC} = \frac{^1/_2 K_{BC}}{K_{AB} + {}^1/_2 K_{BC} + K_{Bu} + K_{Bl}}$$

$$= \frac{^1/_{18}}{^1/_6 + {}^1/_{18} + {}^1/_{40} + {}^1/_{40}} = \frac{10}{49} = 0.204$$

then

$$M_{BC} = 303.8 + 0.204$$

$$\times \left(\frac{2 \times 0.769 \times 135 - 4 \times 168.8}{4 - 0.769 \times 0.612} \right)$$

$$= 276.7 \, \text{kN m}$$

The maximum support moment to be considered at B is thus 276.7 kN m.

According to the criterion specified in BS8110, the maximum moments in the columns at B will occur with sub-frame AB–BC–CD, when BC carries a load of $1.4G_k + 1.6Q_k$ while AB and CD are loaded with $1.0G_k$. The maximum moments are then 31.9 kN m at B.

If these results are compared with those obtained earlier by analysing the floor and adjoining columns as a complete structure it is seen that in this particular example, which may not be typical, the values obtained correspond reasonably well. The maximum support moment at B is overestimated by $276.7 - 219.8$, or about 57 kN m (i.e. 26%), while the maximum moment in span BC is overestimated by about $214 - 200 = 14$ kN m (i.e. 6%). The three-bay sub-frame method underestimates the column moments, as given by more accurate analysis, by $(48.1 - 31.9)/2 = 8.1$ kN m, or about one-third of the true value.

3.9 APPLICATION TO THE DESIGN OF A BUILDING

Some of the methods of calculating the bending moments on beams and slabs described in this chapter are applied to the design of the upper floors of the building in Part Two, as described in detail in Chapter 9 for beam-and-slab construction with the slabs spanning in one direction. Briefly the application is as follows.

3.9.1 Slabs

The bending moments on the floor slabs spanning in one direction are calculated from the bending-moment formulæ on *Data Sheet 7*, as these slabs are continuous over equal spans and carry uniformly distributed loads, the ratio of characteristic imposed to characteristic dead load (excluding partitions) being greater than 1.25.

3.9.2 Secondary beams

The formulae in *Data Sheet 7* are also used to calculate the bending moments on the secondary beams. Since the variation in span length (5.425 m minimum to 6 m maximum) is about 10%, the resulting moments are only approximate and therefore no redistribution is made. The approximate coefficients given in section 3.2 cannot be used as the characteristic imposed load exceeds the characteristic dead load.

3.9.3 Main beams

The bending moments on the main beams supporting the secondary beams are calculated from the formulæ given on *Data Sheet 9* for three-span beams supporting uniform and central concentrated loads. Since these beams have almost exactly equal spans, the resulting moments are sufficiently accurate to undertake some moment redistribution. However, as the resulting beam sections are to be designed by the simplified Code formulae (see section 5.2.3), which assume a value of x/d of 0.5, the amount of this redistribution is limited to 10%. The effect of the beams being monolithic with the exterior columns is dealt with in accordance with the data given on *Data Sheet 14*.

Chapter 4
Material strengths and design stresses

4.1 SPECIFICATION OF CONCRETE

As stated in the Foreword to BS8110, the specifying of concrete mixes is now only dealt with summarily in the Code since detailed information on the specification, production and testing of mixes is provided in BS5328 'Methods of specifying concrete, including ready-mixed concrete', which BS8110 states must be complied with. Two basic types of mixes are defined in BS5328: 1981, namely designed mixes and prescribed mixes, and theoretically each type may be arranged to produce either ordinary structural concrete or special structural concrete. In the case of ordinary structural concrete, the only constituents that may be used are plain, blast-furnace or sulphate-resisting Portland cement, certain types of natural aggregates and water; whereas with special structural concrete, other constituents such as admixtures, and special types of cement or aggregate or both, may be employed.

Rather confusingly, BS8110 lists three types of mix, namely designed mixes, special prescribed mixes and ordinary standard mixes; however, the *Code Handbook* explains that what BS8110 refers to as an 'ordinary standard mix' corresponds to the BS5328 (and CP110) designation 'ordinary prescribed mix'. In the revision to BS5328 currently being prepared, the three types of mix listed in BS8110 will apparently be referred to simply as 'designed', 'prescribed' and 'standard', respectively.

Prescribed (and standard) mixes may be considered broadly as the successors to the standard mixes described in earlier codes such as CP114. With such mixes it is the responsibility of the engineer both to specify the performance required from the concrete and to select mix proportions that will achieve the necessary requirements regarding strength, durability etc. In this case, the sole responsibility of the producer is to achieve a properly mixed concrete containing the specified amounts of constituents. BS5328 provides a table giving the weights of aggregates to be used with 100 kg of cement to produce finished concrete of grades 7.5, 10, 15, 20, 25 and 30. (In BS8110, BS5328 and elsewhere, such as this book, the grade number of a concrete mix represents the characteristic strength at 28 days in N/mm². This is specifically true in the case of designed mixes, and normally – though not contractually – true otherwise.) The amount of constituents given in the Standard are generally appropriate except where a combination of poor control and poor materials occurs, but slight adjustments may be needed in the quantities and proportions of aggregate in order to achieve the strength, workability and cement content required or to take account of the properties of materials obtained locally.

If the proportions tabulated in BS5328 are adopted, all the designer normally need do is state the grade of concrete, type of cement, and type and maximum size of aggregate. Since the cement content of the mixes is high, the need to produce trial mixes or acceptance cubes is eliminated. In special circumstances, however, other information such as workability, nature and sources of materials, and limiting mix temperatures may be specified.

If a prescribed mix other than those tabulated in the Standard is to be used (i.e. a special prescribed mix), the engineer must specify additional factors, for example the permitted types of cement, the minimum cement content, the proportions of materials, the permissible types of aggregates, and so on. Preliminary strength tests will probably be needed, and further tests may also be requested as work proceeds.

Designed mixes place on the supplier of the concrete the responsibility for selecting the materials and proportions necessary to achieve the required strength and properties specified by the engineer. In the case of ordinary structural concrete, all the engineer need do is indicate the permissible type and minimum amount of cement, the permissible type and maximum size of aggregate, and any other requirements that must be met to ensure satisfactory performance. BS5328 draws attention to the fact that, with such mixes, strength testing is an essential factor in judging whether the concrete complies with its specification. Designed mixes for special structural concretes cater for the use of other materials and the need to achieve special properties or finishes, and so on, and additional information may be required for these.

For full details of the specification of concrete according to BS5328, reference should be made to the Standard itself, supplemented by the remarks in section 6 of BS8110 and the *Code Handbook*.

4.2 DESIGN STRENGTHS OF CONCRETE

4.2.1 Characteristic and design strengths

BS5328 defines the characteristic strength f_{cu} of a mix as the value of the strength below which not more than 5% of all possible measurements of the strength of the concrete concerned are expected to fall. This requirement can be more usefully expressed as the need to achieve a 'target mean strength' which exceeds the characteristic strength by a difference which is sometimes known as the 'current margin', which is usually taken as 1.64 times s, the positive square root of the variance, but not less than 3.75 N/mm^2 on tests on at least 100 similar batches of concrete made within 12 months, or 1.64s but not less than 7.5 N/mm^2 on tests on at least 40 similar concrete batches made between 5 days and 6 months.

When sections are analysed according to ultimate limit-state principles, the design strength of a mix is obtained by dividing the characteristic strength f_{cu} by the appropriate partial safety factor for materials γ_m. In many design formulae, however, the appropriate partial safety factor for concrete in bending and direct compression is embodied in the formulae themselves, so that the resulting values are directly related to the characteristic strength of the concrete, thus simplifying the calculations. However, if the section is to be analysed for a less usual ultimate limit-state, such as that resulting from local damage or overloading, BS8110 recommends the adoption of a value of γ_m of 1.3 instead of 1.5, and such an adjustment should also be taken into account when using those design formulæ that do not include γ_m as a variable.

In all limit-state calculations concerning the design of sections which involve the strength of the concrete in bending or direct compression, the appropriate formulae require the direct use of the characteristic strength f_{cu} of the concrete. In earlier codes such as CP114, in the case of slender members such as columns, the load-carrying capacity of such members was obtained by multiplying the calculated capacity of a 'short' member by a reduction factor related to the slenderness; this in effect was equivalent to reducing the design stresses in the concrete and the reinforcement. However, in BS8110 (and its predecessor CP110) the procedure is entirely different, as explained in section 14.8, and consists of designing the section to resist an additional moment relating to the slenderness.

According to BS8110, the limiting shearing stress v_c on a concrete without special shearing reinforcement is not solely based on a simple relationship to the crushing strength of the mix. Instead, the formulae and tabulated values given in the Code have been established from large numbers of tests and depend on the depth of section and on the proportion of main steel provided at, and extending a distance of at least d beyond, the section under consideration, as well as on the cube root of the characteristic strength: see Chapter 6. The values of limiting strength for a concrete with special shearing reinforcement given by BS8110 have been rounded off from those given by the expression $v_{max} = 0.8 \sqrt{f_{cu}}$ (but not more than 5 N/mm^2). Similarly, the limiting ultimate torsional stresses with and without special reinforcement are given by the expressions $v_{tu} = 0.8 \sqrt{f_{cu}}$ (but not more than 5 N/mm^2) and $v_{t min} = 0.067 \sqrt{f_{cu}}$ (but not more than 0.4 N/mm^2). Note that BS8110 specifies the use of a partial safety factor of 1.25 when calculating the shearing resistance of concrete without special shearing reinforcement.

In the case of lightweight-aggregate concrete also, the limiting values of v_c according to the Code are 80% of those of normal dense concrete. Thus for mixes of grade 25 and above, the values of v_{max}, $v_{t min}$ and v_{tu} are obtained by multiplying the values given by the expressions and limits in the previous paragraph by 0.8. For grade 20 lightweight concrete, BS8110 (in Table 5.3 of Part 2) gives values of v_c for different proportions of main steel. However, these values only apply for section depths of 400 mm or greater. For more shallow sections, the correct values can be calculated by using the Code formula (see *Data Sheet 15*) and multiplying the resulting values by 0.8.

Anchorage-bond stresses f_{bu} are also related to the cube root of the concrete strength. The basic relationship is $f_{bu} = \beta \sqrt{f_{cu}}$ where, for type 1 deformed bars in tension, $\beta = 0.4$. For plain bars and type 2 deformed bars, the values of β are 0.28 (i.e. 70%) and 0.5 (i.e. 125%) respectively. With bars in compression, all these values are increased by a further 25%. The Code specifies the use of a partial safety factor of 1.4 for bond strength calculations.

4.2.2 Modification of strength with age

If it is certain that a member will not be required to carry its full design load until some time after it is cast, the member may be designed using a modified characteristic strength, if so desired. The appropriate values given in the Code, which depend on the value of f_{cu} at 28 days and on the age of loading, are shown graphically in *Figure 4.1*.

4.3 CHARACTERISTIC AND DESIGN STRENGTHS IN REINFORCEMENT

As in the case of concrete, the characteristic strength of reinforcement is defined in BS8110 as that value of yield or proof stress below which not more than 5% of the test results fall. Reinforcement complying with the requirements of BS4449, BS4461 and BS4483 has a specified characteristic strength corresponding to the

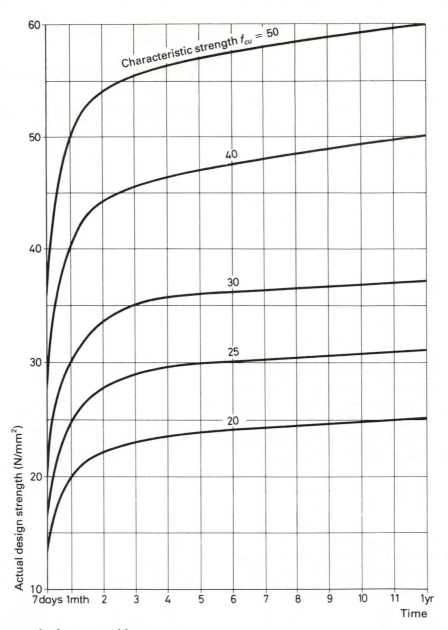

Figure 4.1 Increase in strength of concrete with age.

minimum yield-point stress for the particular type of steel specified. These British Standards thus specify the following characteristic strengths:

Mild steel bars of all sizes $\qquad f_y = 250 \, \text{N/mm}^2$
High-yield steel bars of all sizes $\qquad f_y = 460 \, \text{N/mm}^2$
  (hot-rolled or cold-worked)

The maximum design strength f_{yd2} in the tension steel is obtained by dividing the characteristic strength f_y by the appropriate partial safety factor γ_m, but again many design formulæ embody the partial safety factor in the numerical constants involved, so that the resulting value is directly related to the characteristic strength. This is done, for example, in the series of design formulæ for rectangular beams provided in BS8110. As discussed in section 5.2, if rigorous limit-state analysis is employed, the design stress f_{yd2} is also related to the strain in the reinforcement, and thus to the ratio of x/d. Thus the value of f_y/γ_m is the *maximum* design stress that can be adopted.

In CP110, the maximum design stress f_{yd1} in the compression reinforcement was restricted to about four-fifths of the limiting value in tension, but this restriction has been removed in BS8110. Thus if rigorous limit-state analysis is adopted, the maximum value of f_{yd1} is equal to that of f_{yd2} (i.e. $f_y/1.15$). However, since it is related to the strain in the steel, the actual value that may be adopted for f_{yd1} also depends on the ratio of d', the depth to the compression steel from the concrete surface in compression, to x, the depth to the neutral axis (see section 5.2.1).

In reinforcement that is provided in the form of longitudinal or inclined bars or links to resist direct or torsional shearing forces, the same maximum design stresses may be adopted as those employed for the same types of steel elsewhere. Unlike the situation with CP110, no reduced upper limit is prescribed for the maximum characteristic strength of this steel.

Chapter 5
Members subjected to bending only

BS8110 considers three methods of determining the moment of resistance of a section at the ultimate limit-state. A rigorous analysis from first principles may be made, either utilizing the stress–strain relationship for concrete provided in the Code itself, or by replacing the somewhat complex 'stress-block' that results from this relationship by an equivalent rectangle. Alternatively, BS8110 also provides specific design formulae that are derived from rigorous analysis with a rectangular stress-block, with the addition of further simplifying assumptions. In this chapter, each of these methods is discussed in turn as it is applied to rectangular sections reinforced in tension only, to rectangular sections reinforced in tension and compression, and to sections of other shapes.

For brevity and simplicity the design of the beams and slabs for the building in Part Two is undertaken using only the design formulae given in the Code. Worked examples illustrating all three of the foregoing design methods are given in *RCDH*.

5.1 RECTANGULAR SECTIONS REINFORCED IN TENSION ONLY

If no compression reinforcement is to be provided, BS8110 imposes an upper limit of $\frac{1}{2}d$ on the value of x, the depth to the neutral axis.

5.1.1 Rigorous analysis: parabolic-rectangular concrete stress-block

The stress–strain relationships for the concrete and steel specified in BS8110 are as shown in *Figure 5.1*. Any strength of the concrete in tension is ignored and the maximum compressive strain in the concrete at failure is assumed to be 0.0035. Since it is assumed that sections that are plane before bending remain plane after bending, the strain at any point across the section is linearly proportional to its distance from the neutral axis (*Figure 5.2b*). Thus from the stress–strain diagram for concrete it is clear that the shape of the assumed distribution of compressive stress in the concrete (*Figure 5.2c*) is formed

by the combination of a rectangle and a parabola; in other words, the concrete stress-block is parabolic-rectangular in shape. One point that should be noted is that, because of the way in which the shape of the parabolic portion of the curve is defined, the relative areas contributed to the stress-block by the parabolic portion and the rectangular

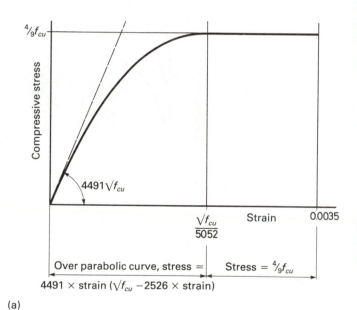

(a)

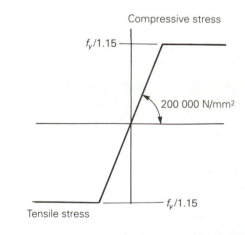

(b)

Figure 5.1 Stress–strain diagrams for (a) concrete (b) reinforcement.

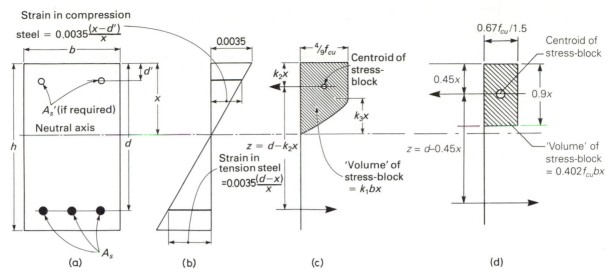

Figure 5.2 (a) Section (b) distribution of strain across section (c) resistance of section assuming parabolic-rectangular stress-block (d) resistance of section assuming uniform rectangular stress-block.

portion depend on the particular value of f_{cu}. Thus the total area does not vary linearly with f_{cu}, and the resulting expressions for the position of the centroid and the lever arm are rather complex. For this reason it will be apparent that doubling the value of f_{cu}, for example, does not double the resulting moment of resistance.

From *Figures 5.2a* and *5.2c*,

total compression on concrete stress-block $= k_1 bx$

where k_1, the factor defining the 'volume' of the concrete stress-block, is equal to $4f_{cu}(3 - k_3)/27$, and k_3, the factor defining the proportion of the compression zone covered by the parabolic curve, is equal to $16\sqrt{(6f_{cu})}/693$.

The depth of the centroid of the parabolic-rectangular stress-block from the top of the section is equal to $k_2 x$, where the factor k_2 is $[\frac{1}{2} - \frac{1}{3}k_3(1 - \frac{1}{4}k_3)]/(1 - \frac{1}{3}k_3)$, and thus the lever-arm between the centres of tension and compression is $d - k_2 x$. Then the moment of resistance of the concrete stress-block M_u is given by

$$M_u = k_1 bx(d - k_2 x)$$

or

$$\frac{M_u}{bd^2} = k_1 \frac{x}{d}\left(1 - k_2 \frac{x}{d}\right) \qquad (5.1)$$

Thus, rearranging,

$$d_{min} = \sqrt{\left\{ M_u \Big/ \left[bk_1 \frac{x}{d}(1 - k_2)\frac{x}{d} \right] \right\}} \qquad (5.2)$$

Now since the total tension must be equal to the total compression,

$$A_s f_{yd2} = k_1 bx$$

where f_{yd2} is the design stress in the tension reinforcement. Then as ϱ, the proportion of tension steel, is equal to A_s/bd,

$$\varrho = k_1 \frac{x}{d}\frac{1}{f_{yd2}} \qquad (5.3)$$

Substituting for x/d in equation (5.1),

$$\frac{M_u}{bd^2} = \varrho f_{yd2}\left(1 - \frac{k_2}{k_1}\varrho f_{yd2}\right) \qquad (5.4)$$

If the actual value adopted for d exceeds d_{min}, expression (5.4) can be rearranged to give

$$\varrho = \frac{1}{2f_{yd2}}\frac{k_1}{k_2}\left[1 - \sqrt{\left(1 - 4\frac{k_2}{k_1}\frac{M_u}{bd^2}\right)} \right] \qquad (5.5)$$

From *Figure 5.1b*, it is clear that the value of f_{yd2} to be used in a particular calculation depends on the corresponding strain in the reinforcement, and it is therefore necessary to sketch the distribution of strain across the section, as shown in *Figure 5.2b*. Alternatively, since the strain in the tension steel depends only on the maximum compressive strain in the concrete and the position of the neutral axis, it is possible to express the strain, and hence the corresponding stress f_{yd2} in the reinforcement, in terms of f_y and the ratio x/d, as shown in the appropriate expressions on *Figure 5.3*. This method is further developed in *RCDH*.

The mathematical derivation of the factors k_1, k_2 and k_3 is given in more detail in *RCDH*, where appropriate design formulae for rigorous analysis with a parabolic-rectangular stress-block are also provided. Such formulae have been used to prepare the design charts forming Part 3 of BS8110. The method is also discussed in detail in *Allen*.

5.1.2 Rigorous analysis: uniform rectangular concrete stress-block

As in the case of a parabolic-rectangular stress-block, it is assumed that plane sections remain plane, that any tensile strength of the concrete is neglected, and that the maximum compressive strain at failure is 0.0035. The distribution of compressive stress in the concrete is

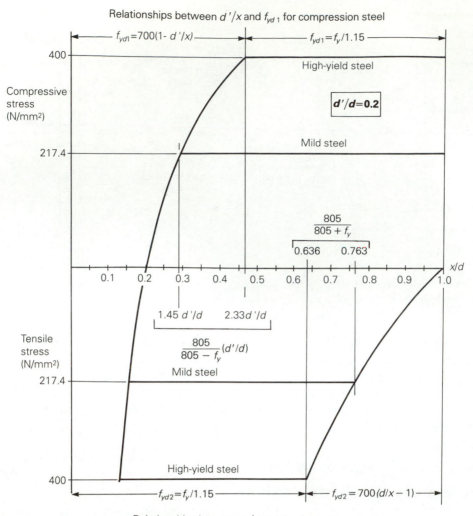

Relationships between d'/x and f_{yd1} for compression steel

Figure 5.3 Relationships between d'/x and x/d, and f_{yd1} and f_{yd2}.

assumed to have a uniform value of $0.444f_{cu}$ down to a depth of 0.9 of that of the neutral axis. The origins of this value are as follows. When defining the stress–strain diagram for normal-weight concrete, BS8110 employs a factor of $0.67/\gamma_m$ to determine the bending strength in flexure from the specified characteristic strength. If, as is usual, $\gamma_m = 1.5$, this factor becomes $67/150 \approx 0.4467$. Elsewhere in the Code, it is simplified to 0.45. For simplicity, throughout the present book, the value employed for this factor is $4/9 \approx 0.444$. In a similar manner, BS8110 sometimes expresses the limiting design strength in the reinforcement as f_y/γ_m (i.e. $f_y/1.15$ if $\gamma_m = 1.15$ as is usual) but frequently substitutes $0.87f_y$, although these expressions are not strictly equivalent.

Thus, from *Figures 5.2a* and *5.2d*,

total compression on concrete stress-block $= 0.4f_{cu}bx$

and since the lever-arm z between the centres of tension and compression is $d - 0.45x$,

moment of resistance of concrete stress-block
$$M = 0.4f_{cu}bx(d - 0.45x)$$

or

$$\frac{M}{bd^2 f_{cu}} = 0.2\frac{x}{d}\left(2 - 0.9\frac{x}{d}\right) \qquad (5.6)$$

Note that, when a uniform rectangular concrete stress-block is used, the ultimate moment of resistance is linearly proportional to the concrete strength, and thus the resistance moment may be expressed in the non-dimensional form of equation (5.6).

Expression (5.6) may be rearranged to give

$$d_{min} = \sqrt{\left\{ 5M \Big/ \left[bf_{cu}\left(2 - 0.9\frac{x}{d}\right)\frac{x}{d} \right] \right\}} \qquad (5.7)$$

Now, since the total tension on the section must equal the total compression,

$$A_s f_{yd2} = 0.4f_{cu}bx$$

where f_{yd2} is the design stress in the tension reinforcement; if x is less than $0.5d$, f_{yd2} is always equal to $f_y/1.15$. Then if the proportion of tension steel $\varrho = A_s/bd$,

$$\varrho = 0.4\frac{x}{d}\frac{1.15f_{cu}}{f_y} = 0.46\frac{x}{d}\frac{f_{cu}}{f_y} \qquad (5.8)$$

Substituting for x/d in equation (5.6),

$$\frac{M}{bd^2 f_{cu}} = \frac{\varrho f_y}{1.15 f_{cu}}\left(1 - \frac{4.5\varrho f_y}{4.6 f_{cu}}\right) \qquad (5.9)$$

If the value selected for d is greater than the minimum value, expression (5.9) can be rearranged as follows:

$$\varrho = \frac{4.6 f_{cu}}{9 f_y}\left[1 - \sqrt{\left(1 - \frac{4.5M}{bd^2 f_{cu}}\right)}\right] \qquad (5.10)$$

or

$$A_s = \frac{4.6 bd f_{cu}}{9 f_y}\left[1 - \sqrt{\left(1 - \frac{4.5M}{bd^2 f_{cu}}\right)}\right] \qquad (5.11)$$

According to BS8110, expressions (5.6), (5.9), (5.10) and (5.11) are only valid provided that z does not exceed $0.95d$. This limit is reached when $M = (0.4 f_{cu} bd/9) \times (0.95d)$, that is when $M/bd^2 f_{cu} = 38/900$. For smaller values of applied moment,

$$\frac{M}{bd^2 f_{cu}} = 0.38\,\frac{x}{d} \qquad (5.12)$$

Substituting this value of x/d in equation (5.8) gives

$$\frac{M}{bd^2 f_{cu}} = 0.95\varrho\,\frac{f_y}{1.15 f_{cu}} = \frac{19\varrho f_y}{23 f_{cu}} \qquad (5.13)$$

so that

$$\varrho = \frac{23}{19 f_y}\,\frac{M}{bd^2} \qquad (5.14)$$

and

$$A_s = \frac{23M}{19 df_y} \qquad (5.15)$$

5.1.3 Code formulae

The formulae given in BS8110 are derived from those obtained by rigorous analysis with a uniform rectangular concrete stress-block by making some simplifications. For sections reinforced in tension only, if the maximum stress in the concrete is taken as $0.67 f_{cu}/1.5$ (rather than $4 f_{cu}/9$), the total force in compression is $0.9 xb(0.67/1.5) f_{cu} = 0.402 f_{cu} bx$, and if the stress in the tension steel is taken as $0.87 f_y$ (rather than $f_y/1.15$), the total force in tension is $0.87 f_y A_s$. Then since the total tension equals the total compression, $0.402 f_{cu} bx = 0.87 f_y A_s$, or $x = 0.87 f_y A_s/ 0.402 f_{cu} b$. Since the lever-arm z is equal to $d - 0.45x$,

$$z = d - 0.45 \times 0.87 f_y A_s/(0.402 f_{cu} b) \qquad (5.16)$$

Taking moments about the centre of compression gives $M = 0.87 f_y A_s z$. Substituting for $0.87 f_y A_s$ in expression (5.16), $z = d - 0.45(M/z)/(0.402 f_{cu} b)$, which leads to the quadratic equation $z^2 - dz + 0.45M/(0.402 f_{cu} b) = 0$. The positive solution of this is

$$z = d\left[0.5 + \sqrt{\left(0.25 - \frac{0.45M}{0.402 bd^2 f_{cu}}\right)}\right] \qquad (5.17)$$

or

$$z = d\left[0.5 + \sqrt{\left(0.25 - \frac{0.45K}{0.402}\right)}\right] \qquad (5.18)$$

where $K = M/bd^2 f_{cu}$. This expression is simplified to $z = d[0.5 + \sqrt{(0.25 - K/0.9)}]$ in clause 3.4.4.4 of BS8110. Now substituting for z in the expression $M = 0.87 f_y A_s z$ and rearranging gives

$$A_{s\,req} = M/\{0.435[1 + \sqrt{(1 - 4.45K)}\, f_y d]\} \qquad (5.19)$$

if the BS8110 expression for z is adopted. As before, this expression is only valid if z does not exceed $0.95d$, which corresponds to a value of $M/bd^2 f_{cu}$ of $38/900 \approx 0.0422$. For lesser values of applied moment, expressions (5.14) and (5.15) remain valid.

5.1.4 Design formulae

For design purposes there is very little advantage in using the expressions provided in BS8110 clause 3.4.4.4 over the most precise formulæ derived in section 5.1.2. For practical use these expressions may be programmed for analysis using a programmable calculator or desk-top computer, or employed to produce design charts or tables. One suitable graphical arrangement is shown on *Data Sheet 13*, the use of which is illustrated in the example below. If the size of the rectangular section is already decided, the corresponding value of M/bd^2 is calculated and, at the intersection on the chart between this value and the appropriate concrete grade, the corresponding value of ϱf_y is read off. Then, by dividing by the characteristic steel strength, the corresponding proportion of tension steel is obtained, and thus the area of reinforcement is given by multiplying by bd. The compressive resistance of the concrete is satisfactory if the intersection point falls below the diagonal chain line.

If the effective depth is not specified, the minimum value that may be adopted is obtained by reading off the value of M/bd^2 corresponding to the intersection of the chain line with the grade of concrete being used. Then

$$d_{min} = \sqrt{[\text{applied ultimate moment}/}$$
$$(M/bd^2 \text{ value given by chart} \times b)]$$

Design charts similar to that on *Data Sheet 13* but for specified values of f_y of 250 and 460 N/mm^2 are given in *RCDH*. Design tables have also been produced by *Allen* (refs 9, 10).

5.1.5 Example

Design a rectangular beam 200 mm wide and reinforced in tension only to withstand an ultimate bending moment of 61 kN m if $f_{cu} = 30$ N/mm^2 and $f_y = 460$ N/mm^2.

From the chart on *Data Sheet 13*, when $f_{cu} = 30$ N/mm^2 the maximum value of M/bd^2 provided by the concrete alone is 4.7. Thus $d_{min} = \sqrt{[61 \times 10^6/(4.7 \times 200)]} = 254.7$ mm. With a total depth h of 300 mm and 25 mm cover to

10 mm links, and assuming that 20 mm bars are to be used, $d = 300 - 25 - 10 - 10 = 255$ mm. Then $M/bd^2 = 61 \times 10^6/(200 \times 255^2) = 4.69$, and if $f_{cu} = 30$ N/mm^2, the corresponding value of ϱf_y is 6.9. Thus $\varrho = 6.9/460 = 0.015$, and $A_{s\,req} = 0.015 \times 200 \times 255 = 765$ mm^2. Provide three 20 mm bars.

5.2 RECTANGULAR SECTIONS WITH TENSION AND COMPRESSION REINFORCEMENT

5.2.1 Rigorous analysis: parabolic-rectangular concrete stress-block

As with the design methods in previous use, the moment of resistance of a section reinforced in tension and compression is found by adding the resistance resulting from the couple due to the concrete in compression and its balancing tension steel, to the couple due to the compression reinforcement and additional tension steel. The first part of the analysis is therefore identical to the case where only tension reinforcement is provided. Then if f_{yd1} is the design stress in the compression steel which is provided at a distance d' from the compression face,

$$\text{total compression} = k_1 bx + A_s' f_{yd1}$$

and the moment of resistance is given by

$$M_u = k_1 bx(d - k_2 x) + A_s' f_{yd1}(d - d')$$

or

$$\frac{M_u}{bd^2} = k_1 \frac{x}{d}\left(1 - k_2\frac{x}{d}\right) + \varrho' f_{yd1}\left(1 - \frac{d'}{d}\right) \quad (5.20)$$

where $\varrho' = A_s'/bd$. Thus

$$A_s' = \frac{M_u - k_1 bx(d - k_2 x)}{f_{yd1}(d - d')} \quad (5.21)$$

or, expressed non-dimensionally,

$$\varrho' = \left[\frac{M_u}{bd^2} - k_1\frac{x}{d}\left(1 - k_2\frac{x}{d}\right)\right] \Big/ \left[f_{yd1}\left(1 - \frac{d'}{d}\right)\right] \quad (5.22)$$

Since the total tension must be equal to the total compression,

$$A_s f_{yd2} = k_1 bx + A_s' f_{yd1}$$

so that

$$A_s = \frac{k_1 bx + A_s' f_{yd1}}{f_{yd2}} = \frac{k_1 bx}{f_{yd2}} + A_s'\frac{f_{yd1}}{f_{yd2}} \quad (5.23)$$

or, expressed non-dimensionally,

$$\varrho = \frac{k_1}{f_{yd2}}\frac{x}{d} + \varrho'\frac{f_{yd1}}{f_{yd2}} \quad (5.24)$$

The values of f_{yd1} and f_{yd2} to be used in any particular case depend on the corresponding strains in the tension and compression steel, and can thus be determined by plotting the strain profile across the section (*Figure 5.2b*). Since

the shape of this profile is determined by the maximum strain in the concrete and the position of the neutral axis, it is possible to express the strain, and hence the design stress f_{yd1} in the compression reinforcement, in terms of f_y and the ratio of x/d', as given by the appropriate expressions on *Figure 5.3*, in a similar manner to which f_{yd2} is related to f_y and x/d.

5.2.2 Rigorous analysis: uniform rectangular concrete stress-block

The analysis here is similar to the foregoing case with a parabolic-rectangular concrete stress-block, the contribution of the compression reinforcement being identical in both cases. Then

$$\text{total compression} = 0.4f_{cu} bx + A_s' f_{yd1}$$

and the moment of resistance is given by

$$M_u = 0.4f_{cu} bx(d - 0.45x) + A_s' f_{yd1}(d - d')$$

or

$$\frac{M_u}{bd^2} = 0.2f_{cu}\frac{x}{d}\left(2 - 0.9\frac{x}{d}\right) + \varrho' f_{yd1}\left(1 - \frac{d'}{d}\right) \quad (5.25)$$

where $\varrho' = A_s'/bd$. Thus

$$A_s' = \frac{M_u - 0.2f_{cu} bx(2d - 0.9x)}{f_{yd1}(d - d')} \quad (5.26)$$

or in non-dimensional terms,

$$\varrho' = \left[\frac{M_u}{bd^2} - 0.2f_{cu}\frac{x}{d}\left(2 - 0.9\frac{x}{d}\right)\right] \Big/ \left[f_{yd1}\left(1 - \frac{d'}{d}\right)\right] \quad (5.27)$$

Since the total tension is equal to the total compression,

$$A_s f_{yd2} = 0.4f_{cu} bx + A_s' f_{yd1}$$

so that

$$A_s = \frac{0.4f_{cu} bx + A_s' f_{yd1}}{f_{yd2}} = 0.4bx\frac{f_{cu}}{f_{yd2}} + A_s'\frac{f_{yd1}}{f_{yd2}} \quad (5.28)$$

or, expressed non-dimensionally,

$$\varrho = 0.4\frac{x}{d}\frac{f_{cu}}{f_{yd2}} + \varrho'\frac{f_{yd1}}{f_{yd2}} \quad (5.29)$$

With either of the foregoing analyses, any value of x may be chosen. However, the choice that is made determines the values of f_{yd1} and f_{yd2} that may be employed, and also the amount of moment redistribution that may be made. This matter is discussed later in this chapter.

5.2.3 Code formulae

The design formulae provided in BS8110 for sections with tension and compression reinforcement are derived from those obtained by rigorous analysis for a uniform rectangular stress-block with slight simplifications. By taking moments about the tension reinforcement, the contribution to the moment of resistance provided by the concrete stress-block is

$$M = 0.402 f_{cu} bx(d - 0.45x) \qquad (5.30)$$

(using the Code factor of 0.402 rather than that of 0.4 employed elsewhere in this book). Now if the ratio of the resistance moment after redistribution to that before redistribution is β_b, the maximum value of x is related to β_b by the expression $x = (\beta_b - 0.4)d$. Substituting for x in expression (5.30) gives $M = 0.402 f_{cu} bd(\beta_b - 0.4) \times [d - 0.45(\beta_b - 0.4)d]$, which may be simplified to $K' = 0.402 \times (\beta_b - 0.4) - 0.1809(\beta_b - 0.4)^2$, where $K' = M/f_{cu} bd^2$. In BS8110, the factor of 0.1809 has been rounded to 0.18.

The resistance required from the compression steel is then obtained by subtracting that provided by the concrete alone from the applied moment. The steel area is determined by dividing this value by the design stress in the steel $(0.87 f_y)$ multiplied by the distance between the centroids of the tension and compression steel, i.e.

$$A'_s = (K - K') f_{cu} bd^2 / [0.87 f_y (d - d')]$$

Tension steel must then be provided to match the compressive force in the concrete plus that contributed by the compression steel, i.e.

$$A_s = K' f_{cu} bd^2 / (0.87 f_y z) + A'_s$$

The Code formulae assume that all the reinforcement is acting at its maximum design strength. However, if the ratio of the depth of the compression steel to that of the neutral axis (i.e. d'/x) exceeds 0.69 in the case of mild steel or 0.43 with high-yield steel, or if the ratio x/d exceeds 0.76 or 0.64 in the case of mild steel and high-yield steel, respectively, this assumption will be incorrect, and the actual design stress corresponding to the strain in the steel at the point concerned must be determined from first principles, as described in *RCDH*. (Note that in clause 4.3.3.3, BS8110 only specifically mentions the limit of $d'/x = 0.43$.)

The chart on *Data Sheet 13* can also be used to design sections with both tension and compression reinforcement provided that the value of M/bd^2 does not exceed 6. The procedure, which is illustrated in the following example, is similar to that for sections reinforced in tension only. Values of ϱf_y and $\varrho' f_y$ corresponding to the intersection of given values of M/bd^2 and f_{cu} are read from the chart and are multiplied by bd/f_y to obtain the areas of tension and compression steel required. The chart has been prepared on the assumptions that $x/d = 0.5$ when compression steel is provided, which implies that moment redistribution is limited to 10%, and that $d' = 0.1d$. If this latter assumption is incorrect, the resulting values can be corrected as follows. Multiply the value of ϱ' obtained by the coefficient relating to the actual ratio of d'/d, which can be read from the scale on the right-hand side of the chart, to obtain the true value of ϱ'. Then, to obtain the adjusted value of ϱ, add or subtract to the value of ϱ

obtained from the chart (depending on whether the true ratio of d'/d is greater or less than 0.1) the same adjustment as is made to ϱ'.

5.2.4 Example

Design a rectangular beam with $b = 200$ mm, $d = 300$ mm, $f_{cu} = 25$ N/mm^2 and $f_y = 460$ N/mm^2 to withstand an ultimate bending moment of 100 kN m.

From the curves on *Data Sheet 13* with $M/bd^2 = 100 \times 10^6/ (200 \times 300^2) = 5.56$ and $f_{cu} = 25$ N/mm^2, $\varrho f_y = 7.8$ and $\varrho' f_y = 2.1$. Thus $\varrho = 0.017$ and $\varrho' = 0.0046$. These values apply when $d'/d = 0.1$, but the actual value of d' is about $25 + 10 + 10 = 45$ mm, so that $d'/d = 45/300 = 0.15$. Thus, with a correction factor of 1.06 from the scale on the right-hand side of *Data Sheet 13*, the actual value of ϱ' required is $0.0046 \times 1.06 = 0.0048$, and $A'_{s\,req} = 0.0048 \times 200 \times 300 = 290$ mm^2: provide two 20 mm bars. Now the actual proportion ϱ required is $0.017 + 0.0002 = 0.0172$ and $A_{s\,req} = 0.0172 \times 200 \times 300 = 1032$ mm^2: provide four 20 mm bars.

5.3 SECTIONS OF OTHER SHAPES

Other than rectangular sections, the most usual shapes encountered in reinforced concrete design are T- or L-sections, where the flange is normally provided by the floor slab. Such sections may be analysed using either of the foregoing rigorous methods. In addition, BS8110 provides design formulae that are applicable in certain conditions only.

Two cases require consideration, namely where the neutral axis falls within the flange (i.e. $x \le h_f$, where h_f is the flange thickness) and where this is not so. In the former case (*Figure 5.4b*), the behaviour is identical to that of a normal rectangular section equal to the flange width. It can thus be designed for bending as such; the required amount of tension steel is merely accommodated within the width b_w of the web. If b is the width of the flange, the maximum moment of resistance that results in this case is, with rigorous analysis with a parabolic-rectangular stress-block,

$$M = k_1 bh_f(d - k_2 h_f)$$

and with rigorous analysis with a uniform stress-block (and the associated BS8110 formulae),

$$M = 0.4 f_{cu} bh_f(d - 0.45 h_f)$$

This relationship is shown by the broken lines on the chart on *Data Sheet 13*. If the value of h_f/d read from the chart corresponding to the given values of f_{cu} and M/bd^2 is less than the true value, the section acts as a rectangular beam and may be designed as such.

If the applied moment exceeds the value of M given by the appropriate expression above, the neutral axis falls below the underside of the flange. Taking moments about

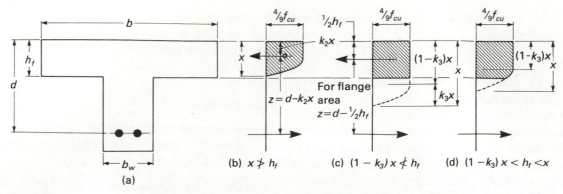

Figure 5.4 (a) Section; (b)–(d) distribution of stress in flange under different conditions.

the centroid of the tension steel gives, with a uniform rectangular stress-block,

$$M = (4/9)f_{cu}(b - b_w)h_f(d - h_f/2)$$
$$+ (4/9)f_{cu}0.9xb_w(d - 0.45x)$$

or

$$M = (4/9)f_{cu}[(b - b_w)h_f(d - h_f/2)$$
$$+ 0.9xb_w(d - 0.45x)]$$

When $x = 0.5d$,

$$\frac{M}{bd^2f_{cu}} = 0.45\frac{h_f}{d}\left(1 - \frac{b_w}{b}\right)\left(1 - \frac{h_f}{2d}\right) + 0.155\frac{b_w}{b} = \beta_f$$

which is equation (2) in clause 3.4.4.4 of the Code, although in BS8110 the factor 0.155 has been rounded to 0.15. This relationship is shown in *Figure 5.5*. Since the full depth of the concrete flange has been assumed to provide compressive resistance, the neutral-axis depth x must be at least $h_f/0.9$, i.e. if $x = d/2$, then $h_f \leq 0.45d$. This restriction is also specified in BS8110.

Taking moments about mid-depth of the flange gives

$$M = 0.87f_y A_s(d - h_f/2) - (4/9)f_{cu}b_w 0.9x(0.45x - h_f/2)$$

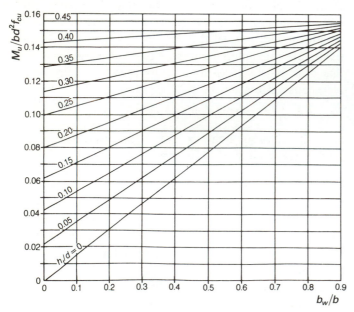

Figure 5.5

and if a value of x of $d/2$ is adopted and the expression rearranged,

$$A_{s\,req} = \frac{M + 0.1f_{cu}b_w d(0.45d - h_f)}{0.87f_y(d - h_f/2)}$$

which is equation (1) in BS8110. These Code equations only apply when $x = d/2$; otherwise rigorous analysis must be employed.

If a parabolic-rectangular distribution of compressive stress in the concrete is adopted and $(1 - k_3)x$ exceeds h_f (*Figure 5.4c*), the distribution of stress over the flange thickness is uniform and equal to $(4/9)f_{cu}$. Thus if no compression reinforcement is provided,

$$\text{total compression on section} = k_1 b_w x + (4/9)f_{cu}(b - b_w)h_f$$

and the moment of resistance is given by

$$M = k_1 b_w x(d - k_2 x)$$
$$+ (4/9)f_{cu}(b - b_w)h_f(d - h_f/2)$$

Then since the total tension is equal to the total compression,

$$A_s = \frac{1}{f_{yd2}}[k_1 b_w x + (4/9)f_{cu}(b - b_w)h_f]$$

However, if x exceeds h_f but $(1 - k_3)x$ does not, the distribution of stress over the depth of the flange takes the form of a combination of a rectangle and a truncated parabola (*Figure 5.4d*). The resulting analysis, which involves a knowledge of the area and the position of the centroid of this somewhat complex shape, although practicable, is thus tedious, and unless a computer or suitable sets of design charts are to hand, it is normally preferable to assume a uniform rectangular distribution of stress instead.

If a uniform rectangular distribution is assumed, if x exceeds h_f, and if no compression steel is to be used, it is necessary to assume a value of x such that

$$x = \left\{ d - \sqrt{\left[d^2 - \frac{4.5M_u}{b_w f_{cu}} \right.} \right.$$
$$\left. \left. + h_f(2d - h_f)\left(\frac{b}{b_w} - 1\right) \right] \right\} / 0.9$$

and

$$A_s = (4/9)f_{cu}[0.9b_w x + (b - b_w)h_f]/f_y$$

and the resulting value of x must not exceed $\frac{1}{2}d$.

If this is not so, some compression steel must be provided. Select a suitable value for x, bearing in mind the resulting values of f_{yd1} and f_{yd2} and the amount of moment redistribution desired. Then

$$A'_s = \frac{M_u - 0.2f_{cu}[h_f(b - b_w)(2d - h_f) + 0.9b_w(2d - 0.9x)x]}{f_{yd1}(d - d')}$$

and

$$A_s = (4/9)\frac{f_{cu}}{f_{yd2}}[0.9b_w x + (b - b_w)h_f] + A'_s\frac{f_{yd1}}{f_{yd2}}$$

Other simple sections comprising a combination of rectangles can be analysed similarly. If a curved section requires analysis, however, or if reinforcement is to be provided at a number of different levels, the analysis becomes complex and the use of the parabolic-rectangular stress-block is out of the question unless sophisticated computer aids are available.

The first step in carrying out such an analysis using a uniform rectangular stress-block is to assume a position of the neutral axis and to divide the resulting concrete compression zone into a number of convenient areas, ascertaining the position of the centroid of each. The strain profile across the section is then sketched from a knowledge of the position of the neutral axis and with the maximum compressive strain of 0.0035 in the concrete (*Figure 5.6b*). The strain at the position of each reinforcing bar is calculated from this diagram, and the corresponding stress in each bar is determined from the appropriate stress–strain diagram (*Figure 5.1b*). Alternatively, the stress in each bar may be calculated from the expressions given on *Figure 5.3*. Then if A_c is the area of an individual element comprising the concrete compression zone, δA_s and f_{yd} are the area of an individual tension bar and the particular design stress in

that bar, and $\delta A'_s$ and f'_{yd} are the area of an individual compression bar and the particular design stress in that bar, since the total tension on the section must be equal to the total compression, it is necessary to satisfy the equation

$$0.4f_{cu}\Sigma A_c + \Sigma(\delta A'_s f'_{yd}) = \Sigma(\delta A_s f_{yd})$$

The position of the neutral axis and the numbers, sizes and positions of the reinforcing bars must be adjusted until this equation is satisfied. When this is achieved, the moment of resistance of the section is then given by the expression

$$M_u = 0.4f_{cu}\Sigma(A_c x') + \Sigma(\delta A'_s f'_{yd} a') + \Sigma(\delta A_s f_{yd} a)$$

where x', a' and a are the distances of each individual concrete element, compression bar and tension bar, respectively, from the neutral axis.

It will be seen that the procedure is lengthy and tedious to undertake by hand, although not difficult to program for computer analysis. In most cases it is possible to analyse first an equivalent rectangle similar in area and having roughly the same overall shape, in order to obtain starting values of x, A'_s and A_s. The results obtained can then be applied to the actual shape to be investigated, thus eliminating most of the tedious preliminary calculation. It is clear, however, that the use of awkwardly shaped sections should be avoided if they require accurate analysis.

5.4 CHOICE OF METHOD OF ANALYSIS AND VALUE OF x

When members with tension and compression reinforcement are designed using rigorous analysis, the choice of the depth x to the neutral axis is left to the designer. As explained above, the design stresses f_{yd1} and f_{yd2} in the compression and tension reinforcement are controlled by the ratios of x/d and x/d', and therefore it is advantageous to choose a value of x such that f_{yd1} and f_{yd2}

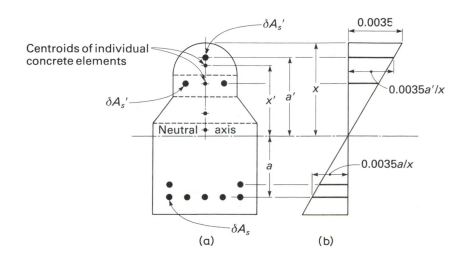

(a) (b)

Figure 5.6

both attain their maximum possible values, while the concrete in compression contributes as much as possible to the resistance of the section. However, a further factor restricting the choice of x is the amount of moment redistribution desired. BS8110 permits the ultimate moment of resistance at any section to be reduced to a maximum of 70% of the value obtained by an elastic analysis (see section 3.3) provided that certain requirements are met. The ratio β_b of the moment after redistribution to that before redistribution is, however, related to the ratio x/d by $x/d \leq \beta_b - 0.4$ or $\beta_b \leq x/d + 0.4$. Where no reduction in moment is made, or where an increased moment occurs after redistribution, β_b is taken as 1. Then, for example, if $x/d = 0.3$ or less a maximum reduction of moment of 30% may be made, but if $x/d = 0.6$ or more no reduction is possible. This restriction must be borne in mind when choosing a value for x.

As indicated by the typical relationships between x/d and f_{yd} shown in *Figure 5.3*, the zone over which both f_{yd1} and f_{yd2} attain their maximum values extends from $x/d = [805/(805 - f_y)](d'/d)$ to $x/d = 805/(805 + f_y)$. Since the resistance provided by the concrete compression zone increases (and thus the corresponding amount of compression steel required decreases accordingly) as the ratio of x/d increases, it can be shown that the most satisfactory section in terms of minimizing the amount of reinforcement required is achieved when the value of x/d is given by the latter expression. However, if compression steel is to be used, BS8110 specifies that a minimum of 0.2% must be provided in rectangular sections and in flanged sections where the web is in compression, and a minimum of 0.4% in other flanged sections. There is no point in choosing a ratio of x/d such that the required amount of compression steel falls below these limits.

Furthermore the relationship between β_b and x/d given in clause 3.2.2.1 of BS8110 seems to imply that the maximum value adopted for x/d should not exceed 0.6, and this is probably a sensible upper limit to adopt.

When designing members reinforced in tension only, the maximum value of x/d that may be adopted is 0.5.

In terms of true economy, of course, it is often advantageous to increase the effective depth of the section still further, thus reducing the amount of reinforcement needed accordingly. However, by so doing the self-weight and hence the moment that the section must be designed to resist are increased, as is the weight transferred to the columns and foundations (and for which these must be designed). In any case, the minimum thickness which may be adopted for a slab or beam is often determined by deflection or fire-resistance considerations (see Chapters 8 and 19).

Because the resistance moment provided by the section does not increase linearly with an increase in concrete strength when a parabolic-rectangular concrete stress-block is adopted, a comparison between both rigorous methods of analysis shows that when f_{cu} is less than about $28 \, \text{N/mm}^2$ the parabolic-rectangular stress-block provides a higher resistance moment than a uniform rectangular stress-block for a given section. For higher concrete strengths the reverse is true; the additional strength provided by the latter assumption becomes more marked as f_{cu} increases. Although a concrete grade of C25 may be adopted in suitable circumstances according to BS8110, a grade of C30 or higher is now more common, and therefore from the theoretical point of view the consideration of the more complex parabolic-rectangular concrete stress-block appears unwarranted.

Chapter 6
Shearing and torsional forces

6.1 SHEARING FORCE

The basic principles of the calculation of the shearing and torsional forces on beams and the provision of reinforcement to resist such forces are discussed in this chapter. These principles and the corresponding tables are applied to the calculations for the beams in the building in Part Two.

As in the case of bending moments (see Chapter 3), various methods of calculating the shearing forces on continuous beams are permitted in BS8110. For systems of beams (a) of at least three spans, none of which differs in length by more than 15% of the longest span, and on which (b) the characteristic dead load is not less than the characteristic imposed load and (c) the loads that are supported are mainly uniformly distributed, the Code gives the following simplified coefficients for shearing force:

At the outer support: $V = 0.45F$
At the penultimate support: $V = 0.60F$
At the interior supports: $V = 0.55F$

For one-way slabs where (a) the area of each bay is greater than $30\,\text{m}^2$, (b) the ratio of imposed load to dead load is not more than 1.25, and (c) the characteristic imposed load (excluding partitions) is not greater than $5\,\text{kN/m}^2$, BS8110 suggests the following simplified coefficients:

At the outer support: $V = 0.4F$
At the penultimate support: $V = 0.6F$
At the interior supports: $V = 0.5F$

In these expressions, V is the total ultimate shearing force and F is the total ultimate load (i.e. $1.4G_k + 1.6Q_k$) on the span.

In the general case of a continuous-beam system, the same arrangement of loading as that described in section 3.1 must be considered, and once again this is most conveniently undertaken by considering instead a dead load of $1.0G_k$ and an 'imposed' load of $0.4G_k + 1.6Q_k$. The upper table on *Data Sheet 14* gives coefficients for the maximum elastic shearing forces on beams that are continuous over equal or nearly equal spans, assuming that free rotation may occur at all the supports. The

coefficients are for three types of loading and are given separately for dead and imposed characteristic loads. Coefficients are also provided for combined ratios of characteristic imposed to characteristic dead load of from 0.5 to 3. These coefficients incorporate the requisite partial safety factor for loads. For example, in the case of a characteristic dead load of say $1\,\text{kN/m}$ and a characteristic imposed load of $2\,\text{kN/m}$, the coefficient corresponding to the appropriate characteristic imposed-to-dead load ratio of 2 takes account of a dead load of $1\,\text{kN/m}$ and an 'imposed' load of $[0.4 + (1.6 \times 2)] = 3.6\,\text{kN/m}$.

The coefficients for most other types of loading can be interpolated from *Data Sheet 14*. The effect of bending moments at end supports can often be assessed by inspecting the free-end and fixed-end coefficients, or *Data Sheet 14* can again be used; the lower table gives coefficients for the shearing forces that must be added to or deducted from the elastic shearing force to allow for a bending moment at an end support. The effect of partial end fixity is, on the end span, to increase the shearing force adjacent to the end support and to reduce the shearing force adjacent to the penultimate support. The shearing force on the other spans is also affected but to a lesser extent, and such variations may be ignored. The bending moments corresponding to the shearing forces on *Data Sheet 14* are given in the same table.

In special beams that support irregular loading or continuous-beam systems with widely differing spans, where bending-moment diagrams are drawn out, the theoretical shearing force at any section can be obtained directly from these diagrams by determining the slope of the tangent to the bending-moment diagram at the section considered.

For the basis of the calculation of the shearing forces to be the same as that adopted for the bending moments, some modification would appear to be necessary to the elastic shearing forces if the bending moments are redistributed within the limits permitted by BS8110. However, provided that the redistribution does not greatly exceed about 15%, such modifications are seldom worth while, as can be seen from the extreme case of a uniformly loaded beam that is continuous at one end and freely supported at the other. The difference between the elastic shearing forces at the outer (free) support and at the inner support is $0.625F - 0.375F = 0.250F$. If the bending moment at the inner support is reduced by 15%, the change in the shearing force is likewise reduced by 15%; i.e. the elastic shearing forces are reduced or increased by $0.0375F$, so that the net shearing forces are $0.625F - 0.037F = 0.588F$ and $0.375F + 0.038F = 0.413F$ at the inner and outer supports respectively, in place of the non-adjusted shearing forces of $0.625F$ and $0.375F$ respectively. These differences are about 6% and 10% respectively; such values are probably negligible compared with some of the uncertainties in the calculation. For redistributions of bending moment in excess of 15% or thereabouts, however, it may be necessary to consider the corresponding adjustments to the shearing forces.

If a beam is fully loaded with a uniformly distributed imposed load (*Figure 6.1a*), there is a section near midspan (under symmetrical restraint conditions it is at midspan) at which there is no shearing force due to this load. Frequently the condition of being fully loaded is the only one that need be considered because it produces the maximum shearing forces on sections towards the ends of a beam in a building. It is possible, however, for the imposed load to occupy only a part of the span, in which case the shearing forces other than at the ends of the span are underestimated. The maximum shearing forces throughout the length of a beam due to a partial imposed load are given in *Figure 6.1c* for symmetrical conditions; the distinction between negative and positive shearing forces should be noted. It is seen that at every section there is a shearing force, and consequently a shearing stress, under some loading condition. In a large beam, especially one with a varying cross-section, it is thus necessary to calculate the maximum shearing force near midspan, taking into account the effect of unequal restraining moments (*Figure 6.1b*) to ensure that the concrete alone (with nominal links) is sufficient to resist this shearing force safely. Otherwise, additional re-

inforcement should be provided for this purpose, as explained later. For a uniformly distributed load the maximum shearing force at midspan, under symmetrical support conditions, is $\frac{1}{8}nl$ and at the supports it is $\frac{1}{2}nl$, i.e. four times as much; so, if the cross-section of the beam is constant, the shearing stress at midspan is one-quarter of that at the supports. The ratio of four is further increased when the effect of the shearing forces due to dead load is combined with those due to imposed load. Therefore, if the shearing stress at the supports as normally calculated does not exceed a stress of $4v$, the shearing stress at midspan will not exceed v and it is clear that the permissible stress in the concrete with nominal reinforcement only is therefore not exceeded.

If the cross-section of the beam is not uniform, or the difference between the negative bending moments at the supports is great (as in the case of an end span), the maximum negative and positive shearing forces and the resulting stresses should be calculated from formulae (6.1) or (6.2) at pertinent points along the span.

6.2 SHEARING STRESS

BS8110 recommends that the shearing stress v at any section should be calculated from the expression $v = V/bd$, in which V is the total shearing force on the section due to ultimate loads, d is the effective depth and b is the width of the section (in the case of a flanged beam, the width of the rib). This expression applies to beams of uniform depth only. When the effective depth varies and the bending moment increases in the same direction as the effective depth (as at a haunch adjacent to the interior support of a continuous beam),

$$v = \frac{V - (M/d)\tan\theta_s}{bd} \qquad (6.1)$$

The angle θ_s is that between the top and bottom edges at the section considered. When the bending moment decreases as d increases, as at a haunch adjacent to a free

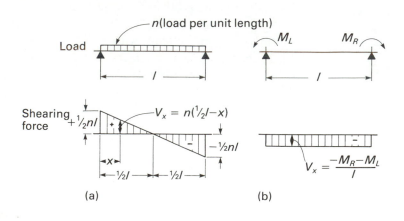
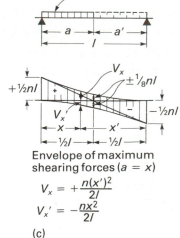

Figure 6.1 Diagrams of shearing force: (a) uniformly distributed load on entire span (b) end restraining bending moments (c) uniformly distributed load on part of span.

support,

$$v = \frac{V + (M/d)\tan\theta_s}{bd} \qquad (6.2)$$

Reinforcement to resist shearing is necessary only when v exceeds the permissible shearing resistance v_c of the concrete alone, given in *Data Sheet 15*. If this is the case, reinforcement must be provided to resist the difference between the applied ultimate shearing force and the resistance provided by the concrete. In no circumstances, however, may the section be subjected to a shearing stress greater than v_{max}; values of v_{max} are also given on *Data Sheet 15*.

It will be observed that the shearing resistance of the concrete alone is related to the proportion of main longitudinal steel which continues for at least a distance d beyond the section under consideration. At supports, however, the entire area of tension steel may be taken into account, provided that anchorage-bond requirements are satisfied. Note that, since the shearing resistance of the concrete rises with the proportion of longitudinal reinforcement, when evaluating the resistance of the concrete alone it is advantageous to consider the actual area of main tension steel provided, as this may substantially exceed the calculated amount required and thus correspond to higher values of v_c.

Where shearing reinforcement is not provided, shear failure usually occurs at an angle of about 30° to the horizontal; in situations where only a more steeply inclined plane of failure is possible, the force required to bring about such a failure increases markedly. Therefore, within a distance of twice the effective depth from a concentrated load or from the face of a support, BS8110 permits the design shear strength v_c to be increased to $v_c 2d/a_v$, where a_v is the distance from the support face or concentrated load to the point being considered. However, in no circumstances may the shear stress exceed the value of v_{max} given on *Data Sheet 15*. In such situations, correct detailing of the steel is essential to attain the optimum shear strength, and so the Code specifies that all tension bars must extend each side of the intersection point with the theoretical shear crack for a distance of not less than the effective depth unless an equivalent anchorage is provided.

6.3 REINFORCEMENT TO RESIST SHEARING

Reinforcement to resist shearing forces may be provided either as links or as inclined bars. The total shearing resistance is the sum of the shearing resistances of the links and the inclined bars, calculated separately if both are provided, but BS8110 specifies that inclined bars may not provide more than one-half of the total shearing resistance required from the reinforcement at any point.

6.3.1 Links

The shearing resistance provided by vertical links is $0.87 A_{sv} f_{yv} d/s_v$, in which A_{sv} is the cross-sectional area

of a single link (both legs), f_{yv} is the characteristic strength of the link reinforcement (which must not exceed $460\,\text{N/mm}^2$), d is the effective depth of the section and s_v is the pitch or spacing between adjacent links, which must not exceed ¾d longitudinally. In addition, the lateral spacing must not be greater than d; nor should any tension bar be more than 150 mm from the vertical leg of a link. It is convenient to express the shearing resistance of a system of links of a given size, spacing and grade of reinforcement by the product $K_u d$ of the effective depth of the section d and the link-reinforcement factor K_u, where $K_u = 0.87 A_{sv} f_{yv}/s_v$. Values of K_u are given in *Data Sheet 16*, as is the resistance of a system of links spaced at $0.75d$. As the spacing may be controlled by the requirements for compression reinforcement, the smallest size of bar in compression corresponding to each spacing is also tabulated; the diameter of the link must be at least one-quarter of the largest compression bar.

Except in minor beams such as lintels, or where the maximum calculated shearing stress is less than one-half of the permissible stress that the concrete can withstand alone, nominal links must always be provided in beams. Instead of specifying a minimum proportion, as has been the case in previous codes, BS8110 requires that sufficient reinforcement is provided to give a design shear resistance of $0.4\,\text{N/mm}^2$. If $K_u d = 0.4 b_t d$ then $b_t = K_u/0.4$, and the requirement can be expressed as a limiting breadth b_t corresponding to various combinations of A_{sv}, s_v and f_{yv}. Appropriate maximum values of b_t for given systems of links are also tabulated on *Data Sheet 16*, and all that is necessary to ensure that a given arrangement meets the Code requirements is to check that the actual value of b_t adopted does not exceed that given in this table.

BS8110 permits the enhanced shearing strength near a support to be considered in two ways. The rigorous method involves calculating the total area of shearing reinforcement ΣA_{sv} needed from the expression

$$\Sigma A_{sv} = a_v b_v (v - 2dv_c/a_v)/(0.87 f_{yv}) \geq 0.4 b_v a_v/(0.87 f_{yv})$$

This reinforcement should be arranged symmetrically over a length of $0.75a_v$. If the ratio a_v/d is less than about 0.6, vertical links tend to be ineffective and horizontal links should be provided instead.

Such rigorous calculations are often somewhat lengthy unless undertaken automatically by computer, and are probably only worth while in special circumstances, such as where a concentrated load acts close to a support. For normal situations the Code provides (clause 3.4.5.10) a simplified approach if the load is mainly uniform or the principal load occurs at a distance from the support of more than twice the effective depth. In such circumstances the Code recommends that the critical shear stress be calculated at a distance equal to the effective depth from the support, ignoring any enhancement that may be possible. The resulting amount of shear steel necessary should then be provided at all sections closer to the support, but no further calculations need be undertaken for these sections.

Links should enclose all the tension reinforcement and be adequately anchored at both ends to develop their resistance. If a sufficient length of anchorage can be obtained in the compression zone of the beam, it is not essential for the link to pass around the top bars, but if links that provide resistance to shearing are also required to bind bars in compression, the links must pass around the compression bars as well as those in tension. Links are usually considered to be anchored effectively if they are bent through an angle of at least 90° around a bar, and then project beyond the end of the bend for a length of at least eight times the diameter of the bar forming the link.

6.3.2 Inclined bars

The principle adopted to consider the action of inclined bars according to BS8110 is that they form the tensile members of an imaginary lattice girder of which the concrete forms the compression members. The shearing resistance at any section is the sum of the vertical components of the inclined tensile and compressive forces crossing the section, and the ultimate shearing resistance V_b of a bar inclined at θ to the horizontal is $0.87 f_{yv} A_{sb} \sin \theta$, where A_{sb} is the cross-sectional area of the bar and f_{yv} is the characteristic strength of the shearing reinforcement, which must not exceed $460 \, \text{N/mm}^2$.

As already stated, not more than one-half of the total resistance required from the shearing reinforcement may be provided in the form of inclined bars. Values of V_i for both grades of steel are given on *Data Sheet 17* for an inclination of 45°, this being the minimum angle permitted by BS8110. If the design stresses in the inclined part of the bar and the horizontal parts of the bar are equal, to achieve equilibrium the spacing of the inclined bars should be not greater than that indicated in the top left-hand part of *Data Sheet 17*, which gives the correct spacing for an angle of inclination of 45°. Unlike its predecessors, BS8110 does not permit bars inclined at less than this angle to be considered, and bars inclined at more than 60° are normally inconvenient. With bars spaced as shown on *Data Sheet 17*, the system of inclined bars is commonly known as a 'single system'. If the spacing is decreased to one-half of this amount, the arrangement is a 'double system' and the resistance V_i is doubled; values for double systems are also given on *Data Sheet 17*. Depending on the spacing adopted, intermediate or multiple systems can also be considered, of course, and data are also given for treble and quadruple systems.

Generally, an inclined bar is one that is employed to resist tension in the bottom of the beam near midspan and is then bent up at 45° into the top of the member, where it may provide resistance over the support. In such a case the force in the horizontal parts of the bar must balance the horizontal components of the force in the inclined part and the complementary compressive resistance. This is ensured in cases in which the design stresses in the horizontal and inclined lengths are equal, if the spacing is as shown in the lower part of *Data Sheet 17* for a single

system or one-half of this spacing for a double system. The spacing of inclined bars in a single system, when the design stress is greater or less than $0.87 f_{yv}$, is given in the top right-hand part of *Data Sheet 17* for design stresses of from $0.615 f_{yv}$ to $0.923 f_{yv}$. The latter value corresponds to a spacing s of $1.5d$, which is the maximum permitted by BS8110. Generally the maximum design stress in bending f_{yd} occurs in the bars in the bottom of the beam only at midspan and reduces towards the supports, near which the inclined part of the bar usually occurs. Therefore the inclined part of a bent-up bar can often be at a closer spacing which corresponds to a reduced design stress in the horizontal part, leading to consequent economies since the corresponding shearing resistance of the system is increased. For example, if the design stress in the horizontal part of the bar adjoining the bent-up portion is $0.8 f_{yv}$, the spacing of the inclined bars may be reduced to $1.3d$, resulting in a shearing resistance equal to about 1.09 times that of a single system.

Inclined bars are seldom used in present-day practice because the cost of bending is not insignificant. They are also less easy to manipulate and fix than straight bars, particularly in congested situations. However, where large concentrated loads must be supported their use should be considered in order to avoid the congestion that can arise when the multiple-link systems, that would otherwise be necessary, are employed. Inclined bars used as shearing reinforcement must be checked for anchorage (see section 7.1) and bearing (see section 7.2).

6.4 CONCENTRATED LOADS ON BEAMS ETC.

If the distance a_v from the face of a support to the nearer edge of a concentrated load is less than twice the effective depth, as explained in section 6.3, the shearing resistance v_c of the concrete alone, as given by *Data Sheet 15*, may be increased to $2dv_c/a_v$ provided that the resulting value does not exceed v_{max}. CP110 specified that this situation applied where the concentrated load caused more than 70% of the total shearing force at that support. The same document specified that in such a case all the main reinforcement must be extended to the support and be provided with an anchorage length equal to 20ϕ. If $a_v < 0.6d$, horizontal rather than vertical links should be provided. Detailed information regarding the design of nibs and corbels supported concentrated loads is given in section 5.2.7 of BS8110 and is summarized in *RCDH*.

6.5 SHEARING FORCES ON SLABS

It is seldom necessary to reinforce slabs to resist shearing forces except where these arise as a result of the application of concentrated loads or around the column heads of flat slabs; the procedure for dealing with such situations is outlined in section 6.5.1. Otherwise, shearing reinforcement is usually only necessary in a slab where an abnormally high load is supported on a fairly short span. The limiting value of v_c should be determined from *Data*

Sheet 15 in the same way as for a beam. If $v > v_c$, shearing reinforcement in the form of links or inclined bars may be provided as in a beam. BS8110 restricts the use of shear steel to slabs not less than 200 mm in overall thickness, owing to the difficulty in bending and fixing such reinforcement accurately in shallow slabs. Unlike the corresponding requirements for beams (i.e. Table 3.8), Table 3.17 of the Code indicates that any system of links only, inclined bars only, or a combination of both may be used, but the table implies that bent-up bars may only be used when the shearing stress v exceeds $v_c + 0.4$. The Code does not define the borderline between a beam and a slab, but the *Code Handbook* suggests generally that a member having a ratio of b/h exceeding 4 should be initially thought of as a slab, although certain design or loading aspects may mean that it should be treated as a beam. Note that the same enhancement of shearing resistance close to the supports described in section 6.3.1 may be employed; with the simplified approach this means that it is only necessary to take account of the shear force at a distance of d from the support face, although this is seldom taken into account in practical design.

6.5.1 Concentrated loads on slabs

The critical perimeter for punching shear due to the action of a concentrated load N on a slab is located at a distance of $1.5d$ from the edge of the loaded area (or the column face in the case of flat-slab design), as shown in *Figure 6.2*. Thus if u is the length of the perimeter of the load, the length of the critical perimeter for shear is $u + 12d$. If the load (or column) is not rectangular, u should be taken as the perimeter of a smallest rectangle touching the loaded area.

The value of v_c, the permissible shearing stress resisted by the concrete alone, may be determined from *Data Sheet 15*. The appropriate value of ϱ can be the average proportion of tension reinforcement for both directions

within a strip of slab $a + 3d$ in width, where a is the width of the loaded area in the direction under consideration. Then if $v_c \geq N/(u + 12d)d$, no shearing reinforcement is needed. Alternatively, it may be advantageous to calculate the value v_c in each direction separately, using the appropriate value of ϱ for the direction concerned. Then if a_1 and a_2 are the widths of the loaded area in each direction and v_{c1} and v_{c2} are the corresponding shearing stresses in these directions, check whether $d[2v_{c1}(a_1 + 3d) + 2v_{c2}(a_w + 3d)]$ is equal to or exceeds N.

If $v_c < N/(u + 12d)d$ (or the more complex expression in the previous paragraph) $< v_{max}$, where v_{max} is the maximum permissible shearing stress in slabs, and providing that h is not less than 200 mm, shearing reinforcement is necessary; successive planes at distances of $0.75d$, commencing at the face of the load, must be investigated. The shearing resistance provided by the concrete may be enhanced as described in section 6.3.1, but the enhancement factor is now $1.5d/a_v$; in no case may v_{enh}, the enhanced value of v_c, exceed v_{max}. At the face of the load $v = N/ud$, and shearing reinforcement having a total area of not less than $(N - v_{enh}ud)/(0.87f_{yv})$ or $0.4ud/(0.87f_{yv})$, whichever is the greater, must be provided. This reinforcement should be evenly spaced along at least two perimeters, the spacing of the links not exceeding $1.5d$.

Perimeters at successive distances of $0.75d$ from the load face must now be checked. Reinforcement having a total area of the greater of $(N - v_{enh}u_nd)/0.87f_{yv}$ or $0.4u_n d/0.87f_{yv}$, where u_n is the perimeter at the plane being considered, must be provided until a plane is reached where v is less than v_c or 0.4. For each plane the shear reinforcement determined from the foregoing expressions should be evenly distributed along the perimeter currently being considered, and a similar amount provided along the perimeter $0.75d$ nearer to the load; this is to ensure that the theoretical plane of

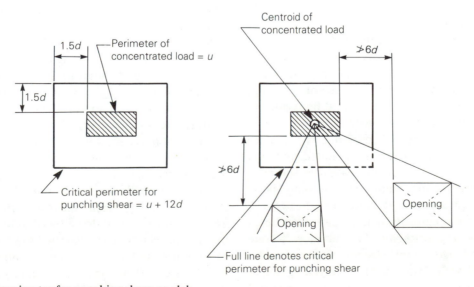

Figure 6.2 Critical perimeter for punching shear on slab.

failure will intersect at least two sets of steel. In calculating the reinforcement required at any perimeter, BS8110 permits the steel already determined by the calculations for a previous perimeter to be taken into account. The link spacing around any perimeter should not exceed $1.5d$. CP110 required that the shear steel should be in the form of vertical or inclined links, securely anchored by both ends being passed around the main reinforcement. BS8110 appears to omit this requirement, although it now permits inclined bars to be used, as noted above.

The design of reinforcement to resist shearing around the column heads and drop panels of flat slabs is undertaken in a similar manner to that described above, each column head being considered as an inverted concentrated load (see section 11.1.2).

Where openings are provided in the slab within a distance of $6h$ from the edge of a concentrated load, the resistance to shearing of the portion of the critical perimeter that lies within the angle subtended by the opening must be omitted in the foregoing calculations, as shown in *Figure 6.2*. However, in the case of a single opening immediately adjoining a column, where its greatest width does not exceed either one-quarter of the column side or one-half of the slab thickness, the opening may be ignored.

6.6 TORSION

One innovation in CP110 was the inclusion for the first time of detailed information concerning the design of sections to resist torsion. In BS8110 this information has been moved to Part 2 of the Code, but although the material has been rewritten, the content remains virtually unaltered, except for a minor change in the relationship between the elastic and shear moduli. According to the Code, if the resistance to torsion or torsional stiffness is ignored when analysing a structure, the nominal shearing reinforcement that is provided will restrict torsional cracking and so detailed investigations of the effects of torsion are generally unnecessary. However, the *Code Handbook* points out that, as well as the need to calculate and provide for any ultimate torsional moments on statically determinate systems, it may also be necessary to investigate the effects of torsion on statically indeterminate structures at the limit-state of cracking, although with such structures the torsional behaviour at the ultimate limit-state is usually satisfactory.

In conventional beam-and-slab design it is therefore generally not necessary to undertake detailed torsional calculations. However, in certain cases a statically indeterminate member may undergo unusually large rotations and thus justify more detailed investigation. A typical example is the case of an L-shaped edge beam having a large opening adjoining the web close to one support. Over the length of this opening the rectangular beam section is liable to serious torsional cracking due to the need of this section (which has a lower torsional

resistance than the L-shaped section provided elsewhere) to transfer the large twisting moment arising from the adjacent slab back to the face of the support.

When calculating the effects of torsion it may be necessary to take account of the torsional rigidity GC of the section. In this expression G is the shear modulus which, according to BS8110, should be taken as 42% of the modulus of elasticity of the concrete, i.e. 0.42 times the value of E_c obtained from *Data Sheet 21*; C is the torsional constant. The Code recommends the use of values of C that are one-half of the Saint Venant values for the corresponding plain concrete section. For rectangular sections, suitable values of C may be obtained by multiplying the values of γ read from the scale on *Data Sheet 15* by $\frac{1}{2}h_{min}^3 h_{max}$, where h_{min} and h_{max} are the lesser and greater overall dimensions of the rectangle. For flanged sections such as T-, L- or I-beams, it is necessary to sum the values obtained for each component rectangle, always taking h_{max} as the larger overall dimension of the individual rectangle. In cases where several ways of subdividing a section are possible, BS8110 recommends the adoption of the arrangement that maximizes $\Sigma(h_{min}^3 h_{max})$, and states that this is normally achieved when the widest rectangle is as long as possible. Thus for a T-section, if $b_w > h_f$ the two rectangles considered should have dimensions h and b_w, and $b - b_w$ and h_f, respectively. In doubtful cases, the value of C resulting from each possible subdivision should be determined and the maximum value adopted.

6.6.1 Torsional stresses

If T is the ultimate torsional moment acting on a rectangular section, the corresponding torsional stress is, according to BS8110, given by the equation

$$v_t = \frac{2T}{h_{min}^2(h_{max} - \frac{1}{3}h_{min})}$$

where, as before, h_{max} and h_{min} are the greater and lesser dimensions of the rectangle. In the case of a flanged section, the total torsional moment should be divided among the component rectangles in proportion to the ratio of $h_{min}^3 h_{max}$ of each individual rectangle to the sum $\Sigma(h_{min}^3 h_{max})$ for all the rectangles.

As in the case of shearing, BS8110 prescribes limiting torsional stresses $v_{t min}$ that concrete of various grades can withstand without the need to provide special reinforcement. However, if v_t exceeds $v_{t min}$, torsional reinforcement is necessary, and unlike the case of normal shear, the *entire* torsional force must then be resisted by the reinforcement. To prevent premature crushing of the concrete, in no circumstances may the sum $v + v_t$ of the stresses resulting from shearing and torsion exceed limiting values v_{tu} laid down by BS8110. Furthermore, in the case of sections in which the larger overall dimension of a reinforcing link is less than 550 mm, v_t may also not exceed $y_1 v_{tu}/550$. Values of $v_{t min}$ and v_{tu} may be read from the scales on *Data Sheet 15*.

6.6.2 Torsional reinforcement

When the torsional stress v_t acting on any component rectangle forming a section, obtained as described above, exceeds $v_{t min}$, reinforcement in the form of a combination of closed rectangular links and longitudinal bars must be provided to resist the entire torsional moment on that rectangle. This reinforcement is additional to any that is provided to resist bending and/or shearing. The link reinforcement provided must be such that

$$0.8x_1\, y_1\, \frac{0.87 f_{yv}\, A_{sv}}{s_v} \geqslant T \qquad (6.3)$$

where A_{sv}, f_{yv} and s_v are the cross-sectional area (both legs), characteristic strength and spacing of the link reinforcement, respectively, and x_1 and y_1 are the lesser and greater dimensions of the rectangular link, respectively; f_{yv} must not exceed $460\,\text{N/mm}^2$.

Since the link-reinforcement factor $K_u = 0.87 f_{yv} A_{sv}/s_v$, equation (6.3) can be expressed as

$$K_u \not< \frac{T}{0.8x_1\, y_1} \qquad (6.3a)$$

Appropriate values of K_u for various grades, sizes and spacings of link reinforcement can be read from *Data Sheet 16*, and it thus becomes a simple matter to select a suitable arrangement of links to give the requisite value of $T/0.8x_1\, y_1$.

The longitudinal steel provided must be such that

$$A_{sl} \not< \frac{x_1 + y_1}{0.87 f_{yl}}\, \frac{0.87 f_{yv}\, A_{sv}}{s_v} \qquad (6.4)$$

where A_{sl} and f_{yl} are the cross-sectional area and the characteristic strength of this reinforcement; again f_{yl} must not exceed $460\,\text{N/mm}^2$. (In BS8110, A_{sl} and f_{yl} are denoted by A_s and f_y.)

As before, since the link-reinforcement factor $K_u = 0.87 f_{yv} A_{sv}/S_v$, equation (6.4) can be rewritten in the form

$$A_{sl} \not< \frac{x_1 + y_1}{0.87 f_{yl}}\, K_u \qquad (6.4a)$$

Now, to meet the requirements of BS8110, all that is necessary to obtain the required amount of longitudinal steel is to multiply the value of K_u corresponding to the system of links required by $(x_1 + y_1)/0.87 f_{yl}$, and to select a suitable arrangement of bars to give this area of steel from *Data Sheet 41*.

The maximum permissible spacing of the closed links is either x_1, $\tfrac{1}{2} y_1$ or $200\,\text{mm}$, whichever is the least, and they must interlock in such a manner that the individual rectangles forming the section are bound together. At least one longitudinal bar must be provided in each corner of each component rectangle requiring torsional steel. The maximum clear distance between the longitudinal bars is $300\,\text{mm}$, and these bars should be so arranged inside the perimeter of the links that the distribution of steel is as uniform as possible. Where the requirement for longitudinal reinforcement coincides with tension or compression bars already provided to resist bending, the additional requirement can be catered for by increasing the sizes of the bars used. Although not stated in BS8110 or the *Code Handbook*, a similar procedure is presumably satisfactory in the case of the link reinforcement. That is to say, the requirements for links to resist shearing and torsion may be calculated separately and summed, and a system chosen that meets the total requirement, taking care to satisfy all the necessary requirements regarding sizes, spacings and so on. Two points should be noted if this procedure is followed. Firstly, since $K_u = 0.87 A_{sv}\, f_{yv}/S_v$, all that is needed is to calculate the required values of K_u for shearing and torsion separately, to sum these, and to select an arrangement of links from *Data Sheet 16* corresponding to this total value of K_u. Secondly, remember that the required area of longitudinal torsional reinforcement A_{sl} is that which is equivalent to the value of K_u for the torsional link reinforcement only.

All torsional reinforcement must extend, beyond the point at which it may theoretically be terminated, for a distance at least equal to the maximum dimension of the section.

In a component rectangle where v_t does not exceed $v_{t min}$, no torsional steel is theoretically required, unless the rectangle forms a major component of the section. The Code does not define 'major' and 'minor' component rectangles, but it seems undesirable to omit torsional reinforcement from any rectangle that defines the overall shape of a flanged section.

Most of the foregoing requirements are illustrated by the worked example given below.

6.6.3 Example

An L-shaped section with $b = 600\,\text{mm}$, $b_w = 200\,\text{mm}$, $h = 350\,\text{mm}$ and $h_f = 150\,\text{mm}$ is subjected to a torsional moment of $15\,\text{kN m}$ and a shearing force of $65\,\text{kN}$. Design suitable shearing and torsional reinforcement if $f_{cu} = 25\,\text{N/mm}^2$, $f_{yv} = f_{yl} = f_y = 250\,\text{N/mm}^2$ and the main tension reinforcement consists of three $20\,\text{mm}$ bars.

Assuming the use of $10\,\text{mm}$ links throughout with $20\,\text{mm}$ cover and the possibility of $25\,\text{mm}$ main bars, $d = 350 - 20 - 10 - 12\tfrac{1}{2} = 307\,\text{mm}$ say. Also for the web rectangle, $x_1 = 200 - (2 \times 20) - 10 = 150\,\text{mm}$ and $y_1 = 350 - (2 \times 20) - 10 = 300\,\text{mm}$. Similarly, for the flange rectangle, $x_1 = 150 - 50 = 100\,\text{mm}$ and $y_1 = 600 - 50 = 550\,\text{mm}$.

Now considering first the shearing force on the web rectangle, $v = 65 \times 10^3/(307 \times 200) = 1.06\,\text{N/mm}^2$. Since $\varrho = 942/(307 \times 200) = 0.0153$ and $f_{cu} = 25\,\text{N/mm}^2$, the resistance to shearing of the concrete alone v_c is $0.77\,\text{N/mm}^2$; as this is less than v, shearing reinforcement would be needed even if no torsional moment were present. Assuming that it is decided not to provide inclined bars, the shearing resistance required from the link reinforcement is thus $65\,000 - (0.77 \times 307 \times 200) = 17\,800$ N. Thus the value of K_u required is $17\,800/307 = 58$.

Next, considering the effects of torsion, if the rectangle containing the web is made as long as possible,

$$\Sigma h_{min}^3 h_{max} = (200^3 \times 350) + (150^3 \times 400)$$
$$= (2.8 + 1.35) \times 10^9 = 4.15 \times 10^9,$$

while if the length of the rectangle containing the flange is maximized, $\Sigma h_{min}^3 h_{max} = (150^3 \times 600) + (200^3 \times 200) = (2.03 + 1.6) \times 10^9 = 3.63 \times 10^9$; thus the former choice is the correct one. Consequently the proportion of the total moment resisted by the web rectangle is $2.8/4.15 = 0.675$. Thus, for the web rectangle,

$$v_t = \frac{2 \times 0.675 \times 15 \times 10^6}{200^2(350 - \frac{1}{3} \times 200)} = 1.79 \text{ N/mm}^2$$

For the flange rectangle,

$$v_t = \frac{2 \times 0.325 \times 15 \times 10^6}{150^2(400 - \frac{1}{3} \times 150)} = 1.24 \text{ N/mm}^2$$

Therefore, since $v_{t\,min} = 0.33$ N/mm² if $f_{cu} = 25$ N/mm², both web and flange rectangles must be reinforced to resist torsion. However, since, for the web rectangle, $v + v_t = 1.06 + 1.79 = 2.85$ N/mm² is less than the limiting value of v_{tu} of 3.75 N/mm² for 25 grade concrete, and also since, as $y_1 = 300$ mm is less than 550 mm, $v_t = 1.79$ N/mm² is less than $v_{tu} y_1/550 = 3.75 \times 300/550 = 2.05$ N/mm², this part of the section is feasible. Furthermore since, for the flange rectangle, $v_t = 1.24$ N/mm² is less than $v_{tu} = 3.75$ N/mm², this rectangle is also feasible.

Now considering first the flange rectangle, from equation (6.3a),

$$K_{u\,req} = \frac{0.325 \times 15 \times 10^6}{0.8 \times 100 \times 550} = 111$$

Provide 6 mm links at 100 mm centres ($K_u = 123$). Also, from equation (6.4a),

$$A_{sl} = \frac{x_1 + y_1}{0.87 f_y} K_u = \frac{(100 + 550) \times 111}{0.87 \times 250} = 332 \text{ mm}^2$$

Provide eight 8 mm bars ($A_s = 402$ mm²).

In the case of the web rectangle, to resist torsion, from equation (6.3a),

$$K_{u\,req} = \frac{0.675 \times 15 \times 10^6}{0.8 \times 150 \times 300} = 281$$

Thus the total value of K_u required to resist the combined shearing and torsional stresses is $58 + 281 = 339$. Provide 10 mm links at 100 mm centres ($K_u = 341$). Now, from equation (6.4a),

$$A_{sl} = \frac{(150 + 300) \times 281}{0.87 \times 250} = 581 \text{ mm}^2$$

Provide six 8 mm bars and increase the diameter of the two outer main bars from 20 to 25 mm ($A_s = 655$ mm²). The reinforcement is arranged in the section as shown in *Figure 6.3*. Note that, in the corner common to the two component rectangles, one 12 mm bar has been substituted for two 8 mm bars.

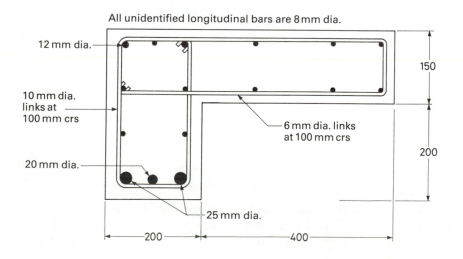

Figure 6.3

Chapter 7
Bond and anchorage

In CP114 the permissible bond stresses were linearly related to the allowable bending stresses and, although the rounding-off of the values tended to mask this linear relationship, a more careful examination of the tabulated values of ultimate bond stress in CP110 indicated the same relationship to the concrete strength. However, in BS8110 the design anchorage-bond stress is now specified as proportional to the square root of the concrete strength.

7.1 ANCHORAGE BOND

If a reinforcing bar is to withstand the very high tensile or compressive stresses for which it is designed, there must be no possibility of significant movement occurring between the bar and the surrounding concrete. To prevent this it is necessary to ensure that the stress over the interface between the steel and the concrete is limited to a safe value, thus enabling the force in the bar to be transferred gradually to the concrete. The permissible stress in this situation is known as the anchorage-bond stress. It varies with the strength of the concrete and the strength and surface characteristics of the reinforcing bar.

BS8110 includes the requirement that every bar must extend beyond any section for a length that is sufficient to develop by bond the force in the bar at that section. Alternatively the Code specifies that the bar must extend to the point where either the shear capacity of the section is twice that for which it is being designed, or the remaining bars provide twice the area of steel required to resist the applied bending moment at that section. The minimum length that may be provided is specified in the Code as equal to the effective depth of the member or twelve times the diameter of the bar, whichever is the greater; this means that a bar must extend a distance of at least 12ϕ or d beyond the point at which there is no theoretical stress in the bar.

It is not essential to provide an anchorage at the end of the bond length, and in present-day practice such anchorages are seldom used when high-yield steel is employed. However, if such an anchorage is provided it is possible to reduce the bond length accordingly. Anchorages in the form of semicircular hooks or right-angled bobs, as shown in *Figure 7.1*, should only be provided where needed for specific design requirements. The internal radius of such hooks or bobs should, according to BS4466, normally be 2ϕ for mild steel bars and 3ϕ for high-yield bars. However, both here and also for bends in reinforcing bars generally, in no case should the radius be less than twice that guaranteed by the reinforcement manufacturer. Provided that this condition is not violated, where a hook fits over a main bar or an anchor bar, the internal diameter of the hook may be made equal to the diameter of the bar around which it fits.

If a semicircular hook is provided then the bond length, measured from the start of the bend to a point 4ϕ beyond the end of the bend, may be considered to provide an anchorage equal to 24ϕ or $8r$, where r is the internal radius of the bend. If the actual length of bar provided, including the straight portion, is greater, the anchorage provided may be considered as this length instead. For a right-angled bob, the anchorage provided is equivalent to the lesser of 12ϕ or $4r$, or to the actual length of the bar if this is greater. Since the actual length of bar needed to form a semicircular hook is $\pi r + 4\phi$, if $r = 3\phi$ a length of bar of about 13.5ϕ will provide an anchorage equivalent to 24ϕ, thus saving a bar length of about 10.5ϕ at the cost of forming the hook. This is the maximum saving of length that can be achieved. For example, if $r = 2\phi$ a bar length of 10.5ϕ will provide an anchorage of 16ϕ, thus saving 5.5ϕ; and with $r = 4\phi$ an anchorage length of 24ϕ requires 16.5ϕ of bar, thus saving only 7.5ϕ. With a right-angled bob, the maximum saving in bar length (of about 4.8ϕ) occurs when $r = 2\phi$; when $r = 3\phi$ the saving is about 3.3ϕ. BS8110 does not recommend the use of hooks of other shapes, such as the 135° bend which was sometimes employed in the past.

Since the total force in the bars must not exceed the bond stress times the contact area between the concrete and the steel,

$$f_{bu}\, l\pi\phi_e \geq (\pi\phi_e^2/4)\,(f_y/1.15)$$

where f_{bu} is the limiting ultimate anchorage-bond stress specified in BS8110 for tension reinforcement, l is the anchorage-bond length and ϕ_e is the effective bar size. If reinforcing bars of size ϕ are arranged in groups in contact (i.e. pairs or bundles) the effective bar size is equal to the diameter of a single bar having an equal total area. In

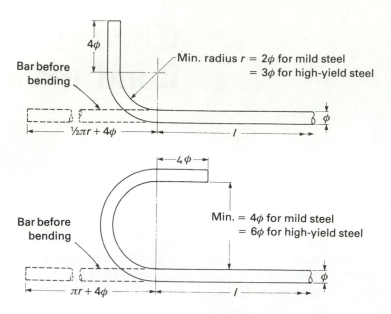

Figure 7.1 Standard hooks for anchorage bond.

other words, if the number of bars forming the group concerned is j, $\phi_e = \sqrt{(j\phi^2)}$. For a single bar, $\phi_e = \phi$. The above expression may be rearranged in the form

$$l \geq 0.2174 f_y \phi_e / f_{bu}$$

Similarly, for compression reinforcement, the foregoing expressions are again valid if f_{bu} now represents the ultimate anchorage-bond stress in compression. In BS8110, as in its predecessors CP114 and CP110, the permissible anchorage-bond stress for bars in compression is 25% higher than for bars of the same type in tension.

The bond resistance of deformed bars is somewhat greater than that of plain bars. For the purposes of bond strength, BS8110 recognizes two types of deformed bars, corresponding broadly to plain or chamfered square-twisted bars and ribbed bars respectively; full details of this classification are given in BS4449 and BS4461. Deformed bars meeting the specified more stringent requirements, either because their deformation pattern conforms to specified characteristics or as the result of performance tests, may be classed as type 2; other deformed bars are known as type 1. The allowable ultimate average bond stresses provided in BS8110 for type 1 deformed bars are about 40% higher than those applying to plain round bars. In addition, the Code permits a further increase of 25% in the values of f_{bu} if the bars are of type 2. However, BS8110 states that where nominal links are omitted from beams, the limiting anchorage-bond stresses considered should be those corresponding to plain round bars, irrespective of the bond type employed.

In *Data Sheet 18* the actual bond lengths required in millimetres are given for three grades of concrete ($f_{cu} = 25$, 30 and 40 N/mm²) for the normal characteristic strengths of reinforcement and for standard bar sizes of from 6 to 40 mm; these lengths have been rounded to the 5 mm dimension above the exact length calculated.

(Unlike its predecessor, BS8110 sets no limit to the increase in allowable bond stress with an increase in concrete grade. However, *Allen* recommends that for values of f_{cu} greater than 40 N/mm² the allowable bond stress employed should not exceed that corresponding to grade 40 concrete.) The lengths are given for straight bars and also with the formation of right-angled bobs and semicircular hooks having internal radii of 2ϕ and 3ϕ for mild steel and high-yield steel bars respectively. Note that all the tabulated values are based on the reinforcement being stressed to its full design value at the point from which the anchorage is measured. If the actual design stress in the reinforcement at the point from which the anchorage is measured is less, the anchorage length may be reduced, in the case of straight bars on a pro rata basis. If an anchorage is to be provided, first calculate the length of straight bar needed by multiplying the no-hook length given in the table by the ratio of the actual design stress to $f_y/1.15$. Then subtract from this value the difference between the length of anchorage required with no hook and that required with the type of hook being used. For example, if $f_{yd1} = 300$ N/mm² with 25 mm high-yield type 2 bars, and $f_{cu} = 25$ N/mm², the anchorage length required at full stress is 1000 mm, so that the actual length required is $1000 \times 300 \times 1.15/460 = 750$ mm. If a semicircular hook of 75 mm internal radius is to be provided, subtract $1000 - 400 = 600$ mm, so that the actual length needed is $750 - 600 = 150$ mm plus the hook.

The bond strength of structural lightweight-aggregate concrete is appreciably lower than that of normal-weight concrete, and BS8110 states that, for both plain and deformed bars, anchorage-bond stresses of 80% of those for corresponding grades of normal-weight concrete should be employed. For practical purposes the appropriate lengths for concrete grades 25, 30 and 40 may be obtained by increasing the lengths given on *Data Sheet 18* for bars without hooks by 25% and then making an appropriate deduction for the hook.

7.1.1 Laps

When bars in tension are lapped, the length of overlap provided must be not less than either the anchorage-bond length appropriate for the smaller of the two bars being lapped, or 15ϕ or 300 mm, whichever is the greater. In addition, where the lap occurs near the top of a concrete section as cast and the minimum cover provided is less than twice the diameter of the bars that are being lapped, the lap length provided must be increased by 40%. If the lap occurs near the corner of a section where the minimum cover to either of the adjoining faces is less than twice the size of the larger bar forming the lap, or where the clear distance between adjacent laps is less than 75 mm or six times the size of the larger lapped bar, a similar increase should be made. If both conditions requiring this 40% increase occur simultaneously, the lap length originally required should be doubled. Anchorage-bond lengths corresponding to these factors of 1.4 and 2 are also tabulated on *Data Sheet 18*.

Where bars in compression are lapped, the lap length should be at least 1.25 times the anchorage-bond length in compression required for the smaller of the lapped bars.

Where both bars forming a lap in either tension or compression are greater than 20 mm in size and the concrete cover provided is less than 50% greater than the size of the smaller bar, transverse links having a diameter of at least one-quarter of the smaller bar must be provided at not more than 200 mm centres throughout the length of the lap. BS8110 also specifies that at any lap the sum of the sizes of all the bars forming a particular layer must not exceed two-fifths of the breadth of the section at that level.

As an alternative to lapping, connections that transfer stress from one bar to another may be formed by proprietary mechanical couplers or sleeves. The bars may also be welded together, although BS8110 states that welding should not be used where the imposed loading is predominantly cyclic. For further details reference should be made to clauses 3.12.8.16 to 3.12.8.21 of BS8110, Part 1.

All laps should, wherever possible, be located at points where the tensile or compressive forces in the bars are low, and they should preferably be staggered.

7.2 BEARING INSIDE BENDS

If a bar does not extend, or is assumed not to be stressed, beyond a distance of four times the size of the bar from the end of a bend (as in, for instance, the case of a standard anchorage), it is not necessary to check the bearing stress inside the bend. Elsewhere, however, the stress on normal-weight concrete within the bend, as given by the expression $F_{bt}/r\phi$, where F_{bt} is the tensile stress in the bar, must not exceed $2f_{cu}/(1 + 2\phi/a_b)$, where a_b is the distance between bar centres measured perpendicularly to the plane of bending. In the case of a bar adjoining the surface, a_b is equal to the concrete cover plus the bar size; if the bars are in bundles, ϕ is the equivalent bar diameter. Now if $F_{bt} = \pi\phi^2 f_{yd2}/4$, these expressions may be rearranged to give

$$r = K_1 \phi = (1 + 2\phi/a_b)(\pi/8)(f_{yd2}/f_{cu})\phi \quad (7.1)$$

where f_{yd2} is the actual design stress in the tension reinforcement. Values of K_1 corresponding to various ratios of ϕ/a_b and f_{yd2}/f_{cu} can be read from the nomogram on *Data Sheet 19*.

For lightweight-aggregate concrete the corresponding limiting stress permitted by BS8110 is two-thirds that for dense concrete, thus resulting in the expression

$$r = K_2 \phi = (1 + 2\phi/a_b)(3\pi/16)(f_{yd2}/f_{cu})\phi \quad (7.2)$$

Values of K_2 can also be read from the nomogram on *Data Sheet 19*.

7.3 LOCAL BOND

Until the drafting of BS8110 it had often been considered possible that, despite a sufficient anchorage-bond length being provided, bond failure between the steel and the surrounding concrete might occur if sufficiently large changes in the tensile force in a bar occurred over a very short distance. Such changes are not unusual where high shearing forces occur in conjunction with low moments (e.g. at span ends that are assumed to be freely supported.) For this reason, up to and including CP110 an additional check for local-bond stress was required. However, it became known that numerous designs had been prepared without this check being taken into account, with no apparent ill-effects. In addition, at those points where a check seemed necessary, the theory relating to local-bond failure required the assumption that flexural cracking had occurred, which was clearly incorrect. Therefore BS8110 simply states that, provided that on each side of any given cross-section a sufficient anchorage-bond length (or alternative method of anchorage) is available to develop the force in the bar concerned, the local-bond stress need not be checked.

Chapter 8
Serviceability limit-states

In addition to having sufficient resistance to failure by bending, shearing, bond etc., a reinforced concrete member must, of course, behave satisfactorily under normal working (i.e. service) loads. The three principal criteria regarding serviceability are that, when a member is subjected to the forces and moments that arise from working loads, deflection, vibration and cracking must not be excessive.

As regards vibration, such as might arise due to the action of machinery or the effects of wind on the structure, the Code states that measures should be taken to prevent discomfort, alarm, and damage to or interference with the proper function of the structure. For acceptable vibration levels, reference must be made to specialist literature (for example ref. 11).

Deflection and cracking are dealt with in some detail in Part 2 of BS8110, and the Code requirements concerning these phenomena are now discussed.

8.1 DEFLECTION: GENERAL REQUIREMENTS

For reinforced concrete members, the general requirements of BS8110 are (a) that the final deflection (including all time-dependent effects such as creep and shrinkage as well as those of temperature) of each horizontal member below the supports must not exceed span/250, and (b) that the deflection occurring after the construction of a partition or the application of a finish should not exceed the lesser of span/350 or 20 mm for non-brittle partitions, and span/500 or 20 mm for brittle materials.

Part 2 of BS8110 gives detailed information concerning the calculation of deflections; these requirements are discussed later in this chapter. However, to minimize the need for such calculations, Part 1 of the Code also provides a series of tables which enable a suitable span/effective-depth ratio to be determined for most normal beams and slabs. Provided that the actual ratio adopted does not exceed the limiting value given by the tables, more-detailed calculation is unnecessary. The values obtained by this simplified procedure may be disregarded, however, provided that detailed calculations show that the resulting deflections meet the foregoing general requirements. It is therefore only necessary to undertake detailed calculations if the designer wishes to exceed the limiting span/effective-depth ratios given by the simplified rules, if the structure is abnormal in some respect, or if particularly careful control of the actual deflection is needed.

8.2 DEFLECTION: SIMPLIFIED RULES

The simplified rules concerning deflection given in the Code consist of selecting a basic span/effective-depth ratio corresponding to the span and the type of fixity of the member. This basic ratio is then modified by multiplying it by various factors corresponding to the resistance moment, the working stress in the tension steel, the amount of compression steel, the type and shape of the section, and the type of concrete used. The reasons why these particular factors influence the span/effective-depth ratio are discussed by *Allen* and in ref. 12.

Since BS8110 prescribes limiting basic span/effective-depth ratios corresponding to specific spans, it is a simple matter to express these requirements in the form of minimum effective depths corresponding to specific spans. Then, by re-expressing the various modifying factors as reciprocals of those given in the Code (which, in the case of some of the factors, leads to considerable simplicity), it is possible to multiply this basic effective depth by the necessary factors concerned to achieve a final minimum value. Provided that the actual effective depth adopted is not less than this minimum value, the design meets the simplified Code requirements concerning deflection.

Scales from which the basic effective depths and appropriate modifying factors may be read are given on *Data Sheet 20*. To illustrate the use of this material, consider the case of a freely supported rectangular beam 400 mm wide spanning 10 m, designed to resist an applied moment of 490 kN m (i.e. $M/bd^2 = 4.9$ if $d = 500$ mm) and reinforced with 1% of compression steel. The basic effective depth for such a beam is 500 mm. For $f_y = 250$ N/mm^2 and $M/bd^2 = 4.9$ the multiplying factor is 0.99 and that for 1% of compression steel is 0.8. Thus the minimum effective depth that may be adopted is $500 \times 0.99 \times 0.8 = 396$ mm.

Since it is necessary to know the section dimensions and the proportion of compression steel provided in order to find the minimum effective depth, it will be appreciated that neither *Data Sheet 20* nor the tables given in Part 1 of BS8110 can be used for direct design to meet the Code requirements concerning deflection, and such tables can thus only be used to check designs already produced. When designing a member it is therefore necessary, as suggested in Chapter 9, to adopt suitable dimensions when calculating the reinforcement needed for ultimate limit-state requirements, which also meet the Code requirements as regards deflection; a certain amount of trial and adjustment may therefore be necessary.

In CP110 the tension steel factor was directly related to the amount of, and stress in, this steel. However, to enable a section to be checked for compliance with deflection requirements at a much earlier stage in the design, in BS8110 this factor is now related to the applied-moment factor (i.e. M/bd^2) and the stress in the tension steel. (Note that clause 3.4.6.5 of BS8110 Part 1 may cause confusion as it refers to the need to modify the basic span/effective-depth ratio according to the area of steel provided rather than the resistance-moment factor. This is because the text was copied directly from CP110, and the revision to the table to which the clause refers was overlooked.) The intention with the BS8110 revision is that, when a value of M/bd^2 has been selected that satisfies ultimate limit-state requirements, an immediate check can be made as to whether the deflection requirements are also met. Since the inclusion of any compression reinforcement is beneficial (i.e. increases the permissible span/effective-depth ratio), its inclusion or otherwise may safely be neglected at this design stage.

8.3 DEFLECTION: ANALYTICAL METHOD

8.3.1 Determination of moments and forces

The first step in the calculation procedure for deflections described in Part 2 of the Code is to determine the actual moments and forces that arise due to working loads by making an elastic analysis: the moments determined from this analysis may not be redistributed, of course. The relative proportions of dead and imposed loads will differ from those considered when undertaking the ultimate limit-state analysis to determine the reinforcement required. It is therefore necessary to make a fresh analysis, unless all the moments have been obtained by factoring the values that result from the application of independent arbitrary loads. If the behaviour of each member is to be characterized by a single stiffness value, BS8110 recommends that this stiffness is based on the gross uncracked concrete section, as it is thought that this represents the true moment and force fields more accurately than the use of values based on transformed cracked sections. However, if a computer analysis that can take account of changes in properties along the lengths of individual members is utilized, a more accurate result is likely if the stiffness of those parts of a member that are stressed more highly is determined on the basis of the transformed cracked section.

In CP110, four principal factors were pinpointed that are difficult to allow for but which considerably influence the accuracy with which deflections may be predicted. These are the actual degree of restraint provided by the supports; the precise loading, and in particular that proportion which is permanent; whether or not a member is cracked; and the effects of finishes and partitions. However, if such points are correctly considered, the method described in the Code has been found to predict both short-term and long-term deflections reasonably accurately. BS8110 suggests that, for normal domestic or office buildings, one-quarter of the imposed load may be considered as permanent, but this proportion rises to at least three-quarters for storage structures when the calculations are made to determine the maximum likely deflection.

8.3.2 Calculation of sectional properties

When the moments and forces have been calculated using working loads of G_k and Q_k, the next step is to evaluate the curvatures. According to BS8110 the way in which this should be done depends on whether or not a particular section is cracked. The procedure involves determining the curvature according to both possibilities and then taking the greater resulting value; this is usually that which occurs when the section is assumed to be cracked.

Cracked section. In this case both the concrete in compression and the steel are assumed to behave elastically and the distribution of strain across the section is assumed to be linear. The elastic modulus of steel is taken as 200 kN/mm^2 and that of concrete in compression for short-term loading is as given on *Data Sheet 21*. For long-term loading, this elastic modulus should be multiplied by $1/(1 + \phi)$, where ϕ is the appropriate creep coefficient (i.e. the ratio of creep strain to initial elastic strain). The appropriate value of ϕ depends on several variables but chiefly on the effective section thickness, the ambient relative humidity and the age at which the concrete is loaded. According to BS8110, for uniform sections the effective section thickness may be obtained by dividing the area of the section by one-half of the perimeter. If the concrete is sealed or permanently immersed in water so that drying is prevented, an effective thickness of 600 mm should be adopted. For ambient relative humidity, BS8110 suggests values of 45% and 85% for indoor and outdoor exposure respectively. Based on these assumptions, the recommended Code values are indicated by the lower graph on *Data Sheet 21*; more-detailed information regarding creep may be obtained from ref. 13. Now to obtain the appropriate value for long-term loading, multiply the short-term elastic modulus $E_{c\,28}$ (given by the expression adjoining the lower graph on *Data Sheet 21*) by the appropriate multiplier read from the lower graph.

With these assumptions the tensile stresses in the concrete may be calculated, assuming a stress distribution that increases linearly from zero at the neutral axis to a value at the centroid of the tension steel that depends on the duration of the load, and is as shown on the scale immediately beneath the lower graph on *Data Sheet 21*.

The stresses in the concrete are thus as shown in *Figure 8.1c*. In order to relate the bending moment to the curvature it is necessary to construct a diagram of the forces on the section (*Figure 8.1b*) and to calculate the moments from this. The resulting analysis becomes rather complex, particularly in the case of flanged sections, and

to simplify this part of the work *Allen* suggests that the sectional properties such as the depth to the neutral axis x, the lever-arm z, and the equivalent moment of inertia I_e are calculated on the assumption that the concrete in the tension zone is unstressed. The depth to the neutral axis is consequently slightly underestimated but comparative calculations show that the effects are negligible.

Uncracked section. If the section is considered to be uncracked, the foregoing assumptions again apply with the exception of that concerning the tensile stress in the concrete, the concrete now being assumed fully elastic in tension (*Figure 8.1d*).

It is now necessary to calculate the properties of the transformed section.

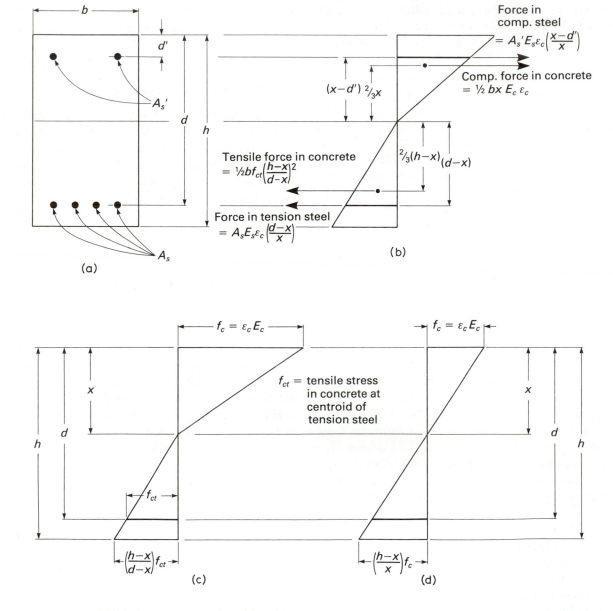

Figure 8.1 (a) Section; (b) forces acting on section; (c) stresses in cracked concrete section; (d) stresses in uncracked concrete section.

Rectangular section. For a rectangular section reinforced in tension and compression,

$$\frac{x}{d} = \sqrt{\left\{ [\alpha_e\varrho + (\alpha_e - 1)\varrho']^2 + 2\left[\alpha_e\varrho + (\varrho_e - 1)\varrho'\frac{d'}{d}\right]\right\}} - [\alpha_e\varrho + (\alpha_e - 1)\varrho'] \tag{8.1a}$$

$$\frac{z}{d} = \frac{3\left(\frac{x}{d}\right)^2 - \left(\frac{x}{d}\right)^3 + 6(\alpha_e - 1)\varrho'\left(\frac{x}{d} - \frac{d'}{d}\right)\left(1 - \frac{d'}{d}\right)}{3\left(\frac{x}{d}\right)^2 + 6(\alpha_e - 1)\varrho'\left(\frac{x}{d} - \frac{d'}{d}\right)} \tag{8.2a}$$

$$\frac{I_e}{bd^3} = \tfrac{1}{3}\left(\frac{x}{d}\right)^3 + \alpha_e\varrho\left(1 - \frac{x}{d}\right)^2 + (\alpha_e - 1)\varrho'\left(\frac{x}{d} - \frac{d'}{d}\right)^2 \tag{8.3a}$$

If the section is reinforced in tension only, $\varrho'(\alpha_e - 1)$ becomes zero and these formulae simplify considerably.

The charts on *Data Sheets 22* to *25* have been prepared to enable values of x/d, z/d and I_e/bd^3 to be determined rapidly for many given combinations of $\varrho\alpha_e$ and $\varrho'(\alpha_e - 1)$. The four charts cover values of d'/d of 0.05, 0.10, 0.15 and 0.20, and linear interpolation may be used between the charts for intermediate cover ratios. Entering the charts with the known reinforcement factors, the neutral-axis factor is given by the broken lines, the lever-arm factor by the chain lines and the moment-of-inertia factor by the full lines.

Flanged section. For flanged sections that are also reinforced in tension and compression, the corresponding formulae for the sectional properties are given by *Allen* as follows:

$$\frac{x}{d} = \sqrt{\left\{\left[\alpha_e\varrho + (\alpha_e - 1)\varrho' + \left(\frac{b}{b_w} - 1\right)\frac{h_f}{d}\right]^2 + 2\left[\alpha_e\varrho + (\alpha_e - 1)\varrho'\frac{d'}{d} + \tfrac{1}{2}\left(\frac{b}{b_w} - 1\right)\left(\frac{h_f}{d}\right)^2\right]\right\}}$$
$$- \left[\alpha_e\varrho + (\alpha_e - 1)\varrho' + \left(\frac{b}{b_w} - 1\right)\frac{h_f}{d}\right] \tag{8.1b}$$

$$\frac{z}{d} = \frac{3\left(\frac{x}{d}\right)^2 - \left(\frac{x}{d}\right)^3 + 2\left(\frac{b}{b_w} - 1\right)\left(\frac{x}{d} - \frac{h_f}{2d}\right)\left[3 - \frac{\left(3\frac{x}{d} - 2\frac{h_f}{d}\right)}{\left(2\frac{x}{d} - \frac{h_f}{d}\right)}\frac{h_f}{d}\right]\frac{h_f}{d} + 6(\alpha_e - 1)\varrho'\left(\frac{x}{d} - \frac{d'}{d}\right)\left(1 - \frac{d'}{d}\right)}{3\left(\frac{x}{d}\right)^2 + 6\left(\frac{b}{b_w} - 1\right)\left(\frac{x}{d} - \frac{h_f}{2d}\right)\frac{h_f}{d} + 6(\alpha_e - 1)\varrho'\left(\frac{x}{d} - \frac{d'}{d}\right)} \tag{8.2b}$$

$$\frac{I_e}{b_w d^3} = \tfrac{1}{3}\left[\left(\frac{x}{d}\right)^3\frac{b}{b_w} - \left(\frac{b}{b_w} - 1\right)\left(\frac{x}{d} - \frac{h_f}{d}\right)^3\right] + \alpha_e\varrho\left(1 - \frac{x}{d}\right)^2 + (\alpha_e - 1)\varrho'\left(\frac{x}{d} - \frac{\delta'}{d}\right)^2 \tag{8.3b}$$

In these expressions the proportions of reinforcement ϱ and ϱ' are based on the width of the web (i.e. $\varrho = A_s/b_w d$ and $\varrho' = A'_s/b_w d$). To avoid the need for a further series of charts, it is possible to utilize those already given for rectangular sections with tension and compression steel by considering the flange as 'compression steel', to obtain the appropriate values of x/d and $I_e/b_w d^3$ for flanged sections which are reinforced in tension only (as is almost always the case). For the rare cases of flanged sections that are also reinforced in compression, the section properties must be evaluated from the above formulae.

The procedure for employing the charts is as follows. Assume first that the neutral axis falls within the flange. If this is so, the section may be considered as a rectangular section of breadth b and depth d; thus determine values of x/d and $I_e bd_3$ by employing steel proportions of $\varrho = A_s/bd$ and $\varrho' = 0$.

If the resulting ratio of x/d exceeds h_f/d, the neutral axis falls within the web. Knowing the values of h_f/d and b/b_w, calculate the equivalent proportion of compression steel that represents the flanges of the section from the expression $\varrho'(\alpha_e - 1) = (h_f/d)[(b/b_w) - 1]$. Then determine $\varrho\alpha_e$ when $\varrho = A_s/b_w d$. With these values of $\varrho\alpha_e$ and $\varrho'(\alpha_e - 1)$, and taking $d'/d = \frac{1}{2}h_f/d$, determine the corresponding factor of x/d for the flanged section from the appropriate chart (interpolating linearly between charts if required). With this revised value of x/d, calculate $(h_f/d)/(x/d)$ and read off the equivalent value of $(d'/d)/(x/d)$ from the scale provided, thus determining an 'equivalent' value of d'/d that represents the position of the centroid of the flange. Then with the value of $\varrho'(\alpha_e - 1)$ obtained originally and the value of d'/d just calculated, use the appropriate chart (or charts) to obtain $I_e/b_w d^3$.

It is not possible to produce a corresponding simple conversion scale to enable z/d to be determined for a flanged section using the charts for rectangular sections. However, the ratio of z/d is affected very little by the exact position of the centroid in the flange, and the assumption of a value of z of $d - \frac{1}{2}h_f$ is very convenient while erring slightly on the safe side.

8.3.3 Evaluation of curvatures

Curvature of rectangular section. When the sectional properties have been determined, the curvature can be calculated from *Figure 8.1b*, using the well-known expression for elastic bending

$$\frac{f_c}{y} = \frac{E_c}{r}$$

Thus the curvature is given by

$$\frac{1}{r} = \frac{f_c}{E_c y} = \frac{\varepsilon_c}{x}$$

since E_c, the short-term or long-term modulus of elasticity of the concrete, is equivalent to f_c/ε_c, and the distance y from the neutral axis to the face considered is equal to x.

Consequently, from *Figure 8.1b*, taking moments about the neutral axis for all the forces acting on the section,

$$M = \tfrac{1}{3}bx^2 f_c + \tfrac{1}{3}b(h - x)^2 f_{ct}\frac{(h - x)}{(d - x)}$$

$$+ \alpha_e \varrho bd(d - x) f_s + (\alpha_e - 1)\varrho' bd(x - d') f'_s$$

where f_c and f_{ct} are the compressive stress and the tensile stress at the centroid of the tension steel respectively, and f_s and f'_s are the stresses in the tension and compression reinforcement respectively. Thus

$$M = \tfrac{1}{3}bx^2 E_c \varepsilon_c + \tfrac{1}{3}b\frac{(h - x)^3}{(d - x)} f_{ct}$$

$$+ \alpha_e \varrho bd(d - x)^2 \frac{E_c \varepsilon_c}{x} + (\alpha_e - 1)\varrho' bd(x - d')^2 \frac{E_c \varepsilon_c}{x}$$

since

$$f_s = E_c \varepsilon_c(d - x)/x \text{ and } f'_s = E_c \varepsilon_c(x - d')/x.$$

Rearranging,

$$M = \frac{E_c \varepsilon_c}{x}[\tfrac{1}{3}bx^3 + \alpha_e \varrho bd(d - x)^2$$

$$+ (\alpha_e - 1)\varrho' bd(x - d')^2] + \tfrac{1}{3}b\frac{(h + x)^3}{(d - x)} f_{ct}$$

But since

$$I_e = \tfrac{1}{3}bx^3 + \alpha_e \varrho bd(d - x)^2 + (\alpha_e - 1)\varrho' bd(x - d')^2$$

for a rectangular section,

$$M = \frac{I_e E_c \varepsilon_c}{x} + \tfrac{1}{3}b\frac{(h - x)^3}{(d - x)} f_{ct} = \frac{I_e E_c}{r} + \tfrac{1}{3}b\frac{(h - x)^3}{(d - x)} f_{ct}$$

Thus

$$\frac{1}{r} = \left[M - \tfrac{1}{3}b\frac{(h - x)^3}{(d - x)} f_{ct} \right] \bigg/ I_e E_c \qquad (8.4a)$$

Expression (8.4a) may be expressed non-dimensionally as

$$\frac{d}{r} = \left[\frac{M}{bd^3} - \tfrac{1}{3}\frac{\left(\dfrac{h}{d} - \dfrac{x}{d}\right)^3}{\left(1 - \dfrac{x}{d}\right)} f_{ct} \right] \bigg/ \frac{I_e E_c}{bd^3} \qquad (8.4b)$$

Curvature of flanged section. For a flanged section, the corresponding expressions are

$$\frac{1}{r} = \left[M - \tfrac{1}{3}b_w\frac{(h - x)^3}{(d - x)} f_{ct} \right] \bigg/ I_e E_c \qquad (8.5a)$$

and non-dimensionally

$$\frac{d}{r} = \left[\frac{M}{b_w d^3} - \tfrac{1}{3}\frac{\left(\dfrac{h}{d} - \dfrac{x}{d}\right)^3}{\left(1 - \dfrac{x}{d}\right)} f_{ct} \right] \bigg/ \frac{I_e E_c}{b_w d^3} \qquad (8.5b)$$

if the contribution of the tensile resistance of the flange concrete below the neutral axis is neglected when the neutral axis falls within the flange. If the neutral axis falls within the web, expressions (8.4) apply.

If the section is assumed to be uncracked, the required values of x/h and I_g/bh^3 for the gross section can be calculated from the following expressions.

Sectional properties of rectangular section

$$\frac{x}{h} = \frac{1 + 2(\alpha_e - 1)\left(\varrho_1 \frac{d}{h} + \varrho_1' \frac{d'}{h}\right)}{2[1 + (\alpha_e - 1)(\varrho_1 + \varrho_1')]}$$

$$\frac{I_g}{bh^3} = \tfrac{1}{3}\left[\left(\frac{x}{h}\right)^3 + \left(1 - \frac{x}{h}\right)^3\right]$$

$$+ (\alpha_e - 1)\left[\varrho\left(\frac{d}{h} - \frac{x}{h}\right)^2 + \varrho'\left(\frac{x}{h} - \frac{d'}{h}\right)^2\right]$$

Sectional properties of flanged section

$$\frac{x}{h} = \frac{\frac{b_w}{b} + \left(1 - \frac{b_w}{b}\right)\left(\frac{h_f}{h}\right)^2 + 2(\alpha_e - 1)\left(\varrho_1 \frac{d}{h} + \varrho_1' \frac{d'}{h}\right)}{2\left[\frac{b_w}{b} + \frac{h_f}{h}\left(1 - \frac{b_w}{b}\right) + (\alpha_e - 1)(\varrho_1 + \varrho_1')\right]}$$

$$\frac{I_g}{bh^3} = \tfrac{1}{3}\left[\left(\frac{x}{h}\right)^3 + \frac{b_w}{b}\left(1 - \frac{x}{h}\right)^3 - \left(1 - \frac{b_w}{b}\right)\left(\frac{x}{h} - \frac{h_f}{h}\right)^3\right] + (\alpha_e - 1)\left[\varrho_1\left(\frac{d}{h} - \frac{x}{h}\right)^2 + \varrho_1'\left(\frac{x}{h} - \frac{d'}{h}\right)^2\right]$$

where $\varrho_1 = A_s/bh$ and $\varrho_1' = A_s'/bh$.

Since the section is uncracked, the curvature may now be calculated directly from the relationship for simple bending:

$$\frac{1}{r} = \frac{M}{I_g E_c}$$

Long-term curvature. According to BS8110, the long-term curvature $1/r_{tt}$ is now obtained as follows:

1. Calculate the instantaneous curvatures due to the total load and to the so-called 'permanent' load ($1/r_{it}$ and $1/r_{ip}$ respectively) and determine the difference between them. The permanent load consists of the dead load plus that due to the partitions and finishes, and includes a proportion of the imposed load; the Code suggests that the proportion of imposed load considered as permanent may range from one-quarter for domestic and office buildings to more than three-quarters for storage structures.
2. Next calculate the long-term curvature $1/r_{tp}$ due to the permanent load; i.e. using the long-term value of E_c.
3. Add to the long-term curvature, the difference between the instantaneous values.
4. Finally, add the curvature $1/r_{cs}$ due to shrinkage, which is calculated from the expression

$$\frac{1}{r_{cs}} = \frac{\varepsilon_{cs} \alpha_e S_s}{I}$$

In this expression, ε_{cs} is the free shrinkage strain, α_e is the appropriate modular ratio, S_s is the first moment of

area of the reinforcement about the centroid of the cracked or uncracked section, and I is the second moment of area of the cracked or uncracked section, depending which has been used elsewhere. The modular ratio α_e is obtained by dividing the elastic modulus for steel by the effective elastic modulus of the concrete, given by multiplying the short-term modulus by the creep multiplier; calculations for I using the transformed section should employ this modular ratio. The shrinkage factors recommended in clause 7.4 of BS8110, Part 2, for indoor and outdoor ambient relative humidities of 45% and 85% respectively and a range of effective section thicknesses, are reproduced in the upper graph on *Data Sheet 21*. More-detailed information regarding shrinkage strain is given in ref. 13.

Thus, to summarize, the total long-term curvature is given by

$$\frac{1}{r_{tt}} = \frac{1}{r_{tp}} + \left(\frac{1}{r_{it}} - \frac{1}{r_{ip}}\right) + \frac{1}{r_{cs}} \qquad (8.6)$$

8.3.4 Relationship between curvature and deflection

By determining the curvature along the member, the deflected shape can be calculated from the expression

$$\frac{1}{r_x} = \frac{d^2 a}{dx^2}$$

where r_x is the curvature at any point x at which the deflection is a. The double integration may be undertaken using a numerical technique such as that due to Newmark (ref. 13), but normally the use of Simpson's rule will be adequate.

The Code also permits the adoption of a simplified approach in which the deflection a is evaluated from the expression

$$a = \frac{Kl^2}{r_b} \qquad (8.7)$$

where l is the effective span, $1/r_b$ is the curvature at midspan for a freely supported or fixed member and at the support for a cantilever, and K is a coefficient depending on the type of loading and the support conditions. The term K is obtained by dividing the numerical coefficient relating to the curvature at the point at which the deflection is calculated (i.e. at midspan for a freely supported or fixed span and at the free end for a simple cantilever) by the coefficient for the *maximum* bending moment on the member, or the maximum positive moment in the case of a fixed member. (Note that Table 3.1 in BS8110, Part 2, which gives values of K corresponding to various bending moment diagrams, appears to contain two errors. The expression for K corresponding to a single concentrated load is only valid if $a > 0.5$, and the expression for the maximum moment due to a trapezoidal load should be $wl^2[(3 - 4a^2)/(1 - a)]/24$ and not as shown.)

For example, in the case of a single concentrated load F acting at a distance αl from the left-hand support of a freely supported member of length l, if $\alpha \geq \frac{1}{2}$,

$$\text{Deflection at midspan} = \frac{Fl^2}{48EI} (1 - \alpha) [3 - 4(1 - \alpha)^2]$$

$$\text{maximum bending moment on span} = F\alpha(1 - \alpha)l$$

Therefore the appropriate value of K is given by

$$K = \frac{3 - 4(1 - \alpha)^2}{48\alpha} = \frac{8\alpha - 4\alpha^2 - 1}{48\alpha}$$

If the curvature is measured at midspan, the resulting deflection given by this method is that at midspan. If, as above, the load is not arranged symmetrically on the span, this will not be the maximum deflection, but the resulting difference is normally negligible.

Since the coefficients K are 'compound' factors it is not strictly possible to consider deflections due to combinations of various types of load simply by summing the appropriate values of K for the individual loadings concerned; instead, a value appropriate to the complete load must be determined. It is possible to establish such a value of K by summing the deflection coefficients due to the various loadings involved, multiplied by the values of these loads; and, after multiplying the resulting sum by the span, dividing by the maximum moment on the span. This procedure is illustrated in the worked example in section 8.4.

The lengthy process of determining the coefficients for the midspan deflections (or tip deflections in the case of cantilevers) and maximum positive moments is considerably simplified by using the fifteen charts given on *Data Sheets 26 to 33*. These charts give the required deflection and bending-moment coefficients due to partial uniform and triangular loading for freely supported and fixed spans, and simple and propped cantilevers. Virtually any arrangement of loading can be broken down into a combination of the loadings given. The procedure is then to multiply the coefficients read from the appropriate chart for each partial load by the load concerned. To obtain K, the sum of the deflection terms multiplied by their individual loads should be multiplied by the span and divided by the maximum positive moment on the span (or the support moment in the case of a simple cantilever). This moment can be determined by multiplying the moment coefficients for the partial loads involved by the

loads in question and, by superposition, sketching the resulting total moment diagram to determine the maximum value. The resulting value of k is then substituted in expression (8.7).

The foregoing deflection coefficients can also be used to calculate deflections using other formulae such as those given by basic structural theory. Then

$$\text{deflection} = \text{coefficient} \times \frac{Fl^3}{EI}$$

where F is the appropriate total load, l is the span, and suitable values are adopted for E and I. It should be remembered, however, that the coefficients give the deflection at midspan and not the maximum deflection. Nevertheless the difference between these values is usually minimal. For example, with a partial uniform load or a concentrated load on a freely supported span, the greatest difference (about 2.5%) between the maximum deflection and that at midspan occurs when the load is at one extreme end of the span, when the deflection values are minimal anyway.

For completeness, on *Data Sheet 31* scales are given which enable the position and maximum deflection of free and fixed spans and propped cantilevers to be determined when loaded with a single concentrated load.

In a similar manner, the coefficients for the maximum bending moments can be used to determine the value and position of the maximum moment due to any partial load on a span. This information enables the bending-moment diagram due to such a load to be sketched very rapidly, as shown in *Figure 8.2*. Most arrangements of load can be broken down into combinations of the partial loads given and the appropriate moment diagrams obtained by superposition. The procedure illustrated in the upper part of *Figure 8.2* is to plot the ordinates at A and B using the formulae given and then to construct from this base line the curved part of the diagram with the aid of the ordinates in the lower sketches. The coefficients given on the appropriate data sheets provide a useful check that the diagram has been drawn correctly.

8.3.5 Rotation of cantilever supports

The coefficients given for simple cantilevers assume that there is no rotation at the support. In practice this is seldom true, because the load on the cantilever or the adjoining span may cause the support to rotate if the supporting structure is not very rigid. The rotation θ at a support B which is free to rotate is given by the expression

$$\theta = \frac{l_s}{E_c I_s} [K_1 M_s - \tfrac{1}{6}(M_A + 2M_B)]$$

where l_s, I_s and E_c are the span, moment of inertia and elastic modulus of the member AB adjoining the cantilever, M_A and M_B are the support moments at the far and near supports of AB respectively, and $K_1 M_s$ is the moment of the area of the free bending-moment diagram due to the applied loads on AB about A, divided by l_s^2.

By writing $1/r_A$ for the curvature at A, and so on, the rotation θ at B can be expressed as

$$\theta = l_s \left[\frac{K_1}{r_s} - \frac{1}{6} \left(\frac{1}{r_A} + \frac{2}{r_B} \right) \right]$$

For partial uniform and triangular loads, values of $K_1 M_s$ can be read from the charts on *Data Sheets 34* and *35*, and values of M_s, the free moment on span AB, can be read from *Data Sheets 26* to *28*. As before, virtually any loading may be considered by subdividing the load into combinations of partial loads, and summing the values of $K_1 M_s$ and of M_s (multiplied by their respective loads) for the individual loads concerned. The value of K_1 appropriate to the total load considered is then obtained by dividing $\Sigma K_1 M_s$ by $M_{s\,max}$, where $M_{s\,max}$ is the maximum free moment on the span and $1/r_s$ is the curvature at the position at which this moment occurs.

When θ has been calculated, the resulting deflection at the tip of the cantilever is θl_c, where l_c is the span of the cantilever. If θ is positive in the above formula this deflection will be upwards, and vice versa.

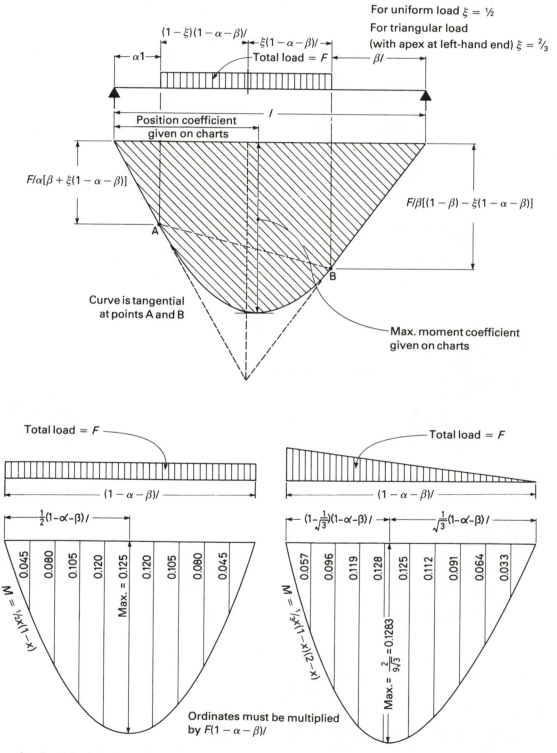

Figure 8.2 Data for sketching bending-moment diagrams.

Owing to limitations of space, such a worked example involving the rotation of a cantilever support cannot be included here, but a valuable one is provided in *Allen*.

8.4 DEFLECTION: WORKED EXAMPLE

The freely supported flanged beam shown in *Figure 8.3* supports a dead load (including self-weight) of 16 kN/m and imposed loads of 11 kN/m (uniformly distributed) and 100 kN (concentrated at 1.5 m from one support). The section is reinforced with six 25 mm bars. If $l = 4.5$ m, $f_{cu} = 25$ N/mm^2, $f_y = 460$ N/mm^2, $\phi = 2.34$ and $\varepsilon_{cs} = 0.0003$, calculate: the instantaneous deflection at midspan due to the permanent load; the long-term deflection at midspan due to permanent load and shrinkage; and the instantaneous deflection due to the total load, assuming that one-half of the imposed loads are permanent.

First calculate the maximum service moments: due to the permanent loads,

$$M_s = [(16 + 5.5) \times 3 \times 1.5 \times \tfrac{1}{2}]$$
$$+ (50 \times 3 \times 1.5/4.5) = 98.4\ \text{kN m}$$

and due to the total load,

$$M_s = [(16 + 11) \times 3 \times 1.5 \times \tfrac{1}{2}]$$
$$+ (100 \times 3 \times 1.5/4.5) = 160.8\ \text{kN m}$$

Next calculate the sectional properties. For short-term loading, if $f_{cu} = 25$ N/mm^2, an appropriate value of E_c (from *Data Sheet 21*) is 25 kN/mm^2. Thus $\alpha_e = 200/25 = 8$. Considering the section as flanged (i.e. the neutral axis falls below the flange), $\varrho = A_s/b_w d = \tfrac{1}{4}\pi \times 25^2 \times 6/$ $(315 \times 200) = 0.0467$ and $\varrho\alpha_e = 0.374$. The flange is equivalent to an area of compression steel of

$$\varrho'(\alpha_e - 1) = \frac{h_f}{d}\left(\frac{b}{b_w} - 1\right) = \frac{125}{315}\left(\frac{960}{200} - 1\right) = 1.508$$

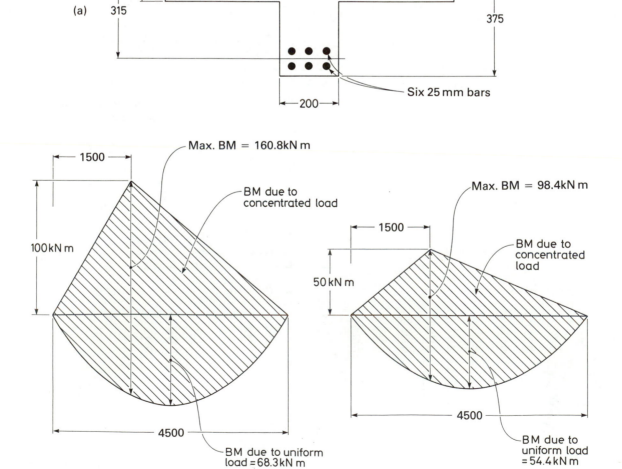

Figure 8.3 (a) Section; (b) envelope of bending moments due to total load; (c) envelope of bending moments due to permanent load.

and $d'/d = 125/(2 \times 315) = 0.198$. Using the chart for $d'/d = 0.2$ on *Data Sheet 25*, with the given values of $\varrho\alpha_e$ and $\varrho'(\alpha_e - 1)$, $x/d = 0.329$. Thus $x = 0.329 \times 315 = 104$ mm; i.e. the neutral axis falls within the flange. The section must therefore be reconsidered as a rectangular section of breadth b (i.e. 960 mm) and depth d. Then $\varrho = 2945/(315 \times 960) = 0.00974$, so that $\varrho\alpha_e = 0.078$; and since the full width of the section has already been considered, $\varrho'(\alpha_e - 1) = 0$. Then from any of the charts on *Data Sheets 22 to 25*, with the given values of $\varrho\alpha_e$ and $\varrho'(\alpha_e - 1)$, $x/d = 0.32$ and $I_e/bd^3 = 0.047$.

For long-term loading, $\phi = 2.34$ so that $E_c = 25/(1 + 2.34) = 7.5$ kN/mm². Thus, considering the section as flanged (i.e. basing the proportion of reinforcement on b_w), $\alpha_e = 200/7.5 = 26.7$ and thus $\varrho\alpha_e = 26.7 \times 0.0467 = 1.249$. Since this value exceeds the range of the charts, in this particular case it is necessary to evaluate x/d and $I_e/b_w d^3$ from the basic formulae for flanged sections given in section 8.3.2. If this is done it will be found that $x/d = 0.514$ and $I_e/b_w d^3 = 0.5103$.

The curvatures and deflections are now calculated as follows.

8.4.1 Instantaneous deflection due to permanent load only

The relationship for a flanged section in which $M = 98.4$ kN m, $f_{ct} = 1$ N/mm², $E_c = 25$ kN/mm², $x = 0.329 \times 315 = 104$ mm, and $I_e = 0.047 \times 315^3 \times 960 = 1.41 \times 10^9$ gives

$$\frac{1}{r_{ip}} = \frac{\left[M - \dfrac{(h-x)^3}{(d-x)} \dfrac{b_w f_{ct}}{3} \right]}{E_c I_e}$$

$$= \frac{\left[98.4 \times 10^6 - \dfrac{(375 - 104)^3 \times 200 \times 1.0}{(315 - 104) \times 3} \right]}{25 \times 10^3 \times 1.41 \times 10^9}$$

$$= 2.61 \times 10^{-6}$$

It is now necessary to calculate the appropriate value of K for the combination of loads. From the charts on *Data Sheet 29*, the midspan deflection coefficient for a uniform load is 0.0130 and that for a concentrated load at the third-point is 0.0177. The maximum free bending moment due to the permanent loads is 98.4 kN m, and thus the required value of K is given by

$$K = \frac{[0.0130 \times (16 + 5.5) \times 4.5 + 0.0177 \times 50]\,4.5}{98.4}$$

$$= 0.098$$

The instantaneous deflection due to the permanent loads is thus

$$a = Kl^2\,\frac{1}{r_{ip}} = 0.098 \times 4500^2 \times 2.61 \times 10^{-6} = 5.18\,\text{mm}$$

since $l = 4500$ mm.

8.4.2 Long-term deflection due to permanent loads and shrinkage

Since, for long-term loading, $E_c = 7.5$ kN/mm², $f_{ct} = 0.55$ N/mm², $x = 0.514 \times 315 = 161.9$ mm and $I_e = 0.5103 \times 200 \times 315^3 = 3.19 \times 10^9$ mm⁴ and with $M = 98.4$ kN m, the long-term curvature due to the permanent loads is

$$\frac{1}{r_{lp}} = \frac{\left[98.4 \times 10^6 - \dfrac{(375 - 161.9)^3 \times 200 \times 0.55}{(315 - 161.9) \times 3} \right]}{7.5 \times 10^3 \times 3.19 \times 10^9}$$

$$= 4.02 \times 10^{-6}$$

To this the curvature due to shrinkage must be added. Since $\alpha_e = 8$, $I = 3.19 \times 10^9$ and $S_s = 2945 \times (315 - 161.9) = 451 \times 10^3$, and since $\varepsilon_{cs} = 0.0003$,

$$\frac{1}{r_{cs}} = \frac{\varepsilon_{cs}\alpha_e S_s}{I} = \frac{0.0003 \times 8 \times 451 \times 10^3}{3.19 \times 10^9} = 339 \times 10^{-9}$$

The total curvature at midspan due to the combined effects of long-term permanent load and shrinkage is thus

$$\frac{1}{r_{tp}} + \frac{1}{r_{cs}} = (4.02 \times 10^{-6}) + (339 \times 10^{-9}) = 4.36 \times 10^{-6}$$

Using the coefficient for K previously determined, the resulting long-term deflection is thus

$$a = 0.098 \times 4500^2 \times 4.36 \times 10^{-6} = 8.65\,\text{mm}$$

8.4.3 Instantaneous deflection due to total loads

With $M = 160.8$ kN m, $E_c = 25$ kN/mm², $x = 104$ mm, $f_{ct} = 1$ N/mm² and $I_e = 1.41 \times 10^9$ mm⁴, the instantaneous curvature due to the total loading is given by

$$\frac{1}{r_{it}} = \frac{\left[160.8 \times 10^6 - \dfrac{(375 - 104)^3 \times 200 \times 1.0}{(315 - 104) \times 3} \right]}{25 \times 10^3 \times 1.41 \times 10^9}$$

$$= 4.38 \times 10^{-6}$$

The relevant value of K for this loading condition is

$$\frac{[0.0130 \times (16 + 11) \times 4.5 + 0.0177 \times 100]\,4.5}{160.8} = 0.094$$

and the resulting deflection is thus

$$a = Kl^2\,\frac{1}{r_{it}} = 0.094 \times 4500^2 \times 4.38 \times 10^{-6} = 8.32\,\text{mm}$$

The maximum deflection is thus obtained by adding the difference between the instantaneous deflections due to the total load and the permanent load to the long-term deflection due to the permanent load and shrinkage. Thus

$$a_{max} = (8.32 - 5.18) + 8.65 = 11.79\,\text{mm}$$

This represents a span/deflection ratio of $4500/11.79 = 381$, which satisfactorily meets the requirements of BS8110.

It will be noted that the maximum deflection has been determined here by summing the individual deflections obtained for the various loading conditions, rather than by summing the curvatures and then calculating the resulting deflection, as suggested in the Code and described in section 8.3.4. Undertaking the calculations as shown in this example avoids the problem of having to decide which particular value of K should be used to evaluate the deflection, once the curvature is known. However, where the same type of loading occurs throughout, as is often the case, the Code method is perhaps simpler to follow.

If the simplified rules for deflection given in BS8110 are followed for this example, the multiplier for the tension steel stress and the resistance-moment factor is 1.0, and that for a flanged beam with a ratio b_w/b of $200/960 = 0.21$ is 1.25. For a freely supported beam spanning 4.5 m, the basic effective depth is 225 mm and thus the minimum final effective depth is $225 \times 1.0 \times 1.25 = 281$ mm. Since the actual effective depth provided is 315 mm, this meets the Code requirements.

8.5 CRACKING: GENERAL RECOMMENDATIONS

BS8110 specifies that, for the requirements of the structure concerned, cracking must not mar appearance or lead to corrosion and loss of performance. As a guide the Code recommends that, unless the environment to which the member is exposed is particularly aggressive, a maximum width of crack not exceeding 0.3 mm will generally be satisfactory. It is understood that, since cracking is a semi-random phenomenon an exact maximum width cannot be predicted with certainty. In situations where the formation of cracks of this width may jeopardize the performance of the structure, for example in the containment of liquids, it may be necessary to limit the specified maximum crack width to a lesser figure, as is done in BS8007 'Code of practice for the design of concrete structures for retaining aqueous liquids'.

As in the case of deflection previously considered, to minimize the amount of calculation that would otherwise be necessary, the requirements contained in Part 1 of BS8110 to prevent the formation of cracks of excessive width are given in the form of simplified rules as regards maximum bar spacing. If these rules, which form clauses 3.12.11.2 of the Code, are complied with, the behaviour as regards cracking under service loads should be satisfactory and further calculation is unnecessary. However, the limiting values given by these simplified rules may be disregarded provided that the designer can still show that the resulting 'designed surface crack widths', when calculated as described in section 3.8 of Part 2 of BS8110, do not exceed the above limit.

8.6 CRACKING: SIMPLIFIED RULES

8.6.1 Beams

Where the concrete cover does not exceed 50 mm and with normal conditions of internal or external exposure,

the maximum clear horizontal distance between bars in tension must not exceed the values given in the graph on *Data Sheet 37*. These values are derived from the expression

clear spacing in millimetres $\leq 75\,000\,\beta_b/f_y$ or 300

where β_b is the ratio, at the section concerned, of the resistance moment after redistribution to that before redistribution. This expression, the derivation of which is explained in the *Code Handbook*, gives the lines on the graph. The values tabulated in BS8110, which are shown by the heavy black circles, have been obtained by rounding off the calculated spacings to the nearest 5 mm value above that actually required. The clear horizontal distance to the corner of a beam must not exceed one-half of the values indicated. As an alternative, the Code allows the limiting clear spacing to be determined from the formula

clear spacing in millimetres $\leq 47\,000\,\beta_b/f_s$ or 300

where f_s is the estimated working stress in the steel, which may be estimated from the expression $f_s = 0.625 f_y (A_{s\,req}/A_{s\,prov})/\beta_b$.

When the overall depth h of a beam is more than 750 mm, longitudinal bars must be provided close to the sides of the beam for at least two-thirds of the depth from the tension face. The spacing of such bars must not exceed 250 mm and their diameter should be at least equal to $\sqrt{(bs_b/f_y)}$, where b is the breadth of the beam at the point at which the bar occurs and s_b is the distance between the bar centres. The maximum breadths of beams corresponding to given spacings for various types and sizes of bar are shown on *Data Sheet 36*.

Apart from the bars in the beam sides, all bars whose diameters are less than 45% of the largest bar used in the section should be ignored when calculating the maximum spacing between bars.

8.6.2 Slabs

The foregoing rules for the spacing of bars in beams are equally applicable to slabs. However, when $f_y = 250$ N/mm^2 and the total thickness is not greater than 250 mm, or $f_y = 460$ N/mm^2 and the total thickness is not greater than 200 mm, or the area of reinforcement provided is less than $0.003bd$, the clear distance between bars is restricted to a maximum of $3d$ but no other check is required. For thicker slabs or greater amounts of steel, if the amount of tension reinforcement provided is less than $bh/100$, the maximum spacing may be increased by multiplying the spacing given by the graph on *Data Sheet 37* by $1/100\varrho_1$, where $\varrho_1 = A_s/bh$. These recommendations are summarized on the flow chart on *Data Sheet 37*. In no case, however, must the maximum clear spacing exceed $3d$ or 750 mm.

For slabs only, if the amount of moment redistribution employed is unknown, values of -15% may be adopted for the support moments and zero for the span moments.

8.7 CRACKING: ANALYTICAL METHOD

The foregoing simplified rules for the maximum spacing of bars may be ignored provided that it can be shown by calculation that the resulting maximum crack widths do not exceed the limiting value of 0.3 mm permitted by BS8110. Since the arbitrary rules described above have been designed to ensure that excessive cracking is prevented in the most extreme practical conditions, detailed calculations will almost always show that wider bar spacings than those permitted by the simplified rules can be adopted. This is particularly likely to be true where the member concerned is fairly shallow, and is normally only untrue for deep beams.

A suitable mathematical procedure for calculating crack widths is described in clauses 3.8 of Part 2 of BS8110. Provided that the strain in the tension steel does not exceed $0.8f_y/E_s$,

$$\text{design surface crack width} = \frac{3a_{cr}\varepsilon_m}{1+\dfrac{1(a_{cr}-c_{min})}{h-x}} \quad (8.8)$$

where a_{cr} is the distance between the point on the surface at which the crack width is being calculated and the face of the nearest longitudinal bar, h is the overall depth of the member, x is the depth to the neutral axis obtained when calculating ε_1, and c_{min} is the minimum cover to the tension bars.

In normal circumstances with a rectangular tension zone, the average strain ε_m in the member at the level at which the crack width is being calculated, taking into account the stiffening effect of the concrete in the tension zone, may be calculated from the expression

$$\varepsilon_m = \varepsilon_1 - \frac{b_t(h-x)(a'-x)}{3E_s A_s(d-x)} \quad (8.9)$$

where A_s and f_y are the area and characteristic stress of the tension reinforcement, b_t is the width of the member at the level of the centroid of the tension steel, a' is the distance from the compression face to the point being considered, and ε_1 is the strain in the member at the point being considered. To allow for creep, when calculating ε_1, the value assumed for the modulus of elasticity of the concrete should be only 50% of the instantaneous value. If shrinkage strains exceeding 0.06% are expected in the concrete, ε_m should be increased by adding one-half to the anticipated shrinkage strain. If tension occurs across the entire section (i.e. $x = 0$), BS8110 recommends that a suitable value for $h-x$ should be interpolated from the limits that $h-x = h$ when the neutral axis coincides with the edge of the section, and $h-x = 2h$ when the loading is purely axial (i.e. there is no applied moment.)

To calculate a design surface crack width, the first step is to determine the moment M_s at the section concerned due to service loads. Calculate (using the expression on *Data Sheet 21*) the modulus of elasticity E_c corresponding to the characteristic strength of the concrete used, and determine the modular ratio α_e by dividing the modulus

of elasticity of the steel ($200\,\text{N/mm}^2$) by $\frac{1}{2}E_c$ (i.e. $\alpha_e = 400/E_c$). The charts on *Data Sheets 22* to *25* may now be used to read off the values of the neutral-axis and lever-arm factors x/d and z/d corresponding to given values of $\alpha_e\varrho$ and $(\alpha_e-1)\varrho'$. Then calculate the average surface strain ε_h at the tension face from the expression

$$\varepsilon_h = \frac{M_s}{A_s z E_s}\frac{(h-x)}{(d-x)} \quad (8.10)$$

The crack width at any given section may then be calculated by using expressions (8.8) and (8.9). The procedure is illustrated by the worked examples given in section 8.8.

The principal criteria determining crack widths are the distances of the point considered from the nearest bar running at right angles to the crack and from the neutral axis, and the average surface strain. Thus across the tension face of a member the width of the crack rises from a minimum value directly above a bar to a maximum at a point midway between bars or at an edge. Over the sides of a beam, the crack width varies from a minimum value at the same level as the tension reinforcement to zero at the neutral axis, attaining its maximum value at a depth equal to about one-third of the distance from the tension steel to the neutral axis.

In the light of the foregoing, unless the exposure conditions come within the 'very severe' category, the task of the designer merely resolves into ensuring that the crack widths (a) in the tension face midway between bars (or at the edge of the member) and (b) at the critical level in the sides, do not exceed 0.3 mm. To facilitate this task it is possible to prepare design charts to simplify the somewhat cumbersome calculations involved once ε_{mh}, the 'adjusted' strain at the tension face, has been determined. Such charts are provided on *Data Sheets 38* and *39*, and their derivation is as follows.

8.7.1 Tension face

Here ε_{mh} equals ε_h and the maximum crack width occurs midway between the bars, the spacing of which is s_b. Thus, for this condition,

$$s_b = 2\sqrt{[(a_{cr}+\tfrac{1}{2}\phi)^2 - (c_{min}+\tfrac{1}{2}\phi)^2]}$$

This requirement may be rewritten as

$$s_b = 2\sqrt{[(a_{cr}-c_{min})(a_{cr}+c_{min}+\phi)]} \quad (8.11)$$

With a crack width of 0.3 mm, expression (8.8) may be rewritten as

$$0.3 = \frac{3a_{cr}\varepsilon_m}{1+2\left(\dfrac{a_{cr}-c_{min}}{h-x}\right)}$$

from which

$$a_{cr} = \frac{\tfrac{1}{2}(h-x)-c_{min}}{5\varepsilon_m(h-x)-1} \quad (8.12)$$

Thus, dividing expressions (8.11) and (8.12) through by c_{min},

$$\frac{s_b}{c_{min}} = 2\sqrt{\left[\left(\frac{a_{cr}}{c_{min}} - 1\right)\left(\frac{a_{cr}}{c_{min}} + \frac{\phi}{c_{min}} + 1\right)\right]} \quad (8.13)$$

where

$$\frac{a_{cr}}{c_{min}} = \frac{\frac{1}{2}\left(\frac{h-x}{c_{min}}\right) - 1}{5\varepsilon_m c_{min}\left(\frac{h-x}{c_{min}}\right) - 1} \quad (8.14)$$

where $\varepsilon_m c_{min}$ is expressed in millimetres. By substituting appropriate values of $(h-x)/c_{min}$, $\varepsilon_m c_{min}$ and ϕ/c_{min}, the curves represented by the full lines on the charts may be obtained.

8.7.2 Sides of members

The precise depth at which expression (8.8) reaches its maximum value depends on a combination of various factors and no simple rule can be devised to give it. However, it has been pointed out by *Allen* that, provided that the thickness of cover is relatively insignificant compared with the depth of the section, in practice the maximum width of crack occurs at about one-third of the distance from the tension reinforcement to the neutral axis. If the crack width at this depth (i.e. $a' = \frac{1}{3}(2d + x)$) is calculated, the difference between the value obtained and the true maximum will be found to be negligible.

Now at a depth a', $\varepsilon_m = (a' - x)\varepsilon_{mh}/(h-x)$. Thus, for a maximum crack width of 0.3 mm,

$$0.3 = \frac{3a_{cr}}{1 + 2\left(\frac{a_{cr} - c_{min}}{h-x}\right)}\left[\frac{(a'-x)}{(h-x)}\varepsilon_{mh}\right]$$

Thus

$$\varepsilon_{mh} = \frac{(h-x) + 2(a_{cr} - c_{min})}{10a_{cr}(a'-x)}$$

and since $a' - x = \frac{2}{3}(d-x)$,

$$\varepsilon_{mh} = \frac{3[\frac{1}{2}(h-x) + a_{cr} - c_{min}]}{10a_{cr}(d-x)} \quad (8.15)$$

Then dividing through by c_{min},

$$\varepsilon_{mh} c_{min} = \frac{3\left[\frac{1}{2}\left(\frac{h-x}{c_{min}}\right) + \frac{a_{cr}}{c_{min}} - 1\right]}{10\frac{a_{cr}}{c_{min}}\left(\frac{d-x}{c_{min}}\right)} \quad (8.16)$$

Assuming that the critical value of a' occurs at $\frac{1}{3}(2d + x)$,

$$\begin{aligned}a_{cr} &= \sqrt{[(c_{min} + \frac{1}{2}\phi)^2 + (d - a')^2]} - \frac{1}{2}\phi \\ &= \sqrt{[(c_{min} + \frac{1}{2}\phi)^2 + \frac{1}{9}(d-x)^2]} - \frac{1}{2}\phi \quad (8.17)\end{aligned}$$

Thus dividing through by c_{min},

$$\frac{a_{cr}}{c_{min}} = \sqrt{\left[\left(1 + \frac{1}{2}\frac{\phi}{c_{min}}\right)^2 + \frac{1}{9}\left(\frac{d-x}{c_{min}}\right)^2\right]} - \frac{1}{2}\frac{\phi}{c_{min}} \quad (8.18)$$

By substituting appropriate values of $(h-x)/c_{min}$, $(d-x)/c_{min}$, ϕ/c_{min} and $\varepsilon_{mh} c_{min}$, the curves represented by the broken lines on the charts may be determined; as before, $\varepsilon_{mh} c_{min}$ is expressed in millimetres.

8.7.3 Use of design charts

The normal method of using the design charts on *Data Sheets 38* and *39* is to calculate ε_{mh}, the 'adjusted' strain at the tensile face of the member, as described earlier. The relevant chart is then entered with the known values of $\varepsilon_{mh} c_{min}$ and $(h-x)/c_{min}$, and the corresponding ratios of s_b/c_{min} and, if appropriate, $(d-x)/(h-x)$ are read off. The former factor gives the maximum spacing between longitudinal bars that may be utilized; the distance from the centre-line of the outermost bar to the edge of a beam must not exceed one-half of this distance. The latter factor gives the maximum ratio of $(d-x)/(h-x)$ that may be adopted.

It must be borne in mind that, in the case of a slab, the spacing of an assumed size of bars cannot be directly increased to the maximum value given by the charts since (in addition to altering the resistance moment of the section) any adjustment to the amount of reinforcement provided will alter the values of x, z, f_s and so on. If the required amount of reinforcement, instead of the amount provided, is substituted for A_s in the calculations to determine ε_{mh}, the maximum spacing given by the chart may be used in conjunction with the appropriate value of A_s to determine a suitable bar arrangement. It must be remembered however that, to be accurate, interpolation between the charts depends on the actual value of ϕ/c, and at this stage ϕ is unknown. Furthermore, since $h = d + \frac{1}{2}\phi + c$, h, d and ϕ are related and it is impossible to evaluate ε_{mh} without assuming some value for ϕ. However, approximate designs can be prepared in this way and the final bar arrangement checked by ensuring that the spacing adopted does not exceed the maximum permissible value given by the charts.

Charts are only provided for the limiting conditions of $\phi/c_{min} = 0$ (i.e. when the bar size is negligible compared with the concrete cover) and $\phi/c_{min} = 1$ (when the bar size is equal to the cover provided), since inspection will show that the actual ratio of ϕ/c_{min} has a relatively minor effect on the values obtained. Since the values of s_b/c_{min} given by the chart for $\phi/c_{min} = 0$ are always lower than those when $\phi/c_{min} = 1$, the former chart may be used directly for preliminary design. For greater accuracy, linear interpolation between the charts should be employed.

As regards cracking in the sides of beams, the ratio of $(d-x)/(h-x)$ is not very sensitive to the actual value of ϕ/c_{min}. Above the chain line on the charts, the chart for $\phi/c_{min} = 1$ actually gives the more critical value for $(d-x)/(h-x)$; below this line, the chart for $\phi/c_{min} = 0$ should be used. Again, for extreme accuracy, linear interpolation between the charts is recommended.

The use of the charts is illustrated in the worked examples that follow. Note that c_{min} relates to the minimum cover of the tension reinforcement to the face of the member being considered. Thus if different amounts of cover are provided to the side and bottom (or top) faces of a beam, when determining s_b the value of c_{min} should be that beneath (or above) the bars, but when checking $(d-x)/(h-x)$ the value taken for c_{min} should be the cover to the sides of the beam.

8.8 CRACKING: WORKED EXAMPLES

8.8.1 Slab

Design the reinforcement for a freely supported 125 mm slab spanning 4.5 m to support an imposed load of 3 kN/m^2 if $f_{cu} = 25 \text{ N/mm}^2$ and $f_y = 460 \text{ N/mm}^2$.

Assuming that no allowance need be made for finishes, the self-weight of the slab is $23.6 \times 0.125 = 2.95 \text{ kN/m}^2$. Thus

ultimate moment $= [(2.95 \times 1.4) + (3.0 \times 1.6)]$
$$\times 4.5^2 \times 0.125 = 22.6 \text{ kN m per m}$$

Using the design chart on *Data Sheet 13*, the required amount of tension reinforcement is 713 mm^2 per metre. It would be convenient to provide 16 mm bars at 275 mm centres, and this would meet the Code requirement that the maximum bar spacing must not exceed $3d + \phi = (3 \times 92) + 16 = 292 \text{ mm}$. However, the simplified rules in Part 1 of BS8110 limit the maximum spacing between high-yield bars to 160 mm.

Using the analytical method described in BS8110, an approximate value of E_c for grade 25 concrete (from the expression on *Data Sheet 21*) is 25 kN/mm^2; thus $\alpha_e = 200/(0.5 \times 25) = 16$. The proportion of reinforcement required is $713/(92 \times 10^3) = 0.00775$. From the charts on *Data Sheets 22 to 25* the corresponding values of x/d and z/d (with $\alpha_e = 16$) are 0.389 and 0.870 respectively. The bending moment due to service loads is $(2.95 + 3.0) \times 4.5^2 \times 0.125 = 15.06 \text{ kNm per m}$, and thus the corresponding stress in the reinforcement is

$$f_s = M_s/A_s z = 15.06 \times 10^6/(713 \times 0.87 \times 92) = 264 \text{ N/mm}^2$$

Since this is less than $0.8 f_y = 368 \text{ N/mm}^2$, the Code analytical method may be used. Then the strain in the steel $= f_s/E_s = 264/200\,000 = 0.00132$. Thus

$$\varepsilon_h = \frac{h-x}{d-x} \varepsilon_s$$

$$= \frac{125 - (0.389 \times 92)}{92(1 - 0.389)} \times 0.00132 = 0.00209$$

$$\varepsilon_{mh} = \varepsilon_h - \frac{b_t(h-x)(h-x)}{3 E_s A_s (d-x)}$$

$$= 0.00209 - \frac{10^3 \times (125 - 0.389 \times 92)^2}{3 \times 2 \times 10^5 \times 713 \times 92(1 - 0.389)}$$

$$= 0.00176$$

Since the maximum crack width occurs midway between bars, by calculation

$$a_{cr} = \sqrt{(137.5^2 + 33^2)} - 8 = 133.4 \text{ mm}$$

and with $c_{min} = 25 \text{ mm}$,

$$\text{maximum crack width} = \frac{3 a_{cr} \varepsilon_{mh}}{1 + 2 \dfrac{(a_{cr} - c_{min})}{h - x}}$$

$$= \frac{3 \times 133.4 \times 0.00176}{1 + \dfrac{2 \times (133.4 - 25)}{125 - (0.389 \times 92)}}$$

$$= 0.205 \text{ mm}$$

Alternatively, by using the charts on *Data Sheets 38* and *39*, with $\phi/c = 16/25 = 0.64$, $(h-x)/c = [125 - (0.389 \times 92)]/25 = 3.57$ and $\varepsilon_h c = 0.00188 \times 25 = 0.047$, it will be seen that the point where the given curves for $(h-x)/c$ and $\varepsilon_h c$ intersect falls outside the charts, but the corresponding limiting value of s_b/c is well above 50. Calculations will in fact show that, from the point of view of cracking, the 20 mm bars could well be spaced at 400 mm centres without a maximum crack width of 0.3 mm being exceeded.

8.8.2 Beam

A beam spanning two 14 m spans has an overall depth of 800 mm and is 400 mm wide. The beam, which carries dead and imposed service loads of 15 kN/m and 10 kN/m respectively, is reinforced over the central support with three 40 mm bars. Check that the Code requirements regarding cracking at this point are not exceeded if $f_{cu} = 25 \text{ N/mm}^2$ and $f_y = 460 \text{ N/mm}^2$.

If a moment redistribution of 20% has been made, the maximum clear spacing between bars given by *Data Sheet 37* is 130 mm. With 40 mm wide cover and three 40 mm bars, the actual clear spacing between the bars is only 100 mm, so that the Code requirements are met in this respect. However, since the beam has an overall depth of more than 750 mm, additional longitudinal bars (say 16 mm at 250 mm centres) must be provided near the sides over a depth of about 535 mm from the tension face unless detailed calculations indicate otherwise.

The maximum moment at the support due to service loads is

$$M_s = (10 + 15) \times 14^2 \times \tfrac{1}{8} = 613 \times 10^6 \text{ kN m}$$

For grade 25 concrete, an appropriate value of E_c (*Data Sheet 21*) is $25\,\text{kN/mm}^2$. Thus $\alpha_e = 400/25 = 16$. The proportion of reinforcement provided is $(3 \times 40^2 \times \tfrac{1}{4}\pi)/(740 \times 400) = 3770/296\,000 = 0.0127$, so that $\varrho\alpha_e = 0.2032$. Thus $x/d = 0.466$ and $z/d = 0.845$ from the chart on *Data Sheet 22*. Now

$$f_s = \frac{M_s}{A_s z} = \frac{613 \times 10^6}{3770 \times 0.845 \times 740} = 260\,\text{N/mm}^2$$

Since this is less than $0.8f_y$ (i.e. $368\,\text{N/mm}^2$), the analytical method is valid. Then

$$\varepsilon_s = \frac{260}{200\,000} = 0.001\,30$$

and

$$\varepsilon_h = \frac{h-x}{d-x}\,\varepsilon_s = \frac{800 - (0.466 \times 740)}{740(1 - 0.466)} \times 0.001\,30 = 0.001\,50$$

giving

$$\varepsilon_{mh} = 0.001\,50 - \frac{400 \times (800 - 0.466 \times 740)^2}{3 \times 2 \times 10^5 \times 3770 \times 740(1 - 0.466)}$$

$$= 0.001\,41$$

Considering the top of the beam, $a_{cr\,max} = \sqrt{(70^2 + 60^2)} - 20 = 72.2\,\text{mm}$. Then

$$\text{maximum crack width} = \frac{3a_{cr}\varepsilon_{mh}}{1 + 2\dfrac{(a_{cr} - c_{min})}{(h-x)}}$$

$$= \frac{3 \times 72.2 \times 0.001\,41}{1 + \dfrac{2 \times (72.2 - 40)}{800 - (0.466 \times 740)}}$$

$$= 0.267\,\text{mm}$$

Considering the sides of the beam, the maximum width of crack will occur at a depth of about $\tfrac{1}{3}(2d + x)$; i.e. about $\tfrac{1}{3} \times 740 \times 2.46 = 607\,\text{mm}$. At this level,

$$\varepsilon_m = \frac{607 - (0.466 \times 740)}{800 - (0.466 \times 740)} \times 0.001\,41 = 0.000\,81$$

Since $a_{cr} = \sqrt{[(740 - 607)^2 + 60^2]} - 20 = 126\,\text{mm}$,

$$\text{maximum crack width} = \frac{3 \times 126 \times 0.000\,81}{1 + \dfrac{2(126 - 40)}{800 - (0.466 \times 740)}}$$

$$= 0.223\,\text{mm}$$

If the chart on *Data Sheet 39* is used instead, with $\phi/c = 1.0$, $(h-x)/c = [800 - (0.466 \times 740)]/40 = 11.4$ and $\varepsilon_h c = 0.001\,41 \times 40 = 0.056$, the chart gives maximum values of $s_b/c = 4.3$ and $(d-x)/(h-x) = >1.0$. Since the actual values of s_b/c and $(d-x)/(h-x)$ are 3.5 and 0.87 respectively, the section is clearly satisfactory, and no adjustment or additional steel need be provided.

Methods of applying the data in the preceding chapters are given on *Calculation Sheets 1* to *9*, in Part Two of this book, which contain the calculations for the slabs and beams of the upper five floors of the building shown in *Drawings 1* and *2*. The dimensions of the members and details of the reinforcement are given on *Drawings 3* to *7*. In this chapter, explanations of the details and calculations are given.

The floors are of ordinary beam-and-slab construction; i.e. they comprise slabs spanning in one direction continuously over several spans, between secondary beams that are themselves supported on columns, or on main beams which transmit the loads to the columns. It is generally more economical for the main beams to span the shorter distance between the columns and for the secondary beams to span the greater distance, as is shown on the general arrangement of the beams in *Drawing 3*. This drawing also gives the reference numbers to the panels of slabs and to the secondary and main beams which relate to the reference numbers in the calculations. Designs for cantilevering slabs and freely supported slabs are also given in this chapter.

9.1 CONTINUOUS SOLID SLABS

Although the design of the floor slab is simple, it is discussed at some length as several features of BS8110 are introduced, and a full treatment here enables the explanations relating to other slabs to be curtailed. By using design tables or charts, no more calculation than that given on *Calculation Sheets 1* is required. The characteristic strengths of the concrete and steel are 30 N/mm² and 250 N/mm² respectively (see section 1.2), and the reinforcement is as shown in *Drawings 3* and *4*.

9.1.1 Loads

The characteristic imposed load is determined as described in section 2.1, and is thus 5 kN/m². For panel P1

on *Calculation Sheets 1*, the dead load comprises the self-weight of the slab, the finishes and the partitions. It is necessary to assume a thickness for the slab in order to determine the self-weight. With not more than about $\frac{1}{2}\%$ of mild steel reinforcement and normal-weight concrete, the minimum effective depth permitted by the Code is $\frac{1}{52}$ (i.e. $\frac{1}{2} \times \frac{1}{26}$) of the effective span if the slab is continuous, and $\frac{1}{40}$ of the effective span if freely supported. The effective depth of the upper floor slabs spanning 2.5 m must not therefore be less than 2500/52 = 48 mm. With a minimum concrete cover of 20 mm to all steel (i.e. 'mild exposure' rating when $f_{cu} = 30$ N/mm², and assuming that a systematic checking regime, as stated in clause 3.3.5.2 of BS8110, Part I, is employed: see *Data Sheet 42*) and 12 mm bars, the minimum possible thickness to meet deflection requirements would be $48 + 20 + (12 \times \frac{1}{2}) = 74$ mm; for the purpose of calculating the dead load, a 100 mm slab is assumed, giving a self-weight of $0.1 \times 24 = 2.4$ kN/m².

The total load on the slab forming the terrace at the fifth floor is virtually the same as that on the floor slabs but differs in detail, comprising the weights of asphalt (0.48 kN/m²), topping (0.24 kN/m²), tiles (0.72 kN/m²) and ceiling finish (0.24 kN/m²) in addition to the self-weight of the 100 mm slab (2.40 kN/m²), thus giving a total characteristic dead load of 4.08 kN/m², with a characteristic imposed load of 5 kN/m², as described in section 2.8.2.

9.1.2 Bending moments

Since the ratio of characteristic imposed load to characteristic dead load (neglecting the weight allowed for partitions) exceeds 1.25, it is not possible to use the approximate bending-moment coefficients given in BS8110; therefore the more-exact formulae for an infinite number of equal spans given on *Data Sheet 7* are employed. The maximum support moment is that obtained by using the appropriate formulae for the internal support, and the maximum span moments are given by the formulae for the moment in the internal span. Although the panels of slab between the front fascia beams and the first line of interior secondary beams, and between the rear fascia beams and the last line of interior secondary beams, are apparently end spans, the fascia beams are sufficiently substantial compared with the thin floor slab to produce fixity of the slab at the outer support. All panels P1 are therefore designed as interior panels, and reinforcement is provided in the top at the outer supports to resist the maximum negative bending moment.

The edges of panels P2, P3 and P4 (see *Drawing 3*) adjacent to the stair-wells and lift-well are nominally freely supported for part of their length and continuous with the stairs, landings or adjacent panels for the remainder of their length. The coefficients for the ultimate bending moments at midspan and at the inner support of each of these panels are intermediate between the corresponding coefficients for an end span and an interior span; i.e. say about $nl^2/12$ for the positive bending moment near midspan and $nl^2/10$ at the inner support. Calculations are not given for these panels but, adopting these coefficients, it will be found that a 100 mm slab will still suffice for panel P2. If the same thickness is provided for panel P4, the torsional resistance of beam S6 is likely to be more than sufficient to compensate for the inequality of the bending resistances of the two slabs meeting at the common support.

9.1.3 Resistance to bending

The thickness of the slab is determined by calculating the minimum effective depth required to resist the greatest ultimate bending moment. The minimum effective depth d_{min} that may be adopted is given by the expression

$$d_{min} = \sqrt{\left(\frac{M}{0.156bf_{cu}}\right)}$$

In the present case the greatest bending moment is that over the supports, and it is seen that the thickness of 100 mm is satisfactory if the cover of concrete is 20 mm. The same thickness is therefore more than sufficient for the smaller ultimate moment at midspan.

The amount of reinforcement required at midspan or support is found by calculating M/bd^2 for the given value of M and, with the aid of the chart on *Data Sheet 13*, reading off the value of pf_y corresponding to the appropriate value of f_{cu} (in the present case, 30 N/mm^2). The actual area of reinforcement required is then obtained by multiplying this value of pf_y by bd/f_y.

On *Data Sheet 40* the maximum ultimate moment of resistance M_u and the corresponding amount of reinforcement A_s required are set out for slabs of various thicknesses with various characteristic steel and concrete strengths. A suggested arrangement of reinforcement relating to each combination is also given. The resistance moments and steel areas are calculated by employing rigorous limit-state analysis with a parabolic-rectangular concrete stress-block, as described in section 5.1.1. A table similar to *Data Sheet 40*, but based on the uniform rectangular concrete stress-block, is provided in *RCDH*. It will be realized, of course, that the amounts of steel given in the table are those required to resist the limiting values of M_u tabulated. If the actual ultimate moment to which the section is subjected is lower, as is usually the case, the amount of steel will be reduced accordingly and should be calculated by the methods described in Chapter 5.

If the bending moments are determined by using the so-called exact methods of structural analysis discussed in Chapter 3, it is often possible to undertake sufficient moment redistribution to obtain equal bending moments at midspan and over the supports; this simplifies the arrangement of the reinforcement. This has not been done on *Calculation Sheets 1* because of the assumptions involved in considering the outer supports to be fully fixed and assuming an infinite number of spans (instead of the true number of six). The actual percentage of redistribution required in the case considered would be about 16%, giving equal span and support moments of 6.63 kN/m per metre, i.e. requiring an area of steel of 428 mm^2. This moment is obtained simply by finding the mean value of the maximum span and support moments; this is a simple and safe approximation. However, it is possible to reduce the moments still further by making up to the maximum permissible redistribution of 30%. As explained in detail in section 3.3, since the maximum moments at midspan and support caused by the imposed loads are due to different arrangements of loading it is normally possible to reduce both simultaneously. In the example considered both span and support moments can simultaneously be reduced to 5.38 kN/m per metre, which would require an area of 348 mm^2 per metre width. When making such a redistribution, however, the corresponding restriction on the maximum value of x/d must be borne in mind.

9.1.4 Size of bars and minimum amount of reinforcement

BS8110 does not specify a minimum size for bars in slabs. Generally, rolled bars smaller than 6 mm are not advisable as main reinforcement in slabs, although smaller bars might conceivably be used as distribution bars. Reference 22 states that the smallest preferred size of bar is 8 mm, since size 6 is not freely available owing to low demand and infrequent rollings. Bars smaller than 5 mm are commonly used in fabricated meshes of hard cold-drawn wire.

The Code recommends that in solid slabs the cross-sectional area of reinforcement provided in each principal direction should not be less than 0.24% of bh if mild steel reinforcement is used, and not less than 0.13% of bh if the reinforcement is of high-yield steel. The maximum spacing of these bars must conform to the requirements described in Chapter 8.

Bars of suitable size and spacing can be selected from *Data Sheet 41*.

9.1.5 Main reinforcement

In clauses 3.12.9 and 3.12.10, BS8110 gives rules for curtailing and anchoring bars in one-way slabs. For slabs that support mainly uniformly distributed loads and where, in the case of a continuous system, the spans are approximately equal and the system is analysed for the single loading condition of maximum load on all spans, simplified rules described in clause 3.12.10.3 of the Code can be employed. These rules for slabs, which are illustrated diagrammatically on *Data Sheet 42*, differ slightly from those provided for beams in clause 3.12.10.2 of BS8110 (and summarized on the same data sheet). In all other cases the general requirements set out in clauses 3.12.9 must be adhered to. These requirements, which are identical for both beams and slabs, are discussed when considering the reinforcement for the secondary beams for the floors in section 9.5.3. For the floor slabs in the example being considered, the analysis is not based on a single-load case and so the simplified detailing rules given in the Code do not apply.

An economical method of reinforcing solid continuous slabs is for the bars near the bottom at midspan to be bent up at the points of contraflexure, so as to pass over the supports in the top of the slab. Alternate bars from the panels on each side of the support may be bent up in this way and thus provide the same area over the support as at midspan. As discussed above, in a slab of uniform thickness more reinforcement is normally required over the supports than at midspan. In heavily reinforced slabs this arrangement tends to be wasteful, and it is preferable to provide only the calculated area of steel at midspan, supplementing the cranked bars over the support by providing additional short lengths of straight bar. In current UK practice the use of such cranked bars is much less than hitherto; separate systems of straight bars are provided to resist the positive moment at midspan and the negative moment over the supports. Although greater anchorage-bond lengths may be necessary, the savings in bending and fixing costs and the ability to select the optimum bar arrangement providing just the area necessary normally outweigh the cost of any extra steel.

The cross-sectional area necessary for such bars is calculated as shown for panel P1 on *Calculation Sheets 1*. These additional bars are not provided at the support afforded by the fascia beam to the terrace at the fifth floor, as the effective span is 2.299 m (i.e. a clear span of 2.225 m plus an effective depth of 0.074 m) and the bending moment is therefore less than that on a normal P1 panel, for which the span is assumed to be 2.5 m. Therefore, as shown on *Drawing 4*, the reinforcement at the top of the fascia beam is 10 mm bars at 400 mm centres plus the links extending from the beam itself; additional 8 mm bars are provided at 200 mm centres over the secondary beam to conform to the requirements of panel P1.

9.1.6 Distribution bars

The minimum proportion of distribution steel required by BS8110 is substantially greater than necessary according to previous Codes; 60% greater in the case of mild steel, although only 8% if high-yield bars are used. To select reinforcement conforming to the requirements of BS8110 regarding the provision of distribution steel, *Data Sheet 41* can be used. Other bars at other spacings (not greater than $3d$) can be used instead, of course. In the case of slabs greater than 250 mm in thickness, note that the Code restricts the maximum clear spacing to 750 mm. In the 100 mm slabs forming the upper floors of the example, 10 mm bars at 300 mm centres are used.

9.1.7 Flanges of T-beams and L-beams

Recommendations are given in BS8110 for the reinforcement of the flanges in compression of T-beams and L-beams. Since this reinforcement is in the slab of monolithic beam-and-slab construction, it is necessary to consider it here. In order to resist the horizontal shearing force, the flange must be reinforced with bars transversely to the length of the beam and extending across the entire effective width of the flange. The greatest effective width is the least of the values of b given on *Data Sheet 41*. The amount of reinforcement that must be provided according to the Code is 0.15% of bh (irrespective of the type of steel used). The transverse bending moment on the slab, where it crosses a beam which is parallel to the direction of the span of the slab, is also resisted by this flange reinforcement; where the slab spans at right angles to the beam beneath, the reinforcement provided over the top of the beam to resist the negative bending moment in the slab is normally more than sufficient to meet this additional requirement. BS8110 does not specify maximum spacings for this reinforcement, but it must clearly meet the requirements of Chapter 8 and thus should not exceed $3d$ or 750 mm. Suitable flange reinforcement for slabs of differing thicknesses is suggested on *Data Sheet 41*.

In the 100 mm slab (*Drawing 4*) the flange reinforcement over the main beam is 8 mm bars at 300 mm centres, which is based on *Data Sheet 41*. If the greatest effective width of flange (*Data Sheet 41*) of the main beam were required to resist compression, the minimum length of the bars comprising the flange reinforcement would be the lesser of $0.2l_e + b_w$, i.e. $(0.2 \times 0.7 \times 5.0) + 0.25 = 0.95$ m, and the distance between the main beams, i.e. 6 m. Thus the bars must be at least 950 mm long. The minimum bond length required by an 8 mm bar in grade 30 concrete, if no hooks are provided, is 285 mm, and at least this distance should be provided on each side of the rib of the main beam. *Calculation Sheets 10* shows that a width of flange of about 387 mm is required to resist the compression in the middle of the interior spans of the main beams; this is less than the maximum permissible width of 950 mm. The flange reinforcement should thus extend for a minimum distance of $(2 \times 285) + 250 = 820$ mm symmetrically about the main beam.

For the secondary beams, *Calculation Sheet 7* shows that in this case also the entire permissible width of flange is not required to resist compression. The reinforcement provided in the slab over these beams, and resisting the negative ultimate bending moment, provides amply for the width required, namely 249 mm. It is also good practice to stagger the ends of the flange bars, as in *Drawing 4*. It will be seen that this is done for the bars over the main beams, and the same result is achieved with bars (1) and (6) over the secondary beams by staggering the positions of stopping-off and bending-down.

In a solid slab of the type shown in *Drawing 4*, bond seldom needs consideration as small-diameter bars are normally used throughout, and there is no difficulty in providing ample lengths for laps and end anchorages without forming hooks, except at the fascia beams. With small ratios of characteristic imposed load to characteristic dead load (about 1.25 in the present case) the possibility of negative bending moments occurring at midspan need not be investigated. Sufficient tensile resistance in the top of the slab is provided up to (or beyond) the points of contraflexure if the top bars are carried slightly beyond the first and third quarter-points which, in this example, are 63 mm from the centres of the supports.

9.1.8 Concrete cover

According to BS8110 the minimum concrete cover required to all steel is as set out on *Data Sheet 42*. It will be seen that the thickness depends on the grade of concrete and the exposure conditions (and also the fire resistance: see *Data Sheet 62*). In addition, the cover provided must be not less than the diameter of the adjoining bar. In the present example, the minimum cover specified for 'mild' exposure conditions with grade 30 concrete is 20 mm, which is adequate for the 10 mm bars employed.

9.2 FREELY SUPPORTED SOLID SLABS

The roof over the rear stair-well is an example of a freely supported slab spanning one way only. The design of this slab is shown on *Calculation Sheet 4* and *Drawing 4*, and differs from the design of the floor slabs in that the positive ultimate bending moment at midspan is $\frac{1}{8}nl^2$, where $n = 1.4g_k + 1.6q_k$. Theoretically there is no bending moment at the supports, which in this example are brick walls. For this reason there appears to be no point in extending all the main bars to the support, or in bending up any bars into the top of the slab. However, as a precaution against any restraint that there may be around the edges of the slab, one-half of the bars are bent up. As asphalt is provided on this roof, the top face is well protected, and cover to meet 'mild' exposure requirements will suffice. To meet deflection requirements, the minimum effective depth of a freely supported slab is one-fortieth of the span, if the percentage of steel does not exceed about $\frac{1}{2}\%$. Thus in the present case d must be

at least $2390/40 = 60$ mm; therefore a 100 mm slab is necessary, although a thinner section would suffice to resist bending.

9.3 CANTILEVERED SLABS

There are two examples of cantilevered slabs in the building in Part Two, namely the canopy at first-floor level and the parapet. The designs are given on *Calculation Sheets 2* and *3* and *Drawings 4* and *6*.

9.3.1 Canopy

The canopy at first-floor level is designed as a simple cantilever, the ultimate bending moment on which is $\frac{1}{2}nl^2$. BS8110 recommends in the simplified rules that the basic ratio of overhang to effective depth of a cantilever should not exceed 7; i.e. in this example the minimum basic effective depth is $1025/7 = 146$ mm, since the overhang is 1 m and the assumed effective depth is 50 mm. As there is no finishing material on the canopy, the structural concrete is exposed to the weather, and the cover of the concrete over the reinforcement should be at least 40 mm with grade 35 concrete (the minimum permissible grade for severe exposure). With 50 mm cover and 10 mm bars the minimum thickness is therefore 118 mm, if less than $\frac{1}{2}\%$ of reinforcement is required. This is a case where, because of the lightness of the imposed loading (1.5 kN/m^2) and the unsightliness of a thick cantilever, it is worth while employing a higher-grade concrete and following through the rigorous calculation procedure for deflections described in Chapter 8. If this is done it will be found that, with grade 40 concrete, a slab 100 mm thick at the root (i.e. with a thickness matching that of the adjoining floor slab) will be satisfactory: see *Calculation Sheet 2*. The canopy depends for its action as a cantilever on the restraining effect of the fascia beam and the adjoining floor slab; therefore the concrete forming the beam and slab must be sufficiently matured before the formwork is removed from below the canopy.

9.3.2 Parapet

As stated in section 2.3.4, the parapet is to be designed as a vertical cantilever subjected to a horizontal load of 740 N per metre acting at a height of 1.1 m. BS6399, Part I, does not state whether this load should be assumed to act inwards or outwards, but it seems reasonable to assume that the parapet should be designed to resist this load acting in either direction: the resulting bending moment at the foot is thus Fl, where $F = 1.6Q_k$. The restraint justifying the assumption of full cantilever action is provided by the slab forming the terrace at the level of the fifth floor and the fascia beam. To ensure stiffness, the thickness should be 100 mm, but in the case of a vertical cantilever it seems unnecessary to meet the Code requirements for limiting the deflection of horizontal members.

The actual effective depth necessary to meet the bending-moment requirements is only 20 mm. Since the parapet is exposed to the weather on both faces the concrete cover (for 'severe' exposure conditions, grade 35 concrete must be used) should be 40 mm, and the minimum thickness required to resist bending is therefore about 75 mm. The provision of adequate thickness thus necessitates a thicker slab than would be required for bending purposes, but for architectural reasons the parapet is required to be 50 mm thicker at the base than at the top; see *Drawing 6*. To enable proper placing of the concrete the least practicable thickness is 100 mm; therefore the parapet is made 150 mm thick at the bottom. The total reinforcement needed to resist bending is 68 mm^2 per metre at both sides of the slab, giving a total of 136 mm^2. However, the minimum permissible amount of main reinforcement in a slab 150 mm thick (i.e. $d = 107$ mm) is $0.24 \times 107 \times 10^3/100 = 257$ mm^2 per metre. Since the effective depth is 107 mm; therefore, from *Data Sheet 40*, 6 mm bars at 200 mm centres near both faces will suffice for the vertical bars and also for the horizontal distribution steel.

9.4 CONCENTRATED LOADS ON SLABS SPANNING IN ONE DIRECTION

Common examples of loads concentrated on small areas or narrow widths of slabs in buildings are computing or data processing equipment, safes, and heavy partitions. There is no simple method of calculating the bending moments produced by such concentrated loads. BS8110 suggests that they be determined by using the elastic methods proposed by Westergaard or Pigeaud, or by collapse theory such as Johansen's yield-line method or Hillerborg's strip method. However, BS8110 also provides a simple empirical method for single-span slabs, based on the results of tests to failure. Concentrated loads on one-way slabs are discussed below: their action on two-way slabs is considered in section 10.7.1.

9.4.1 Freely supported slabs: Pigeaud's method

Bending-moment coefficients based on Pigeaud's method are given on *Data Sheet 43* for a load F that is concentrated on a rectangular area a_x by a_y, symmetrically positioned about the centre of a rectangular panel which is freely supported along the opposite edges only. The coefficients have been derived on the assumption that Poisson's ratio is 0.2, the value recommended for serviceability calculations in BS8110. The bending moment per unit width of slab in the direction of span l_x is then $\alpha_x F$, and that at right angles to the direction of span is $\alpha_y F$. A more detailed discussion of Pigeaud's method, when applied to two-way slabs, is given in section 10.7.1.

9.4.2 Freely supported slabs: Code method

In the case of a freely supported one-way slab, BS8110 recommends that the width y (in metres) of the strip of slab assumed to assist in carrying a load of width a_y, as in *Figure 9.1a*, should not exceed $a_y + 2.4x[1 - (x/l)]$, where l is the span of the slab in metres and x is the distance in metres from the nearer support to the centre of the load. The strip is symmetrical about the centre-line of the load. If the width of $\frac{1}{2}y$ is not available on one or both sides of the centre-line of the load, the width of the strip should be reduced to the width available (*Figure 9.1b*). For example, for a load at the edge of a slab, y is $a_y + 1.2x[1 - (x/l)]$, as in *Figure 9.1c*. If the load acts at the centre of the span, $x = \frac{1}{2}l$ and the width of the strip is the greatest permissible, namely $a_y + 0.6l$, as in *Figure 9.1d*. Other conditions with a central load are given in *Figure 9.1e* and *9.1f*.

The bending moment on each metre width of strip y is given by the expressions in *Figure 9.1*, which are based on the appropriate formulae for a freely supported beam carrying a load distributed over part of the span only. It is interesting to compare these bending moments with those calculated by Pigeaud's method. For the symmetrical conditions in *Figure 9.1d*, the bending moments in kN m per metre of width at midspan, calculated with the expressions given in the diagram, with $y = 0.2l$ are: $0.28F$ if $a_x = 0.2l$; $0.23F$ if $a_x = 0.5l$ and $0.16F$ if $a_x = l$. The corresponding bending moments given by Pigeaud's theory (if Poisson's ratio is assumed to be 0.2 as is recommended in BS8110) are $0.24F$, $0.17F$ and $0.11F$ respectively. That is, the bending moments calculated by the empirical method suggested in the Code are greater than those calculated by a more theoretical analysis, and are therefore acceptable.

The foregoing consideration concerns only the bending moment in the direction of the span. A concentrated load also produces bending in a direction at right angles to the span, but the Code gives no recommendations for the evaluation of this bending moment. If Pigeaud's method is applied to the symmetrical conditions in *Figure 9.1d*, the bending moments at right angles to the span for $a_y = 0.2l$ are $0.19F$, $0.13F$ and $0.09F$ kN m per metre of width for $a_x = 0.2l$, $0.5l$ and l respectively. By inspection, it will be seen that more than the minimum area of reinforcement (0.24% of bh or 0.13% of bh for mild and high-yield steel respectively) is required in the bottom of the slab to resist the bending moments in the direction of the span, calculated by the method recommended in the Code. If the amount of distribution steel under the load is similar to the reinforcement provided in the direction of the span to resist the concentrated load, there is sufficient reinforcement to resist the bending moments at right angles to the span calculated by Pigeaud's method.

9.4.3 Continuous slabs

No specific recommendations are given in the Code for slabs that are continuous over one or both supports. In the absence of more accurate data, it is reasonable to assume that the width y of the strip assumed to carry the load is not less for a continuous slab than for a freely supported

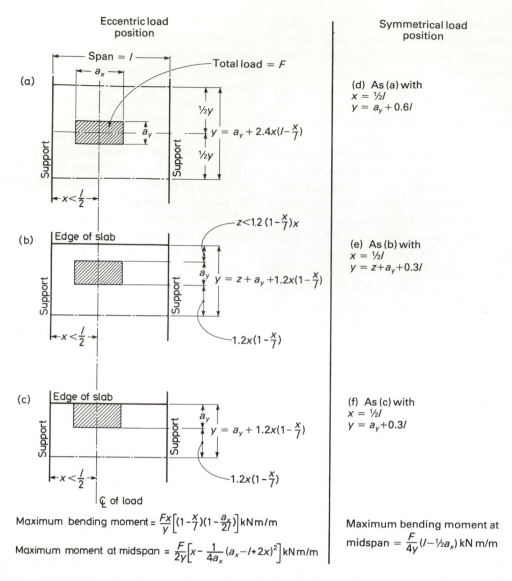

Figure 9.1 Concentrated loads on freely supported one-way slabs (Code method).

slab. Therefore, in the limiting condition of the load extending the entire length of the span ($a_x = l$), the bending moments per metre of width are calculated by one of the methods described in Chapter 3 for a continuous beam carrying a total uniformly distributed load of F/y kN, y being determined as in *Figure 9.1d, e* or *f*, and account being taken of whether the concentrated load occurs on more than one span at a time. In the other limiting condition of the load being concentrated on a small area at midspan (say a_x not exceeding $0.2l$), it is sufficiently accurate to calculate the bending moments as for a continuous beam carrying a total load of F/y concentrated at midspan, using the appropriate tables and other data in Chapter 3. For intermediate conditions, the bending moments can generally be assessed with sufficient accuracy for the design of slabs by considering the actual load in relation to the two limiting conditions. In the case of an important slab carrying a large concentrated load, it would be worth while undertaking a yield-line analysis (ref. 15) or employing Hillerborg's strip method (ref. 16).

The shearing stresses in slabs carrying loads commonly occurring in buildings and uniformly distributed over the entire area are seldom critical, but it is advisable to investigate the resistance to shearing of slabs supporting large concentrated loads: this matter is discussed in detail in section 10.7.2.

9.5 CONTINUOUS SECONDARY BEAMS

The essential calculations to determine the load, bending moments and reinforcement for the secondary beams of the upper floors are given on *Calculation Sheets 5* and *6*, and the details of these beams are given on *Drawing 5*. Each line of secondary beams is continuous over five spans, the effective span being assumed to be the distance (6 m or 5.425 m) between the centres of the supports. The imposed load of 5 kN/m^2 of slab supported, and the dead load of 4.12 kN/m^2 of slab, are taken from *Calculation Sheets 1*, and the weight of the rib of the beam is assumed to be 1.5 kN per metre.

9.5.1 Bending moments

As the characteristic imposed load (12.50 kN/m) exceeds the characteristic dead load (11.80 kN/m) it is not possible to use the approximate bending-moment coefficients provided in the Code (see section 3.2). However, since the difference in length between the end and the interior spans is less than 10%, the formulae given on *Data Sheet 7* can be employed with sufficient accuracy for practical purposes, considering the actual spans in order to obtain the midspan and interior support moments, and an average span of $\frac{1}{2}(6.00 + 5.425) = 5.713$ m to determine the moment over the penultimate supports. No moment redistribution is undertaken since the analysis is only approximate.

It is interesting to compare the bending moments calculated by adopting the foregoing assumptions with those obtained by using the approximate coefficients given in section 3.2, and by a more exact analysis employing precise moment distribution. With the approximate coefficients,

$$F = (11.80 \times 1.4) + (12.50 \times 1.6) = 36.52 \text{ kN/m},$$

and thus

maximum bending moment at midspan of AB
$$= 36.52 \times 5.425^2 \times 0.09 \quad = 96.8 \text{ kN m}$$

Similarly:

maximum ultimate bending
    moment at B   = 118.3 kN m
maximum ultimate bending
    moment at midspan of BC = 75.3 kN m
maximum ultimate bending
    moment at C   = 86.0 kN m

A so-called 'exact' analysis by precise moment distribution gives maximum ultimate bending moments of 100 kN m at midspan in AB, 122 kN m over support B, 85.4 kN m at midspan in BC and 107.1 kN m over support C. Thus the approximate method employed on the calculation sheets results in maximum ultimate moments that are 0.5% lower, 3% higher, 8% higher and 2% higher than the 'true' values, in this particular case.

Examining the moments obtained using the Code coefficients, it will be seen that the values at midspan in AB, at support B, at midspan in BC and at support C are 3% lower, 3% lower, 12% lower and 19% lower than the exact values. In view of the considerable approximation involved in assuming that the continuous beam is of uniform section throughout (when in fact it behaves as a T-beam at midspan and as a rectangular section over the supports), and taking into account the interaction with the supporting columns or beams, it may be assumed that the results obtained by either of the approximate methods considered are sufficiently accurate.

Alternate lines of secondary beams are supported on the main transverse beams and on rows of columns. *Calculation Sheet 5* apply to secondary beams that are supported on the main beams. These secondary beams are assumed to be freely supported at their ends where they are supported on the transverse beams carrying the outer brick walls, and at these supports there is no definite bending moment, but the reinforcement in the top at the end support will resist any small negative bending moment that may develop. A calculation is given in *Calculation Sheet 6* for the modification of the ultimate bending moments and shearing forces due to the ends of the secondary beams being supported by, and being constrained monolithically with, the columns. With uniform loads, the coefficient of the end-restraint bending moment may vary from about 1/120 to 1/12, depending on the relative stiffnesses of the columns and beams. In this example a coefficient of about 1/14 is assumed, which provides a negative ultimate bending moment of 77 kN m. This value may be compared with 47.4 kN m given by the subsequent and apparently more accurate calculation on *Calculation Sheet 20*, but with partial safety factors for load of 1.2; for values of γ_f of 1.4 and 1.6, the corresponding moment would be about 60 kN m. If an outer transverse reinforced concrete wall is provided instead of brick panels on a framework of beams, the ends of all the secondary beams, whether supported on main beams or columns, would be restrained to such a degree that they should be designed to resist the moment due to complete fixity, i.e. $nl^2/12$.

The effects of the restraint bending moment are calculated by means of the coefficients given on *Data Sheet 14*, on the assumption that the spans are of equal length. The negative bending moments calculated from this table could not be reduced by moment redistribution since the spans are not equal. If a more accurate analysis is necessary, the positive and negative ultimate bending moments resulting from the end restraint can be determined quickly by precise moment distribution. The variations due to end restraint in the present example are quite small, and there is no difference in the practical amounts of reinforcement required in the secondary beams supported on the main beams or on the columns. The arrangement of the bars at the end supports is slightly different, as shown on *Drawing 5*, in order to provide tensile resistance to the negative bending moment. If the reduction of bending moment at the penultimate support B were greater, it might be worth while reducing the amount of reinforcement in the top over the support, were it certain that the calculated restraint moment would act.

From a consideration of the shearing forces it is evident that, in this example, shearing resistance does not govern the size of the secondary beams; however, their depths are limited by the specified clearance of 3 m (or 3.75 m in the case of the lowest storey) shown on *Drawing 1*. The maximum allowable overall depth of the beams is therefore $3600 - 3000 = 600$ mm. If the overall depth of the main beams is 500 mm (this allows 100 mm, more than sufficient, for finishes) it is convenient to make the secondary beams slightly shallower, say 400 mm, to avoid the need to crank the reinforcement at the intersections. The rib of the secondary beam therefore projects 300 mm below the slab. If shearing resistance does not govern the size of the rib, its width should normally be from one-third to two-thirds of its depth, and is determined from the consideration of the space required to accommodate the reinforcement, as discussed below.

9.5.2 Resistance to bending

When the ultimate bending moments have been determined, the design of the sections follows the usual procedure. If a ratio of x/d of 0.5 is assumed, the neutral axis falls below the slab at midspan and the entire slab thickness may be assumed to be in compression. If the concrete stress is uniform and the resistance provided by the small area in the web between the underside of the flange and the neutral axis is ignored, the lever-arm is $d - 0.5h_f$ and the breadth b of slab required to resist compression is $5M/(2d - h_f)h_f f_{cu}$. This expression is applicable if the breadth required is less than that permissible. In the end spans, $b = 270$ mm; this is much less than the permissible breadth recommended in BS8110, which is the least of the values of b indicated on *Data Sheet 41*. For the secondary beams, the permissible breadth is controlled by the effective span and is 1122 mm if a rib breadth of 200 mm is assumed. If the value of b given by the approximate expression exceeds the permissible width, the more accurate formulae in section 5.3, which take into account the resistance provided by the web, should be applied.

The recommendation in BS8110 relating to high ratios of the length between the lateral supports of a beam to the breadth of the compression flange must be observed. In the case of a freely supported or continuous beam, the breadth of the compression face b_c midway between the points of lateral restraint must be not less than the greater of $l_e/60$ or $\sqrt{(dl_e/250)}$, where l_e is the clear distance between lateral restraints. For cantilevers restrained at the support only, b_c must be not less than $l_e/25$ or $\sqrt{(dl_e/100)}$. This requirement applies principally to rectangular sections, since the width of flange of a T-beam or L-beam is generally sufficiently large compared with the depth and effective span.

The area of reinforcement required at the middle of the interior spans is given by three 25 mm bars in a single layer. Since the shearing resistance does not determine the size of the rib, the rib width is controlled by the Code requirement that the minimum cover to all steel must be not less than the diameter of the bar or, in the present case, 20 mm, and that the spaces between individual horizontal bars (except at laps) should be the greater of either the bar diameter or the largest size of aggregate plus 5 mm. When different sizes of bar are used, the largest bar determines the spacing. If the maximum size of aggregate is 20 mm, the minimum horizontal distance between the bars according to both requirements is 25 mm. The minimum possible width of beam is thus as

follows: two side covers of 20 mm each (40 mm) plus twice the 10 mm of link assumed (20 mm) plus three 25 mm bars (75 mm) plus two intermediate spaces of 25 mm each (50 mm). The resulting value of 185 mm is less than the width of 200 mm assumed.

When deformed bars are used, the size of the cross-section is influenced by the surface deformations, and the maximum dimension may be up to 14% more than the nominal bar size. This must be taken into account when arranging bars in narrow sections.

The controlling factor in the design of the sections at the supports of the beams is the compressive resistance, and the calculations for these sections determine the amount of reinforcement required in compression if the concrete is assumed to resist part of the compression. In accordance with the the requirements of BS8110, it is necessary to restrict the maximum spacing of the links to twelve times the diameter of the compression bars. Thus if no other consideration were decisive (e.g. shearing resistance) the spacing of the links near the supports for the secondary beams would be 12ϕ (i.e. $12 \times 25 = 300$ mm); spacings corresponding to stated diameters are given on *Data Sheet 16*. Remember that, once compression bars are necessary, according to BS8110 a minimum amount of $0.002bh$ must be provided in rectangular sections and flanged sections where the flange is in tension, and twice this amount in other flanged sections.

The sections are designed as described. At the supports it is necessary to allow for a vertical space of two-thirds of the maximum aggregate size between the layers of reinforcement: here $\frac{2}{3} \times 20 = 13\frac{1}{3}$, say 15 mm. Thus the effective depth to the centroid of two layers of steel is $400 - 20 - 10 - 25 - (\frac{1}{2} \times 15) = 337.5$ mm; in fact, the position of the centroid of a five-bar arrangement would give a slightly greater effective depth, but this is ignored. Since the bars in a single layer are sufficient to provide the compression reinforcement required at the supports, the bars lapping at that point may be in contact vertically for the length of the overlap. If the bars extending from each span were needed to form the compression reinforcement, however, it would be necessary to provide a clearance of 15 mm vertically between them and recalculate d' accordingly. When the cross-sectional areas of steel required have been calculated, the numbers and sizes of bars can be read from *Data Sheet 41*.

9.5.3 Shearing resistance

The calculations for the ultimate shearing force and shearing resistance adjacent to the supports of the secondary beams are given on *Calculation Sheets 5* and *6*. The coefficients for the shearing forces due to dead and imposed loads on beams without end restraint are taken from *Data Sheet 14*, and the appropriate coefficients from this data sheet are used to make the necessary adjustments where required to account for the restraint at the end supports.

The shearing resistance of the secondary beams is provided entirely by links where necessary. In each case the resistance provided by the concrete alone is determined for the relevant amount of main tension steel provided and, where this is less than the applied ultimate shearing force, a suitable size and spacing of links to resist the balance is obtained from *Data Sheet 16* for the required value of K_u. The arrangement of links is shown on *Drawing 5*. Between the points where the shearing resistance provided by the concrete alone is sufficient to withstand the entire shearing force on the section, only nominal links need be provided. The maximum spacing of these nominal links must not exceed $^3/_4d$, i.e. about 250 mm in the present example. Unlike its predecessor, BS8110 does not specify minimum percentages of nominal links, but instead requires sufficient links to be provided to give a design shear resistance of 0.4 N/mm^2. However, this requirement can be expressed as $b_t = K_u/0.4$, and so the maximum b_t can be expressed in terms of A_{sv}, f_{yv} and s_v, as is done on *Data Sheet 16*. In the present example, nominal links of 8 mm diameter at 250 mm centres will suffice.

9.5.4 Details of reinforcement

If the loads supported are substantially uniform and if the spans are approximately equal, BS8110 provides in clause 3.12.10.2 simplified detailing rules; these requirements are summarized on *Data Sheet 42*. Where these rules do not apply, the Code requires that at end supports each tension bar must be provided with an effective anchorage, equal to either 12ϕ beyond the centre-line of the support, or $12\phi + ^1/_2d$ beyond the face of the support (no bend commencing before the support centre-line or at a distance of less than $^1/_2d$ from the support face, respectively). Alternatively, if the design ultimate shear stress at the face of the support is less than one-half of the permissible value (as given by *Data Sheet 15*), a straight length of bar equal to either one-third of the support width or 30 mm, whichever is greater, beyond the support centre-line will suffice. Elsewhere the Code states that each bar must extend beyond the point at which it is no longer required, a minimum distance equal to either the effective depth or 12ϕ, whichever is the greater. Furthermore, a bar may not be stopped off in a tension zone unless one of the following three requirements is met: (a) an anchorage length corresponding to the full design strength of $0.87f_y$ is provided beyond the point at which the bar is no longer required; (b) the shearing resistance provided at the stopping-off point is at least twice the actual shearing force; or (c) the bars that continue provide at least twice the area of steel needed to resist the moment beyond the stopping-off point.

The positions where the bars comprising the tension reinforcement are no longer theoretically required can be determined from a roughly sketched bending-moment diagram. The lengths of the bars beyond these points are then determined as described above; it is often worth while providing lengths greater than the minimum bond lengths permitted in BS8110 if this can be done without causing congestion. Hooks on tension bars are seldom necessary, but it may be thought advisable to hook or bend the ends of plain round bars in beams, especially if the bars terminate in a tension zone. However, it is better to arrange for the bars to be stopped off in a compression zone if this is possible; hooks are not then required.

It used to be thought to be good practice, when detailing beams, to make each bar serve two purposes or more. For example, the bars extending throughout the bottom of the beams in the present example provide tensile resistance at midspan and form the compression reinforcement at the supports. In a similar way, bent-up bars can provide tensile resistance at both midspan and supports, and also resist shear. On the other hand, there are disadvantages in detailing reinforcement in such a way that a single bar contains several bends. The resulting rigidity of the arrangement sometimes makes it difficult to fix the bars easily and no leeway is available to take up constructional tolerances. The designer should consider such points carefully when preparing his details. Bent-up bars are seldom employed in current UK practice, but it has been thought worthwhile to illustrate how they may be used in the design of the main floor beams on *Drawing 7* and *Calculation Sheets 8*.

9.6 FASCIA BEAMS

The fascia beams along the edges of the floors parallel to the secondary beams are designed similarly to the secondary beams, except that the loads differ since only one panel of floor slab is supported by each fascia beam which, however, also carries a wall. The total load, as given in the upper part of *Calculation Sheet 7*, is a little less than that on an ordinary secondary beam; the fascia beams can therefore be the same net size and provided with the same amount of reinforcement as the secondary beams supported on columns. Cross-sections of the fascia beams are given on *Drawing 6*. The shape of the links differs from those in the ordinary secondary beams, in that an arm projecting into the adjacent floor slab assists in providing the necessary resistance to the negative bending moment on the slab. This arm is vertical when the stirrup is bent and fixed in the beam, and therefore does not hinder the fixing of the main bars in the beam. After the bars are fixed in position this arm is bent down into its permanent position. Alternatively (and perhaps preferably) separate U-bars tied to the maink links may be employed. The fascia beam along the front of each of the upper floors (excepting the first and fifth floors) is in the form of a boot lintel supporting the facing of brickwork over the windows. To ensure correct coursing of the inner and outer leaves of brickwork, the depth of the lintel should be a 'brick dimension', i.e. a multiple of 75 mm if 65 mm bricks with 10 mm joints are used. Since the nib is exposed to the weather, ref. 16 recommends that 40 mm minimum cover is provided to all reinforcement, and states that a minimum depth of nib of 140 mm should be provided. The corresponding minimum concrete grade according to BS8110 is C35 (if a 'systematic testing regime' is employed). Useful information is also provided in Chapter 6 of ref. 2.

9.7 MAIN BEAMS

The calculations for the main beams are similar to those for the secondary beams and are given on *Calculation Sheets 8*. Details of these beams are shown on *Drawing 7*.

9.7.1 Loads

The concentrated loads F from the secondary beams are equal to the total load on a single span of a secondary beam, multiplied by suitable coefficients from *Data Sheet 14* to allow for the elastic reactions; inspection shows that the greatest load is carried by the main beams spanning from column A2 to column D2, and for this case the coefficients 0.607 and 0.621 apply to the load on the end span of the secondary beams and 0.536 and 0.603 to the load on the second span. A small proportion of the load is considered to be transferred directly to the main beam as a distributed load; in practice this proportion is clearly much larger, but the assumption made on *Calculation Sheets 8* errs on the side of safety.

9.7.2 Bending moments

As all the main beams are built into the columns at the outer supports, the effect of the restraint at the ends is taken into account in the initial calculation of the bending moment, and not considered separately as for the secondary beams. The ultimate restraint bending moment of 65 kN m, which is assumed to act at both supports simultaneously, should be compared with moments of 72.4 kN m for the upper floors and 107.0 kN m for the first floor obtained by the apparently more accurate calculation on *Calculation Sheets 19*. The bending moments are calculated from the expressions given on *Data Sheet 9*. The total ultimate bending moment at each support and at midspan is the summation of the ultimate bending moments due to the concentrated dead and imposed loads, the uniform dead and imposed loads, and the effects of the bending moments at the end supports. This summation is slightly inaccurate in the case of the positive moments at midspan since the maximum bending moment due to the distributed load does not occur exactly at midspan (except in the case of the bending moment due to dead load on the central span), but in the present example the maximum positive ultimate bending moments are so predominant that this inaccuracy is negligible; in any case, values obtained in this way err on the side of safety. The increase in the positive bending moment on the end span AB due to the restraint at the end support can be readily calculated by considering the

similar triangles comprising the restraint bending-moment diagram. A redistribution of 10% is made to the bending moments obtained by assuming free support at the ends, but no similar adjustment is made to the negative bending moments due to restraint at the ends, because the assumed value of the restraining moment is approximate. No differentiation is made between dead and imposed loads in assessing this restraint bending moment; because of the approximation involved, such a refinement is not worth while at this stage. The effects of such subdivision could, if required, be calculated from the coefficients in *Data Sheet 14*.

9.7.3 Shearing resistance

The shearing-force coefficients for the main beams are taken separately from *Data Sheet 14* for the central concentrated dead and imposed loads, for the distributed dead and imposed loads, and for the end restraint; no adjustment has been made for the 10% redistribution of moment. Alternatively, the approximate static shearing forces might be calculated without considering different arrangements of loaded spans and the effect of end restraint. By this approximate method the calculated ultimate shearing force at each support is the same, and is one-half of the total design load on one span, i.e. $\frac{1}{2}\{1.4[(3.43 \times 5.0) + 73.5] + 1.6[(1.25 \times 5.0) + 83.6]\} = 135.4$ kN, compared with maximum and minimum ultimate elastic shearing forces of 178.7 kN and 125.2 kN respectively.

The calculations for the shearing stresses in the main beams show that reinforcement to resist shear is necessary at each support. Since by far the majority of the shearing force is due to the concentrated loads at midspan, the shearing reinforcement must extend right across the span. In such a case the use of inclined bars to resist shear is only practical if bars already provided for bending can be bent up; it is normally uneconomic to include extra bars solely for the purpose of forming shearing reinforcement. However, as explained below, in the present case the bar lengths needed for stopping-off purposes are such that it is worth while cranking up these bars into the compression zone and utilizing the shearing resistance obtained. If one 25 mm bar is bent up at an angle of 45° at a time, the remaining shearing resistance required at support B, for example, is provided by 8 mm links at 200 mm centres, which is the minimum amount that may be provided according to BS8110.

As already mentioned in the Introduction on page viii, due to the economic circumstances of the 1990s, current UK practice does not favour the use of inclined bars as shearing reinforcement. Not only is extra work involved in bending the bars concerned (which must be done extremely accurately, especially if these bars also form part of the main reinforcement near the bottom of the beam at midspan, or the top reinforcement over the supports, or both) but such steel is more complicated to fix. However, beams that are reinforced by the use of vertical links only have occasionally proved inadequate, particularly where substantial concentrated loads are transferred from secondary beams and form the principal loading on a main beam. In such circumstances, consideration should be given to the provision of inclined bars to contribute some of the shear resistance, particularly where economic circumstances are favourable, as may be true elsewhere in the world.

9.7.4 Details of reinforcement

As in the case of the secondary beams, the simplified detailing rules given in the Code do not apply, as only a small portion of each load is uniformly distributed.

Some of the inclined bars in the main beams resist positive bending moment, shearing force and negative bending moment in turn. Hooks could be provided as anchorages, thus reducing the overall lengths of the bars and minimizing congestion, but this is not done here. At the interior supports no compression bars are actually needed. However, two of the bottom bars from midspan are extended to the supports, where they are lapped. Only a nominal lap length is needed here, and no space need be provided between the lapping bars in the bottom over the supports, since the bars are not required to resist compression.

The cover provided to the main bars throughout is 30 mm (i.e. a minimum of 20 mm to all reinforcement, including links, the size of which is assumed to be 10 mm maximum). However, where the bars in the top of the intersecting secondary beams meet the main beams, the distance from the surface to the bars in the top of the main beam is increased to 55 mm. Beyond the ends of the bars, as at the end supports of the main beams, the cover should be not less than the normal minimum cover nor, for good practice, twice the diameter of the bar; thus for the 25 mm bars, the end cover should be at least 50 mm. The length of the bar embedded in the end support may be insufficient to provide satisfactory bond, and so a transverse anchor-bar is inserted in a 45° bend to provide a more satisfactory anchorage.

The bottom bars are bent up beyond the points at which they are no longer required to resist bending moment; these points are determined by roughly sketching the bending-moment diagrams. Bar 10 is considered as effective in resisting negative bending moment, as it is bent down at some distance from the support. If an inclined bar adjacent to a support forms part of a system of shearing reinforcement, it can seldom be considered as effective reinforcement to resist tension over the support, as it is bent down too close to the support. The spacing of the inclined bars is calculated from $1.42(d - d')$ for a single system at 45°, as in *Data Sheet 17*. For bars 9 and 10, for example, the maximum spacing is $1.42(500 - 62\frac{1}{2} - 67\frac{1}{2}) = 525$ mm, where 500 mm is the overall depth of the beam; $62\frac{1}{2}$ mm consists of the bottom cover (20 mm) plus the link diameter (10 mm) plus the diameter of a bar (25 mm) plus one-half of the vertical space between bars ($7\frac{1}{2}$ mm); and $67\frac{1}{2}$ mm consists of the top cover (45 mm) plus the link diameter (10 mm) plus one-half of a bar diameter ($12\frac{1}{2}$ mm).

9.8 FREELY SUPPORTED BEAM

Beam S6 trimming the lift- and stair-well is considered to be freely supported at both ends. Therefore the bending-moment coefficient is $\frac{1}{8}$ for a load uniformly distributed throughout the span. Details of the beam are given on *Drawing 6* and the design on *Calculation Sheets 9*, The overall depth of this beam is 400 mm and the span 6 m. The loads on the beam comprise the load from panel P4 of the floor uniformly distributed throughout the span of the beam, and the load from the stairs which is distributed over part of the span only (see *Figure 16.2* for details of the latter load).

The maximum shearing force at the right-hand end of the span is 69.4 kN. If the main reinforcement consists of four 20 mm bars and two of these extend to the supports, the shearing resistance of the concrete alone when $\varrho = 1256/(357.5 \times 200) = 0.0176$ is 0.83 N/mm^2, and thus the concrete alone will carry $0.83 \times 357.5 \times 200 = 59.3$ kN. Links are therefore needed to support $69.4 - 59.3 = 10.1$ kN. It is necessary to provide nominal links 8 mm in size at 250 mm centres to meet the requirements for minimum reinforcement; this arrangement will support 31.1 kN and thus be satisfactory.

Checking serviceability requirements, if $f_y = 250$ N/mm^2 and $M/bd^2 = 0.75$ then the multiplier for the tension reinforcement 0.50 from *Data Sheet 20* and the multiplier for a flanged beam (with $b_w/b = 0.25$) is 1.25. Since the basic effective depth for a freely supported beam spanning 6 m is 300 mm, the minimum possible effective depth is $200 \times 0.5 \times 1.25 = 188$ mm, which is satisfactory.

Chapter 10
Beam-and-slab construction: two-way slabs

The general recommendations in BS8110 for the design of slabs spanning in two directions apply to slabs of all shapes and with all types of loading and support fixity. The Code suggests that the bending and torsional moments and shearing forces may be determined either by an elastic analysis, such as those proposed by Westergaard and Pigeaud, or by an analysis considering conditions at failure, such as Johansen's yield-line theory or the strip method of Hillerborg. It seems somewhat doubtful whether the behaviour of a solid reinforced concrete slab, cast monolithically with its supports and continuous over several spans, bears much relation to that predicted by complex elastic analysis; thus the use of an analysis at failure, such as that used to obtain the majority of tabulated coefficients given in BS8110, seems preferable. This method is normally used for the design of floors and roofs of buildings, and in this chapter is employed to design the main roof and some of the subsidiary roofs and floors of the building described in Part Two.

If an elastic analysis is made, the loads considered should presumably, strictly speaking, be service (i.e. unfactored) loads, and the resulting moments and shears should be resisted accordingly. However, it is simpler and probably perfectly satisfactory to factor the loads and design the resulting sections on ultimate limit-state principles, as is done in the case of the coefficients given in Table 3.14 of the Code. In undertaking any elastic analysis, BS8110 recommends that Poisson's ratio should be taken as 0.2. If an analysis based on failure conditions is undertaken, the ratio of support to span moment selected should be comparable with that which would result from an elastic analysis, i.e. normally between 1 and 1.5.

BS8110 tabulates empirical bending-moment coefficients for solid slabs that are uniformly loaded and span in two directions at right angles. Apart from the case of a freely supported slab without torsional restraint, these coefficients have been derived by means of yield-line analysis, and they should be used for all uniformly loaded rectangular slabs. For non-rectangular and partially loaded slabs, an analysis at failure (i.e. using Johansen's or Hillerborg's method) is probably preferable, owing to the complexity of elastic analysis with most non-rectangular slabs. However, rectangular slabs carrying partial loads can be analysed elastically fairly simply by Pigeaud's method, using the coefficients given in this chapter.

As *Allen* points out, although the analysis and design are for the ultimate limit-state, the serviceability limit-state of deflection will often be the ruling criterion, and it is important to estimate accurately an appropriate effective depth before carrying through the entire design calculations in too much detail.

10.1 SLAB FREELY SUPPORTED ON FOUR SIDES

The bending-moment coefficients given in BS8110 for a uniformly loaded rectangular panel, freely supported on all sides and with the corners not reinforced to resist torsion or to prevent lifting, have been obtained by application of the Grashof and Rankine analysis. This assumes that the bending moments in each direction are related to k^4, where k is the ratio between the spans, and the coefficients are derived from the expressions

$$M_{sx} = \frac{nl_x^2}{8}\left(\frac{k^4}{1+k^4}\right) \quad \text{and} \quad M_{sy} = \frac{nl_y^2}{8}\left(\frac{1}{1+k^4}\right)$$

(10.1)

Note that, although this analysis derives from elastic principles, factored loads (i.e. $1.6q_k$ and $1.4g_k$) should be used throughout and the resulting sections designed on ultimate limit-state principles. The coefficients are given on *Data Sheet 44*, and an example of this simple case, which occurs relatively infrequently, is the roof of the lift-motor room above the main roof of the building in Part Two. The motor room is assumed to have brick walls supporting the concrete roof slab, the design of which is given on *Drawings 10* and *11*. The calculations required are on *Calculation Sheet 15* and make use of the coefficients given on *Data Sheet 44*.

BS8110 also gives bending-moment coefficients for uniformly loaded rectangular panels supported on all sides and not continuous over the supports, but with the corners restrained from lifting and containing reinforcement to resist torsion. These coefficients, which are determined as described in the following section, are smaller than those for the case considered previously, and are also given on *Data Sheet 44*. The roof of the tank room in *Drawings 10* and *11* and *Calculation Sheets 14* is an example of the design of a simple slab, the edges of which are prevented from lifting; in this example the edges are held down by a 230 mm brick parapet wall. Reinforcement is provided in accordance with the recommendations of BS8110 to resist the torsion in the corners of the slab, as described below.

10.2 SLABS RESTRAINED ALONG THE EDGES

The Code gives bending-moment coefficients for uniformly loaded rectangular panels that are supported on all sides and are continuous over one, two, three or four edges. The coefficients, which are given on *Data Sheet 44*, have been derived from a yield-line analysis and adjusted to take into account the fact that, as the slab is divided into middle strips and edge strips, the reinforcement is not spaced uniformly across the slab. Arbitrary parameters, such as the ratio of support to span moment, have been selected to give moments that correspond to those which would be obtained from an elastic analysis. The coefficients are applicable only to slabs, the corners of which are prevented from lifting and contain reinforcement to resist torsion. The slab is considered to be divided into middle and edge strips in each direction, as shown on the appropriate diagram on *Data Sheet 44*. The positive and negative bending moments as calculated by the Code coefficients are those that occur in the middle strips, which should then be reinforced accordingly. No reinforcement parallel to the adjacent supports is necessary in the edge strips to resist specific bending moments, but in these strips the amount of reinforcement provided should not be less than $0.0024bh$ if mild steel is used or $0.0013bh$ if high-tensile steel is employed, or as required to resist torsion (see below).

BS8110 states that the foregoing coefficients only apply to cases where the dead and imposed loads on adjoining panels are similar to those on that being considered, and when the direction in which the adjacent panels span is the same. Where the moments at a common support obtained by separate calculations for the two adjoining panels differ significantly, the Code proposes the following procedure. Take the difference in the support moments obtained by using the Code coefficients, and distribute it between the relative stiffnesses of the spans concerned to obtain a new common value for the support moment. Then adjust the adjoining midspan moments accordingly so that the sum of the absolute values of the moments at midspan and support in each span is the same after adjusting the support moment as it was before the adjustment was made. In panels where the support moment has significantly increased as a result of such an adjustment, the bars provided to resist the negative moment over the support must be increased significantly beyond the normal provisions. BS8110 suggests, in effect, sketching the new bending-moment diagram to determine the points of contraflexure and the maximum positive moment in the span. Then one-half of the top steel at the support should extend to twelve times the bar diameter (or a distance equal to the effective depth if this is greater) beyond the point of contraflexure and the remaining bars to at least one-half of this distance.

A negative bending moment occurs at the end of a slab where it is not continuous over the support but is cast monolithically with the support, and in ordinary cases this moment should be assumed to be one-half of the positive bending moment at midspan of the middle strip at right angles to the support. However, if the restraint at the support is considerable, such as will occur when a thin slab is monolithic with a deep beam, a larger negative bending moment should be assumed, as the effect may be equivalent to continuity or even to complete fixity. Examples of slabs spanning in two directions and continuous over one or more edges are given on *Drawings 8* and *9* and *Calculation Sheets 10*.

10.3 REINFORCEMENT AT CORNERS

In accordance with BS8110, the basic cross-sectional area per unit width of the reinforcement provided to resist torsion should be three-quarters of that provided per unit width to resist the maximum positive moment in the middle strip. This amount of reinforcement should be provided in both directions in the top and bottom of the slab in each corner and should cover a square, the length of side of which is at least one-fifth of the shorter span. Any reinforcement provided for any other purpose, and in a suitable position within the square defined, may be considered as part of the reinforcement to resist torsion. It is generally necessary to add only a little more reinforcement in the edge strips to make up the quantity required in the corners. These mats of reinforcement are required only in a corner contained by two edges, over neither of which the slab is continuous. If the slab is continuous over one edge only at a corner, the amount of reinforcement required per unit width in each direction in each layer need only be three-eighths of the amount per unit width necessary at midspan. If the slab is continuous over both edges at a corner, no reinforcement is required to resist torsion. The application of these rules is illustrated in the designs on *Drawings 8* to *11*.

10.4 APPLICATION TO BUILDING IN PART TWO

The main roof of the building in *Drawings 1* and *2* is designed on *Calculation Sheets 10* as rectangular panels spanning in two directions. The arrangement of the beams and references to the panels are shown on the plan of the roof on *Drawing 8*. Details of the slab are given on this drawing and on *Drawing 9*. The method of designing the slab differs from the method for slabs spanning in one direction (section 9.1) only in the calculation of the bending moments. The method of designing the slab section is the same, since the same characteristic strengths are assumed. No distribution bars are required, since the main bars are provided in two directions at right angles. Reinforcement to resist torsion is provided where required, or the minimum amount of main reinforcement is provided in the edge strips.

Interior panels P1 are continuous over all four edges, and coefficients for the bending moments in the middle of the spans and at the supports of the shorter and longer spans are taken from *Data Sheet 44*. Continuity exists along the two transverse edges and the inner longitudinal edge of panel P2, but nominal free support is assumed at the outer longitudinal edge, where the negative bending moment provided for is one-half of the positive bending moment at midspan on the corresponding middle strip. The calculations for panel P1 are given on *Calculation Sheets 10*. The supports of the left-hand panel P3 are similar to the other panels P2, except that along the transverse edge at the stair-well there is continuity for part of the length and nominal free support for the remainder; the bending-moment coefficients are therefore intermediate between those for continuity on two sides and those for continuity on three sides. Panel P4 is similar to panel P2 except that it is discontinuous along one of the shorter edges. Panel P3 is nominally freely supported along two adjacent edges. Complete calculations are not given for panels P2, P3 and P4 on *Calculation Sheets 10*.

Panels P5, P6, P7 and P8 are not designed to span in two directions, but span entirely across the shorter dimension. The method of design is identical to that of the slabs in the beam and slab construction described in section 9.1, but it is necessary to ensure that at a support common to two panels, for example the support of panel P1 and the adjacent panel P5, there is sufficient resistance to the negative bending moment on either panel.

The arrangement of the reinforcement in *Drawings 8* and *9* conforms to the Code recommendations as regards spacing, cover and other matters. With large ratios of the longer to shorter span, the bars in the direction of the longer span correspond to the distribution bars in a slab spanning in one direction. Their area should be not less than $0.0024bh$ in the case of mild steel bars or $0.0013bh$ if high-tensile bars are used. The amount of reinforcement in the slab over the beams along the shorter sides of the panels should conform to the requirements for flange reinforcement, i.e. be not less than $0.0015bh$ (see *Data Sheet 41*). Reinforcement in the panel corners is provided

in accordance with the rules described in section 10.3. Special reinforcement for this purpose is required only in the outer corners of panels P2, P3 and P4 of the main roof.

10.5 LOADS ON SUPPORTING BEAMS

The designs of typical beams supporting the two-way slabs are given on *Calculation Sheets 11* and *12* and *Drawing 9*, but they present only one feature not dealt with already, namely the load carried by the beams supporting rectangular panels. BS8110 improves on its predecessors by providing a table (similar to that giving bending-moment coefficients) which indicates appropriate shearing-force coefficients for panels having differing aspect ratios and support conditions. The load transferred from the slab to the supporting beam is obtained by multiplying the coefficient concerned, read from the appropriate table on *Data Sheet 44*, by the total ultimate load on the slab multiplied by the slab dimension at right angles to the supporting beam being considered. Note that the resulting uniform load per unit length transferred to the beam only acts on a length arranged symmetrically and equal to three-quarters of the span (i.e. corresponding to the breadth of the middle strip of the slab being supported). To this load must be added the uniform load due to the self-weight of the beam plus any other construction supported directly on the beam (and remembering that these loads extend along the full span). To enable the moments resulting from such an arrangement of partial uniform load to be calculated quickly, *Data Sheets 7 to 9* include bending-moment coefficients corresponding to this loading condition.

If the support moments between unequal panels are adjusted, as described in section 10.2, to such an extent that they differ markedly from the values given by the original coefficients, clause 3.5.3.7 of BS8110 states that the loads transferred to the beam at this support, as calculated from the Code coefficients, may require adjustment. However, as *Allen* points out, since the adjustment to the moments simply involves the determining of a mean value between two extremes, any resulting change in the reaction will be small, if not negligible, since any increase in load due to the increase in moment on one side of the support will be counterbalanced by the decrease on the other side.

10.6 DESIGN OF ROOF BEAMS

The design of typical beams supporting the panels of the two-way roof slabs is given on *Drawing 9* and *Calculation Sheets 11* and *12*. The detailed design of the beams to resist bending is as described in Chapter 5 and the shearing resistance is determined as discussed in Chapter 6. The ribs of the beams are all 130 mm wide and therefore are of the same width as the arms of the cruciform interior columns supporting them. This design is suitable for residential structures as, for example, the top storey of the building in Part Two, as there are no projections of beams and columns beyond the faces of the partition walls.

10.6.1 Beams along longer edges of panel

A typical beam providing the support along the longer edges of a rectangular panel is the longitudinal beam at the junction of panels P1 and P2. The load R_l from panel P1 is determined from the factors on *Data Sheet 44* for the condition where all the slab edges are continuous and $l_y/l_x = 1.2$. The equivalent load transferred from panel P2 is obtained by employing the factor corresponding to the same aspect ratio of slab but where one longer edge (i.e. the other to that being considered) is discontinuous. In addition, account must be taken of the continuous uniform load due to self-weight. The coefficients now required to determine the bending moments at the support and near midspan are taken from *Data Sheets 7* and *8* for the penultimate support and an end span of an infinite series of spans. Note that, for the factors relating to the moments induced by the slab loads, the formulae corresponding to partial uniform loads extending across the central 75% length of span must theoretically be utilized, whereas when considering the self-weight the required formulae are those relating to uniform loads throughout the entire span. However, in normal circumstances it is usually sufficiently accurate and simpler to consider all the loads as acting over the central 75% of the span only. If this were done in the present case the calculated moments would be decreased by 0.8 and 0.4 kN m respectively, a discrepancy of only about 0.5%. For even greater accuracy, the self-weight load considered can be increased pro rata, i.e. in the present case to $1.73/0.75 = 2.31$ kN m, if it is considered to act over the central 75% length only, and the resulting differences in moments compared with the theoretically exact values are then virtually nil.

10.6.2 Cantilever beam

Each line of transverse beams comprises a cantilever and two spans of 5 m each. The bending moment on the cantilever is calculated first, because it affects the bending moments elsewhere. The cantilever partially supports panel P5 which spans in one direction only between the fascia beam and the longitudinal beam on column line C. Generally, the load from this panel is considered to be carried equally by these two beams, i.e. one-half of the load acts as a concentrated load at the end of the cantilever where it supports the fascia beam, and this is done on *Calculation Sheets 12*. It may be argued, however, that it is a little more accurate and economical to consider that part of the load is applied throughout the length of the cantilever and that the remainder acts as a concentrated load. If this is done, this division may be effected by imagining lines drawn at angles of 45° from the beam intersections and allocating the loads to the various beams in accordance with the trapezium-shaped areas of slab that result. Thus the shorter cantilevers will support triangular slab areas while the longitudinal beam will carry the remaining trapezoidal area of slab. In the design

of the cantilevers, the critical section is that at the face of the column, but the value of the bending moment at the column centre-line is required for the calculations for the adjacent parts of the continuous beam.

It is also necessary to check the slenderness of the rib of the cantilever. According to BS8110, the clear distance from the end of a cantilever to the support face must not exceed either $25b_c$ (i.e. 3.250 m) or $100b_c^2/d$ (i.e. 3.150 m), both of which exceed the actual value of 2.210 m. *Calculation Sheets 12* also shows that the actual effective depth of at least 537.5 mm is well in excess of the minimum depth required to limit the deflection to an acceptable value.

10.6.3 Beams at shorter edges of panel

The calculations for one of the two spans of a line of transverse beams are given on *Calculation Sheets 12* and the details of both spans are given on *Drawing 9*. The load on the span selected is that from two panels P1; R_s for each panel is determined from *Data Sheet 44* and the equivalent distributed load is calculated as described in section 10.6.1. The bending moments due to these loads are calculated on the basis of a beam continuous over two spans (*Data Sheets 7* and *8*) and are then modified to allow for the effects of the cantilever bending moment acting at support C. The unsymmetrical loading and bending moments on this line of beams will cause bending in the columns, but no relief on this account has been made to the bending moments on the beams. The bending moment on the columns has consequently also been neglected, but as is seen in the case of column B3 (*Calculation Sheets 17*) the ultimate resistance of the column to axial load is so much in excess of the actual ultimate load that the section will be quite adequate to resist combined bending and direct load.

10.7 CONCENTRATED LOADS ON SLABS SPANNING IN TWO DIRECTIONS

10.7.1 Bending moments

The evaluation of the bending moments due to loads concentrated on small areas or on narrow strips of a slab spanning in one direction is described in section 9.4, but for similar loads on slabs spanning in two directions the calculations are not so simple. BS8110 recommends that such problems are dealt with by using methods based on elastic analyses such as those suggested by Pigeaud and others. Alternatively, methods based on a consideration of conditions at failure, such as yield-line analysis or the strip method, may be used. In *Data Sheet 45*, bending-moment coefficients α_x (short span) and α_y (long span) based on Pigeaud's method are given for a load F, concentrated on a rectangular area a_x by a_y that is symmetrically disposed about the centre of a rectangular panel which is freely supported along all edges with the corners restrained. The bending moment per unit width of slab is $\alpha_x F$ on the shorter span l_x and $\alpha_y F$ on the longer span l_y. Poisson's ratio has been assumed to be 0.2, as recommended in BS8110 for serviceability analyses. It should be remembered that the resulting moments are maximum values and do not occur over the whole width of the slab.

It should be observed that when the load covers the entire panel the moments resulting from Pigeaud's theory are rather less than those resulting from the coefficients given in BS8110. For example, when $l_y/l_x = 2$ the Code coefficients are 0.111 and 0.056, while the corresponding coefficients due to Pigeaud (in terms of l_x^2) are 0.102 and 0.038. It is therefore recommended that, with a load covering a substantial area of the panel, the appropriate Poisson coefficients should be increased somewhat, for example by increasing the appropriate Pigeaud coefficients by the difference between the Code coefficient and the Pigeaud coefficient for a fully loaded panel of the shape concerned, times the appropriate ratio of a to l. Again, to relate the Code coefficients (which are in terms of l_x^2) to those of Pigeaud (which are in terms of $l_x l_y$) they must be divided by k. For example, if $k = 2$, $\alpha_x/l_x = 0.8$ and $\alpha_y/l_y = 0.6$, since $\alpha_x = 0.086$ and $\alpha_y = 0.038$, the adjusted Pigeaud coefficients would be $0.086 + (0.111 - 0.102)0.8/k = 0.090$ and $0.038 + (0.056 - 0.038)0.6/k = 0.043$ respectively.

To allow for the effect of continuity with adjacent panels, Pigeaud suggested that the positive bending moment should be reduced by 20% if the negative bending moment is assumed to have the same numerical value as that of the positive bending moment on a freely supported panel. This assumption does not, however, allow for differing support or continuity conditions on various edges of the panel. In the lower part of *Data Sheet 45*, proposed variable adjusting factors k_x and k_y are suggested. The bending moment in any span or at any support is therefore $\alpha_x k_x F$ for the shorter span and $\alpha_y k_y F$ for the longer span. In some cases two limiting values of k_x and k_y are given; the first value (which is not necessarily the lower) is for the case where the load covers only a small area of the panel; the second value is for the case where the load covers the entire area of the panel and represents the ratio between the Code coefficients for the support conditions concerned. For loaded areas that are intermediate in size, a suitable factor can be chosen by inspection.

The foregoing analysis is for a load that is concentrated on an area that is symmetrically disposed about the centre of the panel. For unsymmetrically placed loads, rules can be devised whereby the bending moment, due to a load on a symmetrical area embracing the actual loaded area, can be calculated and then divided to give the bending moment due to the actual load. This procedure is only worth while undertaking for important slabs carrying large eccentric loads, and further details are given in *RCDH*. For the conditions encountered in buildings, it is generally sufficiently accurate to reduce the bending moments by a nominal amount to allow for the lack of symmetry in one or both directions.

10.7.2 Shearing forces

If it is necessary to consider the shearing forces on slabs carrying heavy concentrated loads, the following rules given by Pigeaud are normally sufficient for all conditions of continuity. If a_x is greater than a_y, the shearing force in the middle of length a_x is $W/(2a_x + a_y)$, and in the middle of length a_y is $W/3a_x$. If a_y is greater than a_x, the shearing force at the centre of length a_x is $W/3a_y$, and at the centre of length a_y is $W/(2a_y + a_x)$. If necessary, these expressions could be employed on *Calculation Sheets 13* for panel P9.

10.7.3 Example

An example of a concentrated load on a slab spanning in two directions is panel P9 at the level of the main roof of the building in Part Two. This panel is 5.625 m by 3.75 m, and it forms the floor of the tank room carrying the tank, which is supported on closely spaced bearers such that the loaded area on the panel can be assumed to be the same as the plan area of the tank, i.e. 3.0 m by 2.5 m. The loadings and calculation of the bending moments are given on *Calculation Sheets 13* and the details of the slab are given on *Drawings 10* and *11*. The bending moment coefficients α_x and α_y for the concentrated load are interpolated from *Data Sheet 45* for $l_x/l_y = 5.625/3.75 = 1.5$ and for $a_x/l_x = 2.5/3.75 = 0.67$ and $a_y/l_y = 3.0/5.625 = 0.55$, namely 0.104 and 0.067 respectively. The adjustment factors $k_x = 0.95$ and 0.75, and $k_y = 0.90$ and 0.70, for continuity are estimated from the values on *Data Sheet 45* for the condition of continuity over three edges, one short edge being nominally freely supported; since the tank is symmetrically placed on the panel, no reduction may be made for non-central loading. The gross weight of the tank and its contents is about 230 kN. However, in the calculations for the bending moments due to this concentrated load, the load is reduced to 211.3 kN because a general imposed load of 2.5 kN/m^2 is assumed to act over the entire floor, whereas this load cannot occur under the tank when it is full. The reduction is therefore $2.5 \times 3.0 \times 2.5 = 18.7$ kN.

Chapter 11
Flat slabs

CP 110 permitted flat slabs, i.e. floors or flat roofs without beams, to be designed by either of two quite independent methods. If all the specified limiting criteria were met, the so-called empirical method could be used, whereby each slab panel was considered individually. Alternatively, the structure could be considered as a series of frames in each direction, although this approach normally led to the need to provide more reinforcement and its use was thus unwarranted unless the empirical method criteria could not be met.

In BS8110 two differing analytical methods are still described. Although the procedure involving the analysis of equivalent frames is still retained, the empirical method has now been replaced by a simplified version of the equivalent-frame method. This version does not involve a true frame analysis; instead, coefficients are found from the consideration of a single-load case, and a moment redistribution is undertaken in order to reduce the support moments by 20% (as described in clause 3.5.2.3 of BS8110). This simplified method, which may only be used if certain limiting criteria are met, will in general result in a more conservative design needing more reinforcement than the alternative rigorous procedure.

Computers are now frequently used to analyse flat-slab structures, and grillage analysis is frequently adopted; more detailed information is given in ref. 18. This publication provides comprehensive coverage of all aspects of the design of flat slabs in accordance with the Code, and may be considered an essential reference work if simplified analysis is not employed.

Most of the features of the simplified Code procedure are illustrated in an alternative design for the upper floors of the building in Part Two as shown on *Drawings 12* and *13* and *Calculation Sheets 16* and discussed below. The rigorous analytical method is also illustrated in this chapter.

11.1 SIMPLIFIED METHOD

11.1.1 Dimensions

To use the simplified method, BS8110 requires that there must be at least three panels of approximately equal span in each direction. The upper floors of the building in Part Two could therefore each comprise panels 5 m square, three transversely and six longitudinally, although flat-slab construction is not really economical for such small panels. Combining interior panels 5.0 m by 6.0 m with end panels 5.0 m by 5.51 m (*Drawing 12*) is a better arrangement and fewer columns would be required. Where adjacent spans differ in length, the longer span should be used to calculate the bending moments over the supports. Stability of the structure as a whole must be provided by shear-walls or bracing capable of withstanding all lateral forces.

Flat slabs may or may not have drops. A *drop* is a greater thickness of slab in the area over the head of each column. It provides not only a greater depth of slab to resist the most severe moment acting on the slab, but also a greater area of concrete to resist punching around the column head. In CP110 the negative bending moments were slightly greater and the positive moments slightly less in the case of slabs with drops than otherwise, but this is not true with the coefficients provided in BS8110. Drops may be square or rectangular or, since no upper size limit is specified, may presumably be continuous between columns. The length of each drop in any direction must not be less than one-third of the distance of the shorter span of the panel. For exterior panels the width of drop, measured from the centre-line of the column at the free edge, should be not less than one-half of that at the interior edge of the panel. The floor shown on *Drawings 12* and *13* has drops 2.5 m square, which is greater than the minimum size required. The thicknesses of slab and drop are determined from considerations of ultimate bending moments, shearing forces and deflection. Unlike some design rules for flat-slab construction, those in BS8110 do not limit the amount of reinforcement that may be provided to a lower value than that prescribed for other forms of slab, but the desirability of restricting the proportion to 1% or less should be borne in mind. As explained in section 5.4, a more economical design is achieved; and, although the slab that results is slightly thicker than if a greater amount of steel were provided, the additional self-weight is only small in proportion to the total load supported and the saving in steel is normally well worth while. No specific minimum amount of reinforcement is specified (as in some other codes); therefore the general requirement applies that the least amount of steel in each main direction in a solid slab spanning in two directions should be $0.0024bh$ if mild steel is used or $0.0013bh$ if high-yield steel is employed. No minimum thickness of drop is indicated in BS8110.

Among other factors, the bending moments depend on the size of the column, or the enlarged column head if one is provided. The diameter of such an enlarged column head may not be taken as greater than one-quarter of the shortest span supported by the column concerned, and the angle of splay should be considered as not less than 45° to the horizontal. BS8110 also requires that the diameter of any enlarged column head be measured on a plane 40 mm below the soffit of the slab or drop. This and the other requirements regarding sizes of drop and column heads, and the positions of the critical sections specified for resistance to punching shear, are summarized on *Data Sheet 46*. If the topmost 40 mm of the column head is finished vertically, as shown on *Drawing 13*, the effective diameter corresponds to the actual maximum diameter. The diameter of 1250 mm is then one-quarter of the minimum span of 5 m. In UK design practice, drops and enlarged column heads are now employed much less frequently than used to be the case.

11.1.2 Shearing stresses

If drops are provided, the shearing stresses must be investigated for at least two sections, namely around the column head itself and around the perimeter of the drop. The critical plane around the drop is in the slab, at a distance from the edge of the drop equal to 1.5 times the effective depth of the slab. According to BS8110, in order to take account of the non-symmetrical distribution of the shearing force around this perimeter (brought about by the action of the moment being transferred from slab to column), the effective design shear force to be considered is $V_{eff} = \alpha V_t + 1.5 M_t/x$, where M_t is the moment being transferred from slab to column and x is the length of the perimeter side parallel to the axis of bending. If only the single case of loading on all spans is considered, these values of M_t are those that occur after the 20% reduction of support moments is made; if detailed loading patterns have been taken into account, the maximum values of M_t obtained from these analyses may be reduced by 30%. The Code suggests that if rigorous calculations are not made, V_{eff} is taken as βV_t. At intersections between the slab and internal columns, BS8110 proposes values of $\alpha = 1$ and $\beta = 1.15$. For corner columns and edge columns where the action of the moment in the column is parallel to the free edge of the slab, the Code does not give the factors for the rigorous expression but suggests a value of β of 1.25; for the remaining edge columns (i.e. where the moment acts at right angles to the edge), the Code recommends $\alpha = 1.25$ and $\beta = 1.4$. According to the *Code Handbook*, use of the simplified factors is satisfactory where span lengths differ by no more than one-quarter

and where the imposed load does not exceed the dead load; otherwise the rigorous expressions should be evaluated. This document also states that where a single design load acting on all panels simultaneously is the only loading case considered, for internal columns and edge columns where the moment acts at right angles to the slab edge, the values of β should not be less than 1.1 and 1.35 respectively.

When calculating the shearing force to be considered, remember that there is no need to take account of the loads acting within the perimeter being considered, since these are transmitted directly to the supporting column. Thus for an internal column carrying a total ultimate load of n kN/m^2 on a panel of dimensions l_x by l_y, the shear force acting at the perimeter of a square drop panel of side l_d is $1.15n[l_x l_y - (l_d + 3d)^2]$, and thus the shear stress is

$$1.15n[l_x l_y - (l_d + 3d)^2]/[4(l_d + 3d)]$$
$$= 0.2875n\{[l_x l_y/(l_d + 3d)] - (l_d + 3d)\}$$

The critical shearing force around the column head itself is considered to occur on a rectangular plane at a distance of $1.5d$ from the column face, where d is now the effective depth of the bars within the drop panel if one is provided. Again, to obtain the value of the shearing stress acting on the section, the load per unit area of slab multiplied by the area of panel outside the perimeter under consideration and by the appropriate factor β is divided by the perimeter times the effective depth. If the resulting stress exceeds the shear resistance provided by the concrete alone, either the slab must be increased in thickness or specific reinforcement must be provided to resist the excess shear force, as described for concentrated loads on slabs in section 6.5.1. Note, however, that shearing resistance cannot be provided in slabs that are less than 200 mm in thickness.

Remember, when determining the shear resistance contributed by the concrete, that the Code requires the reinforcing bars providing the negative resistance in the column strip over the supports to be more closely spaced within the central half of the strip; this respacing will occur within most, if not all, of the most critical perimeter. In cases where only part of the perimeter being considered falls within the part of the column strip where the bars are more closely spaced, it is often worth while evaluating the resistance of the differently reinforced parts of the perimeter separately and summing them, as calculation will normally show that the shear resistance resulting from evaluating the resistance of the two parts separately and summing them exceeds that obtained by considering the same amount of reinforcement spaced uniformly.

The shear stress around the face of the column itself should also be checked to ensure that it does not exceed the limiting value of $0.8(f_{cu})^{0.5}$ or 5 N/mm^2 specified in the Code. If this unlikely situation should occur, the column dimensions or the slab thickness must be increased.

Typical calculations to determine the shearing forces at the critical planes in a slab with drop panels are given on *Calculation Sheets 16*.

11.1.3 Bending moments

To calculate the ultimate bending moments, each panel is divided into column strips and middle strips. The width of each column strip is one-half of the shorter dimension of the panel except where drops are provided, when it is equal to the width of the drop. The width of the middle strip is that of the entire panel, less the column strip. The bending-moment coefficients given in BS8110 apply to the column strips and middle strips when the width of each is one-half of the width of the panel. When the width of the column strip is less than one-half of the width of the panel, the total bending moment on the middle strip must be increased in proportion to its increased width, and the bending moments on the column strip so decreased that the total ultimate bending moments resisted by column strip and middle strip together remain unchanged. Coefficients based on the values given in BS8110 and which enable the ultimate bending moments to be calculated directly are given on *Data Sheet 46*.

The coefficients k for the total ultimate positive and negative bending moments for the column strips and middle strips of rectangular panels, with and without drops, must be substituted in the expression $M_{ds} = knl_1 l_2 (l_1 - \frac{2}{3}h_c)$ to obtain the bending moment. Here l_1 is the distance between the centres of the columns in the direction of the span being considered; l_2 is the distance between the centres of the columns at right angles to the direction of span; n is the total ultimate load and is equal to $1.4g_k + 1.6q_k$; and h_c is the effective diameter of the column head. For square panels of span l, $l_1 = l_2 = l$, which is substituted in the general expression for the bending moments. If the column-head diameters are unequal, h_c should be the average of the two diameters.

As already stated, when the moments over supports between unequal spans are calculated, the longer of the two spans should be used in the expressions. It is often advantageous to have end panels that are slightly shorter than the interior panels, since the bending moments then become less unequal and the same thickness of slab may be maintained throughout the floor, and possibly the same bar arrangement also. BS8110 states that the moment and shearing-force coefficients that it provides may be employed as long as there are at least three rows of panels of approximately equal span in the direction being considered. Although it does not specify the limitations of of 'approximately equal', it seems reasonable to adopt the same criteria that control the similar approximate coefficients given for continuous beams in clause 3.4.3 of BS8110, i.e. that variations in span length must not exceed 15% of the longest span. These coefficients already take account of the 20% reduction of support moments and the corresponding increase in span moments that may be adopted if a more thorough analysis from first principles is undertaken. Note that the BS8110 coefficients must be multiplied by the product of the total load on the panel and the effective span $(l_1 - h_c/1.5)$; this

is more onerous than the similar requirement in the preceding Code.

The requirements of BS8110 are much more specific regarding the transfer of moment from the slab to the supporting edge and corner columns. Unless an edge beam (or equivalent slab edge strip) is specifically designed to transfer the moment from slab to column by means of torsion, the actual moment that can be transferred directly into these columns will be much less than indicated by the Code coefficients. BS8110 states that the maximum moment that can be transferred in such a way is $0.15 b_e d^2 f_{cu}$, where d is the effective depth for the column-strip bars in the top of the slab and b_e is a breadth of slab determined by the column dimensions and the proximity of the slab edge, but in no case may exceed the width of the column strip. Where an edge column intersects the slab edge, b_e is the sum of the width of the column parallel to the slab edge plus the contact length between column and slab at right angles to the edge; for corner columns, b_e is the sum of the shorter contact length plus one-half of the longer contact length. If the moment resulting from the above expression is less than that determined from the Code coefficients, this lower value must be substituted and the moment near the middle of the end span must be increased accordingly.

Although not considered in the Code, it seems reasonable to assume that the need to consider end panels can be avoided if the peripheral columns are set in from the edges of the floor. Panels that would otherwise be end panels then become interior panels, if the part of the floor that cantilevers beyond the outer row of columns is sufficiently wide to produce negative bending moments that are similar to those that occur in an interior panel.

11.2 UPPER FLOORS AS FLAT SLABS

Calculation Sheets 16 and *Drawings 12* and *13* give the designs for two panels of an upper floor of the building in Part Two. The imposed load is 5 kN/m^2, as for the upper floors designed in Chapters 9 and 10. As explained in section 2.8.2, an additional allowance of 1 kN/m^2 is made for partitions in excess of the minimum of 1 kN/m^2. Of the two panels for which calculations are given, P1 is an internal panel and the bending moments are based on *Data Sheet 46*. Panel P2 is an end panel and would be designed for the appropriate bending moments on *Data Sheet 46*, but for the fact that the edge is stiffened by a fascia beam; since the depth of the beam exceeds $1\frac{1}{2}$ times the thickness of the slab, the fascia beam, as recommended in BS8110, is designed to support the wall and window and one-quarter of the total load on the panel, all this load being considered to be distributed uniformly along the beam. The bending moments on the half column strip adjacent to the beam are then taken as one-quarter of those on *Data Sheet 46*. The end walls are of brick and therefore, along these transverse edges, beams are provided to carry the walls. A substantial edge beam may, by virtue of its torsional resistance, offer sufficient edge restraint to make it necessary to design an end panel as an interior panel as regards negative bending moments, with a consequent reduction in the positive bending moment at right angles to the edge. The Code does not consider this case.

The thickness of the slab is sufficient to resist the greatest negative or positive bending moment on the middle strip or the greatest positive bending moment on the column strip, and the thickness of the drop is sufficient to resist the greatest negative bending moment on the column strip, in all cases without the need to use compression steel. The tension reinforcement is varied at all critical sections to provide sufficient resistance to the varying bending moments.

The calculations and details of corner panel P4 and end panels P3 are not given, but the calculations would be similar to those for end panel P2 because the depth of the beams along the non-continuous edges exceeds $1\frac{1}{2}$ times the thickness of the slab. The basic bending moments would be computed for an interior panel 5.510 m by 5.000 m with column strips 2.500 m wide in both directions. The middle strip is 3.010 m wide across the shorter span and 2.500 m wide across the longer span.

11.3 THICKNESS OF SLAB AND DROP

Although the slab thickness is influenced by the ultimate moments, the normal limiting criteria are deflection and shearing. In clause 3.7.1.6, BS8110 states that the slab thickness is generally controlled by deflection considerations, but this may only be true where drops are provided, where the size of the column head is large or where special steel to resist shear around each column head is employed. Unlike its predecessor, in the present Code the minimum permissible effective depth is related to the *longer* of the spans and the bending-moment factor M/bd^2 relating to that direction (according to the *Code Handbook* this should be determined from the values of M and b relating to the full panel width and not simply the middle strip). Details of the requirements of BS8110 concerning deflection (and also cracking) are given in Chapter 8. In no case may the thickness of a flat slab be less than 125 mm. Remember that if drops of at least one-third of the shorter span in either direction are not provided, the minimum effective depths determined from *Data Sheet 20* must be increased by $\frac{1}{9}$; if a slab with a non-uniform section (such as a waffle slab) is adopted, a further increase of $\frac{1}{4}$ must be included, making a total adjustment of almost 39%.

Unfortunately, too many simplifying assumptions have to be introduced to make it possible to produce simple practical basic rules for selecting an initial slab thickness to meet deflection and shearing requirements. Having assumed trial slab and drop thicknesses, it is advisable to calculate first the positive moment in the end span in the longer direction and to check that the span/effective-depth ratio that is proposed is satisfactory. Next rapidly estimate the amount of steel that will be provided in each

direction over the internal supports, and hence determine the maximum permissible shearing stress provided by the concrete alone. The points to remember here are that, as regards shear around the column head, the steel proportion that determines the shear resistance of the concrete is that relating to the more closely spaced bars in the central half-width of the column strip; however, if large drops are used a check may have to be made around the edge of the drop as to which bars extend at least distance equal to the effective depth beyond the section considered, since only those bars may be taken into account when determining the shear resistance of the concrete. Remember too that at all internal columns the effective shear stress is calculated by multiplying the actual shear force by a factor of 1.15, and greater factors are applicable for edge and corner columns. However, only the slab loads acting further from the column than the perimeter being considered need be taken into account. If the shearing stress due to the loading exceeds the resistance provided by the concrete, the slab thickness must be increased unless special shear links are provided.

11.4 ARRANGEMENT OF REINFORCEMENT

The predecessors to BS8110 gave specific details for the arrangement of reinforcement if the so-called empirical design method was employed. Although CP114 sanctioned a four-way arrangement of bars, this was seldom practicable and was omitted from CP110. BS8110 does not include similar explicit requirements, although the simplified detailing rules for slabs given in clause 3.12.10 of the Code are broadly similar to those prescribed for flat slabs in CP110, and these are illustrated by the arrangements adopted in *Drawings 12* and *13*. The current Code states that 40% of the reinforcement required in each column strip and middle strip to resist positive bending moments must extend beyond the centre-line of each support, while all reinforcement resisting positive moments must extend for a distance of at least 60% of the span on each side of the panel centre-line. All bars provided to resist the negative bending moment in the top of the slab must extend into adjacent panels for a distance of at least 15% of the span (or 45 times the bar diameter if this is the greater), and at least one-half of the reinforcement must extend to within $0.2l$ of the centre-line of the span. In addition, for flat slabs BS8110 states specifically that two-thirds of the reinforcement provided in the column strip to resist the negative moment over the supports should be concentrated within one-half of the width of the strip. If an average spacing has been determined for these bars, this requirement is equivalent to reducing this spacing to three-quarters of its initial value over the central half-width and increasing it to 1.5 times elsewhere. Where the edges of the panel are discontinuous, all bottom bars at right angles to the edge should extend to a distance of within $0.1l$ of the support, while top steel equivalent to at least $0.0013bh$ for slabs reinforced with high-yield steel or $0.0024bh$ where mild steel is employed, and extending a distance of at least 0.1*l* into the slab, must be provided.

When arranging the bars in a panel, it may at first seem advantageous to make the maximum use of the effective depth in the direction in which the bending moments are greatest (i.e. the longer span), arranging for these bars to pass below the transverse steel at midspan and above it at the supports. However, such an arrangement often presents difficulties, particularly if cranked bars are used, when it will be found well-nigh impossible to thread one set of bars through the other. In such circumstances it is therefore preferable to arrange that the steel having the greater effective depth at midspan is at right angles to that having the greater effective depth at the support. If the cranked bars are of the same diameter and have the same cover in each direction, and if drops are not provided, this arrangement has the further advantage that the overall depth of crank is the same in both directions. This arrangement is employed in the flat slab shown on *Drawings 12* and *13*.

Such considerations are less important if cranked bars are not used, as is now normally the case in UK practice. It is then possible to select spacings for top and bottom bars independently, and the amount of steel saved as a consequence balances the additional lengths required to provide sufficient anchorage, with the additional advantages of simplicity of steel fixing and a saving in bending costs. It is interesting to note that none of the examples of the detailing of slabs provided in ref. 2 illustrates the use of cranked bars.

Since the permissible span/effective-depth ratio is normally the critical factor in determining the minimum thickness of a flat slab, and this is related to the bending-moment factor in the longer direction, it is sensible to provide the greater effective depth to these bottom bars.

11.5 OPENINGS

Openings of limited size may be formed in flat slabs, but they must not encroach on the column head or drop. Openings must be trimmed on all sides with beams that transfer the loads to the columns unless, in the case of openings in an area common to two middle strips, the greatest dimension of the opening parallel to the centre-line of the panel does not exceed 40% of *l*. Openings in an area common to two column strips need not be framed if their aggregate length or width does not exceed 10% of the width of the strip in question, provided that the perimeter along which the shearing stress is evaluated is reduced accordingly. Similarly, openings without beams may be formed in an area common to one column strip and one middle strip, if the aggregate lengths or widths do not exceed 25% of the width of the column strip. In all cases the reduced width of the slab on each side of the openings must be sufficiently strong to resist the total ultimate negative and positive bending moments.

Since the opening for the stairs and lift in the floor on *Drawing 12* exceeds the permissible dimensions, the panel in which it occurs is supported on beams, as indicated on *Drawing 12*, and the slabs within the framework of the beams are designed as ordinary solid slabs 175 mm in thickness, to conform to the thickness of the drops in the adjacent flat-slab panels.

11.6 FLAT SLABS DESIGNED AS FRAMES

The recommendations of BS8110 relating to the design of flat slabs as a series of intersecting frames include the same recommendations as for the simplified method as regards dividing the panel into column strips and middle strips, minimum thickness, shearing stresses, openings, edge beams and panels adjacent thereto, and column heads. The structure is considered to be divided longitudinally and transversely into frames consisting of a row of columns and strips of slab. In an analysis for the effects of vertical loading only, the width of each slab strip should be taken as the distance between the panel centre-lines on each side of the column row, but when considering the effects of combined vertical and horizontal loads this width should be halved. A full elastic analysis of the frame may be undertaken; alternatively, for vertical loads only, each individual floor may be analysed as a separate frame, the far ends of the columns above and below being assumed fixed in position and direction. In addition, the Code permits the simplified three-bay sub-frame described in section 3.8.2 to be utilized if desired. To determine the stiffnesses, the moments of inertia of the members may be assumed to be those due to the gross cross-section of the concrete, ignoring the stiffening effects of reinforcement, drops and splayed column heads, although the *Code Handbook* recommends that the additional stiffness contributed by drops should be taken into account where they extend into the span by a distance of more than 0.15*l*. The stiffening effect resulting from the small area of solid slab adjoining the column in a floor that is otherwise formed by recessed or coffered construction may also be neglected provided that the solid area does not extend beyond the column centre-line by more than 15%.

Provided that, in the direction being considered, there are at least three rows of panels approximately equal in span, that the characteristic imposed load does not exceed the characteristic dead load by more than 25%, and that the latter (excluding partitions) is not greater than 5 kN/m^2, BS8110 states that only the single load condition of all spans loaded with the maximum ultimate load need be considered. The resulting moments may now be redistributed by reducing the negative support moments by 20% and increasing the span moments accordingly, provided that at no point in the resulting moment envelope is the final value less than 70% of the moment at that point before redistribution. In addition, BS8110 states that any moments which are greater than those that occur at a distance of $0.5h_c$ from the centre-line of the column need not be considered if the sum of the *maximum* positive and *average* negative design moments across the

Flat slabs

entire panel width in any individual span is at least $0.125nl_2(l_1 - h_c/1.5)^2$. According to the *Code Handbook*, if the supporting column is rectangular, h_c should not exceed 1.5 times the length of the lesser dimension.

These bending moments must now be divided between the column strips and the middle strips in the following proportions: negative moments, 75% on column strip and 25% on middle strip; positive moments, 55% on column strip and 45% on middle strip. These proportions assume that the widths of the column strip and the middle strip are equal. As with the simplified method, however, if the width of column strip is equal to that of the drop or the panel is not square, the moments resisted by the middle strip should be increased in linear proportion to its increased width. The moments in the column strip are then reduced to such an extent that the total positive and negative ultimate bending moments carried by the entire panel width remain unchanged. The requirements for panels supported by marginal beams or walls correspond to those already described for the simplified method. Supporting columns must be designed, using the method described in Chapter 14, to resist the most severe combination of bending moment and direct force.

The critical sections for shear around the columns and drops must be checked as described in section 11.1.2, for various arrangements of imposed load giving rise to differing combinations of V and M.

11.6.1 Buildings not subjected to sway

If a multistorey building is stiffened by bracing or shear-walls so that it cannot deform transversely or longitudinally due to the action of unsymmetrically arranged floor loads or wind (in other words, if the building is not subject to 'sway'), it should be sufficiently accurate to calculate the ultimate bending moments in flat slabs by the simplified sub-frame method described in section 3.8.2, using the appropriate formulae or the chart on *Data Sheet 12*. These formulae give the ultimate bending moments at the ends of the end and interior spans of continuous beams that are monolithic with the supporting columns. When applied to flat slabs, the formulae give the total maximum negative ultimate bending moments on end and interior panels. The fixed-end moments F_{LR}, F_{RL} etc. are, in the case of flat slabs, the total ultimate bending moments on a strip of floor of a width equal to the distance between the centres of the columns, assuming that the panel is fixed at each end of the span. The total negative maximum bending moments calculated from the appropriate formulae for M_{LR} and M_{RL} are then divided in the proportion of 75% on the column strip and 25% on the middle strip (assuming that these are equal in width; otherwise the adjustments described earlier must be made). The negative moment resisted by the column strip need only be that at a distance of $\frac{1}{2}h_c$ from the column centre-line. To calculate the total maximum positive ultimate bending moment, deduct from the total 'free' bending moment on the panel the total negative bending

moment at about midspan. The total maximum positive ultimate bending moment is then divided in the proportions of 55% on the column strip and 45% on the middle strip (again assuming equal widths). The sum of the total maximum positive bending moments and the average of the total negative bending moments on one span must be not less than the value determined, as described in the foregoing section.

The coefficients in *Data Sheet 12* can also be used to determine the ultimate bending moments M_{LG} and M_{LH} on the external columns, by dividing the maximum possible bending moment M_{LR} at L between the upper and lower columns in proportion to their stiffnesses K_{LG} and K_{LH} respectively and taking one-half of the true stiffness for LR; i.e.

$$M_{LG} = \frac{K_{LG}}{\frac{1}{2}K_{LR} + K_{LG} + K_{LH}} M_{LR}$$

and

$$M_{LH} = \frac{K_{LH}}{\frac{1}{2}K_{LR} + K_{LG} + K_{LH}} M_{LR}$$

In these formulae, M_{LR} is the sum of the ultimate bending moments M_{LR} at L when all panels are loaded with $1.4G_k + 1.6Q_k$. As with the simplified method described in section 11.1.3, the maximum moment that may be transferred from the slab to the edge columns must now be checked. If it is less than the previously calculated value, the lower limit must be adopted and the positive moment in the end span increased accordingly.

The bending moments M_{RO} and M_{RP} on the upper and lower internal columns can also be determined approximately by dividing M_{RL} between the upper and lower columns and the adjacent panel RL, in proportion to the stiffnesses of these members, taking only one-half of the true stiffness for RL and RS. Then

$$M_{RO} = \frac{K_{RO}}{\frac{1}{2}K_{RL} + \frac{1}{2}K_{RS} + K_{RO} + K_{RP}} M_{RL}$$

$$M_{RP} = \frac{K_{RP}}{\frac{1}{2}K_{RL} + \frac{1}{2}K_{RS} + K_{RO} + K_{RP}} M_{RL}$$

Alternatively, the ultimate moments on the columns may be calculated by using the expressions

$$M_{RO}\frac{K_{SO}}{\Sigma K_s - K_{RL}} M_{RL} \quad \text{and} \quad M_{RP} = \frac{K_{SP}}{\Sigma K_s - K_{RL}} M_{RL}$$

where M_{RL} is as stated above and $\Sigma K_s = \frac{1}{2}K_{RL} + \frac{1}{2}K_{RS} + K_{RO} + K_{RP}$. It is also possible to take account of the effect of the imposed load on the adjoining panels (i.e. JK and ST) by employing a more complex moment-distribution analysis. Yet another possibility is to calculate the ultimate bending moments on the columns and flat slabs in the same way as described below for a building subject to sway, but omitting the part of the analysis relating to the sway.

The foregoing calculations give the bending moments on the columns in the direction of the spans of the panels being considered (i.e. transversely). The corresponding moments at right angles to those already determined (i.e. longitudinally) must also be calculated, but care must be taken to ensure that panels that are wholly or partially loaded, or not loaded, to give the greatest transverse bending moments, remain in this condition of loading when calculating the longitudinal bending moments. The effects of transverse and longitudinal bending on the columns are then combined with the appropriate axial loads on the columns to determine the most adverse conditions.

The foregoing method is not applicable to the design shown in *Drawings 12* and *13* in Part Two because there is no transverse bracing, the transverse end walls being of brickwork. The effects of sway must therefore be taken into account.

11.6.2 Buildings subjected to sway

The effects of sway in buildings formed with flat-slab floors and without stiffening walls can be dealt with by the common methods of moment distribution. Alternatively, the Code states that the analysis for the effects of vertical loading may be undertaken by considering each floor separately (the columns concerned being assumed fixed at their far ends). This appears to indicate that it is normally not necessary to consider the effects of sway due to vertical loading only. Moment distribution may then be used to calculate the moments due to lateral (i.e. wind) loading. The use of general formulae is too cumbersome for practical design.

The application of the methods to the particular case of one floor of the building in Part Two is described in the following. The design of a flat-slab floor of this building by such a rigorous method should be compared with that obtained by the simplified method, as illustrated on *Drawings 12* and *13*. An exact comparison is not possible because the design shown in *Figure 11.1a* assumes that the slab is of uniform thickness throughout (i.e. is without drops), that no column heads are provided, and that the columns are circular. Only transverse frame action is considered; longitudinal frame action would be dealt with similarly, making due allowance for the different lengths and numbers of panels. The analysis may be undertaken conveniently by using the precise moment-distribution method described in Chapter 3, but instead, to illustrate the work involved, conventional moment distribution has been employed.

The first stage in the analysis, as for all moment-distribution calculations, is to calculate the stiffness factors K, for which purpose sizes must be assumed for the columns and thickness of the slab. To avoid unnecessarily cumbersome arithmetic, the units of K in this example are $mm^4/mm \times 10^{-5}$; the units do not matter so long as the same units are used for every member. Remember that for vertical loading the width of slab used to determine the stiffness should be that between the column centre-lines,

but when analysing the effects of wind this stiffness should be based on only one-half of this width.

The distribution factors are next calculated from the stiffness factors, and are as shown in *Figure 11.1b*. The following step is to assume an arbitrary load on each span, and to calculate by moment distribution (neglecting any sway for the time being) the bending moments on the slabs and columns, as shown in *Figures 11.1c* and *11.1f*. In this example the arbitrary load considered is a uniform load such that the fixed-end moments F_{AB}, F_{BA} etc. that result are each equal to 100 kN m (of appropriate sign). As already explained, it is only necessary to consider a frame comprising one floor and the corresponding upper and lower columns; the columns are assumed fixed at the ends remote from the floor being considered.

So far the analysis is applicable to frames whether subject to sway or not. If sway due to vertical loading is to be considered, the next design stage is to add correcting moments to counteract the unbalanced shearing forces at the further ends of the columns. Since the load on the central panel in *Figure 11.1f* is symmetrically disposed on a symmetrical frame, no sway will result and the bending moments originally obtained can here be used without correction. The load on each end panel in *Figure 11.1c* is unsymmetrically disposed on the frame and, although the frame is symmetrical, correction is therefore necessary. The unbalanced shearing force on the upper columns (that on the lower columns is identical) is

$$\frac{1}{3.6}[(33.1 + 16.5) + (-35.3 - 17.6)$$
$$+ (3.9 + 1.9) + (-0.9 - 0.4)] = 0.33 \text{ kN units}$$

Now proceed as shown in *Figure 11.1d*. The correcting moments applied to the frame must result in an unbalanced shearing force of -0.33 kN on both the upper and lower columns. Apply arbitrary bending moments of -100 kN/m on each column at the intersection of the slab and the column, and distribute these bending moments as shown in *Figure 11.1d*. The resulting unbalanced shearing force on the upper columns is

$$\frac{1}{3.6}[(-46.4 + 26.8) + (-59.0 + 20.5)]2$$
$$= -32.28 \text{ kN units}$$

compared with that of -0.33 kN units required. Therefore the correcting bending moments required are $0.33/32.28 = 0.0103$ times the bending moments given in *Figure 11.1d*. Adding these reduced correcting moments to those shown in *Figure 11.1c* gives the resultant moments shown in *Figure 11.1e*, which produce no unbalanced shearing forces because

$$\frac{1}{3.6}[(32.6 + 16.8) + (-35.9 - 17.4) + (3.3 + 2.1)$$
$$+ (-1.4 - 0.1)] = 0$$

The actual bending moments are now calculated. If the necessary conditions are satisfied, it may only be

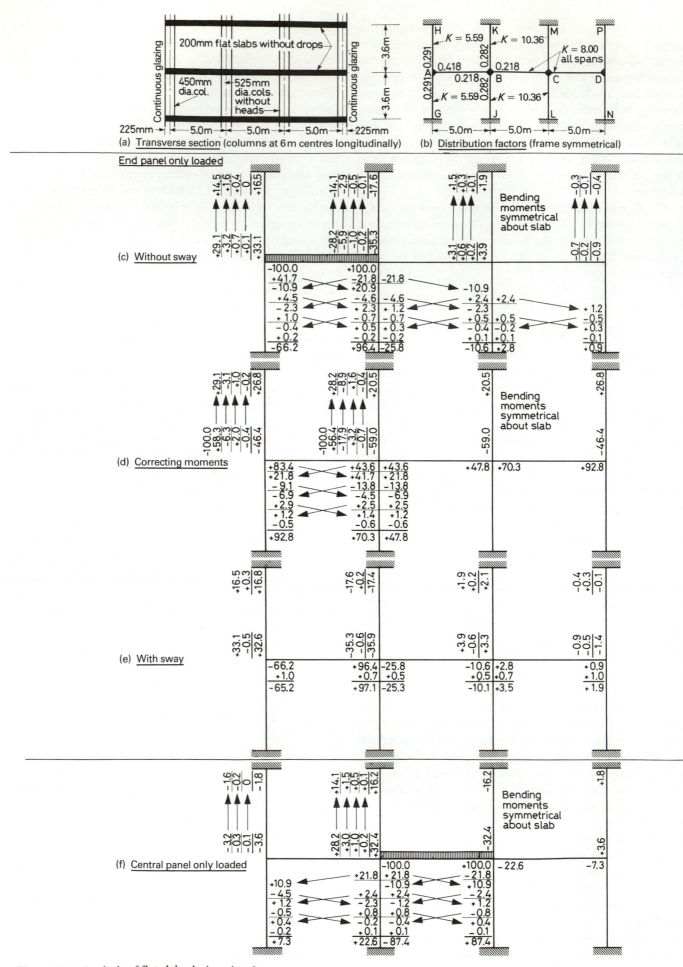

Figure 11.1 Analysis of flat slabs designed as frames.

Load	Fixed-end moment ratio	Loaded panel	Bending moment at B (kN m on 6 m width)		Notes
			Basic (100 units)	Bending moment	
Dead	0.94	AB BC CD	−25.3 −87.4 +10.1	— — —	Algebraic signs of bending moments adjusted to comply with flat-slab sign convention
	Bending moment due to total dead load		−102.6	−96.5	
'Imposed'	1.38	AB BC CD	−25.3 −87.4 +10.1	−34.8 −120.3 +13.9	
Total maximum negative ultimate bending moment over column centre-lines at B				−237.7	Imposed load on all panels
Total negative ultimate bending moment to give maximum positive bending moment at midspan of panel BC				−216.8	Imposed load on panel BC only

Figure 11.2 Flat slab analysed as a frame: transverse span of an interior panel BC (see *Figure 11.1*).

necessary to consider the single load case of all spans loaded with the maximum ultimate load. Otherwise (for example, for storage structures, according to the *Code Handbook*) the relevant loading patterns to obtain maximum positive and negative moments described in section 3.1 must be employed. An example of how this may be done in the case of the interior panel of the slab being considered is shown in *Figure 11.2*; a similar tabulation must be made for the end panels and the columns. The bending moments for the actual ultimate total load on each panel are determined by multiplying the bending moments in *Figures 11.1e* and *11.1f* by the ratio of the actual fixed-end moments to 100 kN/m.

Considering panel BC, the weight of a 200 mm slab (assumed) is 4.8 kN/m^2 and therefore the total dead load is 4.80 + 0.72 (finishes) + 2.00 (partitions) = 7.52 kN/m^2 (compare with *Calculation Sheets 16*). The fixed-end moment for the dead load on a panel 6 m wide is 6.0 × 7.52 × 5.0^2/12 = 94 kN/m, and the ratio is therefore 94/100 = 0.94; similarly for the 'imposed' load of (5.0 × 1.6) + (7.52 × 0.4) = 11.01 kN/m^2, and the ratio is 1.38.

The problem of algebraic signs must now be considered. In the analysis shown in *Figure 11.1* the conventional signs are those adopted in ordinary moment-distribution operations, and they can be interpreted as follows. A bending moment producing tension in the top of a beam at the right-hand side of a support is negative, and a bending moment producing tension in the top at the left-hand side of a support is positive. However, the convention for flat slabs, as for ordinary continuous slabs or beams, is that a bending moment producing tension in the top of the slab at any section is a negative bending moment, and one producing tension in the bottom at any section is positive. The necessary adjustment in signs in the bending moment over the column centre-lines is made when transferring the basic bending moments from *Figure 11.1* to *Figure 11.2* to suit the convention of signs for flat slabs.

The maximum ultimate negative bending moment at any section of the panel or on a column is, according to BS8110, the algebraic summation of the ultimate bending moments due to all panels being loaded with $1.4G_k + 1.6Q_k$. The maximum negative bending moments over the column centre-lines at B and C if calculated in this way are each − 237.7 kN m (see *Figure 11.2*). The negative ultimate bending moments used in the design calculations may be less than this value if advantage is taken of the recommendation in BS8110 that only the bending moment at a distance of $0.5h_c$ from the column centre-line need be provided for; a nominal reduction to 235 kN m is made in this example. The maximum positive moment is calculated in a similar way. As shown in *Figure 11.2*, the negative bending moments over the column centre-lines at B and C to produce the maximum positive moment occur when span BC is loaded with $1.4G_k + 1.6Q_k$ while the adjoining spans support loads of only $1.0G_k$, leading to support moments of −216.8 kN m. If the panel were freely supported at B and C the positive moment at midspan would be $0.125 \times [(1.4 \times 7.52) + (1.6 \times 5.0)] \times 6.0 \times 5.0^2 = 347.4$ kN m. Thus the net maximum positive ultimate bending moment is 347.4 − 216.8 = 130.6 kN m. The sum of the maximum negative and positive ultimate bending moments for which the slab is to be designed is therefore 235.0 + 130.6 = 365.6 kN m. Since this is not less than $0.125 \times [(1.4 \times 7.52) + (1.6 \times 5.0)] \times 6.0 \times [5.0 - (0.525/1.5)]^2 = 300.5$ kN m, the requirements of BS8110 concerning the limitation of negative design moments are met.

The maximum bending moments on the slab are now divided between the column strip and the middle strip of each panel in the proportions previously described. Thus, since drop panels are not provided and the width of each column strip and middle strip is therefore 6.0/2 = 3 m:

Bending moment on column strip:

negative	0.75 × 235.0/3 = 58.8 kN m/m
positive	0.55 × 130.6/3 = 23.9 kN m/m

Bending moment on middle strip:

negative $0.25 \times 235.0/3 = 19.6$ kN m/m

positive $0.45 \times 130.6/3 = 19.6$ kN m/m

A similar tabulation should be made to determine the transverse bending moments on the end panels and on the columns. The longitudinal bending moments should then be calculated by a similar analysis. The design of the slab then proceeds as for a building without sway or as a flat slab, and the columns are designed by the simplified method, except that it may be necessary to take into account the bending moments due to wind, as described in the case of a framed building in section 14.10.

Chapter 12
Other designs
of floor

In this chapter several types of floors are considered and are applied to the design of the ground floor of the building on *Drawings 1* and *2*. Some of the designs are equally suitable for upper floors if modifications are made.

12.1 GROUND FLOOR: GARAGE

12.1.1 Slabs

It is shown in section 2.8.3 that the floor slab of the garage should be designed for imposed loads in excess of the minimum of $5\,kN/m^2$ given in BS6399, Part 1, if it may be subjected to effective wheel loads of 12.5 kN each. The equivalent uniform loads are $22.0\,kN/m^2$ for the bending moment at the support and $27.4\,kN/m^2$ for the bending moment at midspan. In this calculation it is assumed that only one wheel can occur on a span at any one time, but the possible arrangement of vehicles on the ground floor as a whole is such that the sequence of loading necessary to produce the maximum bending moment (i.e. *Figure 12.1*) can be realized. The self-weight of the slab and finishes is about $4.2\,kN/m^2$ and therefore the maximum ultimate bending moment, which occurs at the supports, is about

$$[(0.116 \times 22.0 \times 1.6) + (0.116 \times 4.2 \times 0.4)$$
$$+ (0.106 \times 4.2 \times 1.0)]2.5^2 = 29.5\,kN\,m$$

The bending-moment coefficients that are used here are taken from *Data Sheet 9*. With grade 30 concrete and mild steel, the minimum effective depth required is 80 mm. A 150 mm slab will suffice and also provide sufficient fire resistance. A check should be made of the shearing stress around the periphery of the concentrated load. With an effective depth of 115 mm, the resulting shearing stress along a perimeter $1\frac{1}{2}d$ from the edge of the load is given by the expression $N/(u + 12d)d$, i.e. $12\,500 \times 1.6/[(400 \times 4) + (115 \times 12)]115 = 0.06\,N/mm^2$, which is insignificant.

12.1.2 Secondary beams

The load on the beams supporting the floor should be investigated to determine the most adverse combination of loads. For a wheel in the middle of a fully continuous secondary beam spanning 6 m, it is shown in section 2.8.3 that the equivalent uniform loads are $4.94\,kN/m$ and $3.71\,kN/m$, compared with $12.5\,kN/m$ from the imposed

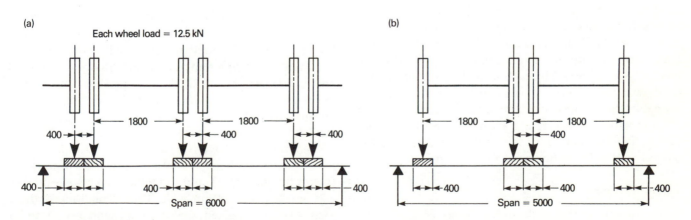

Figure 12.1 Garage floor: critical loadings on (a) secondary beam, interior span (b) main beam, interior span.

load in the case of a 2.5 m panel of slab carrying the minimum load of 5 kN/m². Any beam may, however, support several wheels simultaneously, and *Figure 12.1a* represents the probable worst condition. The total load resulting from this arrangement of wheels is $6 \times 12.5 = 75$ kN. Introducing a factor of, say, 1.5 to allow for the non-uniform distribution of the loads shown in *Figure 12.1a*, the total equivalent uniform imposed load is 113 kN, or 18.75 kN per metre, which exceeds the minimum uniform imposed load of 12.5 kN/m.

On a span of 6 m the total dead load is about 80 kN. The maximum bending moments can be calculated by using the coefficients in *Data Sheet 9*. The maximum positive ultimate bending moment in the end spans is then about

$$[(80.0 \times 1.0 \times 0.078) + (80 \times 0.4 \times 0.100)$$
$$+ (113.0 \times 1.6 \times 0.100)]6 = 165.1 \text{ kN m}$$

and the maximum negative ultimate bending moment at a penultimate support is

$$[(80.0 \times 1.4 \times 0.106) + (113.0 \times 1.6 \times 0.106)]6.0$$
$$= 186.2 \text{ kN m}$$

If the beam is 300 mm wide and 400 mm deep below the 150 mm slab, calculations similar to those on *Calculation Sheets 5* for the beams of the upper floors show that four 25 mm bars are required in the bottom of the end spans and five 25 mm bars are necessary at the penultimate supports.

The load in *Figure 12.1a* also produces the maximum shearing forces; this can be checked by assuming other wheel arrangements. The maximum ultimate shearing force is $\frac{1}{2}(80.0 + 113.0) = 96.5$ kN. The shearing stress is about 0.65 N/mm². With the reinforcement provided (i.e. $\varrho = 0.0161$) the concrete alone will withstand a stress of about 0.78 N/mm² and therefore only nominal shearing reinforcement is required.

12.1.3 Main beams

The imposed load on the main beams may be (a) when the entire floor is covered by the heaviest vehicles likely to use the garage, say 4 tonnes each on about 10 m², which is less than the minimum load of 5 kN/m² or a central concentrated load of 75 kN applied by the secondary beams; or (b) when the wheels are concentrated as adversely as possible on the beam, as shown in *Figure 12.1b*, the total load being about $4 \times 12.5 = 50$ kN. This load is not concentrated at one point on the beam as is the load of 38 kN imposed by each secondary beam. Loading (a) determines the greatest ultimate bending moment and probably the greatest ultimate shearing force. A beam 300 mm wide and projecting 450 mm below the 150 mm slab will be suitable and does not exceed the limiting overall depth of 600 mm shown on *Drawing 1*.

12.1.4 Columns

It is shown above that, if the entire floor is covered by the heaviest vehicles, the corresponding uniform load is

about 4 kN/m². Therefore the columns supporting the garage portion of the ground floor should be designed to carry the minimum specified load for this class of floor, namely 5 kN/m². This is therefore the load taken into account in the design of the interior columns between the basement and the ground floors, as given on *Calculation Sheets 17*.

12.2 GROUND FLOOR: RETAIL SHOPS

As explained in section 2.8.3, although the ground floor is to be occupied partly by a garage and partly by retail shops, there is some advantage in designing the entire floor for the imposed load of 5 kN/m². For the purpose of an alternative example, however, the part of the ground floor that is intended for retail shops shown on *Drawing 2* is now designed for an imposed load of 4 kN/m². If a minimum additional dead load of 1 kN/m² for partitions is included and 3.75 kN/m² is allowed for the weight of the slab and finishes, the total dead load is 4.75 kN/m². In the following, three designs are given for the floor, namely: a solid slab spanning 2.5 m; a ribbed hollow-block slab spanning 5 m; and a floor composed of precast slabs with a cast *in situ* concrete topping. For all these types of floor the minimum thicknesses necessary to meet fire-resistance regulations form an important consideration, and reference should be made to the requirements set out in Chapter 19.

12.2 Solid slab

A total load of 8.75 kN/m² and a span of 2.5 m results in the negative ultimate bending moment on the slab being, from the coefficients in section 3.2, $[(4.75 \times 1.4) + (4.0 \times 1.6)] \times 2.5^2 \times 0.086 = 7.01$ kN m per metre width, if all the spans of the slab are considered as fully continuous. This assumption is reasonable because the outer support of the end span is the basement retaining wall, with which the slab is formed monolithically. As is shown in section 15.7, the basement retaining wall acts as a propped cantilever, the ground floor providing the prop, and therefore the deformation at the top of the wall is such that the corresponding deformation produced at the outer support of the end span of the ground-floor slab is equivalent to imposing a negative bending moment on the slab, which is the condition required for continuity. For the bending moment calculated above, if $f_{cu} = 30$ N/mm², a 100 mm slab reinforced with 8 mm mild steel bars at 100 mm centres will suffice, and for convenience this reinforcement is provided in the bottom of the slab at midspan as well as over the supports. The distribution steel should consist of not less than 8 mm mild steel bars at 200 mm centres, as given on *Data Sheet 41*. A suitable size for the secondary beams is 400 mm deep and 300 mm wide (below the slab soffit), and a suitable main beam would be 500 mm deep and 300 mm wide. The calculations for these members are not given, since they are similar to those for the beams forming the upper floors. The arrangement of

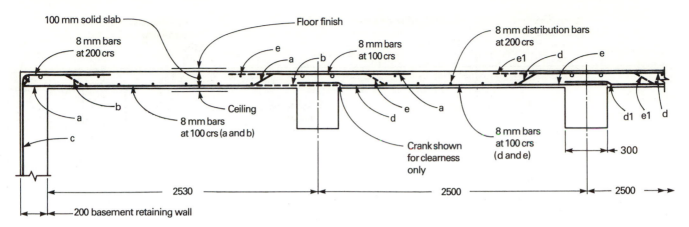

Figure 12.2 Solid slab.

reinforcement in the slab is shown in *Figure 12.2*, in which bars (a), (b), (d) and (e) are each placed at 200 mm centres.

12.2.2 Ribbed slab with hollow clay blocks

The appropriate section in BS8110 (clauses 3.6) covers three main types of construction: namely, slabs formed from permanent blocks with intervening cast *in situ* concrete ribs and with or without topping; slabs formed by the use of forms that are removed after the concrete has hardened; and slabs that have a flat soffit but incorporate voids of various shapes. In this section, the design of the first type of slab is dealt with in some detail, and notes are also given where the requirements differ for the other types.

For the initial calculation of the ultimate bending moment on a ribbed slab forming the retail shop part of the ground floor, it may be assumed that the total dead load is the same as the load for the solid slab, i.e. 4.75 kN/m². Although the thickness of a ribbed slab spanning 5 m (*Figure 12.3*) is greater than 100 mm, which is the thickness of the corresponding solid slab spanning 2.5 m, the voids formed by the hollow clay blocks so lighten the slab that there is normally very little difference in weight. The economy is in the reduction of the number of beams required. In this example only half as many secondary beams are required. In addition, the main beams as such are omitted and are replaced by much smaller members which act as ties between the columns.

Assuming complete continuity of the slab over the longitudinal supporting beams and with the front retaining wall, the ultimate bending moment at the midspan and at the supports of the slab is [(4.75 × 1.4) + (4.0 × 1.6)] × 5² × 0.063 = 20.55 kN m per metre width. If it is difficult to provide sufficient resistance to the negative ultimate bending moment, as is often the case when a ribbed slab is used, BS8110 requires the slab to be designed as freely supported, and for reinforcement, equal in area to one-quarter of the area of steel at midspan, to be provided in the top of the slab over the

supports and to extend from the support a distance of one-tenth of the clear span. This requirement may at first be thought a little severe because, if such resistance to negative bending moment is provided, it seems reasonable to reduce the midspan bending moment accordingly; suitable bending-moment coefficients might then be 1/10 at midspan and 1/32 at the supports. However, although safe, this method of design is anyway somewhat unsatisfactory as regards serviceability, since the redistribution of moment assumed at the supports means that it is likely that the concrete will crack, even under dead loads. In this example, the width of the compression flange required for the longitudinal beams is investigated, to show that it provides sufficient compressive resistance for the negative bending moment in the hollow-block slab when calculated on the basis of full continuity.

If $f_{cu} = 30$ N/mm², the effective depth of the slab at midspan should not be less than $\sqrt{[(20.55 \times 10^6)/(4.69 \times 10^3)]} = 66$ mm. A somewhat greater effective depth than this must be provided because the relatively thin compression flange of the ribbed slab generally gives a compression area that is less than that of the solid rectangular section assumed. At the supports, where the slab section is solid, the effective depth should again be not less than $\sqrt{[(20.55 \times 10^6)/(4.69 \times 10^3)]} = 66$ mm.

According to CP110, if slip-tiles 10 mm thick were provided beneath the ribs, the minimum thickness beneath the top of the tile and the underside of the bar could be reduced to 10 mm (presumably irrespective of the bar size). However, this reduction is not sanctioned by BS8110, and accordingly the full cover to meet durability requirements must be provided. There is therefore now no advantage from this point of view in providing such slip-tiles, and in fact their use is now positively discouraged as they prevent the quality of rib casting from being checked visually.

If a slab having a total thickness of 200 mm (*Figure 12.3*) is assumed and 20 mm bars are employed, the effective depth at both midspan and supports will thus be 200 − 20 − 10 = 170 mm. At both positions this depth exceeds the minimum required to resist bending. The

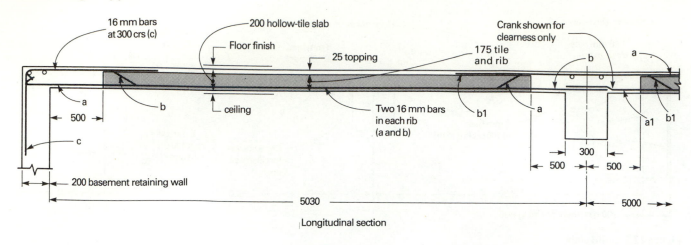

Longitudinal section

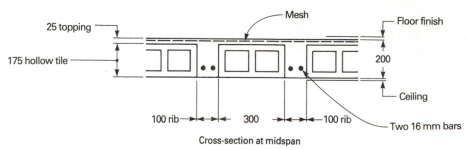

Cross-section at midspan

Figure 12.3 Hollow clay-block floor.

minimum depth necessary to meet serviceability (i.e. deflection) requirements will be checked when the reinforcement required in the ribs has been ascertained.

The minimum effective thickness of concrete required over the blocks between the ribs (i.e. the topping) is 25 mm, but this is only true (a) if the crushing strength of the hollow burnt clay (or concrete) blocks is not less than 14 N/mm² (in the direction of the line of action of the compressive stress in the slab) and the top part of the block forms part of the compression area of the section; (b) if the clear distance between the ribs does not exceed 500 mm; and (c) if the slabs are jointed with cement/sand mortar not weaker than 1 : 3 and having a strength of at least 11 N/mm². These conditions are assumed to apply to the following example. If condition (c) does not apply, the effective thickness of the topping should be not less than 30 mm. If the blocks are insufficiently strong to contribute to the strength of floor, condition (a) will not apply and the topping thickness must not be less than 40 mm or one-tenth of the clear distance between the ribs. Where permanent blocks are not provided, the topping thickness should be not less than 50 mm or one-tenth of the clear distance between the ribs. The effective thickness of the topping is the actual thickness less an allowance for wear. In the present example, a boarded or some other finish is assumed to be laid on the topping (and has been allowed for in the dead load), so that the actual thickness of 25 mm can be adopted. If the contribution to the compressive resistance provided by the hollow blocks is neglected, the lever-arm, i.e. the distance between the centroid of the flange and the bars in tension, is about $170 - 12.5 =$

157.5 mm. If the ribs are provided at 400 mm centres, the area of tension reinforcement required at midspan and over the supports at each rib is $20.55 \times 10^6 \times 0.4/(250 \times 0.87 \times 157.5) = 240$ mm². This area is adequately provided by two 16 mm bars per rib (i.e. 402 mm²). The minimum rib width is determined by cover and bar spacing considerations, but must also be not less than one-quarter of the depth below the topping. To accommodate two 16 mm bars, the width required must be at least $(3 \times 20) + (2 \times 16) = 92$ mm; a suitable rib width would therefore be 100 mm. One-half of the reinforcement (i.e. one bar) must extend in the bottom of the rib as far as the supports. The other bar may be bent up into the top of the solid slab near the supports to provide, with the corresponding bar from the adjacent span, the tensile resistance over the supports. Details of the resulting slab are shown in *Figure 12.3*. The weight of a 200 mm ribbed slab (including the weight of the hollow blocks) will be about 2.75 kN/m², which is adequately covered by the assumed dead load of 3.75 kN/m².

A check must now be made to ensure that serviceability requirements are met. If $f_y = 250$ N/mm² and the ratio $M/bd^2 = 20.55 \times 10^6/(400 \times 170^2) = 1.78$, from *Data Sheet 20* the modifying coefficient for the tension reinforcement is 0.65. Thus, since the basic effective depth corresponding to a continuous span of 5 m is 192 mm, the modified minimum effective depth is $192 \times 0.65 = 124$ mm. This in turn must be multiplied by a further factor, corresponding to a flanged section with a ratio b_w/b of $100/400 = 0.25$. Since this is less than 0.3, the factor is 1.25, giving a final minimum effective depth of 155 mm. Thus the effective depth provided is adequate. For simplicity, in the foregoing calculations it has been assumed that the service stress f_s (on which the modifying factor for tension steel is actually based) is $0.625f_y$. However, to estimate the actual value of f_s, BS8110 permits the value of $0.625f_y$ to be multiplied by the ratio of reinforcement required divided by that provided. If this is done in the present example, the minimum permissible depth is reduced to 138 mm.

The compressive stress in the flange should now be checked. The average compressive stress does not exceed, and may well be much less than, the stress obtained by dividing the ultimate moment by the lever-arm multiplied by the compression area of the flange, i.e. $20.55 \times 10^6/(157.5 \times 1000 \times 25) = 5.22\,\text{N/mm}^2$. The greatest ultimate compressive stress in the top face of the clay block is therefore not more than $5.22\,\text{N/mm}^2$. If the top face of the clay block is to contribute to the compressive resistance of the section, the minimum crushing strength of the block must be at least four times this value, i.e. $21\,\text{N/mm}^2$. According to BS8110, if it can be shown that the strength of not more than one block in twenty falls below the specified minimum strength, the maximum ultimate strength adopted for the blocks may be increased to 30% of the specified strength.

It is now necessary to consider the extent of solid slab needed for the flanges of the supporting beams. The total load on one beam of 6 m span is $8.75 \times 5 \times 6 = 263\,\text{kN}$, say 280 kN including the self-weight of the beam itself. The ultimate bending moment at midspan does not exceed $[(4 \times 1.6 \times 5 \times 6^2) + (4.75 \times 1.4 \times 5 \times 6^2) + (17 \times 1.4 \times 6)] \times 0.07 = 175\,\text{kN m}$. If the total depth of the beam is 575 mm, the approximate lever-arm from the centre of the solid 200 mm slab to the centre of the reinforcement is 420 mm. The width of flange required is therefore about $175 \times 10^6/(420 \times 200 \times 0.4 \times 30) = 174\,\text{mm}$, assuming the formation of a uniform stress-block with an ultimate resistance of $0.4f_{cu}$ in the flange, i.e. less than the rib width for the secondary beams. The point of contraflexure, when the maximum ultimate negative bending moment occurs, is assumed to be at one-fifth of the span, i.e. 1 m from the support. Since the maximum compressive strength of each 100 mm wide rib (neglecting the reinforcing bar in compression) is $4.69 \times 100 \times 170^2 = 13.5\,\text{kN m}$ per rib, or 33.8 kN m per metre width, this is more than adequate to resist the negative moment at the centre-line of the supporting beam by itself. Therefore the solid area of slab should be extended to a nominal distance of say 0.5 m from the beam centre-line and from the centre-line of the retaining wall. Alternatively, the negative bending moment may be ignored and the slab designed as freely supported, with a consequent increase in the thickness and possibly also in the amount of reinforcement required; in the present example an excess of reinforcement is already provided at the midspan of the rib.

It is reasonable to require the topping to be sufficiently strong to span between the ribs if it is subjected to concentrated loads, such as those due to partitions parallel to the ribs. Attention should also be paid to the stresses induced in the topping due to shrinkage and changes in temperature, and possibly to the transverse stress caused by the longitudinal stresses. It is not possible to calculate any of these stresses with even approximate accuracy, and each case must therefore be considered on its merits. The sturdy slab in the present example, comprising 100 mm ribs at 400 mm centres and 25 mm topping (with at least a 15 mm thickness of block below) spanning 300 mm only, should be quite capable of withstanding any concentration of light partitions that do not impose an equivalent distributed load in excess of $1\,\text{kN/m}^2$. Where a partition of clinker blocks or brickwork occurs parallel to the ribs, it is preferable to omit a line of blocks and to make the slab solid immediately below the partition. If the partition runs at right angles to the ribs, there is no need to make any special provision so long as the slab as a whole is capable of carrying the concentrated (knife-edge) load of the partition. When considering such concentrated loads, the *Code Handbook* states that the width of slab assumed to support the load should be the loaded area itself plus the lesser of *either* twice the rib spacing *or* $\frac{1}{4}l$ on *each* side of the loaded area. To cater for the miscellaneous stresses in the topping, BS8110 suggests (but does not insist upon) the inclusion of a single layer of mesh reinforcement having an area of at least 0.12% of the topping in each direction, with wires spaced at not more than one-half of the distance between rib centres. Thus in the present example the cross-sectional area of mesh required, if one were to be provided, should be not less than $0.0012 \times 25 \times 1000 = 30\,\text{mm}^2$ per metre.

All that remains now is to consider the shearing resistance of the slab. The maximum shearing stress occurs at the edge of the solid area of the slab where, in the present example, the shearing force is $0.5(5-1) \times [(4.75 \times 1.4) + (4.0 \times 1.6)] = 26.1\,\text{kN}$ per metre width or 10.44 kN per rib. With one 16 mm bar in a 100 mm wide rib, $A_s/b_w d = 201/(100 \times 170) = 0.012$. From *Data Sheet 15* with $d = 170\,\text{mm}$ and $f_{cu} = 30\,\text{N/mm}^2$, the corresponding value of v_c is about $0.88\,\text{N/mm}^2$. Since BS8110 permits the width of the rib to be increased by the thickness of one wall of the hollow block (*Allen* points out that this presumably only applies if structural-type blocks are used), the shearing resistance of the slab, without the need for special shearing reinforcement, is $0.88 \times 115 \times 170 = 17.2\,\text{kN}$ per rib. Thus special shearing reinforcement is not required.

If the topping of a hollow-block floor does not contribute structurally (e.g. if the crushing strength is less than $14\,\text{N/mm}^2$) of if none is employed, the blocks used must be of the structural type (i.e. comply with BS3921 if not of concrete). A minimum thickness of material of at least 20 mm or 10% of the horizontal width of the voids (whichever is greater) must be provided over the voids, and the distance between ribs must not exceed five times the overall thickness of the slab. When calculating the effective rib width of such sections, a rectangular section,

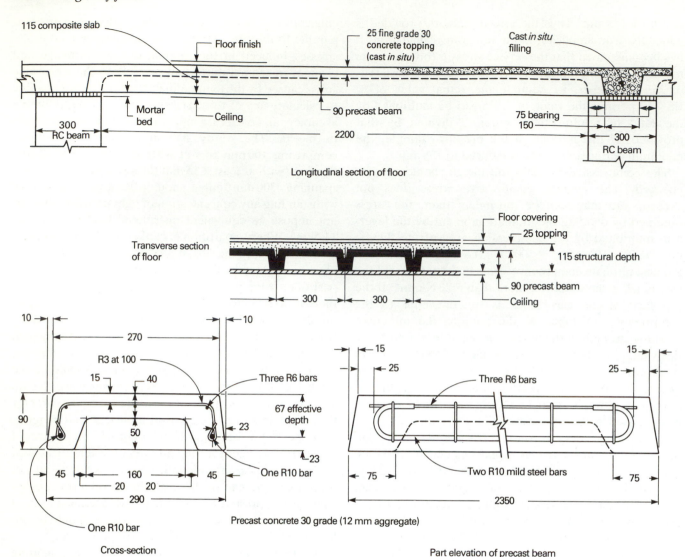

115 composite slab

Floor finish

25 fine grade 30
concrete topping
(cast *in situ*)

Cast *in situ*
filling

Mortar
bed

Ceiling

90 precast beam

75 bearing
150

300
RC beam

2200

300
RC beam

Longitudinal section of floor

Floor covering

25 topping

115 structural depth

Transverse section
of floor

300

300

90 precast beam

Ceiling

10

270

10

R3 at 100

15

40

Three R6 bars

67 effective
depth

90

50

23

45

160

45

20

20

290

One R10 bar

One R10 bar

23

15

25

Three R6 bars

15

25

75

Two R10 mild steel bars

75

2350

Precast concrete 30 grade (12 mm aggregate)

Cross-section

Part elevation of precast beam

Figure 12.4 Precast slab construction.

equal in area and depth to all the material below the top flange of the unit, should be assumed.

12.2.3 Precast floor slabs

The design of the precast concrete floor slabs given in *Figure 12.4* for the ground floor (retail shops) complies with the requirements of BS8110.

The slabs are supported on cast *in situ* secondary beams 300 mm wide at 2.5 m centres. A 75 mm bearing is provided at each end, and the slabs are laid side by side at 300 mm centres with a small space between, which is filled with fine concrete. The cross-section of each unit is in the form of an inverted channel, which results in a light slab that is simple to cast. The depth of the unit is assumed to be 90 mm and the thickness of the top flange is 40 mm; these dimensions are checked in the subsequent calculations. A ceiling is provided below the units, and a topping of fine concrete not less than 25 mm in thickness is laid over the units after they are placed. If the faces of the units in contact with the filling and topping are sufficiently

roughened to effect a bond between the precast units and the cast *in situ* concrete, the topping may be assumed to contribute to the strength of the floor.

Each unit is designed for the following conditions: (a) the temporary condition of the unit acting alone and supporting itself, the newly placed filling and topping, and the loads due to construction; (b) the permanent condition in which the unit, filling and topping act compositely to support the weight of unit, topping, floor covering and ceiling, partitions and imposed load; and (c) the stresses that the unit is subjected to while being handled, transported and erected.

It is assumed that the precast concrete is of grade 30 to minimize the thickness of cover that must be provided. The reinforcement per unit is one mild steel bar near the bottom of each rib and three bars in the top flange. The bars, for which $f_y = 250 \, \text{N/mm}^2$, are kept in position while the unit is cast, by 3 mm links at 100 mm centres. The required minimum cover of 15 mm to all steel is obtained if the dimensions of the ribs of the slab are as shown in *Figure 12.4*. The inclusion of wooden plugs or strips in the bottom of the ribs, to enable the ceiling to be attached, would make it necessary for the bottom bar to have, say, 30 mm cover above the soffit of the rib, thus considerably reducing the effective depth. Thus it is preferable to provide an alternative means of fixing the ceiling, as is assumed in the present case. The weight of the reinforced concrete is taken as 23.6 kN/m^3, and thus the total weight of each precast unit is 1.13 kN. The effective span is the distance between the centres of the bearings, i.e. 2.275 m.

For the temporary condition (a), the load (calculated on the effective span) to be carried by one unit is the weight of the unit (1.13 kN), the 25 mm topping 300 mm wide (0.43 kN) and the filling between the units (0.11 kN), i.e. a total dead load of 1.67 kN. In addition there will be a constructional load of, say, 1.50 kN/m^2 or 1.13 kN per unit. The ultimate bending moment at midspan is thus $\frac{1}{8} \times [(1.67 \times 1.4) + (1.13 \times 1.6)] \times 2.275 = 1.18$ kN m. Assuming 10 mm bars are used, the effective depth is 67 mm. The ultimate moment of resistance of the concrete section is thus $0.156 f_{cu} b d^2 = 0.156 \times 30 \times 270 \times 67^2 = 5.67$ kN m, which is well in excess of that required. The moment of resistance could be reduced by using a lower-grade concrete (though this would involve increasing the cover required to all reinforcement and thus considerably increase the self-weight of the unit), or by decreasing the depth. However, the present depth gives a final span/effective-depth ratio of 34 which is, strictly speaking, slightly excessive for a freely supported flanged beam of the dimensions shown; it seems acceptable, however, for temporary conditions during construction. Thus no further reduction of depth is practicable. Now

$$\frac{M}{bd^2} = \frac{1.18 \times 10^6}{270 \times 67^2} = 0.97 \, \text{N/mm}^2$$

Hence $\varrho f_y = 1.19$, and so

$$A_{s \, req} = \frac{1.19 \times 270 \times 67}{250} = 86 \, \text{mm}^2$$

Two 10 mm bars will be more than sufficient.

The greatest ultimate shearing force on each rib under the temporary condition is $\frac{1}{2} \times \frac{1}{2} \times [(1.67 \times 1.4) + (1.13 \times 1.6)] = 1.04$ kN. Since $A_s/bd = 78/(55 \times 67) = 0.021$, the allowable shearing stress without the need for special reinforcement is 1.35 N/mm^2 for grade 30 concrete, and thus the strength of the section in shearing is $1.35 \times 55 \times 67 = 4.96$ kN, which is well in excess of that required.

For the permanent condition (b), the additional dead load to be carried by the slab is the weight of the partitions (1 kN/m^2) and the floor finish and the ceiling (say 0.72 kN/m^2), and the imposed load is 4 kN/m^2. The plan area of each unit is 0.75 m^2.

The ultimate bending moment is now $\frac{1}{8} \times [(2.96 \times 1.4) + (3.0 \times 1.6)] \times 2.275 = 2.55$ kN m. Since the 25 mm cast *in situ* topping is now considered to contribute to the resistance of the section, the effective depth is increased

to 92 mm. The lever-arm is given by expression (5.16), i.e.

$$d - \frac{0.45 \times 0.87 f_y A_s}{0.402 f_{cu} b}$$

$$= 92 - \frac{0.45 \times 0.87 \times 250 \times 157}{0.402 \times 30 \times 300} = 87.3 \, \text{mm}$$

and thus the depth to the neutral axis is only about 10 mm, i.e. the neutral axis falls well within the depth of the topping. The composite precast slab and topping may thus be assumed to act as a rectangular slab 300 mm wide and with an effective depth of 92 mm. With two 10 mm bars, the ultimate moment of resistance of the section is now 2.98 kN m, which is still in excess of that required.

Checking serviceability requirements, the ratio of M/bd^2 is $2.55 \times 10^6/(300 \times 92^2) = 1.0$. From *Data Sheet 20*, the modifying coefficient for this proportion of steel is 0.51 and the multiplier for a flanged section, with $b_w/b = 100/300 = 0.33$, is 1.24. Thus, since the basic effective depth for a 2.5 m freely supported span is 125 mm, the final minimum effective depth that may be adopted is $125 \times 0.51 \times 1.24 = 79$ mm, so that the actual depth provided is satisfactory.

The total ultimate shearing force per rib is now $\frac{1}{2} \times \frac{1}{2} \times [(2.96 \times 1.4) + (3.0 \times 1.6)] = 2.24$ kN. The corresponding value of A_s/bd is now $78/(55 \times 92) = 0.015$, and thus the appropriate value of v_c is 1.12 N/mm^2. Consequently, the section will withstand $1.12 \times 55 \times 92 = 5.67$ kN without special shearing reinforcement being needed, i.e. more than double the actual shearing force applied.

It is now necessary to consider the horizontal shearing stress v_h at the interface between the precast and the cast *in situ* concrete. The requirements in section 5.4 of BS8110, Part 1, in this respect have been completely rewritten and considerably simplified in comparison with the somewhat theoretical serviceability treatment presented in CP110. All that is needed now is to check that the value of horizontal shearing stress due to ultimate load does not exceed the limiting value given in Table 5.5 of the Code, which is based on the concrete grade, the finish of the precast surface and whether or not links or connectors are provided that extend into the topping. According to BS8110, if the interface occurs within the compression zone of the section, the horizontal shearing force at the interface is equal to the compression acting above the interface and resulting from the ultimate bending moment. However, in the present case, since the depth to the neutral axis is only about 10 mm, the interface is in tension; in such circumstances the Code states that the horizontal shear force corresponds to the total tensile or compressive force (and thus may be obtained by dividing the ultimate moment by the lever-arm). The horizontal shear is thus $2.55 \times 10^6/87.3 = 29.2$ kN, giving a stress of $29.2 \times 10^3/(300 \times 1000) = 0.1$ N/mm^2. If grade 30 concrete is used for the cast *in situ* topping, if links are not provided, and if the surface of the precast concrete is not specially treated in any way to increase the adhesion with the topping, the allowable value of v_h is 0.55 N/mm^2, which is greatly in excess of the applied value.

For condition (c) the greatest stresses during transportation and erection occur when the slab is the right way up and is supported only at the middle. In such a case the ultimate cantilever moment due to the self-weight of the unit is $\frac{1}{8} \times 1.13 \times 1.4 \times 2.5 = 0.49$ kN m. Neglecting the tensile resistance of the concrete itself (which in the present case might well be sufficient to resist such a small moment) the area of reinforcement required is only about 36 mm^2; three 6 mm bars (85 mm^2) are provided near the top of the slab.

Chapter 13
Columns subjected to axial loads only

The design of reinforced concrete columns subjected to concentric or axial ultimate loads without bending is dealt with in this chapter. Analytical principles are used that are similar to those employed when designing sections to resist bending with or without direct force. According to BS8110, axially loaded columns may be considered as either 'short' or 'slender', the former being those members for which the ratio of the effective height to the corresponding overall thickness is less than 15 for braced columns of normal-weight concrete or 10 for all other columns. A braced column is defined as one where, in the plane considered, the lateral stability of the structure is provided by walls, bracing or buttressing which is specifically designed to resist all the lateral forces acting in that plane. If the ratio of l_e/h is equal to or exceeds these limiting values, the procedure for designing slender columns described in the next chapter must be employed. In no case should the clear distance between the end restraints exceed 60 times the least lateral dimension of a column. Nor, in the case of a cantilevered or similar unbraced column with one end unrestrained, should its clear height in any given plane exceed $100b^2/h$, where b and h are the 'breadth' and 'depth' of the cross-section relative to the plane considered. In addition, to prevent excessive deflection, in unbraced columns in normal cases the *average* ratio of the effective height to the corresponding overall thickness for all columns in each direction at each floor level in a structure should not exceed 30; this particular restriction may be waived in the case of a single-storey structure where no damage to finishes is likely to result for any excessive deflection that may occur.

The minimum dimension of a column depends, among other factors, on the period of fire resistance required; for details see Chapter 19.

13.1 EFFECTIVE HEIGHT

When assessing the effective height l_e of a column, BS8110 defines four types of *end condition*:

1. Column connected monolithically to beams at both sides, the depths of which are at least as great as the overall column dimension in the plane considered. Where connection is to a foundation, this must be specifically designed to resist the moment transferred from the column.
2. As condition 1 but with shallower beams.
3. Column connected to members not specifically designed to resist rotation but which do provide some restraint.
4. Column end unrestrained against both lateral movement and rotation.

Figure 13.1 may then be used to find the factor β by which the clear height must be multiplied in order to obtain the effective length.

If the columns form part of a framed structure, clauses 2.5 in Part 2 of BS8110 give the following alternative

Fixity at end 2	Fixity at end 1					
	Braced column			Unbraced column		
	Type 1	Type 2	Type 3	Type 1	Type 2	Type 3
Type 1	0.75	0.80	0.90	1.20	1.30	1.60
Type 2	0.80	0.85	0.95	1.30	1.50	1.80
Type 3	0.90	0.95	1.00	1.60	1.80	—
Type 4	—	—	—	2.20	—	—

Figure 13.1 Factor β for finding effective length of column.

expressions for determining the effective height of a column:

Braced columns:

$$l_e \text{ is lesser of (a) } \left(\frac{14 + \alpha_{c1} + \alpha_{c2}}{20} \right) l_0;$$

$$\text{(b) } \left(\frac{17 + \alpha_{c\,min}}{20} \right) l_0;$$

$$\text{or (c) } l_0$$

Unbraced columns:

$$l_e \text{ is lesser of (a) } \left(\frac{20 + 3\alpha_{c1} + 3\alpha_{c2}}{20} \right) l_0;$$

$$\text{or (b) } \left(\frac{20 + 3\alpha_{c\,min}}{20} \right) l_0$$

where l_0 is the clear height between end restraints; α_{c1} and α_{c2} are the ratios of the sum of the column stiffnesses to the sum of the beam stiffnesses at each end of the column under consideration; and $\alpha_{c\,min}$ is the lesser of these two ratios. The stiffnesses should be evaluated by dividing the second moment of area of a member by its actual length, but only those members that frame into the column in the appropriate plane of bending may be considered. For column-to-base or beam-to-column junctions designed to resist nominal moments only, a value of α_c of 10 should be taken. For column-to-base junctions designed to resist the full column moment, a value of α_c of 1 should be taken. With flat-slab construction, the beam stiffness should be taken as that of a beam equal in size to the slab forming the column strip.

13.2 REINFORCEMENT IN COLUMNS

A minimum cross-sectional area of not less than 0.4% of longitudinal reinforcement should normally be provided in a column. However, where the concrete section provided is of such a size that the ultimate load may be resisted by the concrete alone, the section may be designed as a plain concrete column. In such circumstances the minimum steel areas needed are 0.3% and 0.25% for mild steel and high-yield reinforcement, respectively.

At least four bars should be provided in rectangular columns and at least six bars in circular columns; the size of these bars must be at least 12 mm. Links, having a diameter of at least one-quarter of the size of the largest bar to be tied, must be provided at a spacing of not more than 12 times the size of the smallest bar tied. At each corner of the section and at alternate bars in the outermost layer of steel, a link should pass round the bar with an included angle not exceeding 135°, and all other bars must not be further than 150 mm from the restrained bars. In the case of circular sections, however, adequate lateral support will result if circular links are provided that pass round the longitudinal bars, or groups of bars, that lie near the periphery of the circle.

Except at laps, not more than 6% of longitudinal reinforcement should be provided in columns that are cast vertically, although this limit may be increased to 8% if the columns are cast horizontally. In neither case should the proportion of steel where the laps occur exceed 10%.

At a splice in the longitudinal reinforcement, the length of the overlap provided must be at least 15 times the diameter of the smaller bar at the splice, or 300 mm, whichever is the greater. The length of lap must also be at least 25% more than that required to enable the full compressive stress to be developed in the bar. Thus in the case of mild steel, the minimum lap length is 39 diameters in normal-weight grade 25 concrete, 36 diameters in grade 30 concrete, and 31 diameters in grade 40 concrete. For high-yield bars, the corresponding lap lengths for the same grades of concrete are 40 diameters, 37 diameters and 32 diameters respectively, if the reinforcement consists of type 2 deformed bars. Appropriate lap lengths for lightweight and other grades of normal-weight concrete can be calculated from the expressions given in section 7.1.1.

The cover of concrete over all reinforcement (including links) should be not less than the values given in *Data Sheet 42*. If both lapping bars are greater than 20 mm in size and the cover provided is less than 50% more than the diameter of the smaller bar, BS8110 specifies that links not less than one-quarter of the diameter of the smaller lapping bar and spaced at not more than 200 mm must be provided throughout the entire length of lap.

13.3 SHORT AXIALLY LOADED COLUMNS

According to BS8110, when a column section is subjected to pure axial load, its ultimate resistance N_z is given by the expression

$$N_z = 0.45 f_{cu} A_c + 0.87 f_y A_{sc} \qquad (13.1)$$

The factor of 0.45 corresponds to the value of 4/9 shown on *Figure 5.1a*. The factor of 0.87, relating to the resistance provided by the reinforcement, corresponds to the reciprocal of the partial safety factor for steel (i.e. 1/1.15), and thus converts the characteristic strength of the reinforcement to the corresponding maximum design strength. Note that this expression, and those that follow, incorporate the normal partial safety factors for materials, and if the designer wishes to consider different values of γ_m they should be adjusted on a pro rata basis. Values of N_{uz}/bh corresponding to various characteristic strengths of steel and concrete and various proportions of reinforcement can be read from the nomogram on *Data Sheet 60*.

Pure axial load is extremely rare, except in theory, and expression (13.1) is normally only used (as described in Chapter 14) in the design procedures for slender columns and columns subjected to biaxial bending. For short braced columns that are loaded axially and cannot, owing to the type of structure adopted, be subjected to significant moments, the appropriate design expression given in BS8110 is

$$N = 0.4f_{cu}A_c + 0.75f_yA_{sc} \qquad (13.2)$$

This expression thus gives values that are about 10% less than the pure axial resistance of the section. The resulting margin of safety allows for an eccentricity due to possible constructional tolerances of the order of $h/20$.

In the case of short braced columns that support an arrangement of beams that is approximately symmetrical, the corresponding design expression provided in BS8110 is

$$N = 0.35f_{cu}A_c + 0.67f_yA_{sc} \qquad (13.3)$$

The Code states that this design condition is applicable to cases where beams supporting uniform loads have spans that do not differ by more than 15% of the greater of the spans. It will be seen that the values given by expression (13.3) are slightly more than 20% lower than the pure axial resistance of the section, and thus cater for the greater uncertainty regarding the eccentricity of the applied loading. The *Code Handbook* advises that this expression should not be used to design the peripheral columns of a structure unless the main beams span in the direction parallel to the edge under consideration.

Short unbraced columns that are not designed to resist a specific bending moment must be designed as sections subjected to axial load combined with a minimum moment of $Nh/20$ (but not more than $0.02N$ in kN/m), using the design procedures described in the next chapter.

In expressions (13.1) to (13.3), f_{cu} and f_y are the characteristic strengths of the concrete and steel, respectively, and A_{sc} is the area of longitudinal reinforcement provided. According to BS8110, A_c represents the net area of concrete (i.e. deducting the area of concrete displaced by the longitudinal compression reinforcement). However, in preparing the design charts for sections subjected to combined axial load and bending that are given in Part 3 of BS8110, the usual assumption has been adopted that no reduction is made for the area of concrete displaced by the reinforcement. Since these charts allow for a minimum eccentricity of load of $h/20$, the maximum ultimate loads read from them (i.e. when the moment is not greater than $Nh/20$) should be similar to those given by expression (13.2), i.e. those for a short braced axially loaded column. However, if A_c is taken as the *net* area of concrete in expression (13.2), the values given by this expression will clearly be rather less than those read from the appropriate charts in the Code. Further, since short unbraced columns must be designed using the Code design charts, it may therefore be thought advantageous to ignore the effect of any bracing and to consider the columns as unbraced, so that they may be designed using these charts. Such a stratagem is obviously illogical, and it may be thought preferable for uniformity to take A_c as the gross area of concrete. The maximum difference between the results obtained by the two differing assumptions occurs with the maximum percentage of mild steel reinforcement and the highest possible grade of concrete. If $f_y = 250\,\text{N/mm}^2$, $f_{cu} = 50\,\text{N/mm}^2$ and $\varrho_1 = 0.08$, the value of N obtained assuming that A_c is the gross area of concrete is only about 5% higher than that obtained if A_c is taken as the net area.

As explained in Chapter 14, valid arguments also exist for taking A_c as the gross concrete area when following through the design procedures for slender columns and columns subjected to biaxial bending.

To design a short column of any shape to support a specified ultimate axial load, it is necessary to assume a size and then to calculate the amount of reinforcement required from the following transposed versions of expressions (13.2) and (13.3):

$$A_{sc} = \frac{N - 0.4f_{cu}A_c}{0.75f_y} \qquad (13.2a)$$

$$A_{sc} = \frac{N - 0.35f_{cu}A_c}{0.67f_y} \qquad (13.3a)$$

These design formulae assume that A_c may be taken to be the gross area of the concrete. They are not suitable for direct design if A_c must be considered as the net concrete area, and if this is the case then the following expressions should be employed:

$$A_{sc} = \frac{N - 0.4f_{cu}A_c}{0.75f_y - 0.4f_{cu}} \qquad (13.2b)$$

$$A_{sc} = \frac{N - 0.35f_{cu}A_c}{0.67f_y - 0.35f_{cu}} \qquad (13.3b)$$

In all cases the resulting section must be checked to ensure that the amount of steel determined falls within the prescribed limits.

Alternatively, the nomogram on *Data Sheet 47* can be used, if A_c is assumed to represent the net concrete area. This nomogram gives the ultimate resistances per unit area of a section of a column for any given combination of concrete and steel characteristic strengths, and any percentage of reinforcement corresponding to design expressions (13.2b) and (13.3b). The design procedure is then to assume a column size, to divide the given ultimate load by the cross-sectional area, and to read off from the nomogram the percentage of reinforcement required, by laying a straight-edge on it in such a way that it intersects the ultimate resistance per unit area needed and the given characteristic strengths.

Care must be taken when employing a 'projection' nomogram of this type, especially when using it to determine the percentage of steel required (i.e. rather than the resistance corresponding to a given percentage), since any slight discrepancy in aligning the straight-edge through the given points is multiplied when reading off the answer. Nevertheless, despite this limitation the values obtained should be within 2% or 3% of those given by exact calculation. Even if it is preferred to evaluate the exact amounts of steel from formulae (13.2) and (13.3) and their variants, use of the nomogram on *Data Sheet 47* gives a rapid independent check that a numerical slip has not occurred.

The nature of the specific problem will determine whether the design adopted should be the most economical (in which case the highest-grade concrete and the minimum permissible amount of reinforcement should generally be used), whether it is to have the smallest overall dimensions (i.e. to utilize the strongest concrete mix and the maximum amount of steel), or whether some intermediate combination should be employed. The size may be specified to enable standard forms to be used or to conform to architectural requirements, or the grade of concrete may be specified to avoid using a mix that differs from that used in adjacent floors or other work. If one or more of these factors is specified, the number of possible designs that may be adopted will clearly be less than if there are no restrictions.

13.4 EXAMPLES OF AXIALLY LOADED SHORT COLUMNS

In the following examples, a column section is designed to withstand a total ultimate load of 2 MN in conformity with each of the conditions stated. Mild steel reinforcement ($f_y = 250 \, \text{N/mm}^2$) is used throughout. The above formulae are used for design, and the nomogram on *Data Sheet 47* is employed for checking purposes. The cross-sectional areas of appropriate numbers of round bars are read from *Data Sheet 41*.

13.4.1 Smallest square column

For this column the strongest concrete (say $f_{cu} = 50 \, \text{N/mm}^2$) and the largest amount of reinforcement (i.e. $\varrho_1 = 6\%$) should be used. Then, if the column cannot be subjected to significant moment, and assuming that A_c represents the net area of concrete,

$$A_c = \frac{N}{0.75\varrho_1 f_y + 0.4(1-\varrho_1)f_{cu}}$$

$$= \frac{2 \times 10^6}{(0.75 \times 0.06 \times 250) + (0.4 \times 0.94 \times 50)}$$

$$= 66\,560 \, \text{mm}^2$$

Thus a column 260 mm square will suffice.

The area of main reinforcement should be $66\,560 \times 0.06 = 4000 \, \text{mm}^2$. Four 32 mm bars plus four 16 mm bars will provide 4022 mm².

If the loading is approximately symmetrical and A_c is again the net concrete area,

$$A_c = \frac{N}{0.67\varrho_1 f_y + 0.35(1-\varrho_1)f_{cu}}$$

$$= \frac{2 \times 10^6}{(0.67 \times 0.06 \times 250) + (0.35 \times 0.94 \times 50)}$$

$$= 75\,480 \, \text{mm}^2$$

and a column 275 mm square should be provided. The area of steel required is $75\,480 \times 0.06 = 4530 \, \text{mm}^2$: provide eight 25 mm bars plus four 16 mm bars (i.e. 4731 mm²).

If the nomogram on *Data Sheet 47* is used, with $f_y = 250 \, \text{N/mm}^2$, $f_{cu} = 50 \, \text{N/mm}^2$ and 6% of reinforcement, the corresponding ultimate resistances per unit area for the two loading conditions are $30.0 \, \text{N/mm}^2$ and $26.5 \, \text{N/mm}^2$. Thus, to sustain an ultimate load of 2 MN, areas of 66 667 mm² (i.e. 260 mm square) and 75 470 mm² (i.e. 275 mm square) are necessary, as already calculated.

If it is assumed that A_c represents the gross concrete area, the calculations are as follows. If the loading is axial,

$$A_c = \frac{N}{0.75\varrho_1 f_y + 0.4 f_{cu}}$$

$$= \frac{2 \times 10^6}{(0.75 \times 0.06 \times 250) + (0.4 \times 50)} = 64\,000 \, \text{mm}^2$$

Employ a column 255 mm square. The required area of steel is $0.06 \times 64\,000 = 3840 \, \text{mm}^2$, for which eight 25 mm bars will suffice ($A_{sc} = 3927 \, \text{mm}^2$).

For approximately symmetrical loading,

$$A_c = \frac{N}{0.67\varrho_1 f_y + 0.35 f_{cu}}$$

$$= \frac{2 \times 10^6}{(0.67 \times 0.06 \times 250) + (0.35 \times 50)} = 72\,600 \, \text{mm}^2$$

The resulting column should be 270 mm square and reinforced with $0.06 \times 72\,600 = 4356 \, \text{mm}^2$ of steel: provide four 25 mm bars plus eight 20 mm bars ($A_{sc} = 4476 \, \text{mm}^2$).

13.4.2 Most economical column

The cheapest column normally results from a combination of the strongest concrete (say $f_{cu} = 50 \, \text{N/mm}^2$) and the least amount of steel (i.e. 0.4%). Assuming A_c represents the net concrete area, if the loading is axial,

$$A_c = \frac{2 \times 10^6}{(0.75 \times 0.004 \times 250) + (0.4 \times 0.996 \times 50)}$$

$$= 96\,760 \, \text{mm}^2$$

Provide a 315 mm square column. The reinforcement needed is $0.004 \times 96\,760 = 387 \, \text{mm}^2$; four 16 mm bars will provide 804 mm². Although four 12 mm bars would provide a sufficient area of steel, in normal circumstances in order to form a robust cage the minimum size of bars used for main steel in columns should be 16 mm. This complies with the recommendations in the *Code Handbook*.

Note that, in such a case as this, since the first practicable bar arrangement provides an area of steel considerably in excess of the minimum amount of 0.4% required, it is worth while recalculating the minimum size of concrete section required with this arrangement of bars to support the specified load. For example, if A_c

represents the gross area of concrete and the loading is axial,

$$A_c = \frac{N - 0.75 f_y A_s}{0.4 f_{cu}}$$

$$= \frac{(2 \times 10^6) - (0.75 \times 250 \times 804)}{0.4 \times 50} = 96\,470\,\text{mm}^2$$

i.e. a 305 mm square column will actually suffice.

If the loading is only approximately symmetrical,

$$A_c = \frac{2 \times 10^6}{(0.67 \times 0.004 \times 250) + (0.35 \times 0.996 \times 50)}$$

$$= 110\,500\,\text{mm}^2$$

Thus the column should be 335 mm square and be reinforced with four 16 mm bars ($A_{sc} = 804\,\text{mm}^2$) to cater for the required steel area of $0.004 \times 110\,500 = 442\,\text{mm}^2$.

As before, these calculations can be checked by using the nomogram on *Data Sheet 47*. This indicates that the ultimate resistances per unit area are $20.8\,\text{N/mm}^2$ and $18.0\,\text{N/mm}^2$ respectively, leading to areas of $96\,160\,\text{mm}^2$ and $111\,100\,\text{mm}^2$ to resist 2 MN.

The calculations involved if A_c represents the gross concrete area are similar to those given in section 13.4.1.

13.4.3 Column 250 mm by 500 mm of C25 grade concrete

Assuming that the load is axial and that A_c represents the net concrete area,

$$A_{sc} = \frac{N - 0.4 f_{cu} A_c}{0.75 f_y - 0.4 f_{cu}}$$

$$= \frac{(2 \times 10^6) - (0.4 \times 25 \times 250 \times 500)}{(0.75 \times 250) - (0.4 \times 25)} = 4225\,\text{mm}^2$$

Provide four 20 mm bars plus four 32 mm bars ($A_{sc} = 4474\,\text{mm}^2$). This represents 3.6%. Checking by means of the nomogram, the ultimate resistance per unit area is $2 \times 10^6/(250 \times 500) = 16\,\text{N/mm}^2$ and, with stresses of $f_y = 250\,\text{N/mm}^2$ and $f_{cu} = 25\,\text{N/mm}^2$, the corresponding percentage is also given as 3.6%.

If the load is distributed approximately symmetrically and A_c is as before,

$$A_{sc} = \frac{N - 0.35 f_{cu} A_c}{0.67 f_y - 0.35 f_{cu}}$$

$$= \frac{(2 \times 10^6) - (0.35 \times 25 \times 250 \times 500)}{(0.67 \times 250) - (0.35 \times 25)} = 5709\,\text{mm}^2$$

This represents 4.6%: twelve 25 mm bars ($A_{sc} = 5890\,\text{mm}^2$) will suffice. Again, the nomogram indicates 4.6%, corresponding to stresses of $f_y = 250\,\text{N/mm}^2$ and $f_{cu} = 25\,\text{N/mm}^2$, to give a required ultimate resistance per unit area of $16\,\text{N/mm}^2$.

The calculations are similar if A_c represents the gross concrete area, but formulae (13.2a) and (13.3a) should be employed instead of those used above.

Other examples of the design of axially loaded short columns are given on *Calculation Sheets 17*.

13.5 APPLICATION TO INTERIOR COLUMNS

Drawing 14 gives a design, in accordance with the requirements of BS8110, for interior column B3 on *Drawing 2*. The calculations for a column square in cross-section are given on *Calculation Sheets 17*, and the calculations for other interior columns which have slightly different loadings would be similar. As these interior columns support a symmetrical arrangement of beams (although the arrangement of imposed load need not be symmetrical), they are designed for axial load only, using the expression for short braced columns supporting an approximately symmetrical beam arrangement given in the Code. Since most of the loading on the main beams is in the form of central concentrated loads, it may be argued that the Code expression is therefore not applicable, since this is stated to relate to columns supporting beams designed for uniform loads only. However, since the concentrated loads in question are merely the reactions from the secondary beams due to the uniform loading on the slab they carry, it seems reasonable to assume that the expression employed is applicable here.

The imposed loads on the roof and upper floors are the same as are assumed for the beams and slabs. The imposed load on the ground floor is assumed to be $5\,\text{kN/m}^2$ over the area of floor supported by column B3, although only part of this floor is used as a garage. The reductions of the imposed load on the first to fifth floors have been made in accordance with *Data Sheet 4*. No similar reduction is made for the ground floor, although this is presumably permitted by the Code, because it is thought that this floor could be almost fully loaded at one time. The ultimate dead loads on all floors, including the load from partitions, are based on the calculations for the beams and slabs on *Calculation Sheets 1 to 9*.

In calculating the floor loads, no allowance is made for elastic reactions, i.e. the load on any column is the static load on the area of floor supported by that column. The employment of elastic reactions generally results in a greater load on the interior columns compared with the loads based on static reactions, but an advantage is gained, since the bending moments on the external columns will be combined with lower direct loads than those given by the static reactions. The load-carrying capacity of the columns is obtained from the nomogram on *Data Sheet 47*.

The sizes of the columns depend in some cases on considerations other than the ultimate load. Between the fourth and fifth floors the columns must be sufficiently wide to accommodate the roof beams, although the column provided is much larger in size than is needed to support the load. A single size of column should be used for as many storeys as possible, and normally for at least two storeys, so as to enable the formwork to be reused without alteration. Additional load-carrying capacity for a column of a given size can be achieved by increasing the amount of longitudinal reinforcement, by using steel with a higher characteristic strength, or by increasing the cement content to produce a higher-grade concrete; the last method is normally the most economical. It is advantageous to use the same grade of concrete and size of column for all the interior columns in one storey; variations can be made in the amounts of longitudinal reinforcement to provide for small differences of load from column to column.

The columns in the top storey are cruciform in section and the arms are made as narrow as practicable (130 mm minimum) so that there is no projection into the rooms of the residential premises in the top storey, i.e. the widths are the same as the thickness of the partition walls and the breadth of the beams.

Between the ground and first floors the columns in this example should be as small as practicable. A column 350 mm square is provided. In comparison, the use of an octagonal or circular column would be more expensive as, unlike the requirements of CP114, the present Code allows no increase in load-carrying capacity due to the employment of helical binding, and the cost of formwork would be substantially greater. However, a column of this shape might be more convenient in a shop or garage.

The arrangement of links in each column depends on the number of vertical bars, in order to comply with the requirement of the Code that every longitudinal bar must be prevented from buckling. The arrangement can be adjusted where the bars are restrained by intersecting beams and slabs and, above the soffits of beams, links are only provided to hold the corner bars in position, the number of such links being a minimum since they may cause inconvenience when fixing the bars in the beam. The spacing of the links must not exceed twelve times the size of the smallest longitudinal bar, as required by the Code.

The splices in the longitudinal bars in these designs occur just above floor level, as is common practice. Where bars have to be cranked for lapping purposes, the lower bars are carried up straight through the beam/column intersection and lapped with the lower cranked ends of the upper bars. An alternative arrangement (and one that used normally to be adopted at one time) is to provide the cranks at the upper ends of the lower bars, immediately below the beam/column intersection. This method has the disadvantage that the lever-arm of the steel, and therefore the resistance moment of the column section, is considerably reduced just at the point where any bending moment in the column will be a maximum. It is also not unknown for the cranks in the column bars to be oriented incorrectly in plan owing to errors made when erecting the reinforcement; such mistakes may not come to light until the subsequent floor construction is undertaken, when they are extremely difficult to remedy. Cranking the lower bars does sometimes make it simpler to avoid the beam bars passing through the beam/column intersection, but astute detailers can normally avoid such

difficulties by providing 'loose' bars in the beams at troublesome intersections.

The laps in the column bars are in general not less than 1.25 times the anchorage-bond lengths required by BS8110 for bars in compression, as described in section 7.1.1. When a length of 15 diameters or 300 mm cannot be provided, for example at the heads of top-storey columns, a lower characteristic strength should be adopted for the steel. In the present example the characteristic strength of the mild steel reinforcement is taken as $250 \, N/mm^2$ and therefore bond lengths of 36 diameters are provided, since the concrete is of grade 30 throughout. When the bars in the columns in consecutive storeys are of different diameters, the bond length may be related to the diameter of the smaller bar. Between the ground and first floors the bars in column B3 are 32 mm in diameter, and above the first floor 20 mm and 25 mm. Since a length of 890 mm is sufficient to transfer the ultimate load from one 25 mm bar to a bar of the same size in grade 30 concrete, this length is all that is required to transfer the ultimate load from a 25 mm bar to a larger one. It is necessary, however, to ensure that the larger bars have a sufficient bond length above the highest section of maximum stress, which is usually at the beam soffits. For this reason it is not possible for any of the eight 32 mm bars below the first floor to end immediately below the first floor, although less than this amount of reinforcement is required above this level. At 36 times the diameter, the 32 mm bars require a bond length of 1150 mm, and the distance from the level of the floor to the soffit of the beam is only 500 mm. Some of the bars could therefore terminate $1150 - 500 = 650$ mm above the first floor, but for practical reasons all bars in each storey are made the same length.

It is usual to provide a concrete 'kicker' 50 mm or 75 mm high immediately above the top surface of the floor slab, against which the forms for the next lift of columns may be butted. If this is done, bar lengths and laps should be adjusted accordingly when detailing the column reinforcement. These kickers may be cast immediately after the floor is concreted or, preferably, monolithically with it. The concrete forming the kicker should in theory be of the same grade as that used in the rest of the column although, unless there is a wide variation in strength between the mix used in the columns and in the floor construction, it is difficult to imagine failure occurring due to a slight reduction in strength over such a short length. However, if large bending moments in the columns are anticipated, it is essential that a special batch of column-grade material should be employed to concrete the kickers. Because of various difficulties in forming them, kickers are sometimes prohibited, although their use is sanctioned in ref. 2.

The details of suitable foundations for interior columns are shown on *Drawing 14* and the design of such foundations is described in section 17.1.

Columns forming part of a monolithic frame of a building, and exterior columns in particular, are subjected to bending. It is not always necessary to design interior columns to resist specific bending moments, particularly if they support symmetrical arrangements of beams and loads, although in some cases the effects due to the variation of the position of the imposed load should be investigated. The requirements of BS8110 relating to the interior and exterior columns of buildings that have beam-and-slab floors, and the bending moments in columns supporting flat slabs, are considered in this chapter. The general information regarding the limiting amounts of reinforcement in and the effective heights of columns etc. given in the preceding chapter is also applicable here, of course.

14.1 BENDING MOMENTS ON COLUMNS

Apart from carrying out a full elastic analysis of the complete structural frame using computer techniques or such a method as moment distribution, BS8110 in effect recognizes three particular methods of determining the moments in the columns of a frame subjected to vertical loads only. Firstly, each individual floor with its adjoining columns may be analysed as a separate system, the far ends of the columns being assumed to be fully fixed. Secondly, each individual floor may be divided into a series of three-bay sub-frames in which the far ends of the adjoining columns are assumed fully fixed, and the stiffnesses of the outer beams are taken as one-half of their true values. In both cases, to evaluate the maximum moment on any column the span to one side of the column carries a load of $1.4G_k + 1.6Q_k$, while that on the other side is loaded with $1.0G_k$, the actual disposition of loads being such that the unbalanced moment at the support is a maximum.

As explained in more detail in section 3.8.2, the final moment M in any column at L adjoining a central span LR

is given by the expression

M = distribution factor for column concerned

$$\times \frac{2D_{RL}F_R' + 4F_L'}{4 - D_{LR}D_{RL}}$$

where D_{LR} and D_{RL} are the distribution factors, and F_L' and F_R' are the out-of-balance fixed-end moments for the central span, at L and R respectively.

Since both of the foregoing methods involve 'exact' analysis in terms of BS8110, the column moments that have been obtained may be redistributed by up to 30%. However, such redistribution is seldom possible because the corresponding ratio of x/h that must be adopted when designing the column section is normally too high to permit any redistribution to be made. Nevertheless, see the discussion on unsymmetrically disposed re-inforcement in section 14.4.

The third method of determining column moments is simply to distribute the unbalanced *fixed-end* moments resulting from a loading of $1.4G_k + 1.6Q_k$ on the span to one side of the column, and of $1.0G_k$ on the span to the other side, in proportion to the ratio of the column stiffness to that of the total stiffness of the members meeting at that support. This method, which is almost identical to that familiar to those designers who were conversant with CP114 (clause 322b), is discussed in more detail below. In view of the approximations involved, however, no redistribution of moments is possible.

14.1.1 Exterior columns

The final method mentioned above complies with the requirements of clause 3.2.1.2.5 of BS8110. All the members concerned are assumed fully fixed at their far ends, and the stiffnesses of the beams adjoining the intersection being considered should be taken as one-half of their true stiffnesses. Then the bending moment M_l at the top of the lower column and M_u at the bottom of the upper column are given by

$$M_l = \frac{K_l}{K_l + K_u + \tfrac{1}{2}K_b} FEM$$

and

$$M_u = \frac{K_u}{K_l + K_u + \tfrac{1}{2}K_b} FEM$$

$$(14.1)$$

in which K_l and K_u are the stiffnesses of the lower and upper columns respectively, K_b is the true stiffness of the beam, and FEM is the fixed-end moment at the end of the beam adjoining the column, assuming complete fixity at both ends of the beam. For this condition, with any arbitrary load on the beam $FEM = 2A(2l_b - 3z)/l_b^2$, while if the load is symmetrical $FEM = A/l_b$, where A is the area of the bending-moment diagram for the beam of span l_b assumed to be freely-supported at both supports, and z is the distance from the column to the centroid of the free bending-moment diagram. For a given total load F these formulae can be reduced to $FEM = CFl_b$. Values of the fixed-end moment coefficient C for some common loads are given in *Data Sheet 48*, and most coefficients likely to be encountered in practice can be read from Tables 29, 30 and 31 of the tenth edition of *RCDH*.

If there is no upper column, i.e. where the roof beams are monolithic with the columns in the top storey, K_u is zero and the formula for the bending moment at the top of the column is

$$M_l = \frac{K_l}{K_l + \tfrac{1}{2}K_b} FEM \qquad (14.1a)$$

These formulae give approximate values of the column bending moments at the rigid junctions of columns and beams. In some cases more precise analysis may lead to smaller bending moments being obtained, but on *Calculation Sheets 19* and *20* expressions based on the requirements of clause 3.2.1.2.5 of BS8110 are used to design some of the exterior columns of the building on *Drawings 15* and *16*. The results of more-exact mathematical analysis may be thought more accurate, but such accuracy is largely fictitious because the assumptions made in a detailed analysis may bear little resemblance to actual conditions. For example, in normal structural theory it is assumed that the structural members are mathematical lines, that intersections are points instead of blocks, and that the relationship between stress and strain is linear. An even more serious implicit assumption is that the unloaded structure is not subjected to strain until completed, whereas in practice the construction storey by storey results in initial bending moments (and therefore stresses) being set up in the new concrete that are not represented in ordinary analyses. Similarly, the effects of shrinkage and temperature, particularly on long framed buildings, may be appreciable, especially on end columns. These discrepancies are additional to the variations from actuality that are assumed (generally knowingly) when analysing members subjected to combined bending and direct load, as described later.

There is therefore a great deal in favour of using approximate methods for analysing ordinary framed buildings, provided that such methods result in safe and economical designs. So long as the sum of the ultimate moments of resistance at the end of the beam and at midspan for each condition of loading sufficiently exceeds the ultimate free bending moment on the beam, there is little danger that structural failure will occur; creep under excessive stress tends to cause adjustments of deformation that, within limits, equalize the bending moments to the resistances provided. In view of the many factors (the magnitude of most of which are not known accurately) that enter into the analysis of a framed structure, it is only coincidence if the calculated and measured moments and forces in an actual structure are similar.

14.1.2 Interior columns

If the arrangement of, or the load on, beams supported by interior columns is unsymmetrical, it is necessary to consider the effect of the bending moments on the columns. The magnitude of the bending moments can be calculated approximately from the expressions

$$M_l = \frac{K_l}{K_l + K_u + \tfrac{1}{2}K_{b1} + \tfrac{1}{2}K_{b2}} F'_{b1b2}$$

and

$$M_u = \frac{K_u}{K_l + K_u + \tfrac{1}{2}K_{b1} + \tfrac{1}{2}K_{b2}} F'_{b1b2}$$

$$(14.1b)$$

The symbols in these expressions are the same as in the formulae for exterior columns; in addition, K_{b1} and K_{b2} are the stiffnesses of the two beams meeting at the column, and F'_{b1b2} is the greatest difference between the fixed-end moments on the two beams when one carries a load of $1.4G_k + 1.6Q_k$ and the other carries a load of $1.0G_k$. When calculating the forces on the columns due to these bending moments, it should be remembered that the load on the column is reduced by the amount of load (i.e. $0.4G_k + 1.6Q_k$) omitted from one of the beams. In the case of a floor supporting a large ultimate imposed load, it may be necessary to consider the effect on interior columns supporting a symmetrical arrangement of beams, if one beam is loaded by such a load and the other is not, in which case the above formulae also apply.

14.2 STIFFNESSES OF COLUMNS AND BEAMS

The stiffness of a structural member is defined by the bending moment which, if applied at one end of the member, produces unit rotation at that end, assuming that end to be supported and the other end to be fixed. For a column or beam having a uniform moment of inertia, the numerical value of the stiffness is obtained by dividing the moment of inertia by the length of that member. In practice it is more convenient to express the factors in the formulae for the bending moments in terms of the ratio of the stiffnesses of the members concerned. The relative stiffness of a column to a beam is therefore

$$\frac{\text{moment of inertia of column} \times \text{span of beam}}{\text{moment of inertia of beam} \times \text{height of column}} = \frac{I_c l_b}{I_b l_c}$$

which, for an exterior column, is

$$\frac{\text{stiffness of upper column}}{\text{stiffness of beam}} = \frac{K_u}{K_b} = \frac{I_u l_b}{I_b l_u} = \bar{K}_u$$

$$\frac{\text{stiffness of lower column}}{\text{stiffness of beam}} = \frac{K_l}{K_b} = \frac{I_l l_b}{I_b l_l} = \bar{K}_l$$

Similarly, for interior columns,

$$\frac{K_u}{K_{b1}} = \frac{I_u l_{b1}}{I_{b1} l_u} = \bar{K}_u \quad \text{and} \quad \frac{K_l}{K_{b1}} = \frac{I_l l_{b1}}{I_{b1} l_l} = \bar{K}_l$$

These formulae are only correct when the end-fixing conditions of both members are identical. If the end of the column remote from the junction with the beam is equivalent to 'hinged' while the remote end of the beam is 'fixed', the ratio of the stiffnesses is $(3/4)(I_c l_b/I_b l_c)$. If the end-fixing conditions are reversed, the ratio of stiffness becomes $(4/3)(I_c l_b/I_b l_c)$. The factors 3/4 and 4/3 correspond to the coefficient k on *Data Sheet 49*, where values are given for extreme combinations of end conditions. In ordinary building frames k is unity, because the ends of the beams and columns remote from the monolithic connection are generally continuous with adjacent members and, although the condition at the ends is not that of absolute fixity, the degree of restraint is approximately the same at the ends of both members. In cases where this approximate equality is not obtained, a suitable value of k should be substituted into the expressions for $\bar{K}_u$ and $\bar{K}_l$, to give the general formulae for the ratios presented on *Data Sheet 49*. If the ratios $\bar{K}_u$ and $\bar{K}_l$ are substituted in the bending-moment formulae, the more simple formulae for these bending moments, as given in the table at the bottom of *Data Sheet 49*, are obtained for exterior and interior columns.

The formulae derived from the requirements of clause 3.2.1.2.5 of BS8110 for bending moments on columns apply to specific cases where the only likely instance of end conditions being other than fixity or continuity is where a column in a bottom storey is considered as 'hinged' at the foot because the base is assumed to be a concentrically loaded foundation. This is the case with the interior columns in the building in Part Two; it does not occur with the exterior columns, which are monolithic with the basement walls (see section 15.7.4).

It is presumably acceptable for the lengths of the column and beam used in the calculation of the stiffness factor to be the height of the column from floor to floor and the effective span of the beam as defined in BS8110. Three bases are suggested in the Code (clause 2.5.2) for calculating the moment of inertia: (a) the entire concrete area, ignoring the reinforcement but including the concrete in tension; (b) the entire concrete area, including the concrete in tension and allowing for the reinforcement

on the basis of the modular ratio; and (c) the area of the concrete in compression only, together with the reinforcement on the basis of the modular ratio. Method (a) is clearly generally the simplest to apply and is normally used but, as pointed out in the *Code Handbook*, the other methods are applicable when assessing the ability of existing structures to support revised loadings (i.e. where the areas of reinforcement are already known). Method (a) is used to calculate the stiffness factors on *Calculation Sheets 19* and *20*. The same basis must normally be used to determine all the moments of inertia when calculating the stiffness factors of all members meeting at one junction.

A different method to that used to analyse the frame can, of course, be used to calculate the moment of inertia in order to determine the stresses in the members if this is required, for example methods (b) and (c). Although all three methods give widely differing results, when the resulting values are expressed as ratios of moments of inertia of one member to another, the differences in practical terms are far less marked. When calculating the moment of inertia of a flanged beam such as a T-beam, the breadth of the flange should be that given by the rules on *Data Sheet 41*; these breadths are included in the calculations of the moments of inertia of beams on *Calculation Sheets 19* and *20*.

On *Data Sheet 49*, curves are given to simplify the calculation of the moments of inertia of flanged beams and of beams and columns of other common cross-sections, evaluated on the basis of the gross concrete area and ignoring the reinforcement (i.e. method (a)). Alternatively, the moments of inertia of rectangular and flanged sections determined on the basis of the area of concrete in compression only, and allowing for the reinforcement on the basis of the modular ratio (method (c)), may be found by using the charts and procedures described in Chapter 8. BS8110 suggests that a modular ratio of 15 is used, unless the actual elastic modulus of the concrete is known and hence the true modular ratio can be calculated.

Since the bending moments on a column depend on the stiffness ratio, the units in which the moments of inertia are expressed must be the same. The units for the length of the beams and columns must be the same, but they need not necessarily be the same as those adopted for the moments of inertia. It is generally more convenient to consider the moments of inertia in $\text{mm}^4 \times 10^9$ units, the lengths in metres, and the bending moments in kN m. In all the formulae in *Data Sheet 49*, $\text{mm}^4 \times 10^9$ units can be used for the moments of inertia and metres for the lengths.

14.3 ANALYSIS OF RECTANGULAR SECTIONS

Unlike its predecessor CP110, which also provided simplified design expressions that could be used when analysing the effects of combined bending and direct force at the ultimate limit-state, BS8110 requires that in all cases a rigorous analysis from basic principles must be made, assuming a parabolic-rectangular or uniform rectangular distribution of stress in the concrete.

With any analytical method, because of the number of variables involved it is, generally speaking, not possible to devise practicable design expressions from which the area of reinforcement required to resist a given combination of bending moment and direct force can be calculated directly for a given section. Instead, each basis may be used to prepare sets of tables or, more usually, graphs from which the appropriate proportions of steel may be read. Alternatively, programs are available for microcomputers and programmable pocket calculators that give directly the steel area needed to resist the application of a given force and moment to a section of given dimensions.

14.3.1 Rigorous analysis: parabolic-rectangular stress-block

In this case the behaviour of the section is similar to the corresponding case described in section 5.1.1. By resolving forces axially and then taking moments about the centre-line of the section, the following equations are obtained (*Figure 14.1*):

$$N = k_1 bx + A'_{s1} f_{yd1} - A_{s2} f_{yd2} \qquad (14.2a)$$

$$M = k_1 bx(\tfrac{1}{2}h - k_2 x) + A'_{s1} f_{yd1}(\tfrac{1}{2}h - d')$$
$$+ A_{s2} f_{yd2}(d - \tfrac{1}{2}h) \qquad (14.3a)$$

in which A'_{s1} and A_{s2} are the amounts of 'compression' and 'tension' steel respectively, the design stresses are f_{yd1} and f_{yd2} respectively, and k_1 and k_2 are factors defining the 'volume' and position of the centroid of the concrete stress-block. These expressions can be rewritten as

$$\frac{N}{bh} = k_1 \frac{x}{h} + \tfrac{1}{2}\varrho_1 f_{yd1} - \tfrac{1}{2}\varrho_1 f_{yd2} \qquad (14.2b)$$

$$\frac{M}{bh^2} = k_1 \frac{x}{h}\left(\tfrac{1}{2} - k_2 \frac{x}{h}\right) + \tfrac{1}{2}\varrho_1 f_{yd1}$$

$$\times \left(\tfrac{1}{2} - \frac{d'}{h}\right) + \tfrac{1}{2}\varrho_1 f_{yd2}\left(\frac{d}{h} - \tfrac{1}{2}\right) \qquad (14.3b)$$

when the reinforcement is disposed symmetrically and thus $A'_{s1}/bh = A_{s2}/bh = \tfrac{1}{2}\varrho_1$. By substituting appropriate values of f_y, f_{cu}, ϱ_1, d/h and x/h, a series of design charts can be prepared: such a series forms a portion of Part 3 of BS8110. Note that the terms 'compression' and 'tension' are used in this chapter to distinguish the two areas of reinforcement; it the position of the neutral axis falls below the area of steel farther from the line of action of the resultant load, this 'tension' steel will actually act in compression.

As noted earlier, the manner in which the stress–stress diagram for concrete is defined in BS8110 means that the relationship between the concrete strength and the resistance of a section is not linear. In other words, if the strength of the concrete is doubled, the calculated resistance of the section to bending and axial force does not correspondingly double. For this reason it is, strictly speaking, necessary to prepare separate sets of design charts for each concrete grade, as is done in Part 3 of the Code. However, if the above expressions are divided through by f_{cu} and used to prepare a set of design charts for grade C30 concrete it can be shown that, if these charts are used for concrete strengths ranging from 25 to 50 N/mm², the resulting differences are negligible and well within the limit of accuracy with which the charts can be read by eye. Such a series of charts is provided in *RCDH*.

One point should be noted. When $x/h = 1$ (i.e. the position of the neutral axis coincides with the bottom face of the section: *Figures 14.1b* and *14.1c*), the shape of the concrete stress-block and the position of its centroid are such that they still provide some resistance to applied moment or, in other words, to an eccentrically positioned load. As the eccentricity of the load decreases, the 'volume' of the stress-block rises until, when the eccentricity (i.e. the applied moment) is zero, a uniform stress of $0.45f_{cu}$ extends over the entire section. In most design charts (e.g. those in Part 3 of BS8110) the values obtained when $x/h = 1$ and $M/bh^2 = 0$ are joined by straight lines. In practice this point is of mainly academic interest, since over most of this area of the charts the shape of the design curves is dictated by the criterion that a minimum eccentricity of load of $h/20$ or 20 mm, whichever is the lesser, must be considered.

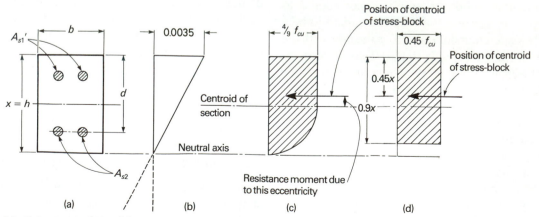

Figure 14.1 (a) Column section; (b) distribution of strain when $x = h$; (c) parabolic-rectangular stress-block; (d) uniform rectangular stress-block.

14.3.2 Rigorous analysis: uniform rectangular stress-block

Here too the analysis closely resembles the corresponding case of a section subjected to bending only, dealth with in section 5.1.2. For example, for a rectangular section the following expressions are obtained:

$$N = 0.402 f_{cu} bx + A'_{s1} f_{yd1} - A_{s2} f_{yd2} \qquad (14.4a)$$

$$M = 0.201 f_{cu} bx(h - 0.9x) + A'_{s2} f_{yd1}(\tfrac{1}{2}h - d')$$
$$+ A_{s2} f_{yd2}(d - \tfrac{1}{2}h) \qquad (14.5a)$$

With symmetrically disposed reinforcement, these expressions may be rearranged as follows:

$$\frac{N}{bhf_{cu}} = 0.402 \frac{x}{h} + \tfrac{1}{2} \frac{\varrho_1}{f_{cu}} f_{yd1} - \tfrac{1}{2} \frac{\varrho_1}{f_{cu}} f_{yd2} \qquad (14.4b)$$

$$\frac{M}{bh^2 f_{cu}} = 0.201 \frac{x}{h}\left(1 - 0.9\frac{x}{h}\right) + \tfrac{1}{2}\frac{\varrho_1}{f_{cu}} f_{yd1}$$
$$\times \left(\tfrac{1}{2} - \frac{d'}{h}\right) + \tfrac{1}{2}\frac{\varrho_1}{f_{cu}} f_{yd2}\left(\frac{d}{h} - \tfrac{1}{2}\right) \qquad (14.5b)$$

Now by substituting appropriate values of $f_{yd1}, f_{yd2}, \varrho_1$ and x/h, sets of design charts can be prepared. The charts for symmetrically reinforced rectangular columns, provided on *Data Sheets 50 to 54*, have been produced from these expressions.

Unlike the situation when a parabolic-rectangular distribution of stress in the concrete is assumed, with a uniform stress-block the relationship between the concrete strength and the load-carrying capacity of the section is linear. Thus a single set of design charts will cater for any concrete grade.

14.4 UNSYMMETRICAL REINFORCEMENT

Most sets of design charts prepared by adopting one of the foregoing analytical bases are for symmetrically disposed reinforcement, and although sets catering for rectangular columns with unsymmetrically arranged reinforcement were produced for CP114 and CP110, a similar set does not seem to be currently available for the present Code. Positioning the reinforcement symmetrically within a

section often means that it is not used to its greatest advantage if the applied moment is such that it cannot reverse in direction (e.g. if it results from the floor loading on an adjoining beam). For example, if M/N is equal to $\tfrac{1}{2}h - d'$, the line of action of the resultant direct load coincides with the position of the 'compression' steel, and thus no 'tension' reinforcement is theoretically required. The amount of reinforcement needed when an unsymmetrical arrangement is adopted is never more, and often considerably less, than that required when equal amounts are provided near both faces.

If M/N was not less than $\tfrac{1}{2}h - d'$ (i.e. if the steel further from the line of action of the resultant direct load was not in compression), CP110 permitted the section (*Figures 14.2a* and *14.2b*) to be designed by ignoring the axial load completely and considering instead an increased moment of $M + (\tfrac{1}{2}h - d')N$. Then, when the section had been designed on this basis, the area of tension steel determined could be reduced by $N/0.87 f_y$. This technique, well known to engineers who are familiar with elastic design methods, consists in effect of introducing equal and opposite forces along the line of the tension reinforcement, as shown in *Figure 14.2c*. These forces are then considered as two separate systems, one (*Figure 14.2d*) consisting of the moment M together with forces N acting over a lever arm of $d - \tfrac{1}{2}h$, and the other consisting of a compressive force N acting along the line of the tension steel (*Figure 14.2e*).

Although CP110 did not specify restrictions on the method to be used to design the section for bending only, it was clear that if rigorous analysis were employed, the maximum ratio adopted for x/d should be such that f_{yd2} were not less than $f_y/1.15$; alternatively, if a higher ratio were used, the final reduction in the amount of steel should theoretically be N/f_{yd2}. However, since it was obviously more economical to use the greatest allowable design stresses, it was desirable to adopt a ratio of x/d such that both f_{yd1} and f_{yd2} attained their maximum values. Thus, as in the case of simple bending, the ideal ratio for x/d to achieve optimum economy was the maximum which might be employed without reducing f_{yd2} below $f_y/1.15$.

One disadvantage of such a method of design is that it is, of course, impossible to choose the relative proportions of reinforcement required near each face.

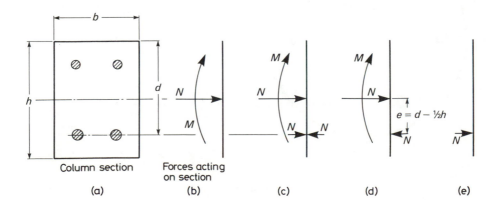

Column section	Forces acting on section			
(a)	(b)	(c)	(d)	(e)

Figure 14.2

The amounts that may actually result may thus be inconvenient and it is sometimes difficult to be certain whether, if the amounts are adjusted to achieve a convenient layout, the strength of the resulting section is sufficient. It should also be remembered that an unsymmetrical arrangement of steel can only be employed if the bending moment cannot reverse in direction. If it can, (for example, if it results from wind loading), equal steel must be provided near both faces. A further disadvantage of this method was that it ignored the true values of strain (and hence, possibly, stress), in the materials, and for this reason it was not included in BS8110.

14.5 CIRCULAR SECTIONS

The analysis of circular sections is basically similar to that given above for symmetrically reinforced rectangular sections. However, if a parabolic-rectangular concrete stress-block is adopted, complex calculations are necessary to determine the contribution of the concrete; because a circular segment is involved, the concrete stress, which varies with the depth from the compression face, acts on a section of varying width. Similarly complex calculations are needed to determine the position of the centroid through which this resistance acts.

Although such an analysis is not difficult if a computer is used to consider the stress-block as a number of strips of equal depth, to find the stress at the midpoint of each, and hence to sum the contribution of the individual strips, it is mathematically simpler to employ the alternative uniform rectangular stress-block. Furthermore, because the uniform stress of $0.4467f_{cu}$ applies to the full depth of $0.9x$, where the section is widest (whereas the corresponding stress given by the alternative method at this point is zero), the resistance provided by the concrete is theoretically greater if this analytical method is adopted.

Referring to *Figure 14.3*, the basic equations are

$$N = 0.45f_{cu}A_c + A_s(f_{A1} + f_{A2} + f_{A3})/3$$

$$M = 0.45f_{cu}A_c(R - x_c) + A_s(f_{A1} - f_{A3})(R\sin 60°)/3$$

where A_c is the area of concrete in compression (i.e. to a depth of $0.9x$ from the compression face) and x_c is the depth to the centroid of this area; A_s is the total area of reinforcement provided; and f_{A1}, f_{A2} and f_{A3} are the stresses in the reinforcement at the three positions shown, and are assumed positive if the bar concerned is in compression. The resistance provided by the section in bending differs depending on the number and orientation of the reinforcing bars provided, but it can be shown that the minimum value of this resistance usually occurs when only six bars are provided and they are arranged as shown.

Unlike its predecessor, Part 3 of BS8110 does not contain design charts for circular columns, but such charts are included in the *Joint Institutions Design Manual*. No explicit explanation of the derivation of these charts is included, but as this manual only endorses the use of the uniform rectangular concrete stress-block, this is the assumption made. In order to reduce the number of charts necessary to a minimum, the charts included in the *Joint Institutions Design Manual* are plotted in such a way that the characteristic steel strength f_y is included as a variable. By so doing, only a single set of charts for a range of concrete cover ratios is needed to cater for any value of f_y. This simplification has a negligible effect in the case of rectangular members. However, with circular sections this is not true, and it can be shown that in the case of mild steel reinforcement (i.e. where $f_y = 250\,\text{N/mm}^2$) the moment-carrying capacity of a section may be underestimated by up to nearly 20% if the charts in the *Manual* are used. Looked at another way, they may indicate that

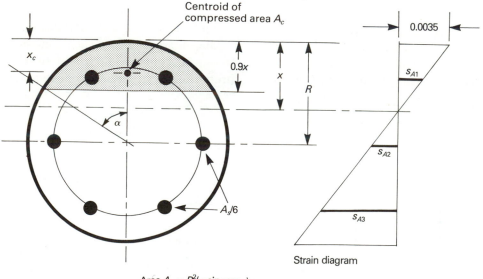

Area $A_c = R^2(\alpha - \sin\alpha\cos\alpha)$
$x_c = R\{1 - 2\sin^3\alpha/[3(\alpha - \sin\alpha\cos\alpha)]\}$

Figure 14.3

up to nearly 20% more steel is needed than is actually required.

On *Data Sheets 55* to *59*, separate sets of charts are provided for circular columns with various cover ratios and for values of f_y of 250 and 460 N/mm². These have been produced using the expressions given above.

14.6 SECTIONS ON OTHER SHAPES

The basic principles outlined above also apply to sections that are neither rectangular nor circular, although in practice the analysis of such sections is a complex matter unless suitable design charts or computer programs are handy. Reference 19 contains a simple computer program for determining the axial and bending resistance of a non-rectangular plane section that is symmetrical at right angles to the plane of bending, and includes an example of the use of this program to analyse a square column bending about a diagonal. Reference 20 includes charts for annular sections, based on the assumption of a uniform rectangular stress-block and a more simplified stress–strain relationship for the reinforcement than that specified in BS8110.

To design any section of arbitrary shape it is necessary to employ rigorous analysis. Clearly the choice of a uniform rectangular stress-block is simplest, and a suitable analytical method is outlined in *RCDH*.

14.7 BENDING ABOUT TWO AXES

Unlike its predecessors, CP110 included recommendations for designing rectangular sections that are subjected to biaxial bending and direct load. Unfortunately, however, these requirements had the considerable disadvantage that they could not be used to develop a direct design procedure giving the required area of steel to reinforce a given column section subjected to given moments and forces. Instead, a trial-and-adjustment procedure was necessary. For this reason, BS8110 has substituted instead a much simpler approach whereby the biaxial bending moments are converted into an equivalent uniaxial moment. With h' and b' as shown in *Figure 14.4* and $\beta = 1 - 7N/6bhf_{cu}$ (but not less than 0.3), then if M_x/h' exceeds M_y/b' the section should be designed to resist an axial load N together with a moment of $M_x + \beta M_y h'/b'$; otherwise the section should be designed to resist an axial load N together with a moment of $M_y + \beta M_x b'/h'$. The use of these expressions is illustrated on *Calculation Sheets 20*.

These expressions are only valid if all the bars are located near the corners of the sections, and BS8110 itself gives no indication of what action should be taken if this arrangement is impossible. The *Code Handbook* suggests, in effect, that the column section is divided into quarters, and that in each quadrant a single bar of equivalent area and positioned at the centroid of all the bars (or parts of bars) located within the quadrant should be considered instead. The *Code Handbook* states that

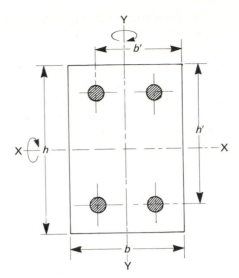

Figure 14.4

this method is conservative but is not usually excessively so.

Alternatively (and especially if suitable computer facilities are available) the trial-and-adjustment method specified in CP110 is still valid and may be used. It should be noted that this method provides a better approximation to a rigorous computer analysis (in which the strain, and hence the stress, in each individual bar is determined separately) than the simple method described in BS8110, and indeed the *Code Handbook* concedes that it is 'probably technically superior'. For this reason the method is described in detail in the following section. Alternatively the amount of reinforcement necessary can be determined using the BS8110 method, and the resulting design then checked (and adjusted if necessary) using expression (14.6).

14.7.1 Biaxial bending: CP110 design procedure

Research has shown that if a given section is subjected to ultimate bending moments of M_x and M_y about the two principal axes and to an axial load N, the section will be satisfactory provided that

$$\left(\frac{M_x}{M_{ux}}\right)^{\alpha_n} + \left(\frac{M_y}{M_{uy}}\right)^{\alpha_n} \not> 1 \qquad (14.6)$$

In this expression M_{ux} and M_{uy} are the ultimate uniaxial bending-moment capacities of the given section about the respective axes, when the axial load N acts on the section. In other words, if the section contains say 4% of reinforcement, M_{ux} is the ultimate bending-moment (read from the chart for combined bending and direct force) corresponding to an axial load N and 4% of steel. $\alpha_n = \frac{1}{3}(2 + 5N/N_{uz})$, where $N_{uz} (= 0.45f_{cu}A_c + 0.87f_yA_{sc})$ is the resistance of the given section to pure axial load, with the proviso that $1 \not> \alpha_n \not> 2$.

The foregoing relationships can be expressed graphically as shown on *Data Sheet 60*, and the value of

Columns subjected to bending

N_{uz} can be read from the adjacent nomogram. In the latter, an allowance has been made for the area of concrete displaced by the reinforcement. If such an allowance is not made, the resulting value of N_{uz} is, of course, slightly higher and thus the corresponding value of α_n is lower, in turn increasing the resulting values of $(M/M_u)^{\alpha_n}$. Thus for safety it would appear preferable to adopt a higher rather than a lower value of N_{uz}, and it is therefore perhaps better in practice to calculate N_{uz} on the assumption that A_c represents the gross concrete area. However, since the column design tables (based on a uniform rectangular concrete stress-block) accompanying this book tend to underestimate the strength of a section compared with those based on a parabolic-rectangular stress-block, it has been thought satisfactory to adopt the alternative assumption on *Data Sheet 60*. Data based on the assumption that A_c represents the gross concrete section are given in the *RCDH*.

It will already be clear that, as with many design methods, the foregoing requirements make it impossible to produce a direct design procedure that gives the required area of steel for a given column section subjected to given forces. Instead, a trial-and-adjustment procedure, such as that now described, must be adopted.

The first step is to assume suitable dimensions for b and h if these are not specified and, by selecting d', to calculate d'/b, d'/h, $M_x/bh^2 f_{cu}$, $M_y/b^2 hf_{cu}$ and N/bhf_{cu}. Next a suitable ratio must be chosen for M_y/M_{uy} and, after calculating $M_{uy}/b^2 hf_{cu}$ the column design charts are used to determine the value of ϱ_y corresponding to these given values of $M_{uy}/b^2 hf_{cu}$ and N/bhf_{cu}. The next step is to calculate (or to use the nomogram on *Data Sheet 60* to obtain) N_{uz} and thus determine the ratio N/N_{uz}. Then with the given ratios of M_y/M_{uy} and N/N_{uz}, the graph on *Data Sheet 60* can be used to read off the corresponding value of M_x/M_{ux}, thus enabling $M_{ux}/bh^2 f_{cu}$ to be evaluated. The column design charts are now used once again to find the value of ϱ_x corresponding to the given values of $M_{ux}/bh^2 f_{cu}$ and N/bhf_{cu}.

It is obviously unlikely that ϱ_x and ϱ_y will be equal, and the actual area of reinforcement required will be somewhere between these two proportions. By taking a proportion midway between these two values and rounding up slightly, the new value of ϱ_{tot} can be used with N/bhf_{cu} to obtain revised values of M_{ux} and M_{uy} from the column design charts, and of N_{uz} from the nomogram on *Data Sheet 60*. Then, with the actual values of M_x, M_y and N, the graph on *Data Sheet 60* can be used to ensure that the original Code criterion is met. If not, a further trial-and-adjustment cycle is called for. This procedure is illustrated in the example in section 14.7.2.

The foregoing procedure is only valid if the resulting reinforcing bars, or groups of bars, are located near the corners of the section and thus contribute to the strength of the section in bending in both directions. If this is not so, and if the various areas of reinforcement are as designed in *Figure 14.5*, then $\varrho_x = \varrho_1 = \varrho_2$, $\varrho_y = \varrho_1 + \varrho_3$ and $\varrho_{tot} = \varrho_1 + \varrho_2 + \varrho_3$. Thus, by rearranging these

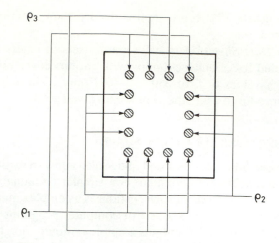

Figure 14.5

equations,

$$\varrho_1 = \varrho_x + \varrho_y - \varrho_{tot}$$
$$\varrho_2 = \varrho_{tot} - \varrho_y \qquad (14.7)$$
$$\varrho_3 = \varrho_{tot} - \varrho_x$$

The modified design procedure is now as follows. First, choose a ratio of M_y/M_{uy} and, after calculating $M_{uy}/b^2 hf_{cu}$, use the column design charts to find the value of ϱ_y corresponding to N/bhf_{cu}. Similarly, choose a value for M_x/M_{ux} and use the charts to determine the proportion of reinforcement ϱ_x corresponding to $M_{ux}/bh^2 f_{cu}$. Finally, with the chosen ratios of M_x/M_{ux} and M_y/M_{uy}, use the graph on *Data Sheet 60* to read off the corresponding minimum ratio of N/N_{uz}. With this value and the given value of N/bhf_{cu}, the nomogram on *Data Sheet 60* may be used to determine ϱ_{tot} and expressions (14.7) can then be employed to find ϱ_1, ϱ_2 and ϱ_3. If the resulting steel areas (principally that corresponding to ϱ_1) are impracticable, a different choice should be made for the ratios M_x/M_{ux} and M_y/M_{uy}. This procedure is illustrated in the example in section 14.7.3.

One point to be observed is that the ratios of M/M_u can be higher if the corresponding ratio of N/N_{uz} is higher; to put it another way, the limiting ratios of M/M_u correspond to the minimum ratio of N/N_{uz}, and thus to the minimum value of N commensurate with the values of M_x and M_y. Thus, any condition in which relatively high moments occur in conjunction with a substantially reduced axial load should be investigated, if this is not the condition for which the section has been designed.

14.7.2 Biaxial bending: worked example 1

Determine suitable reinforcement for a 500 mm by 250 mm column of grade C30 concrete which is subjected to ultimate moments of 100 kN/m and 45 kN/m about the major and minor axes respectively, together with an ultimate axial load of 750 kN, using both the formulae in BS8110 and the rigorous method outlined in CP110. The exposure conditions are moderate, and mild steel reinforcement is to be used.

With grade C30 concrete ('a systematic checking regime' is assumed) in moderate exposure conditions, BS8110 requires 35 mm cover to all reinforcement. If 8 mm links are provided to a maximum size of bar of 32 mm, d' will be $35 + 8 + (32 \times 0.5) = 59$ mm and thus $d/h = (500 - 59)/500 = 0.882$ and $d/b = 191/250 = 0.764$. With the BS8110 simplified method, since $\beta = 1 - (7N/6bhf_{cu}) = 0.767$ and M_y/b' $(= 45 \times 10^6/191 = 235 \times 10^3)$ exceeds M_x/h' $(= 100 \times 10^6/441 = 227 \times 10^3)$, the section must be designed for a modified uniaxial moment of $M_y + \beta M_x b'/h' = (45 \times 10^6) + (0.767 \times 100 \times 10^6 \times 191/441) = 78.2$ kN m acting about the minor axis. Thus interpolating between the charts for $d/h = 0.75$ and $d/h = 0.80$ with $M/b^2 hf_{cu} = 78.2 \times 10^6/(250^2 \times 500 \times 30) = 0.0834$ and $N/bhf_{cu} = 750 \times 10^3/(250 \times 500 \times 30) = 0.20$, $\varrho_1/f_{cu} = 0.00055$, so that $\varrho_1 = 0.0165$ and $A_{s\,req} = 2065$ mm^2.

With the CP110 analytical procedure, first assume that $M_{ux}/M_x = 2$. Then interpolating between the upper charts for $d/h = 0.85$ and $d/h = 0.90$ on *Data Sheets 51* and *52*, with $M_{ux}/bh^2 f_{cu} = 2 \times 100 \times 10^6/(500^2 \times 250 \times 30) = 0.107$, $p_1/f_{cu} = 0.0007$ and thus $p_x = 30 \times 0.0007 = 0.021$.

From the nomogram on *Data Sheet 60*, with this proportion of steel and with $f_{cu} = 30$ N/mm^2 and $f_y = 250$ N/mm^2, $N_{uz}/bh = 17.8$ N/mm^2. Thus $N/N_{uz} = 0.20 \times 30/17.8 = 0.337$, and from the chart on *Data Sheet 60* with $M_x/M_{ux} = 0.5$ the corresponding value of M_y/M_{uy} must not exceed 0.64, so that $M_{uy} = 45 \times 10^6/0.64 = 70.3$ kN m.

Now interpolating for $d/b = 0.764$ between the charts for $d/h = 0.75$ and $d/h = 0.80$ gives $M_{uy}/b^2 hf_{cu} = 70.3 \times 10^6/(250^2 \times 500 \times 30) = 0.075$, and with $N/bhf_{cu} = 0.20$ as before, $\varrho_1/f_{cu} = 0.0004$ and so $\varrho_y = 0.0004 \times 30 = 0.012$.

Now the actual value of ϱ_1 required will lie somewhere between the values obtained for ϱ_x and ϱ_y: try $\varrho_1 = 0.017$, so that $\varrho_1/f_{cu} = 0.00057$. Then with $\varrho_1/f_{cu} = 0.00057$ and $N/bhf_{cu} = 0.20$, $M_{ux}/bh^2 f_{cu} = 0.096$, so that $M_x/M_{ux} = 0.053/0.096 = 0.552$; and $M_{uy}/b^2 hf_{cu} = 0.085$, so that $M_y/M_{uy} = 0.048/0.085 = 0.565$. Then since $N_{ux}/bh = 17$ N/mm^2 (from the nomogram on *Data Sheet 60* with $\varrho_1 = 0.017, f_{cu} = 30$ N/mm^2 and $f_y = 250$ N/mm^2), $N/N_{ux} = 6.0/17.0 = 0.354$ and thus $\alpha_n = [(5 \times 0.354) + 2]/3 = 1.256$. Now

$$\left(\frac{M_x}{M_{ux}}\right)^{\alpha_n} + \left(\frac{M_y}{M_{uy}}\right)^{\alpha_n} = 0.552^{1.256} + 0.565^{1.256}$$
$$= 0.962$$

which is satisfactory. The area of reinforcement required is $0.017 \times 500 \times 250 = 2125$ mm^2: provide four 32 mm bars, one near each corner of the section.

As can be seen by comparing the above results, in the present example the areas of steel required by the two design methods are very similar.

14.7.3 Biaxial bending: worked example 2

Design a column section to withstand an axial ultimate load of 1250 kN together with ultimate bending moments of 105 kN m and 40 kN m acting at right angles. Assume grade C30 concrete, 40 mm cover to all steel and $f_y = 460$ N/mm^2.

Try a section 400 mm by 250 mm in size. Then if 25 mm bars are to be used with 8 mm links, d' will be $40 + 8 + 12.5 = 60.5$ mm, so that $d/h = 339.5/400 = 0.85$ and $d/b = 189.5/250 = 0.758$.

Next assume that $M_{uy} = 1.6M_y$ say. Then from the lower charts on *Data Sheets 53* and *54*, with $M_{uy}/b^2 hf_{cu} = 1.6 \times 40 \times 10^6/(250^2 \times 400 \times 30) = 0.085$ and $N/bhf_{cu} = 1250 \times 10^3/(250 \times 400 \times 30) = 0.417$, $\varrho_1/f_{cu} = 0.00065$ so that $\varrho_y = 0.0195$. Also assume that $M_{ux} = 1.6M_x$, so that, from the lower chart on *Data Sheet 52*, with $M_{ux}/b^2 hf_{cu} = 1.6 \times 100 \times 10^6/(250 \times 400^2 \times 30) = 0.133$ and $N/bhf_{cu} = 0.417$ as before, $\varrho_1/f_{cu} = 0.00095$ so that $\varrho_x = 0.0285$.

Now with $M_x/M_{ux} = M_y/M_{uy} = 0.625$, from the chart on *Data Sheet 60*, the corresponding minimum value of $N/N_{uz} = 0.49$. Thus the target value of N_{uz} is $0.417 \times 30/0.49 = 25.5$ N/mm^2. From the nomogram on *Data Sheet 60*, with $f_{cu} = 30$ N/mm^2 and $f_y = 460$ N/mm^2, $\varrho_{tot} = 0.0285$. Equations (14.7) are now as follows:

$$\varrho_1 = 0.0195 + 0.0285 - 0.0285 = 0.0195$$

$$\varrho_2 = 0.0285 - 0.0285 = 0$$

$$\varrho_3 = 0.0285 - 0.0195 = 0.009$$

so that $A_{sc1} = 0.0195 \times 250 \times 400 = 1950$ mm^2, $A_{sc2} = 0$ and $A_{sc3} = 0.009 \times 250 \times 400 = 900$ mm^2. These areas may be provided by employing four 25 mm bars (i.e. 1963 mm^2) for A_{sc1} and two 25 mm bars (i.e. 982 mm^2) for A_{sc3}.

If six bars, so arranged that three are near each shorter face, are to be used and the method suggested in the *Code Handbook* is employed to estimate the contribution of the two bars that are not located near the corners of the section, the procedure would be as follows. The centroid of the 1.5 bars located in each quadrant occurs at $(125 - 60.5)/1.5 = 43$ mm from the minor axis, and thus $b' = 125 + 43 = 168$ mm. Now since M_x/h' $(= 100 \times 10^6/339.5)$ exceeds M_y/b' $(= 40 \times 10^6/168)$, the modified bending moment that must be considered is $M'_x = M_x + \beta M_y h'/b' = (100 \times 10^6) + (0.514 \times 40 \times 10^6 \times 339.5/168) = 141.6 \times 10^6$, since $\beta = 1 - 7N/6bhf_{cu} = 0.514$. Then from the lower chart on *Data Sheet 52*, $\varrho_1/f_{cu} = 0.0008$, so that $\varrho_1 = 0.024$ and thus $A_{s\,req} = 2400$ mm^2. According to these calculations, six 25 mm bars would be insufficient.

If the values obtained for ϱ_1, ϱ_1 and ϱ_3 do not lead to a convenient bar arrangement, it is possible to adjust them by altering the ratio of N/N_{uz} (remembering that the figure read from the chart on *Data Sheet 60* is a *minimum* value). A higher ratio will give a correspondingly lower value of N_{uz} and hence require a lower value of ϱ_{tot}. Alternatively, different ratios of M_x/M_{ux} and M_y/M_{uy} may be selected.

14.8 SLENDER COLUMNS

When the ratio of effective height to overall thickness about either axis is equal to or more than the limiting value of 15 for braced columns of normal-weight concrete, or 10 for unbraced and lightweight concrete columns, the member must be considered as a long or 'slender' column. In CP114 and earlier codes such members were designed by multiplying the load-carrying capacity of an equivalent short column by a reduction factor related to the slenderness ratio. However, in both CP110 and BS8110 an entirely different method of dealing with slender columns has been introduced which is based on European Concrete Committee (CEB) recommendations; the background to the method is described in detail in ref. 21. Known as the *additional-moment* concept, the method is also applicable to axially loaded members that are not initially subjected to bending, as well as to members that support an axial load combined with bending about either or both principal axes.

In outline, the method consists of determining an additional moment corresponding to the theoretical deflection resulting from the slenderness of the column, and adding this to the initial applied ultimate moment, if any. The column section is then designed to resist the resulting combination of the total bending moment and axial load, using the normal sets of design charts prepared for this purpose. When a preliminary value has been determined for the amount of reinforcement necessary, provided the actual axial load exceeds that corresponding to the so-called *balanced* situation (see below), the value of the additional moment may now be reduced, if desired, by multiplying it by a further factor K, which relates to the load and the properties of the designed section. The reason why this reduction may be made is that, as the loading condition of the column more nearly approaches that of axial load only, the less likelihood there is of the column buckling, and the additional moment may therefore be reduced accordingly. The revised additional-moment value is now added to the initial applied ultimate moment, and the entire design cycle is repeated until any adjustments become negligible.

The additional moment corresponding to the slenderness of the member is specified in BS8110 (equation 35) as $M_{add} = Na_u$, where a_u is the calculated deflection. For rectangular or circular columns only, the Code states (equation 32) that a_u may be taken as $\beta_a Kh$. According to BS8110 (equation 34), for normal-weight concrete $\beta_a = (l_e/b')^2/2000$, while for lightweight concrete $\beta_a = (l_e/b')^2/1200$; these equations are represented by the scales on *Data Sheet 48*. Thus, for normal-weight concrete,

$$M_{add} = \frac{Nh}{2000}\left(\frac{l_e}{b'}\right)^2 \tag{14.8}$$

and for lightweight concrete

$$M_{add} = \frac{Nh}{1200}\left(\frac{l_e}{b'}\right)^2 \tag{14.9}$$

According to the Code, with bending in one direction only, l_e is the effective height of the column in the plane of bending considered (note that this definition differs from that in CP110) while b' is the smaller cross-sectional dimension of the column.

To determine the maximum initial moment M_i, BS8110 provides diagrams (*Figures 3.20* and *3.21* in the Code) illustrating the envelopes of final design moments for braced and unbraced columns respectively. With braced columns the Code states that M_i may be assumed to act near mid-height and may be taken as $0.4M_1 + 0.6M_2$, where M_1 and M_2 are the smaller and larger initial end moments induced by the design loads respectively. If, as is usual with fixed ends, the column is bent in double curvature, M_1 should be taken as negative. In addition, M_i must never be less than $0.4M_2$: this corresponds to the situation where $M_1 = 0.5M_2$.

The section must now be designed to resist a total moment M_t which, from the diagrams given in BS8110, will be the maximum of the following four values: M_2; $M_i + M_{add}$; $M_1 + 0.5M_{add}$; and $e_{min}N$. In the case of columns bending about a single axis, provided that the longer dimension of the section h is less than three times the shorter b, and, for columns bent about their major axis, if l_e/h does not exceed 20, the section may now be designed for the values of N and M_t thus obtained. If these two conditions are not met, the Code recommends that the section should be designed for biaxial bending with no initial moment about the other principal axis.

The Code states that when bending is significant about both axes, additional moments must be calculated as described above for both directions in which bending occurs; when calculating the additional moments, the value taken for b' in expressions (14.8) and (14.9) must now be the depth of cross-section in the plane being considered. These moments are then added to their respective initial moments to obtain total design moments in both directions, and these are finally combined as described in section 14.7. BS8110 does not make clear what moment should be considered as 'significant', but *Allen* suggests that if the calculated moment in either direction is sufficiently small for $e_{min}N$ to be the determining moment in that direction, then this moment may be considered to be insignificant and the section designed for uniaxial bending only. Alternatively, the CP110 design procedure for biaxial bending may be utilized, although this is extremely protracted with slender sections unless computer analysis is available.

Unlike the previous situation where the deflection of each individual column may be considered individually, with unbraced columns it is necessary to examine the behaviour of all columns at a given storey, as these are usually constrained to deflect laterally by a similar distance. BS8110 therefore requires the average ultimate deflection a_{uav} to be assessed by dividing the sum of the individual deflections a_u for each column involved by the number of columns. If for any individual column $a_u > 2a_{uav}$, BS8110 states that the value of a_u for this column should be disregarded and a modified average, based on the adjusted number of columns, be calculated.

As indicated by the appropriate diagram in the Code, unbraced columns will be bent in single curvature and thus the additional moments resulting from deflection will increase the initial moments at the column ends. The full additional moment M_{add} may thus be considered to act at the column end having the greater restraint (i.e. the stiffer joint), and the additional moment acting at the opposite end may be reduced in proportion to the ratio of the joint stiffnesses at the two ends.

As with braced columns, in the case of a column bending about a single axis, provided that the longer dimension h is less than three times the shorter b for a column bending about the major axis with $le/h \leqslant 20$, and if $M_2 > M_1$, the maximum design moment $M_t = M_2 + M_{add} \geqslant Ne_{min}$. If $h \geqslant 3b$ or $l_e/h > 20$, design the section for biaxial bending with an initial moment of zero about the other principal axis. BS8110 also sets out absolute slenderness limits. For any column the clear distance between end restraints must not exceed 60 times the smaller column dimension b'; nor in the case of an unbraced unrestrained column may it exceed $100b'^2/h'$, where h' is the greater column dimension.

One important point to which attention is not drawn in either the Code or the *Code Handbook* is that, when designing slender columns subjected to biaxial bending, care must be taken when combining the moments in the two directions if the effective lengths or end fixities or both differ about the principal axes. It is not unusual to encounter a situation where the columns are braced in one direction but unbraced in the other, and the case of columns being restrained at mid-height by fascia beams but entirely unrestrained at right angles is not unknown. In such circumstances the designer must be clear about which values of M_t relating to each principal axis must be combined, particularly if several differing combinations of load and moment (or moments) must be considered as well. In such a situation it may be simpler to combine the maximum values of M_t in the two directions irrespective of the actual position along the column at which each occurs.

As mentioned earlier, when the section has been designed, the additional moment M_{add} may be reduced by multiplying it by a factor K, where $K = (N_{uz} - N)/(N_{uz} - N_{bal}) > 1$. In this expression N is the ultimate axial load for which the section is being designed; N_{uz}, the resistance of the section to pure axial load, is equal to $0.45f_{cu}A_c + 0.87f_yA_{sc}$. N_{bal} is defined as the design axial load capacity of a balanced section. This corresponds to the axial load at which, in the section considered, a maximum compressive strain of 0.0035 in the concrete and a tensile strain of 0.002 in the outermost layer of tension reinforcement are attained simultaneously. By sketching a diagram of the distribution of strain across the section, it can be seen that this situation occurs when $x = 7(h - d')/11$ and, provided that d'/h does not exceed 3/14, the tensile stress in the tension reinforcement is then equal to the compressive stress in the compression

steel. Consequently, if equal amounts of reinforcement are provided at equal distances from the respective faces of a rectangular section, the forces in the two areas of steel are equal and opposite, and thus the resistance of the section to axial force corresponds to that provided by the concrete section alone. In such a case the *amount* of steel provided does not influence the value of N_{bal}, although the *position* of this reinforcement does. If the section is not rectangular, the reinforcing bars are disposed unsymmetrically; or if $d'/h > 3/14$, N_{bal} must be calculated from first principles.

With a uniform rectangular concrete stress-block, with equal amounts of symmetrically disposed reinforcement and with $d'/h \leqslant 3/14$, it can be shown that $N_{bal} = 14bdf_{cu}/55$, assuming that $d = h - d'$. Values of N_{bal}/bhf_{cu} derived from this relationship may be read from the appropriate scale on *Data Sheet 54*. For simplicity, BS8110 states that for symmetrically reinforced rectangular sections N_{bal} may be taken as $0.25f_{cu}bd$.

Allen was the first to draw attention to the fact that the appropriate values of K can actually be plotted on the charts that are used to design sections to resist combined bending and direct thrust, and this has been done on the charts provided in Part 3 of BS8110 and on *Data Sheets 50 to 59*. However, in the case of the Code charts, there is a slight error for those cases (nos 21, 26, 31, 36, 41 and 46) where $d/h = 0.75$ since, with such a ratio, d'/h exceeds 3/14 and the so-called 'balanced' load is no longer independent of the amount of steel provided.

In evaluating N_{uz}, the question yet again arises as to whether the gross or the net area of concrete should be substituted for A_c in the relevant formulae. Although BS8110 specifically states that A_c represents the net cross-sectional area of concrete, in preparing the design charts to resist combined bending and direct thrust in Part 3 of BS8110 no allowance was made for the area of concrete displaced by the reinforcement. Consequently, it seems desirable to make the same assumption when calculating values of N_{uz} to be used in conjunction with the charts. Furthermore, it will be noticed that if N_{uz} is taken to be lower than its true probable value, as may be the case if A_c is considered to be the net area of concrete, the resulting value of K will be lower than it should be, leading in turn to a lower additional moment than should be considered. Thus it is recommended that A_c should be taken as the gross area of concrete, and this has been done in the expression given above to calculate values of K. However, for uniformity with the nomogram for designing axially loaded columns, in the nomogram giving N_{uz} on *Data Sheet 60* an allowance has been made for the area of concrete displaced by the reinforcement (graphical data based on the alternative assumption are included in *RCDH*). Such an allowance has not been made in calculating the values of K on the charts on *Data Sheets 50 to 59*, because this cannot be done when giving curves for ϱ_1/f_{cu}. In practical design it makes no significant difference whether the net or gross concrete section is considered, but such differences become important if precise mathematics is involved, as happens if the equations are being programmed for solution using a computer.

The additional-moment concept employed in BS8110 (and CP110) displays various anomalies and has been widely criticized. For example, where a square column section resists uniaxial bending, if l_e/h exceeds 20, BS8110 requires the section to be designed for either uniaxial or biaxial bending depending on which of the two principal axes is considered the major, and a considerable difference in the amount of steel needed results from the choice made; in this instance it seems that the intention of the Code is that biaxial bending with zero moment about the minor axis should be considered. The principal problem is that as the limits set by the ratios $l_e/h = 20$, $h/b = 3$ etc. are crossed, the calculated steel areas may change dramatically; in extreme circumstances the load-carrying capacity of a given section bending about the major axis may be virtually halved instantaneously as the value of l_e/h exceeds 20. Critics have suggested that the additional-moment concept should be replaced by an updated version of the method employed in earlier codes, where the moment-carrying capacity of a slender column was determined by multiplying that of an equivalent short column by a reduction factor relating to the slenderness ratio. This is the method set out in the current permissible-stress code drafted by the Institution of Structural Engineers.

The BS8110 procedure for designing slender columns subjected to axial load and uniaxial bending is illustrated in the following example.

14.8.1 Slender columns: worked example

Design a column pinned at both ends to resist an ultimate bending moment of 140 kN m about the major axis and a direct force of 2 MN, if the effective length of the column about both axes if 8.750 m. Assume that grade C30 normal-weight concrete and mild steel are to be used in moderate exposure conditions.

For grade C30 concrete and moderate exposure, a minimum cover to all steel of 35 mm is required. Thus, assuming that 8 mm links are used with 32 mm main bars, d' will be $35 + 8 + (0.5 \times 32) = 59$ mm. Try a section with $h = 550$ mm and $b = 350$ mm, so that $d = 550 - 59 = 491$ mm and $d/h = 0.893$. Since $l_e = 8750/350 = 25$, the corresponding additional-moment factor read from the appropriate scale on *Data Sheet 60* is 0.313, and thus $M_t = M_i + \beta_a Nh = (140 \times 10^6) + (0.313 \times 2 \times 10^6 \times 550) = 484 \times 10^6$ N mm. Now with $M_t/bh^2 f_{cu} = 484 \times 10^6/(350 \times 550^2 \times 30) = 0.152$ and $N/bhf_{cu} = 2 \times 10^6/(350 \times 550 \times 30) = 0.346$, interpolation between the upper charts for $d/h = 0.85$ and $d/h = 0.90$ on *Data Sheets 51* and *52* indicates that $\varrho_1/f_{cu} = 0.0013$ and $K = 0.84$, so that $\varrho_1 = 0.0013 \times 30 = 0.039$.

The modified value of M_t is thus $(140 \times 10^6) + (0.84 \times 0.313 \times 2 \times 10^6 \times 550) = 429 \times 10^6$ N mm. Then

$M_t/bh^2 f_{cu} = 429 \times 10^6/(350 \times 550^2 \times 30) = 0.135$ and, with $N/bhf_{cu} = 2 \times 10^6/(350 \times 550 \times 30) = 0.346$ as before, interpolation between the charts indicates that $\varrho_1/f_{cu} = 0.0012$ and $K = 0.83$.

If a further cycle is undertaken, it will be found that the adjusted value of M_t is therefore $(140 \times 10^6) + (0.83 \times 0.313 \times 2 \times 10^6 \times 550) = 425 \times 10^6$ N mm, and further use of the charts gives a final value of ϱ_1/f_{cu} of 0.0011, i.e. $\varrho_1 = 0.033$. Thus the area of steel required is $0.033 \times 350 \times 550 = 6352$ mm^2. Provide eight 32 mm bars ($A_{sc} = 6434$ mm^2).

14.9 APPLICATION TO EXTERIOR COLUMNS

The design of two exterior columns, D3 and C1, of the building in *Drawings 1* and *2* in accordance with BS8110 is given on *Calculation Sheets 19* and *20* and *Drawings 15* and *16*. Some of the methods of calculation described in this chapter are used. The calculations for the remaining columns are similar. Column C1 is not considered at every storey, but the calculations that are omitted do not demonstrate different principles. The loads on each column are based on simple 'static reactions', since continuity with the beams reduces the difference between such reactions and 'elastic reactions' (see section 6.1) to such an extent that the apparently more accurate calculations are not worth while undertaking. The self-weights of the columns are calculated on the floor-to-soffit height of the columns, as the self-weights of the beams are determined from the gross lengths of these members.

To calculate the stiffnesses of columns and beams the moment of inertia is based on the gross concrete section only, the reinforcement being ignored. In general, the cover of concrete over the main reinforcement is 30 mm in the upper storeys of column D3, since the outer face of this column is protected by brickwork, but 50 mm of cover is provided between the ground floor and first floor, to ensure that there is at least 40 mm of concrete over all the reinforcement (including the links) where the concrete is exposed to the weather. Similarly, 60 mm of concrete cover is provided outside the main bars throughout column C1 since, although this column when first constructed may be protected by the walls of the adjacent building, it is not known if this condition will apply throughout its life. The width of column C1 is 400 mm in each storey to ensure that each brick panel is the same length, i.e. 4.600 m. Metal ties are built into the column to secure the brickwork. A small rebate is provided on the beams and column C1 to ensure that the joint between the brickwork and the concrete is lightproof and weatherproof. The small increase in width at the inner face of column C1, due to the projections forming this rebate, is neglected in the calculations. A similar rebate is not required on column D3 because the outer leaf of the brick wall passes in front of the column.

The calculation of the bending moments due to the monolithic construction of the columns and beams follows the methods described in this chapter. The factors of $\frac{1}{8}$ and $\frac{1}{12}$, employed in calculating the fixed-end moments for column D3, are for a central concentrated load and distributed load respectively and are taken from *Data Sheet 48*. The corresponding coefficient of $\frac{1}{10}$ for column C1 is for the trapezoidal load on the beam supporting part of the roof slab spanning in two directions; the coefficient for a triangular load is $\frac{5}{48}$ (*Data Sheet 48*) and the trapezium in this case is nearly a triangle.

The maximum bending moments on a column occur theoretically at the intersection of the axes of the beam and column. The critical planes of the column are at the level of the soffit of the beam and at the top of the floor slab, at which sections the bending moments are less than the maximum bending moments. To make the necessary reduction, it can be assumed that the point of contra-flexure is at the mid-height of the column, except where there is a large difference between the bending moment at the top and bottom of the column in a single storey. From the bending-moment diagrams on *Calculation Sheets 19* it is seen that this assumption is fairly accurate except in the top and bottom storeys. On *Calculation Sheets 19* and *20* the ratios 1.55/1.80 and 3.20/3.60 are approximate reduction factors to determine the ultimate bending moment at the soffit of the beam in the inter-mediate storeys; the denominator in each fraction is one-half of the storey height. It is necessary to calculate more accurately the reduction in bending moment at the top of columns C1 and D3, as is done on *Calculation Sheets 19* and *20*. Likewise, column D3 requires separate consideration between ground and first floors. Compared with the size of the column, the more-massive construc-tion of the basement wall, combined with the ground floor, produces conditions that are almost equivalent to complete fixity at ground-floor level; the bending moment on the column at this level is therefore about one half of the bending moment that is applied at the upper end of the column in this storey. The diagram on *Calculation Sheets 19* explains the reduction factor of 2.65/2.90 on this calculation sheet. The bending moment at the bottom of this column is taken into account in the calculations for the basement retaining wall in section 15.7.4.

The column section in any storey should be designed for the ultimate bending moments which may occur at the top or bottom of the column, and which are determined from the separate calculations for the upper and lower junctions with the beams. It is generally more convenient when designing columns to proceed downwards from the top of the building, so that reinforcement required to resist the combined bending and axial load at the top of the column in any storey is determined before the requirement at the bottom of the same storey is investigated. When the bending moment at the bottom has been determined, it should be compared with that at the top, and if it is greater the section provided at this point should be checked. An example of this occurs on column A1 between the fourth and fifth floors. The

ultimate bending moment at the fourth floor is 47.88 kN and at the fifth floor the bending moment is 42.47 kN. As the effects of the smaller bending moment are small, recalculation is unnecessary. If the difference in the bending moments at the top and bottom and the pro-portion of the moment to the axial load at the top of this storey make it advisable to investigate conditions at the bottom, the fact that tension occurs in the outer column face at the top and at the inner column face at the bottom must not be overlooked.

In buildings it is not generally necessary to investigate the case of the minimum axial load (i.e. omitting the imposed load) combined with the corresponding bending moment. It may, however, be necessary to consider this case if the maximum axial load is very little or no greater than the so-called balanced load.

The bending moment on the beam at each intersection is the sum of the ultimate bending moments on the lower and upper columns at the intersection. Although not required for the design of the columns, the bending moment is given for comparison with the ultimate bending moments assumed in designing the beams (see *Calculation Sheets 19* and *20*).

Since the bending moments are greatest at the top and bottom of each storey height of column, less reinforce-ment may theoretically be required at about the middle of this height because the axial load is combined with a diminishing bending moment. In large columns it is sometimes economical to provide fewer or smaller-diameter bars over the middle part of the shaft than at the ends. If this is done in the case of a slender column, however, it should be remembered that the additional moments to be considered to cater for the slenderness must be based on the additional-moment diagrams provided in BS8110. In comparatively small columns such as those shown in *Drawings 15* and *16* it is clearly not worth while reducing the reinforcement provided at the ends of the column.

Where the bending moments on a column section cannot reverse in direction (e.g. such as where they result from the loading on the adjoining floor beams), it may be advantageous to arrange the reinforcement unsym-metrically within the column section. For example, where the ultimate bending moment divided by the ultimate axial load is equal to the distance of the 'compression' steel from the centre-line of a rectangular section, a theoretical saving of 50% of the reinforcement can be made, compared with that needed with a symmetrical arrangement. However, the *Code Handbook* states that there are 'very strong practical and behavioural reasons for employing symmetrically reinforced sections' (although it does not explain what these are).

The length of the overlap of the bars at each floor must be sufficient to transfer the stress from one bar to another (as discussed in section 13.2), but taking into account whether the bar is in compression or tension. Generally the length corresponding to the maximum design stress is provided, in spite of the fact that the calculated design stress may actually be less (because more reinforcement is provided than the minimum needed); by so doing, some allowance is made for any incidental stress in excess of that calculated. Since the bars must be in the positions assumed in the calculations, the cranks should therefore be in the direction that maintains all the bars at a splice in this position relative to the direction of the moment (i.e. the direction of the crank should be at right angles to the moment).

The shearing forces due to the ultimate bending moments on exterior columns are generally so small that it is unnecessary to calculate the shearing stresses. Consider, for example, the bottom storey of exterior column D3. According to *Calculation Sheets 19*, the bending moment at the level of the first floor is 84.8 kN/m, changing to a reverse moment of 42.4 kN/m at the ground floor. Since the shearing force is equal to the rate of change of the bending moment, it is therefore $(84.8 + 42.4)/4.35 = 29.3$ kN. On the gross cross-section of the column this force produces a shearing stress of $(29.3 \times 10^3)/(500 \times 300) = 0.19$ N/mm², which is negligible, especially as the principal tensile stress due to the shearing stress and the direct stress may be considerably reduced by the compressive stress in the column. BS8110 states that there is no need to check the shear resistance of rectangular columns where M is less than $0.75\,Nh$ (provided that the shear stress does not exceed the lesser of $0.8\sqrt{f_{cu}}$ or 5 N/mm²). This relationship is indicated by a chain line on the charts on *Data Sheets 50* to *54*.

14.10 EFFECTS OF WIND

As explained in section 2.8.6, the columns of the building in Part Two need not be designed to resist the effects of wind if the building is stiffened by transverse end walls of concrete. This condition is assumed in the design of exterior column D3.

If concrete end walls are provided, there are no separate columns along lines 1 and 6, and each wall is designed as a load-carrying wall, as described in section 15.5. However, if brick transverse walls are provided, all the columns, including columns C1, must be designed to withstand the bending moments produced by the wind. The bending moments resulting from the floor loading act in a plane parallel to the longitudinal axis of the building, whereas the greatest bending moments due to wind action act in a plane at right angles to this. Consequently, the columns must be designed to resist biaxial bending, as shown on *Calculation Sheets 20*.

The bending moments on columns due to wind can be calculated by analysing the entire structural frame, either by hand (using a method such as moment distribution) or by computer. Building frames of one or two storeys can in general be readily analysed by hand methods, but for multistorey buildings such calculations are tedious and time-consuming, even when relatively simple procedures such as moment distribution are employed.

BS8110 describes a simplified analytical method that it states may be used for most frames subjected to vertical and lateral loading to obtain the moments, loads and forces for which the individual frame members must be designed. The frame should first be divided into a series of sub-frames, each consisting of the beams comprising one floor together with the columns above and below, these columns being assumed fully fixed at their far ends. All the beams forming each sub-frame should carry a load of $1.2(G_k + Q_k)$ and all lateral loading is ignored.

Next, an analysis of the complete frame should be made, assuming points of contrflexure (i.e. zero moment) at the midspan of each beam and the mid-height of each column, under the action of a lateral load of $1.2W_k$, all vertical loads being ignored. Unless absolute rigidity of the column footing is assured, it may be more realistic to assume that the ground-floor columns are pinned at their feet, rather than to adopt points of contraflexure at mid-height. Where it is possible that overturning may occur, such as in narrow tall buildings or cantilevered frames where tension may arise in the windward columns, an alternative combination of loading of $1.0G_k$ and $1.4W_k$ should be considered when making these analyses.

The maximum moments obtained by summing the results of the two foregoing analyses (remembering that a reversal of the wind direction will reverse the signs of the values obtained in the second analysis) should be compared with the corresponding values obtained by carrying out a simplified analysis for vertical loads of $1.4G_k + 1.6Q_k$ and $1.0G_k$ only, as described in section 3.8.1, the members then being designed to withstand the greater of the values obtained. The moments resulting from these analyses may be redistributed by up to 10%, provided that the conditions described in section 3.3 governing such redistribution are complied with; the restriction on the ratio of x/d normally prevents the possibility of any redistribution being made on column sections unless the corresponding axial load is low.

In the above approximate analysis of the entire frame under lateral load only, the wind pressure on the wall of each storey is divided into two equal forces acting horizontally at the two floors comprising the storey. The total horizontal shearing force on all the columns in any storey is the sum of all the horizontal forces above the columns concerned, including the horizontal force acting at the floor forming the ceiling of the storey. This shearing force should theoretically be divided between the columns in proportion to their moments of inertia (since the lateral deformation is assumed to be constant). This precise division may not be worth while if the effects of wind are small compared with the effects of dead and imposed loads, as in the building considered in Part Two, and a practical method is to assume that the shearing force on all interior columns is the same and is equal to twice that on the two exterior columns. The maximum bending moment at the top and bottom of any column is then equal to the shearing force on the column multiplied by one-half of the storey height, the bending moments being such that tension tends to be produced on the windward face at the base of the column and the leeward face at the head of the column in any storey. The direction of the wind giving the most adverse combination of moment, when combined with those due to vertical loads, must therefore be considered. For example, for a column such as D3, wind on the back of the building produces bending moments that increase the total bending moments if the building is not stiffened. For a column such as C1 in an unstiffened building, wind on either the front or the back of the building produces the same additional stresses, although on different faces of the column. Increases or decreases in the axial load on any column are secondary effects caused by wind, and although they may be important in one- or two-bay frames (especially in the case of exterior columns) they can normally be ignored for buildings having three or more bays (as with that in Part Two), except in the case of very tall structures. The wind also induces additional moments in the floor and roof beams and, with the same assumptions made above, the appropriate moment is equal to one-half of the sum of the bending moments on the columns above and below the beam/column intersection concerned.

For very tall buildings, the bending moments and shearing forces due to wind may be of equal or greater importance than those due to dead and imposed loads; such frames must be analysed more accurately, usually using a computer.

The bending moments due to wind are then calculated by the following approximate method. The maximum characteristic wind load on the building, as described in section 2.8.6, is 760 N/m^2, and thus $1.2w_k = 912$ N/m^2. Then

One-half total pressure on top storey:
½ × 29.250 × 3.150 × 0.912   = 42.02 kN
Total number of columns in top storey = 18
Ultimate shearing force on each column
    in top storey = 42.02/18 = 2.34 kN
Ultimate bending moment on each column
    in top storey
    = ½ × 3.0 × 2.34 = 3.50 kN/m

One-half total pressure on top storey
    (as before)   = 42.02
One-half total pressure on fifth storey:
½ × 29.250 × 3.600 × 0.912   = 48.02
                                       _____
Total shearing force on columns in fifth
    storey   = 132.06 kN
Total number of columns in fifth
    storey = 24
Ultimate bending moment on each column
    in fifth storey

$$= \frac{132.06}{24} \times \frac{3.6}{2} = 9.90 \text{ kN m}$$

One-half total pressure on fifth storey
    (as above)   = 48.02

One-half total pressure on fourth storey
    (as fifth storey)   = 48.02
                                       _____
Total shearing force on columns in fourth
    storey   = 228.10 kN
Ultimate bending moment on each column
    in fourth storey

$$= \frac{228.10}{24} \times \frac{3.6}{2} = 17.11 \text{ kN m}$$

One-half total pressure on fourth storey
    (as above)   = 48.02
One-half total pressure on third storey
    (as fifth storey)   = 48.02
                                       _____
Total shearing force on columns in third
    storey   = 324.14 kN
Ultimate bending moment on each column
    in third storey

$$= \frac{324.14}{24} \times \frac{3.6}{2} = 24.31 \text{ kN m}$$

and so on.

This simplified calculation considers only a single (maximum value of w_k, i.e. that corresponding to the top of the building. The wind forces, and hence the moments, may be reduced somewhat by determining the appropriate value of S_2 and thus w_k at each individual storey level (see section 2.6). Then, for the present example, the wind pressures are as follows:

	Height (m)	S_2	V_s (m/s)	w_k (kN/m^2)
At roof	21.90	0.76	33.44	0.686
At fifth floor	18.75	0.735	32.34	0.641
At fourth floor	15.15	0.69	30.36	0.565
At third floor	11.55	0.64	28.16	0.486
At second floor	7.95	0.59	25.96	0.413
At first floor	4.35	0.54	23.76	0.346

If these values are employed in a similar analysis to that given above, the resulting ultimate bending moments are as follows:

On each column in top storey	2.63 kN m
On each column in fifth storey	7.27 kN m
On each column in fourth storey	12.03 kN m
On each column in third storey	16.18 kN m

and so on. These values are thus about two-thirds to three-quarters of the more approximate ultimate bending moments obtained earlier. However, reference to *Calculation Sheets 20* shows that the actual values of the wind forces have no effect on the area of reinforcement required in this particular column, since the minimum permissible amount of steel is sufficient to withstand without difficulty the greater moments obtained by the approximate calculations.

These ultimate bending moments are then combined with those that result when each floor is analysed as a separate structure under the action of vertical loads of $1.2G_k + 1.2Q_k$ on all spans. Thus an analysis such as that shown in *Figure 3.11* must be undertaken for each floor in turn, using the precise moment-distribution method described in Chapter 3, or some similar method. If this is done (spaces does not permit these calculations to be included here), the moments shown in *Figure 14.6b* are obtained. Then, if these are added to the ultimate moments due to wind action as calculated above (*Figure 14.6a*), the final maximum ultimate moments in the columns are as shown in *Figures 14.6c*.

These moments (which for convenience can be termed as due to loading case 1) should be compared with those obtained by considering the action of maximum and minimum vertical loadings of $1.4G_k + 1.6Q_k$ and $1.0G_k$ respectively, so arranged that they give the most severe moments at each critical section (case 2). Then for each individual section, the moments obtained by case 1 and case 2 loading should be compared, and each section designed to resist the greater. Since the maximum moments in the exterior columns and the corresponding beam junctions will normally result from case 1, only this case is considered in the present example. However, for interior columns and beam junctions, provided that the beams forming adjacent spans do not differ greatly in stiffness and length, the question of whether case 1 or case 2 loading will produce the greater moments will depend on whether the effects due to a so-called 'imposed' vertical load of $1.6Q_k + 0.4G_k$ outweigh those that result from a lateral load of $1.2W_k$.

The foregoing analysis assumes that the effects of sway due to the unsymmetrical disposition of the vertical loads can safely be neglected. If this is uncertain, the complete frame must be analysed as a whole, and for such an analysis access to some form of computer is well-nigh essential.

To design a column such as C1, the ultimate moments in the transverse plane obtained above must be combined with those in the longitudinal plane due to frame action. As shown on *Calculation Sheets 20*, these are obtained in the same way as those for column D3. However, in this case the loading on the floor considered is $1.2(G_k + Q_k)$, as this is the vertical loading that has already been considered in order to obtain the moments in the lateral plane. The column sections are then designed, as shown on *Calculation Sheets 20*, to resist combined biaxial bending and axial load as described in section 14.7. Normally, biaxial bending proves to be the most onerous condition on such a column as C1, but in certain cases the effects of bending in one direction due to a vertical loading of $1.4G_k + 1.6Q_k$ should also be checked.

From *Figure 14.6* it is clear that the corner columns (i.e. A1 and D1) are subjected to greater lateral moments due to the combined action of wind and floor loading than columns B1 and C1. However, since the corresponding edge beams running longitudinally support only one-half of the area of slab carried by the secondary beams framing into columns B1 and C1, the resulting moments applied to the corner columns in this direction will be less. Furthermore, the axial load on these columns will also be less owing to the smaller floor area being supported by them. Consequently, the resulting conditions due to biaxial moments combined with axial load are probably no more critical for columns A1 and D1 than for column C1, which is considered in detail here. However, in practice, detailed calculations would need to be prepared for each column or type of column.

14.11 COLUMNS SUPPORTING FLAT SLABS

Interior and exterior columns supporting flat-slab floors or roofs are subjected to bending moments transferred from the slabs and must be designed accordingly. The ultimate bending moments on the columns are related to the ultimate bending moments in the slab and may be calculated by using the coefficients accompanying the simplified analytical method described in BS8110 and considered in section 11.1. Alternatively, if the flat slabs are analysed rigorously as frames acting in conjunction with the columns, the ultimate bending moments must be determined by an analytical method such as that outlined in section 11.6.2. If the interior columns are provided with enlarged heads (see section 11.1.3), the size of these heads should be taken into account when calculating the bending moments. The heads of the exterior columns should also be sufficiently large to justify this method of calculation, and therefore such columns should have as much as possible of a complete column head.

In the simplified Code method is adopted, the bending moment on an internal column is $0.022nl_1 l_2(l_1 - 2h_c/3)$, where l_1 and l_2 are the panel dimensions parallel to and at right angles to the direction of span currently being considered, h_c is the effective diameter of the column head (see section 11.1.3) and n is the total ultimate load on the panel (i.e. $1.4G_k + 1.6Q_k$). For edge and corner columns the corresponding expression is $0.04nl_1 l_2(l_1 - 2h_c/3)$, and this latter value may be limited by the moment that may be considered to be transferred from the adjoining column strip at the edge of the slab (see section 11.1.3). In all cases the total bending moment should be divided between the upper and lower columns at any floor in proportion to their relative stiffnesses.

The critical conditions concerning an internal column must now be considered. The critical section is immediately above the first floor, where the greatest ultimate load is due to the roof, four fully loaded floors, and five storeys of column. The ultimate bending moment is that due to the action of the slab on the first floor. Conditions are less severe at the top of the column, since the bending moment is less because the critical section here occurs at the underside of the column head. At this section the load is also less, and is due to the columns of four storeys, to three fully loaded floors, and to one partially loaded floor (the second) because, in order to exert a bending moment on an interior column, the second-floor panels supported by the columns cannot be

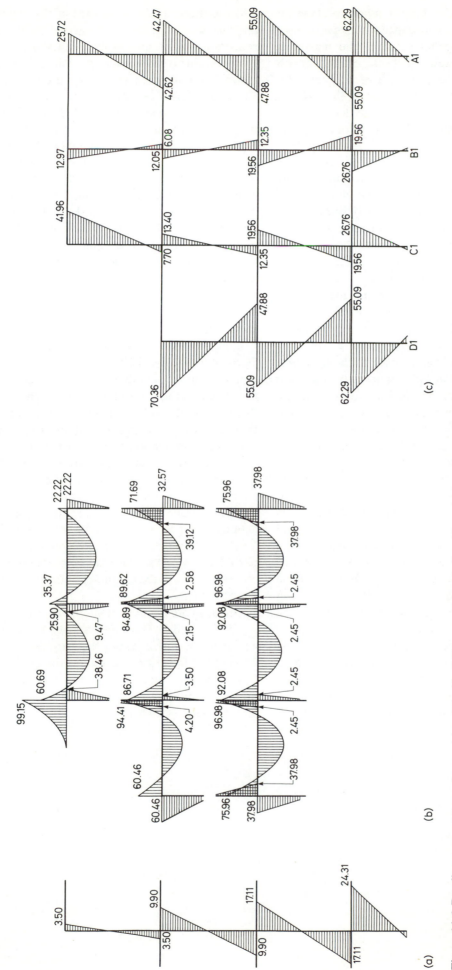

(a)

(b)

(c)

Figure 14.6 Bending moments on columns A1 to D1 of building considered in Part Two: (a) in columns due to wind load of $1.2W_k$ only; (b) in individual floor systems due to vertical loads of $1.2G_k + 1.2Q_k$ only; (c) in columns due to combined loading (moments in beams omitted for clarity).

fully loaded. A reduction of load is made to allow for multistorey construction in accordance with *Data Sheet 4*.

In the case of an external column, the maximum bending moment is caused by both of the panels supported by the column being fully loaded. Therefore, for an external column in the second storey, the load at a critical section immediately above the first floor is due to the weight of the column and walls in five storeys, and to the four fully loaded floors and roof. BS8110 does not give coefficients for bending moments acting on corner columns, but reasonable rules can be formulated from the requirements for edge columns. The bending moment on a corner column can be divided into two, one moment acting in the plane of the longitudinal centre-line of the column and the other in a plane at right angles thereto. Each bending moment can be determined from the expression $0.04nl_1 l_2(l_1 - 2h_c/3)$, where l_1 and l_2 are transposed as necessary. The column section must then be designed for biaxial bending as described in section 14.7.

The greatest bending moments due to wind are produced on the columns of the building in Part Two when the wind is blowing on the front or back of the structure, and the plane of bending is at right angles to the direction of the column strip; the bending moment on the column strip determines the bending moment transferred to the column due to the loads on the floor. On *Calculation Sheets 16*, only the effects caused by the moment due to the floor loads are considered. Since the strength of the section is amply sufficient, the effects of wind need not be investigated specifically, as the resistance to the combined forces (taking account of the reduced partial safety factors for load that are required with dead + imposed + wind loading) will clearly still be well within the capacity of the section. If it were necessary to carry out more-detailed calculations for the effects of wind action, these should be made as shown on *Calculation Sheets 19* and *20*, using once again the methods for dealing with biaxial bending described in section 14.7.

If the floor cantilevers beyond the outer row of columns, as in the type of construction described in section 11.2, it seems reasonable to reduce the bending moments on the edge columns by the amount of negative ultimate bending moment (using a partial safety factor of 1.0) produced by the edge load of the cantilevered slab and the weight of any wall carried by the slab; such a reduction was specifically sanctioned by CP110. However, in no case should the reduction be such that the resulting moment is less than would occur if the column were considered as internal. As an example, consider column D3 (*Drawing 12*) at the level of the first floor, where there is a canopy as shown on *Drawing 4*. Using the BS8110 coefficients, the ultimate bending moment transferred to the column from the slab is $0.04 \times 17.18 \times 6.00 \times 5.00 \times [5.00 - (2 \times 1.25/3)] = 85.9$ kN m. The cantilever bending moment due to the dead load of the canopy only is $2.1(\text{av.}) \times 1.00^2 \times 0.5 = 1.05$ kN m per metre run, i.e. $1.05 \times 6 = 6.3$ kN m total. The bending moment for which these columns should be designed is therefore 79.6 kN m, which is divided between the upper and lower columns in proportion to their relative stiffnesses. If the upper and lower columns are of the same cross-section, the stiffnesses are inversely proportional their lengths of 3.6 m and 4.35 m respectively, i.e. the bending moment at the base of the upper column is $79.6 \times 4.35/(4.35 + 3.60) = 43.55$ kN m. This bending moment must be combined with a direct load equivalent to the sum of the loads and weights of the terrace at fifth-floor level; the fully loaded second, third and fourth floors; and four storeys of column. The forces resulting from this combination must not exceed the resistance of the section. If the bending moment due to wind, which is of course the same as that calculated for the exterior column D3 when supporting beam-and-slab floors and acts in the same plane as the bending moment due to the action of the floor slab, is algebraically combined with the latter bending moment and corresponding direct load, the partial safety factors for load employed should be those corresponding to the condition of dead plus imposed plus wind loading.

The ultimate bending moment at the top of the column is $79.6 \times 3.60/(4.35 + 3.60) = 36.05$ kN m due to the action of the flat slab, but this can be reduced to about 27 kN m at the critical section at the base of the column head. The corresponding direct load is the same as for the upper column plus the load and weight of the first floor and the weight of a 6 m length of the canopy.

Chapter 15
Walls

The four principal types of concrete wall that occur in buildings are (a) load-bearing walls that support floors or other structural parts, and which are subjected to vertical loading with or without bending; (b) panel walls that are merely partitions which carry no vertical loads, but which may be subjected to lateral loads due to wind pressure or suction; (c) basement walls or retaining walls which normally carry some vertical load and which, in addition, have to resist the pressure exerted by retained ground; and (d) non-structural walls which do not stiffen a building or support any load, and which must be carried by the floors as described in Chapter 2. The latter may be formed of blocks of normal or lightweight concrete. In this chapter, the design of walls of types (a), (b) and (c) is dealt with.

15.1 LOAD-BEARING WALLS: DEFINING REQUIREMENTS

Three principal criteria control wall design according to BS8110, namely the proportion of reinforcement provided, whether the wall is short (the term actually used throughout this section of the Code is 'stocky') and whether the wall is braced or unbraced.

15.1.1 Proportion of reinforcement

According to the Code, a reinforced concrete wall is one in which the area of vertical reinforcement provided is not less than $0.004bh$; if less steel is provided, the wall is assumed to be of plain concrete. It is assumed that the breadth b exceeds four times the thickness h; if this is not so, the section is considered to be a column and the corresponding minimum reinforcement requirements apply (see section 13.2).

The minimum thickness and minimum cover to reinforcement necessary for fire resistance also depend on the area of vertical steel provided. Table 4.6 in Part 2 of BS8110 specifies differing criteria for normal-weight concrete with amounts of vertical reinforcement of less than $0.004bh$, more than $0.1bh$, and between these two values, and also for lightweight concrete (where the amount of steel does not influence the values that must be adopted).

15.1.2 Bracing

BS8110 defines a braced wall as one where the lateral stability of the entire structure at right angles to the plane of the wall being considered is provided by walls (or other means) which are arranged to resist all the forces acting in that direction. If this stability is not provided, the wall must be considered as unbraced. Note that in no direction may the overall stability of a multistorey structure depend on the resistance provided by the unbraced columns or walls alone. According to the *Handbook* to CP110, at least one-quarter of the lateral load had to be resisted by columns or walls provided to assist the unbraced members. The current *Code Handbook* merely states that in no direction may lateral stability be provided solely by walls bending about their minor axes, and that the wall arrangement must be such that a stiff structure to resist lateral load results.

15.1.3 Slenderness

If the effective height l_e is not greater than 15 times the thickness h of a wall of normal-weight concrete, the wall may be considered as short. If the wall is constructed of lightweight concrete, the corresponding limiting ratio is 10. The definition of the effective height of a reinforced concrete wall is identical to that of a column (see section 13.1), provided that the wall and the adjoining members are constructed monolithically. If the floors transmitting load to the wall are assumed to be freely supported, however, the effective height should be determined as for a plain concrete wall. For such a wall, BS8110 defines the effective height somewhat differently, as explained below.

For greater ratios of l_e/h the section is considered as slender. Maximum ratios of l_e/h are prescribed in the Code, as noted later.

If the loads are such that the centroid of the resulting vertical load occurs at a different point along the wall to that of the wall section itself, the intensity of loading will be greater at one end of the wall than at the other. The resulting distribution of loading can then be determined by simple statics. Where the wind or some other horizontal load is resisted by the combined action of

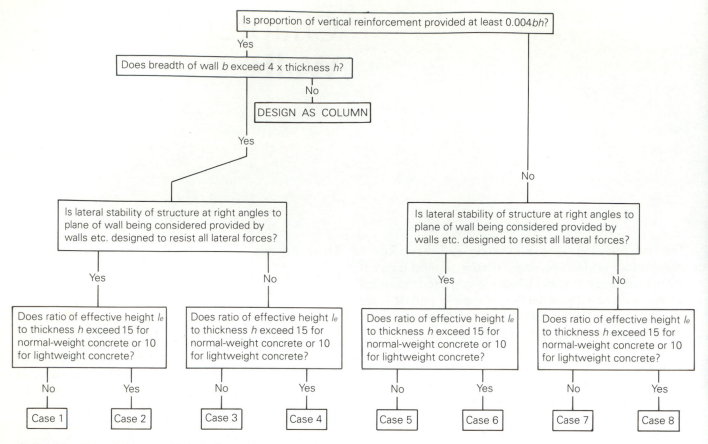

Figure 15.1 Flow chart for determining category of wall.

several walls, it is suggested that the total load is divided in proportion to the relative stiffness of each wall. In the case of plain concrete walls, however, the resulting eccentricity in any individual wall must not exceed one-third of the wall length.

The defining criteria are summarized in *Figure 15.1*, and the eight cases (and appropriate design expressions) that result are now discussed in turn.

15.2 REINFORCED CONCRETE WALLS

15.2.1 Case 1: short braced reinforced concrete wall

In this case the appropriate design formulae correspond to those for columns. If the wall supports an approximately symmetrical arrangement of loading, the ultimate load n_w supported by a unit length of wall is given by the expression

$$n_w = [0.35f_{cu}(1 - \varrho_1) + 0.67f_y\varrho_1] h \qquad (15.1)$$

where ϱ_1 is the proportion of reinforcement provided (i.e. $\varrho_1 = A'_s/bh$, where A'_s is the area of vertical steel). Expression (15.1) assumes that, as the Code specifies, a deduction is made for the area of concrete displaced by the reinforcement; if this is not so the term $(1 - \varrho_1)$ should be omitted. This expression is only applicable if the slabs supported by the wall support uniform loads only and if the lengths of adjacent spans do not differ by more than 15% of the longer span.

Where a wall is subjected to a combined axial load and bending moment it is necessary to design the section as a column of unit breadth as described in section 14.3, using the design charts provided on *Data Sheets 50 to 54*, or those in Part 3 of BS8110.

If the eccentricity of the resultant vertical load is significant in both directions, the wall will be subjected to combined axial load and biaxial bending. To avoid complex calculation, BS8110 permits a simplified approach in which the moment in the plane of the wall is first considered alone. By a simple elastic analysis, assuming that the concrete has no resistance in tension, the resulting distribution of tension and compression along the wall is found. This axial force (or tension) is then combined with the lateral moment alone, using appropriate design charts for bending and thrust.

15.2.2 Case 2: slender braced reinforced concrete wall

If A'_s is not greater than $0.01bh$, the effective height l_e of the wall is restricted by BS8110 to 40 times the thickness h; for larger areas of reinforcement this limiting ratio of l_e/h is increased to 45. The section must be designed in the same manner as a slender column, using the additional-moment method and trial-and-adjustment procedure described in section 14.8. Thus, if the wall is subjected to no initial moment it must be designed to resist a moment relating to its slenderness combined with the applied axial load, using the charts on *Data Sheets 50 to 54*, or those in

Part 3 of the Code. If there is an initial moment, this is added to that due to slenderness. In both cases the initial trial design is refined using the cyclic procedure described for columns.

15.2.3 Case 3: short unbraced reinforced concrete wall

In the case of an unbraced wall, the section must be designed to withstand the combination of an axial load and a moment. If no applied moment is present, a minimum moment corresponding to an eccentricity of $h/20$ or 20 mm, whichever is less, must be considered.

15.2.4 Case 4: slender unbraced reinforced concrete wall

With a slender unbraced wall, the maximum effective height of the wall must not exceed 30 times the wall thickness h. The design procedure is identical to that for a slender braced reinforced concrete wall (i.e. case 2; section 15.2.2).

15.3 PLAIN CONCRETE WALLS

The centroid of a load transferred from an adjoining concrete floor or roof is assumed to act at a distance from the loaded face equal to one-third of the breadth of the bearing. For unbraced walls, the resulting eccentricity at right angles to the wall of the total vertical load must be calculated by taking into account the eccentricities of all the vertical loads and moments acting above that level. For braced walls, however, the resulting eccentricity may be evaluated on the assumption that the eccentricities of all the vertical loads above that point are zero.

Note also that in all the following design expressions for plain concrete walls the resultant eccentricity of load e_x at right angles to the wall may not be taken as less than $h/20$. Although not explicitly stated in BS8110, the maximum value of e_x that need be considered is presumably 20 mm.

BS8110 also sets out restrictions to limit bending and shearing stresses and deflections in plain concrete walls; for details of these reference should be made to the Code itself.

If no reinforcement is provided to withstand handling or prevent cracking, the durability provisions for concrete not containing embedded material, given in clause 6.2.4.2 of BS8110 and the associated Table 6.2, apply rather than those for reinforced concrete. Plain walls are akin to those constructed of masonry, and reference to BS5628 Parts 1 to 3 'The use of masonry' may be useful.

15.3.1 Case 5: short braced plain concrete wall

The effective height l_e of a braced wall of plain concrete depends on its lateral support. If these supports can resist both lateral displacement and bending, l_e may be taken as 0.75 times the distance between the supports, or twice the distance from a support to a free edge. If they resist

displacement only, l_e should be taken as equal to the distance between supports, or 2½ times that from a support to a free edge. With horizontal lateral supports such as floors, the foregoing distances are measured vertically, of course, and vice versa.

According to BS8110 a lateral support must be able to transfer, to the walls or columns providing lateral stability to the whole building in that direction, 1/40 of the total vertical ultimate load that the wall is designed to support at that point, plus the reactions due to the total horizontal ultimate forces. Such a support can only be considered to resist rotation if the intersection between the braced wall and the lateral support is reinforced to provide this restraint, or if the bearing from a floor extends over at least two-thirds of the width of the wall, or where adequate bending restraint is provided by the connection.

The ultimate load n_w per unit length supported by the wall is then given by the expression

$$n_w = 0.3(h - 2e_x) f_{cu} \qquad (15.2)$$

where e_x, the eccentricity (if any) of the load at right-angles to the plane of the wall, may range from a minimum of $h/20$ to a maximum of $h/2$ (when the wall carries no load).

15.3.2 Case 6: slender braced plain concrete wall

The effective height l_e of a slender braced wall of plain concrete must not exceed 30 times the thickness h, the effective height being as defined for a short braced wall (i.e. case 5; section 15.3.1). According to BS8110, the ultimate load n_w per unit length of wall is then still as given by expression (15.2) but, in addition, must now not exceed, for normal-weight concrete,

$$n_w = 0.3\left[h - 1.2e_x - 0.0008\left(\frac{l_e^2}{h}\right) \right] f_{cu} \qquad (15.3)$$

or for lightweight concrete,

$$n_w = 0.3\left[h - 1.2e_x - 0.0012\left(\frac{l_e^2}{h}\right) \right] f_{cu} \qquad (15.4)$$

15.3.3 Case 7: short unbraced plain concrete wall

For an unbraced wall, the effective height l_e is twice the actual height of the wall, unless a floor or roof is provided at the top spanning at right angles to the wall. If this is so, l_e may be taken as only 1.5 times the actual height. If the top of the wall slopes uniformly, the height should be measured to the midpoint of the slope.

The ultimate load n_x supported by a unit length of wall is then as given by the lesser of the two appropriate expressions in the following:

Normal-weight concrete only:

$$n_w = 0.3\left[h - 2e_{x2} - 0.0008\left(\frac{l_e^2}{h}\right) \right] f_{cu} \qquad (15.5)$$

Lightweight concrete only:

$$n_w = 0.3 \left[h - 2e_{x2} - 0.0012 \left(\frac{l_e^2}{h} \right) \right] f_{cu} \qquad (15.6)$$

Both types of concrete:

$$n_w = 0.3[h - 2e_{x1}] f_{cu} \qquad (15.7)$$

In these expressions, e_{x1} and e_{x2} are the resultant eccentricities at the top and bottom of the wall respectively; neither may be taken as less than $h/20$.

15.3.4 Case 8: slender unbraced plain concrete wall

As in the case of a braced plain concrete wall, the maximum effective height is limited to 30 times the thickness h. The appropriate design expressions are then the same as those given for short unbraced plain concrete walls (i.e. case 7; section 15.3.3).

15.4 LIMITING AMOUNTS OF REINFORCEMENT

As already stated, the minimum area of vertical steel that may be provided per unit length in a reinforced concrete wall of thickness h is $h/250$. In addition, the area of vertical steel employed must not exceed $h/25$. An area of at least $h/400$ of high-yield steel or $h/333$ of mild steel must also be provided horizontally. The diameter of these bars must be at least one-quarter that of the vertical bars or 6 mm, whichever is the greater; ref. 2 recommends a minimum bar size of 8 mm. Although not explicitly stated in BS8110, one-half of the steel in each direction should be located near each face.

The *Joint Institutions Design Manual* recommends in addition that the minimum size for vertical bars should be 10 mm and that the maximum spacings of these bars should not exceed 250 mm for mild steel and 200 mm for high-yield steel. The maximum spacing of horizontal bars should not exceed 300 mm.

If the area of vertical reinforcement exceeds $h/50$, BS8110 (clause 3.12.7.5) requires links having a diameter of not less than one-quarter of the largest vertical bar or 6 mm to be provided through the wall. These links must be spaced at not more than $2h$ in either direction and at not more than 16 times the bar size (this presumably means the size of the vertical bars that are being tied) vertically. Any bar resisting vertical force that is not enclosed by such a link must be within 200 mm of a bar that is restrained.

Appropriate areas of reinforcement corresponding to these requirements for reinforced concrete walls of various thicknesses are given on *Data Sheet 61*, together with suggested bar spacings and sizes.

In the following instances, reinforcement should be provided in plain concrete walls to resist the effects of tension, shrinkage or temperature change. Such reinforcement should consist of an area of not less than $0.0025bh$ of high-yield steel or $0.003bh$ of mild steel. If, owing to the loading having an eccentricity in the plane of

the wall exceeding one-third of the length, calculations show that more than one-tenth of the length of the wall is in tension, vertical reinforcement must be provided in two layers, the bar spacing not exceeding 300 mm for mild steel bars and 160 mm otherwise.

If thought necessary by the designer, cast *in situ* concrete plain walls more than 2 m long should be reinforced with closely spaced small-diameter bars in both directions. In walls exposed to the weather this steel should be placed near to the exposed face over the whole surface, but in internal walls it may only be necessary to provide reinforcement adjoining junctions with beams and slabs. In such a case, one-half of the reinforcement should be disposed near each face.

BS8110 suggests that any openings should be trimmed with nominal bars. Owing to the stress concentrations that occur at the corners of such openings, the *Joint Institutions Design Manual* recommends that the area of reinforcement A_s provided at each corner should be calculated from the expression $2Q/0.87f_y$, where Q is the horizontal shear force in the wall adjoining the opening. This steel should be arranged diagonally across the corners and consist of not less than two 16 mm bars, i.e. corresponding to values of Q of 87 kN and 160 kN for mild and high-yield steel respectively.

15.5 APPLICATION OF RECOMMENDATIONS TO DESIGN

The foregoing recommendations are now applied to the cast *in situ* monolithic reinforced concrete transverse walls at the ends of the building in *Drawing 1*. As an example, the wall between the ground and first floors is designed, as this carries the greatest load and is the tallest panel. Since the design is seldom critical, the calculations need not normally be so precise as those for beams and slabs. The thickness of 225 mm provides the high degree of fire resistance desirable for a party wall. The load on a 1 m length is that from a strip about 2.7 m long and 1 m wide of the roof and five upper floors, plus the self-weight of the wall. Making the normal allowance for the reduction of the imposed loads (*Data Sheet 4*) permitted for walls acting as supports, the total ultimate load is not more than 580 kN per metre. The ultimate bending moment from one secondary beam is assumed to be 61 kN m, i.e. 24.4 kN m per metre of wall, which exceeds the minimum value of $Nh/20$ that must be considered.

Since the wall is adequately restrained in position and direction at each end, the effective height may be taken as $\frac{3}{4}l_e$, i.e. as $0.75 \times 4.35 = 3.26$ m. Then since h is 225 mm, $l_e/h = 14.5$, and thus $\alpha = 0.105$ (from the scale on *Data Sheet 48*). Consequently

$$M_t = M_i + \alpha Nh = (24.4 \times 10^6) + (0.105 \times 580 \times 10^3 \times 225)$$
$$= 38.1 \times 10^6 \text{ kN m}$$

If 40 mm cover is to be provided to all reinforcement, $d' = 40 + 6$ and thus $d/h = 179/225 \approx 0.80$. Then, if $f_{cu} = 30 \text{ N/mm}^2$ and $f_y = 250 \text{ N/mm}^2$, from the upper chart on *Data Sheet 53*, since $M/bh^2 f_{cu} = 39.3 \times 10^6/(10^3 \times 225^2 \times 30) = 0.026$ and $N/bhf_{cu} = 580 \times 10^3/(10^3 \times 225 \times 30) = 0.086$, the corresponding value of ϱ_1/f_{cu} is less than zero, and thus only the minimum reinforcement required in a reinforced concrete wall need be provided. Thus, from *Data Sheet 61*, 12 mm bars at 225 mm centres in each face will suffice.

The foregoing calculations for a reinforced concrete wall assume that the building is adequately braced in both principal directions by some other means; for example, by the walls surrounding the stairs or lift shafts acting as vertical cantilevers. Quite frequently, however, the lateral stability of a structure such as the building considered in Part Two is provided by arranging for the floor slabs to act as horizontal beams transferring the wind force back to end shear-walls of reinforced concrete. If this is done in the present case, each wall will be subjected to horizontal loads of $3.6 \times (½ \times 29.25) \times 912 \, \text{N/m}^2 \approx 48 \, \text{kN m}$ at each typical floor level. These loads produce an ultimate moment on each end wall at ground-floor level of about $3.15 \, \text{MN m}$. Since $f = ½ Ml/I$ where $I = hl^3/12$, the stresses produced at the ends of the walls by this moment are $\pm ½ \times 3.15 \times 10^9 \times 12/(15\,450^2 \times 225) = \pm 0.35 \, \text{N/mm}^2$. Even if the permissible reductions in partial safety factors for load are neglected, this stress will only increase the value of N on the end metre of wall to about 660 kN, which has no effect at all on the design.

Bracing in the longitudinal direction must still be provided as already described.

If the end walls are to be considered as unreinforced and the entire horizontal stability of the structure is provided by other means, the effective height of the walls from ground-floor to first-floor level may be taken as $¾ \times 4.35 = 3.26 \, \text{m}$, since both ends are fixed in position and direction, and thus $l_e/h = 14.5$ as before. The wall therefore acts as a slender braced plain wall. Assuming that the ultimate load of 87.8 kN transferred from the first floor acts at a distance of $^1/_3 \times 225 = 75 \, \text{mm}$ from the face of the wall (i.e. at an eccentricity of 37.5 mm), the total ultimate load of 580 kN acts on the wall at an equivalent eccentricity of $87.8 \times 37.5/580 = 5.68 \, \text{mm}$. Then the axial load per unit length n_x must not exceed that given by expression (15.2), i.e.

$$n_x = 0.3[225 - (2 \times 5.68)]\,30 = 1923 \, \text{kN per metre of wall}$$

since fully compacted grade C30 concrete is used. In addition, the ultimate axial load per unit length must not exceed that given by equation (15.3), i.e.

$$n_x = 0.3 \left[225 - (1.2 \times 5.68) - 0.0008 \times \frac{3260^2}{225} \right] 30$$

$$= 1623 \, \text{kN/m}$$

These maximum values of n_x are both well in excess of the actual value of 580 kN/m.

If the wall is assumed to be unbraced, it is not possible to assume that the resulting eccentricity of all the vertical loads transferred from the floors above that being

considered is zero, as has been done above. With a total ultimate loading of 415 kN from the floors, the maximum eccentricity is $415 \times 37.5/580 = 26.9 \, \text{mm}$ and thus equations (15.7) and (15.5) give maximum values of n_x of 1542 kN/m and 1202 kN/m respectively. These are still well in excess of the actual value of 580 kN/m. If the end walls provide lateral stability against wind, the additional vertical force per metre of wall at the ends of the wall due to wind loading is only about 80 kN, and thus the maximum vertical ultimate load is about 660 kN/m, without taking into account the reduced partial load factors of $1.2G_k$ and $1.2Q_k$ that may be adopted when (dead plus imposed plus wind) loading is considered.

Thus plain concrete walls would be quite satisfactory. However, it would be desirable to provide some reinforcement in such large areas of concrete to prevent the formation of shrinkage cracks. In addition, some steel would be essential to prevent cracks forming where the floors frame into the walls. BS8110 recommends the provision of not less than 0.3% of reinforcement in each direction in both these cases, and it may therefore be thought worth while to increase this amount to 0.4%, which is the minimum structural requirement for a reinforced concrete wall. For fire-resistance purposes, increasing the amount of steel to 0.4% would double the fire-resistance period to 3 hours. Even this may be insufficient for a party wall, and an increase of thickness to 240 mm, giving a period of 4 hours, would then be necessary.

15.6 PANEL WALLS

A reinforced concrete panel wall, which is an infilling in the structural frame and is considered not to carry any vertical load, must withstand the pressure or suction due to wind. The panel must be sufficiently strong to resist the bending moments due to its spanning between the members of the frame, and the connections to the frame must be strong enough to transfer the pressures on the panel to the frame either by bearing, if the panel is set in rebates in the members of the frame, or by the resistance to shear of reinforcement that projects from the frame into the panel. A bearing is preferable, since the panel can then be completely free from the frame and is therefore not subjected to secondary stresses due to any deformation of the frame; nor is the connection between the panel and the frame subjected to tensile stress resulting from contraction of the panel caused by shrinkage of the concrete or by temperature changes. By setting the panel in a chase, the connection is also lightproof. The wind pressure for which such a panel must be designed is discussed briefly in sections 2.6 and 2.8.6.

15.7 BASEMENT RETAINING WALLS

The method of designing a retaining wall in a basement depends on the method of construction to be adopted. There are three cases to investigate: (a) when the wall only has been built, at which time the wall will be subjected to the least vertical load and, unless propped, the greatest overturning moment due to cantilever action; (b) when the ground floor (but nothing above) has been constructed, in which case the floor acts as a prop at the head of the wall and the only vertical load, in addition to the weight of the wall itself, is the weight of the ground floor; and (c) when the building is completed and fully loaded.

The conditions in cases (a) and (c) produce the greatest pressure on the ground, and in either of these cases the compressive stress in the concrete may be greatest. Case (a) determines the maximum area of vertical reinforcement required at the back of the wall. In case (b), vertical reinforcement may be required at the front of the wall, which is unnecessary to resist the actions due to either of the other cases.

Typical values for the safe bearing pressure on various types of ground are given in BS8004: 1986 'Foundations'. It is important to note that the permissible values given in this document correspond to *service* loading conditions, and the forces and loads transmitted from the structure, which are used to calculate the pressures on the soil, should also thus be service (i.e. unfactored) values; that is, those that result from applied loads of $1.0G_k$, $1.0Q_k$ or $1.0W_k$. In designing the sections forming the substructure itself, however, ultimate forces and moments are required (i.e. those that result from loads of $1.6Q_k$, $1.4G_k$ or $1.0G_k$, and $1.2W_k$). In the example that follows, the safe bearing pressure beneath the foundation is assumed to be $0.4\,MN/m^2$. (This corresponds to an imperial value of slightly less than $4\,tons/ft^2$.)

As an example of a basement retaining wall, the front wall of the building in *Drawing 1* is considered below. The bearing of the ground floor on the wall is also shown. No special waterproof course is provided, as it is assumed that the ground is either not water-bearing or is well drained. If waterproofing is necessary, a layer of asphalt on a 115 mm brick wall should be placed between the retaining wall and the earth. The asphalt should be continued under the basement floor slab, where it would be applied to a 75 mm layer of plain concrete. To guard against damp, greater impermeability of the wall will be ensured by employing grade 35 concrete. According to *Data Sheet 42*, the bars near the face of the wall must then have a cover of concrete of at least 30 mm, while the cover should be at least 20 mm at the inner face.

15.7.1 Horizontal pressure

The horizontal pressure from the earth may be determined by various methods. Although several other expressions are preferable for special conditions, Rankine's formula is probably the simplest and is commonly used in normal cases, and this is employed here. If the angle of repose of the soil is assumed to be 35° and the weight of dry gravel or earth is taken as equivalent

to $16\,kN/m^3$, the value of the Rankine factor $(1 - \sin\theta)/(1 + \sin\theta)$ is 0.271, and the intensity of active pressure at any depth h is $0.271 \times 16h = 4.34\,kN/m^2$.

With a depth of ground of 3.45 m, as shown in *Figure 15.2a*, the pressure at the bottom of the wall is $4.34 \times 3.45 = 14.97\,kN/m^2$, and decreases uniformly until there is no pressure at the top. An addition must be made to allow for a surcharge of, say, $10\,kN/m^2$ on the ground behind the wall. The surcharge increases the pressure uniformly throughout the height of the wall by $0.271 \times 10 = 2.71\,kN/m^2$. It is assumed that the ground is dry, but to guard against unforeseen water-logging it is desirable to investigate the possible water pressures. If the whole height of the wall may be subjected to water pressure, the greatest hydrostatic pressure at the bottom is $3.45 \times 9.81 = 33.84\,kN/m^2$. As this condition is most unlikely to occur, a partial safety factor of not less than one-half of that adopted for the ordinary condition of pressure on dry earth should be satisfactory. If the ground is continuously or intermittently water-bearing, the factor of safety relating to water pressure should not be reduced. Therefore, for comparative purposes in the present example, the reduced water pressure to be considered is $0.5 \times 33.84 = 16.92\,kN/m^2$ compared with the earth pressure of $17.68\,kN/m^2$. Designing for the ordinary earth pressure therefore allows for the (in this example improbable) condition of water-logging. Water can be prevented from collecting against the wall by providing land drains or by incorporating weep-holes in the wall. For a building in a town, it may not be possible to arrange for an adequate system of land drains, as can be done on an open site; also weep-holes are likely to prove objectionable in a basement. For these reasons it is preferable to investigate the pressures that may result from probable water-logging. If such water-logging is a practical possibility, consideration should be given to designing the wall sections to meet the requirements of BS8007 'Design of concrete structures for retaining aqueous liquids'.

Rankine's (and other) methods of determining the active earth pressure assume that the unsupported end of the cantilever wall is free to displace and rotate. Terzaghi (ref. 22) has shown that the displacement necessary to reduce the pressure from the soil to the active value is in the order of one millimetre for each metre of height of the wall. When a basement wall is restrained in position (and possibly direction) at the top owing to the construction of the ground floor slab, this displacement cannot occur and as a result the pressure exerted by the retained soil is greater than the active value.

A specific value is difficult to estimate since much depends on the constructional method adopted. If a basement is constructed in existing stable ground, the normal procedure is to excavate a trench in which the wall and base are formed, and the space between the soil and the wall is then backfilled. This procedure minimizes the horizontal pressure from the soil since the undisturbed

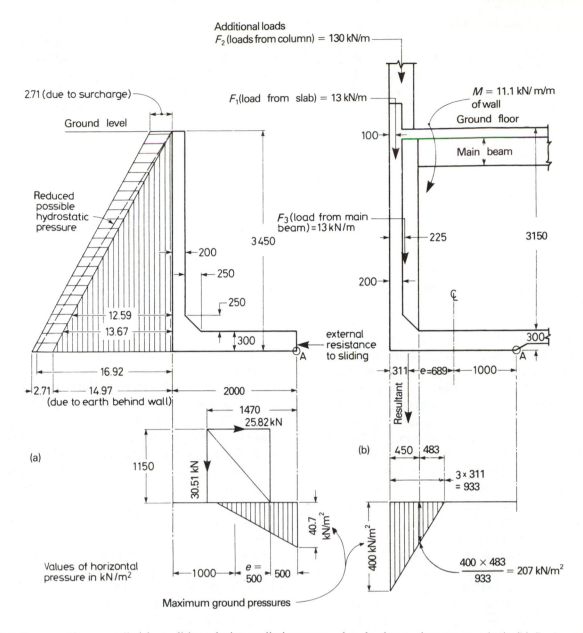

Ground level

2.71 (due to surcharge)

Additional loads
F_2 (loads from column) = 130 kN/m

F_1 (load from slab) = 13 kN/m

Reduced possible hydrostatic pressure

200

250

250

12.59

13.67

300

16.92

2.71

14.97
(due to earth behind wall)

2000

M = 11.1 kN/m/m of wall

Ground floor

100

Main beam

F_3 (load from main beam) = 13 kN/m

3450

225

3150

200

300

external resistance to sliding

A

311   e=689   1000

1470
25.82 kN

1150

30.51 kN

40.7 kN/m²

Resultant

450   483

3 × 311 = 933

400 kN/m²

(a)

(b)

Values of horizontal pressure in kN/m²

1000   e = 500   500

Maximum ground pressures

$\dfrac{400 \times 483}{933}$ = 207 kN/m²

Figure 15.2 Forces acting on wall: (a) conditions during preliminary stage (no shoring against overturning); (b) final conditions.

ground is allowed to expand laterally while construction is in progress. If, however, large areas adjoining the wall are backfilled and well compacted after construction, very high earth pressures may result.

If movement of the wall and lateral strain are both prevented, the resulting earth pressure p corresponds to the so-called 'at rest' condition where $p = kwh$, in which k depends on the type of soil and other factors and w is the unit weight. For normally consolidated soils k is often taken as $1 - \sin \theta$, but it may range from 0.45 for loose sand to 6 for a clay that is machine compacted over the entire backfill. In the present case, with a soil having an angle of internal friction of 35°, $1 - \sin \theta = 0.426$ and the intensity of pressure at a depth h is $0.426 \times 16h = 6.82h$ kN/m². With a depth of wall of 3.45 m, the pressure at the bottom is now $6.82 \times 3.45 = 23.53$ kN/m²,

decreasing uniformly as before, and the additional pressure due to surcharge becomes $0.426 \times 10 = 4.26$ kN/m².

15.7.2 Wall as simple cantilever

During the first stage of construction, the wall acts as a cantilever unsupported at the top. It is necessary to investigate the stability and the resistance to sliding of the cantilever wall, and to calculate the ground pressures and ultimate bending moments. Assume that a portion of base 2 m in length and 300 mm in thickness is constructed at the same time as the wall, that there is a 250 mm splay at the foot of the wall, and that the wall is 200 mm in thickness, as shown in *Figure 15.2a*. The calculation of the stability of a 1 m length of wall is as follows:

			Load		Moment about A
Horizontal forces:	Normal	$14.97 \times 3.45 \times \frac{1}{2}$	$= 25.82\,\text{kN} \times 3.45 \times \frac{1}{3}$		$= 29.70\,\text{kN}\,\text{m}$
	Surcharge	2.71×3.45	$= 9.35\,\text{kN} \times 3.45 \times \frac{1}{2}$		$= 16.13\,\text{kN}\,\text{m}$
			$36.17\,\text{kN}$		$45.83\,\text{kN}\,\text{m}$
Vertical forces:	Base	$2 \times 0.3 \times 24$	$= 14.40\,\text{kN} \times 1.0$		$= 14.40\,\text{kN}\,\text{m}$
	Stem	$3.2 \times 0.2 \times 24$	$= 15.36\,\text{kN} \times 1.9$		$= 29.18\,\text{kN}\,\text{m}$
	Splay	$0.25^2 \times \frac{1}{2} \times 24$	$= 0.75\,\text{kN} \times 1.72$		$= 1.29\,\text{kN}\,\text{m}$
			$30.51\,\text{kN}$		$44.87\,\text{kN}\,\text{m}$

During the first stage, it is normally safe to assume that the pressure due to the surcharge does not act owing to the position of the hoarding around the site, and precautions can be taken to prevent loads being deposited on the ground immediately behind the wall. In these circumstances the unfactored overturning moment is 29.70 kN m, which is resisted by a counter-moment of 44.87 kN m, giving a factor of safety of 1.51, which is satisfactory for a temporary condition. If precautions cannot be taken to prevent loads being placed on the ground behind the wall during this preliminary stage, the corresponding overturning moment is 45.83 kN m, which is in excess of the counter-moment. It would then be necessary to provide shores against the wall until the ground floor were constructed.

Without external aids, the resistance to sliding is, say, $0.4 \times 30.51 = 12.2$ kN, compared with the minimum horizontal force causing sliding of 25.82 kN. It is therefore necessary to provide struts to prevent the possible forward movement of the wall. If shores are provided to resist overturning, these can be arranged to give resistance against sliding. The earth in front of the base slab cannot be considered as providing any resistance, as it may not be in contact with the concrete and, in any case, it will be largely removed when the excavation for the remainder of the basement floor is made. The coefficient of friction of 0.4 used in the calculations for resistance to sliding is reasonable for dry earths and gravels, but a lower coefficient should be taken for wet earths, and the coefficient may be zero for wet clay.

The next step is to calculate the maximum ground pressure due to service loading (assuming there is no external restraint against overturning) caused by the horizontal force of 25.82 kN acting at a height of $3.45/3 = 1.15$ m above the base of the wall, and the vertical load of 30.51 kN acting at a distance of $44.87/30.51 = 1.47$ m from point A (*Figure 15.2a*). The point at which the resultant of these two forces intersects the underside of the base is at a distance from point A equal to $1.47 - (25.82 \times 1.15/30.51) = 0.5$ m. Thus the eccentricity e is $(0.5 \times 2) - 0.5 = 0.5$ m and, as this is greater than one-sixth of the width of the base (i.e. 0.33 m), the maximum

ground pressure is

$$\frac{4F}{3(l - 2e)} = \frac{4 \times 30.51}{3(2.0 - 1.0)} = 40.7\,\text{kN/m}^2$$

which is much less than the permissible pressure of $0.4\,\text{MN/m}^2$.

All that now remains is to calculate the critical ultimate bending moments required to design the concrete sections. The foregoing forces and moments have all been determined from characteristic (i.e. unfactored) loads, but to determine the design moments it is necessary to take account of the appropriate partial safety factors. According to BS8110 the partial safety factor for earth pressure is 1.4. For the self-weight of the concrete structure the Code gives dead-load factors of 1.4 (adverse) and 1.0 (beneficial), and the question arises as to which parts of the wall/base structure each of these factors should be applied in order to investigate the most adverse situation. To avoid such an extremely complex analysis, and remembering that the resulting bearing pressures are also determined by the same factors adopted, a single analysis with a partial safety factor for dead loads of 1.4 will be sufficiently accurate in almost all circumstances. A partial safety factor of 1.6 must be adopted when considering the surcharge load.

Considering first the base slab, with the centre of pressure at 0.5 m from A, this corresponds to a distance of $1.55 - 0.55 = 1.05$ m from the end of the splay. The ultimate bending moment at a section at the end of the splay is thus

$$-(30.51 \times 1.4) \times 1.05 = -44.85$$
$$+0.3 \times 24 \times 1.4 \times 1.05^2/2 = +5.56$$
$$\overline{\qquad -39.29\,\text{kN}\,\text{m}}$$

The pressure at the level of the top of the splay is $4.34 \times 2.90 = 12.59\,\text{kN/m}^2$, excluding the pressure due to surcharge. The ultimate bending moment at this section is thus $(12.59 \times 1.4) \times 2.90^2/6 = 24.71$ kN m. Due to the surcharge alone, the ultimate bending moment is $(2.71 \times 1.6) \times 2.90^2/2 = 18.24$ kN m, and thus the total ultimate bending moment is 42.95 kN m. These ultimate bending moments are shown in *Figure 15.3a*.

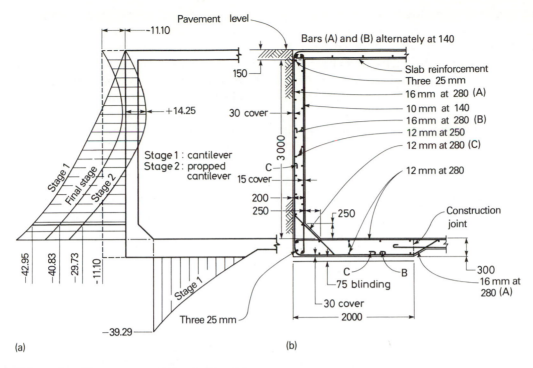

Figure 15.3 (a) Ultimate bending moments (kN m); (b) reinforcement.

15.7.3 Wall as propped cantilever

When the ground floor only has been constructed, the floor acts as a prop at the head of the wall and the wall supports the floor. Considering the cross-section of the basement shown in *Drawing 1*, the minimum vertical load on the front retaining wall is the dead weight only of the 150 mm slab spanning 2.5 m, i.e. a characteristic dead load of $\frac{1}{2} \times 2.5 \times 0.15 \times 24 = 4.5$ kN/m. The effect of this load on the total vertical force is

	Load	Moment about A
Ground floor	4.5 kN	$\times 1.9 =$ 8.55 kN m
Wall (as before)	30.51 kN	$= 44.87$ kN m
	35.01 kN	53.42 kN m

The resistance offered to sliding is $0.4 \times 35.01 = 14$ kN. At this stage it is assumed that the full pressure of the surcharge acts, but that the basement floor has not been constructed. Owing to the horizontal propping effect of the floor, the approximate horizontal force acting at the bottom of the wall is $(0.8 \times 23.53 \times 3.45 \times 0.5) + (0.625 \times 4.26 \times 3.45) = 41.66$ kN, which exceeds the calculated resisitance to sliding. Hence the props provided during the first stage to resist sliding must be retained until the basement floor has been constructed.

The ground pressures and ultimate bending moments on the base slab need not be considered at this stage. The ultimate bending moments on the wall, assuming an effective span of 3.25 m, are due to a total pressure, triangularly distributed, of $6.82 \times 3.25^2 \times \frac{1}{2} = 36.02$ kN and a total uniform pressure of $4.26 \times 3.25 = 13.85$ kN. The ultimate bending moment at the top of the splay is

about

$$-\left(\frac{36.02 \times 3.25}{7.5} + \frac{13.85 \times 3.25}{8} \right) \times 1.4 = -29.73 \text{ kN m}$$

At a point slightly above the mid-height of the wall there is a maximum ultimate positive bending moment of

$$-\left(\frac{36.02 \times 3.25}{16.7} + \frac{13.85 \times 3.25}{14.2} \right) \times 1.4 = -14.25 \text{ kN m}$$

The ultimate bending moments are shown in *Figure 15.3a*.

15.7.4 Permanent conditions

When the building is complete, the wall is subjected to vertical forces due to the imposed load on the ground floor and to the loads on the columns, and there is an additional bending moment from the restraint at the end of the ground floor and the columns. The following approximate values are assumed for the loads and bending moments; although these do not conform exactly with the corresponding data in section 14.9 which were calculated subsequently, they are sufficiently accurate to illustrate the design methods involved.

The total characteristic load (excluding the effects of wind) from an external column is, say, 130 kN per metre of wall, and is assumed to act at a distance of 180 mm from the outer face of the wall. Reference to *Calculation Sheets 19* shows that the more accurate and later calculation gives a load of 145 kN acting at a distance of 250 mm from the outer face of the wall, which may be a less severe condition. The total load from the ground-floor slab, including the weight of the shop front, is assumed to be

13 kN per metre of wall. The total load from the main beam of the ground floor is also about 13 kN per metre of wall. The negative ultimate bending moment from the ground-floor slab is 7.2 kN m per metre of wall, and the ultimate bending moment on the wall due to the restraint of the main beam is assumed to be 3.9 kN m per metre of wall.

The total negative ultimate bending moment from the ground floor is therefore 7.2 + 3.9 = 11.1 kN m per metre of wall and is constant throughout the height of the wall, as shown in *Figure 15.3a*. The external vertical loads act at the positions shown in *Figure 15.2b*, and are combined with the weight of the wall as follows:

	Load		Moment about A
Column (F_2)	130 kN	× 1.82 =	236.60
Slab (F_1)	13 kN	× 1.90 =	24.70
Beam (F_3)	13 kN	× 1.775 =	23.08
Wall (as before)	30.51 kN	=	44.87
	186.51 kN		329.25 kN m

The centre of action of the vertical forces is 329.25/186.51 = 1.77 m from A, and the eccentricity of the load about the centre of the 2 m base slab is therefore 0.77 m, causing a positive ultimate bending moment of 186.51 × 0.77 × 1.4 = 201.1 kN m on the base. Combined with the negative ultimate bending moment of 11.1 kN m, the resulting positive ultimate bending moment on the base is 190.0 kN m. At the top of the splay the ultimate bending moment due to the earth pressures, as calculated in the previous stage, is 14.25 kN m, and to this extent the positive bending moment on the base due to the eccentricity of the vertical load may be relieved. As the pressure due to the surcharge, and perhaps part of the pressure due to the earth, may not be exerted all the time, an erroneous view of the conditions may be obtained if the full ultimate bending moment is deducted from the positive ultimate bending moment. Suppose that the positive ultimate bending moment is reduced to only 180 kN m, which, in order to calculate the maximum ground pressure, must be combined with a vertical load of 186.51 kN. The resulting eccentricity is 180/(186.51 × 1.4) = 0.689 m. As this exceeds 2.0/6 = 0.33 m, the greatest ground pressure is

$$\frac{4 \times 186.51}{3[2 - (2 \times 0.689)]} = 400 \text{ kN/m}^2$$

which is equal to the permissible pressure of 0.4 MN/m². From the distribution of ground pressure shown in *Figure 15.2b*, the ultimate bending moment at the edge of the splay on the base slab is 207 × 0.483² × ⅙ × 1.0 = 8.05 kN m less 0.3 × 24.0 × 1.55² × ½ × 1.4 = 12.11 kN m due to the self-weight of the slab, which results in a small net bending moment of about 4 kN m.

The ultimate bending moments in the wall are the same as calculated for the intermediate stage, combined with the effect of the negative ultimate bending moment transferred from the ground floor, as shown in *Figure*

15.3a. The eccentricity (relative to the centre of the wall) of the loads from the column and beam can be neglected because these members are immediately over the pilaster in the wall, and their loads are transmitted directly to the base slab through the pilaster. The critical ultimate bending moment contributed by the permanent condition is therefore only the negative ultimate bending moment on the upper part of the wall.

15.7.5 Details of wall reinforcement

The resistance of the wall is investigated by calculating the minimum effective depth and the amount of reinforcement required. *Figure 15.3a* gives the ultimate bending moments that may occur under the three conditions.

According to the *Joint Institutions Design Manual*, the exposure rating for a basement wall exposed to the earth should be 'moderate' unless the adjoining soil is aggressive. Accordingly, the minimum cover to all reinforcing steel corresponding to this exposure condition permitted by BS8110 (Table 3.4) is 35 mm if grade C35 concrete, the minimum allowable for such an exposure rating, is used. According to clause 3.3.5.2 of the Code, it would theoretically be possible to reduce the concrete grade to C30 provided that a systematic checking regime is established to ensure that the BS8110 limits on the free-water/cement ratio and cement content are strictly observed, but in view of the importance of achieving impermeable concrete this is not done here.

Among the other requirements mentioned in the *Joint Institutions Design Manual* regarding the design of basement walls, it is recommended that the thickness of such walls should be at least 300 mm, but this suggestion is not adopted in the example that follows. The manual also states that adequate durability and fire resistance should be ensured and that the recommendations of ref. 23 should be followed.

Considering the base first, the greatest ultimate bending moment, which occurs in the preliminary stage, is 39.29 kN m. The minimum effective depth required with grade C35 concrete is 86 mm, but the effective depth of the 300 mm slab, with 35 mm cover to 20 mm bars, is 255 mm. The area of mild steel reinforcement required is thus 717 mm². The reinforcement provided consists of 16 mm bars at 280 mm centres (A_s = 718 mm²) and comprises bars (a) in *Figure 15.3b*.

The critical ultimate bending moment for determining the reinforcement at the back face at the top of the splay of the wall is 42.95 kN m, requiring an effective depth of about 89 mm. With a 200 mm slab having an effective depth of 157 mm, the area of reinforcement required is 1339 mm². The arrangement of bars shown in *Figure 15.3b*, which provides 16 mm bars (a) and (b) at 140 mm centres (A_s = 1436 mm²), is therefore suitable. The smaller negative bending moment at the rear face at the top of the wall is resisted by 16 mm bars at 280 mm centres.

During the second stage a positive ultimate bending moment of 14.25 kN m is developed. At an approximate effective depth of 174 mm, the vertical reinforcement required on the inner face is 383 mm^2, which is provided by 10 mm bars at 140 mm centres.

The resistance of the wall to the combined ultimate bending moment and direct force which acts in the final stage can be checked by using the charts for columns on *Data Sheets 50 and 51*. At the top of the splay the ultimate bending moment is about 40.83 kN m, and the ultimate axial load can be calculated approximately by multiplying the sum of the weight of the wall ($2.95 \times 0.2 \times 24 = 14.16$ kN per metre) and the load from the ground-floor slab (13.0 kN per metre) by the partial safety factor for imposed loads of 1.6, resulting in an ultimate axial load of 43.6 kN per metre. With $N/bhf_{cu} = 0.006$ and $M/bh^2f_{cu} = 0.029$ it is clear that only nominal steel is required, and therefore the reinforcement already calculated will suffice. The minimum total area of vertical steel permissible in the wall is 0.4% or 800 mm^2: 10 mm bars at 140 mm centres near one face, and 16 mm bars at 280 mm centres near the other, provide 1279 mm^2 (i.e. 0.64%).

The pilasters under each of the columns transmit the load from the columns and main beams to the base slab which, with the wall, acts as a beam to distribute the load to the ground. It can be shown that the pilaster is able to support the vertical load without assistance from the wall, although the load is, of course, actually carried by the pilaster and the adjoining part of the 200 mm wall. The longitudinal ultimate bending moment resulting from distributing the concentrated load of the column and beam over 6 m is about 800 kN m. The effective depth of the wall acting as a beam is say 3 m, and the compressive ultimate moment of resistance due to the 200 mm wall alone is 8100 kN m, which is ample without considering the compressive resistance of the ground-floor slab at sections near the piers or of the base slab at sections midway between the piers. The longitudinal reinforcement required is $800 \times 10^6/(0.8265 \times 250 \times 300) = 1291$ mm^2, which is provided by three 25 mm bars at the top and bottom of the wall at the critical sections. The horizontal bars in the wall can be provided in accordance with the requirements of BS8110 for distribution bars (*Data Sheet 41*); however, the longitudinal bars should be not less than 12 mm in size and spaced at not more than 250 mm near both faces of the wall.

15.7.6 Arrangement of wall reinforcement

Various schools of thought exist as to the most suitable way of arranging the reinforcement in a retaining wall. As regards steel fixing, the normal method is to erect the vertical bars first, securing these to the steel projecting from the base slab. The horizontal bars are then tied in position. From the point of view of the fixer, it is probably simpler if the horizontal bars are on his side of the vertical steel, since otherwise these bars must be lifted over the vertical reinforcement or placed on the ground on the far side of the starter bars before the vertical steel is erected. If all the steel fixing for the basement is taking place from within the excavation, the most desirable sequence (working from the outside inwards) from the point of view of steel fixing might thus be: outer vertical bars; outer horizontal bars; inner vertical bars; and finally inner horizontal bars.

On the other hand, it is argued that providing the horizontal bars nearest the forms may tend to segregate the concrete, the larger stones in the mix tending to wedge between the lateral bars and the framework. This is more likely to be true where the concrete cover is small, if the concrete is placed from a height, if the height of each lift of concrete is considerable, or if thorough compaction is difficult.

From the structural point of view it is, of course, preferable to provide the maximum effective depth in the direction in which the moments are greatest. Thus if the wall is designed to span vertically, as in the example considered here, the outer bars should run vertically. When detailing, however, care should be taken to ensure that steel fixing will be as simple as possible. If the wall is designed to span horizontally between counterforts, consideration should be given to running the horizontal reinforcement outside the vertical bars.

Chapter 16
Stairs

Stairs not infrequently form one of the most prominent visual features of a building, and as such present a challenge to both engineer and architect. Unlike the normal floor or roof slab where a slight reduction in thickness can seldom be seen, the provision of stairs having the maximum possible slenderness is often visually desirable, making a vast difference between clumsiness and grace. Any consequent increase in the amount of reinforcement required to compensate for restricting the effective depth to the minimum possible value is insignificant in relation to that required for the building as a whole, and is clearly outweighed by the enhanced appearance achieved.

Many types of stair design can be utilized, ranging from a simple flight spanning between supporting beams, to free-standing 'scissor' flights with landings not independently supported, and helicoidal stairs turning through 360° and supported at top and bottom only.

Advice on the analysis of the simplest types only is given in BS8110. To analyse more-complex designs, reference must be made to the many specialist articles that have appeared: see, for example, refs 24–26 (design information drawn from these references is given in *RCDH*).

The normal terms used in describing stairs, listed in *Allen*, are shown in *Figure 16.1a*, together with the commonly adopted limiting dimensions. To reduce the waist thickness to the minimum, if a reasonable thickness of finish is to be applied to the concrete it is possible to adopt the concrete profile shown in *Figure 16.1b*, although such an arrangement is clearly slightly more difficult to form and concrete. The rules for slenderness given in BS8110 will normally determine the minimum waist thickness that can be used. Note that, provided the length of flight forms not less than 60% of the span, the Code permits the minimum basic permissible ratio of span

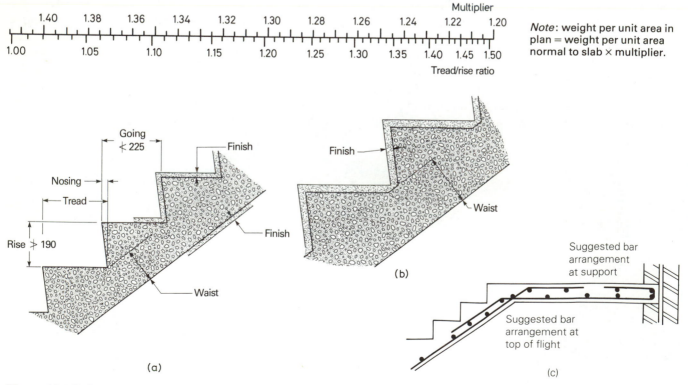

Figure 16.1 Stairs.

138

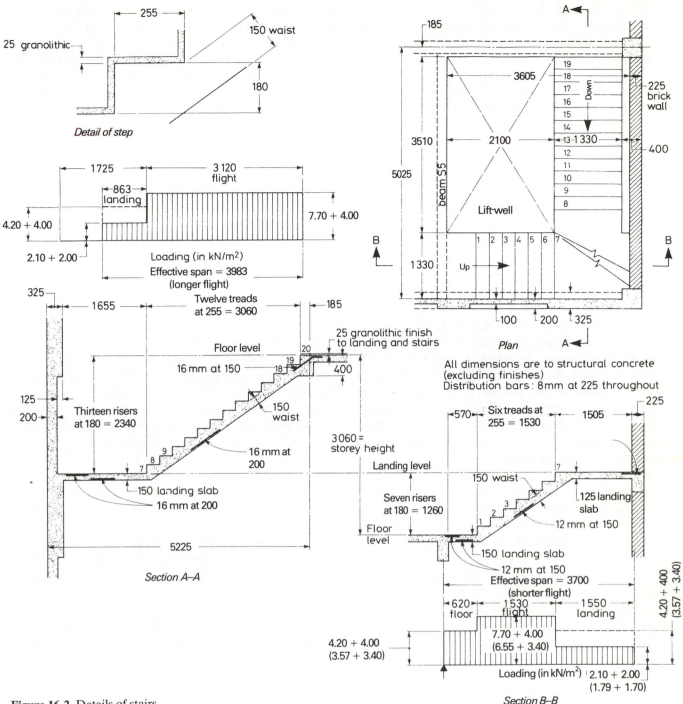

Figure 16.2 Details of stairs.

to effective depth, determined as described in section 8.2, to be increased by 15%. This takes some account of the additional stiffening contributed by the treads. Since stairs are lightly loaded, it may be worth while undertaking a rigorous deflection analysis, although it is then not possible to take account of the stiffening provided by the treads. The interaction of inclined flights at right angles also obviously considerably restricts the deflections that actually occur. At the same time, however, such intersection points are very vulnerable to the development of cracks, and suitable anti-cracking steel must be detailed at these points.

16.1 DESIGN OF STAIRS

As with its predecessors, the requirements of BS8110 relating to stairs are brief. They are best illustrated by designing the principal stairs adjacent to the lift-well in the upper storeys of the building in *Drawing 2*. The dimensions of the stairs are given in *Figure 16.2*, and the characteristic load imposed on the stairs and landings is, according to BS6399, Part 1 'Design loading for buildings', 4 kN/m² (see section 2.3.1).

If the stairs are around an open well (as in this example), BS8110 recommends that the loads on landings

139

common to two spans that are mutually at right angles be divided equally between the two spans. In addition, some relief of load is also permitted if the stairs and landings spanning in the direction of the flight are built into the wall along at least part of their length. The load on a strip 150 mm in width adjacent to the wall may be neglected if the stair is built in at least 110 mm. CP110 also stated that the effective breadth of the section may be increased by two-thirds of the thickness that it is embedded, up to a maximum of 80 mm. This concession is not mentioned in BS8110, but it seems likely that its omission is accidental rather than deliberate.

The span of stairs that are supported not on stringer beams, but on landings or structural members spanning at right angles to the direction of the span of the stairs, may be assumed to be the horizontal distance between the supports plus one-half of the width of each landing (or member), or plus 900 mm if a landing is more than 1.8 m in width. This condition occurs in the stairs being considered (and also in the back stairs), as the longer flight (section A–A on *Figure 16.2*) spans from the longitudinal beam at the upper floor to the intermediate half-landing. The short flight, intermediate half-landing, and small landing at the lower flight (section B–B in *Figure 16.2*) span from the beam in the end wall to the transverse beam trimming the lift-well. The full load must be taken on the lower landing, but the load on the intermediate landing can be divided equally between the two spans.

The effective span of the upper flight is the clear span plus one-half of the effective depth plus one-half of the half-landing width, since the latter is less than 900 mm, and is thus 3.965 m. The effective span of the shorter flight is given by the clear span plus the effective depth, i.e. say 3.7 m.

It is assumed that the thickness of the landings and the waist of the stairs is 150 mm. The characteristic loads carried by the landings are therefore 3.6 kN/m^2 for the self-weight of the slab, 0.6 kN/m^2 for the granolithic finish and 4.0 kN/m^2 for the imposed load. For the intermediate half-landing, one-half of each of these loads is carried on each span. The dead weight of the steps and the finish on the treads and risers (including the nosings) is about 7.7 kN/m^2. The distribution of characteristic loads is as shown in *Figure 16.2*. Because of the reduction in the loaded breadth and the increase in the effective breadth described above, on the lower flight the equivalent characteristic load per metre width of stair or landing is less than the actual characteristic load. The net width of the flight is 1.33 m and therefore the reduced loaded width is 1.18 m and the effective breadth is 1.40 m. The equivalent loads per square metre are therefore the actual loads per square metre multiplied by $(1.18 \times 1.33)/(1.33 \times 1.40) = 0.85$, and the design loads are therefore 3.57 kN/m^2 for the self-weight of the landings (including finishes), 6.55 kN/m^2 for the self-weight of the flight (including finishes), and 3.4 kN/m^2 for the imposed load. The Code does not give bending-moment coefficients for stairs, but it is reasonable to allow for continuity where

this occurs. In the present example, the monolithic construction at the top and bottom of each flight assures continuity. For the longer flight, ultimate bending moments of two-thirds of the free bending moment may therefore be assumed at the three critical sections. For the shorter flight, there is less fixity at the wall beam than at the lower landing beam, so that the bending moment at the lower support and at midspan may be assumed to be 80% of the free bending moment and, say, 40% at the upper support. The average maximum negative and positive ultimate bending moments on the longer flight are therefore about 21.36 kN m per metre width, to resist which a 150 mm slab reinforced with 16 mm bars at 200 mm centres will suffice. The corresponding maximum ultimate bending moments on the shorter flight are about 15.21 kN m per m, requiring 12 mm bars at 150 mm centres.

The section should be checked to confirm that it is satisfactory as regards slenderness. The basic effective depth corresponding to a continuous span of 3.983 m is $153 \times 0.85 = 130$ mm and, with $f_y = 250$ N/mm^2 and a value of M/bd^2 of 1.44, the multiplier for the tension reinforcement is 0.59. Thus the minimum effective depth possible is $0.59 \times 130 = 77$ mm, which is satisfactory. Note that, if the upper flight were not supported at the intermediate half-landing but spanned across to the front wall, the resulting minimum effective depth required to meet deflection requirements would exceed that provided and a thicker slab would be essential.

The longitudinal force in a stair, or in the sloping part of a combined landing and stair considered as a cranked beam, is generally neglected as it is very small. Since it is commonly assumed that the direction of the reactions at the upper and lower supports is vertical, the axial force at any section of the inclined part is the longitudinal component of the vertical shearing force at the section. The component at right angles to the slope is the shearing force acting parallel to a normal cross-section. In the case of the longer flight of stairs in *Figure 16.2*, the average vertical ultimate shearing force at the bottom of the stairs is about 15.3 kN/m. Since the slope of the stairs is about 1 : 1.42, the normal force is about $(15.3 \times 255)/313 = 12.5$ kN/m, and the longitudinal thrust is $(15.3 \times 180)/313 = 8.8$ kN/m, which is insignificant compared with the cross-sectional area of the 150 mm waist of the stair.

Bars in stairs may be curtailed according to the simplified requirements described for slabs in clause 3.12.10.3 of BS8110 (see section 9.1.5). At the re-entrant angle where the bars from the flight pass into the upper landing to prevent cracking, the bars must be arranged as shown in *Figure 16.1c*. It is also good practice to form U-shaped bends where bars terminate at the edges of landings that are built into the supporting walls and where some fixity may thus be expected to occur even if, during the analysis, these points are nominally considered as freely supported; see *Figure 16.1c*.

Chapter 17
Foundations

The principal types of concrete foundations for buildings are as follows:

1. Separate column bases of reinforced concrete or plain concrete.
2. Reinforced concrete or plain concrete strip footings for walls or for two or more columns in line, if the permissible bearing pressure is small or if one or more of the columns is located close to the boundary of the site.
3. A reinforced concrete raft foundation for a group of columns or other supports, when the permissible bearing pressure is low or if the columns are closely spaced.
4. A piled foundation, where the bearing pressure immediately below ground level is too low, but where good ground occurs at a considerable depth below the surface. Even if the bearing capacity of the deeper ground is insufficient to allow end-bearing piles to be employed, it may still be possible to carry the building on piles that support their load by surface friction.

Except in the case of piled foundations, the area of a base loaded concentrically must be not less than the unfactored total load on the ground, divided by the average intensity of pressure that can safely be imposed on the soil. This pressure depends among other factors on the type of soil and the depth of the foundation, but the choice of the correct value in any particular situation is a specialized task, often requiring the judgement of a foundations engineer, assisted by the results of soil mechanics investigations. Presumed bearing values under static loading, which are intended to aid preliminary design, are provided in BS8004: 1986 'Foundations', and in practice these would be confirmed by site investigation. In the following, it is assumed that the designer has obtained reliable information of the bearing pressure that the ground can safely withstand.

Alternative designs for foundations of type 1 for one of the interior columns of the building illustrated in *Drawings 1* and *2* are given in this chapter. Examples of foundation of types 2, 3 and 4 are also given for other structures. All the designs are in accordance with BS8110 where relevant requirements exist. This document gives no guidance on the design of plain concrete bases; since they sometimes prove an economic alternative to reinforced concrete bases, some details are given of their design, which in this case is in accordance with the general requirements of the *Joint Institutions Design Manual* and BS8103: Part 1: 1986 'Code of practice for stability, site investigations, foundations and ground floor slabs for housing'.

In designing any base, an important point to note is that the permissible bearing pressure on the ground corresponds to a service stress, and is often the maximum value that can be supported without excessive deformation taking place. Thus, in determining the corresponding area of base required, the applied load should be the unfactored value (i.e. that corresponding to service loading). However, when designing the concrete sections, ultimate values must be employed, i.e. those resulting from maximum loadings of $1.4G_k + 1.6Q_k$ or $1.2(G_k + Q_k + W_k)$. When designing isolated footings it is not difficult, as is done on *Calculation Sheets 17* to *20*, to determine both the unfactored and factored column loads, using the former to calculate the area required and the latter to design the sections, but with more complex combined bases this refinement may cause problems. If so, the use of a single partial safety factor for loads of 1.5 to convert actual to ultimate loads is convenient and normally accurate enough for practical purposes, bearing in mind the uncertainties involved in assessing an appropriate permissible bearing capacity for the soil. However, it may possibly be argued that, if the proportion of imposed to dead loading is high, the ultimate moments may be underestimated as a result. An overall partial safety factor of 1.6 can never be exceeded and, although its use errs on the side of safety, the excess reinforcement provided as a result is negligible and is useful to cater for possible irregularities in the distribution of pressure beneath the base.

The *Joint Institutions Design Manual* suggests that in many cases it is sufficiently accurate to determine unfactored loads by dividing the factored values by 1.45.

17.1 INDEPENDENT COLUMN BASES

The base of a reinforced concrete column is generally a reinforced concrete block, or it may consist of a

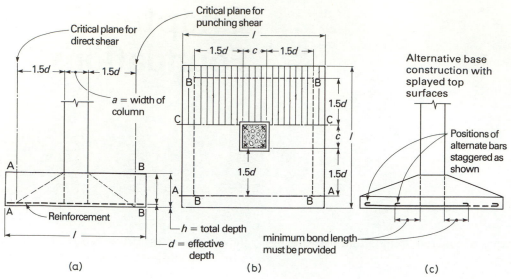

Figure 17.1 Reinforced concrete bases.

rectangular block of plain concrete. Reinforced bases are sometimes splayed at the top as shown in *Figure 17.1*, although owing to the cost of the extra labour involved compared with the saving in materials achieved, this operation is much less common than used to be the case. An example of a reinforced concrete base is given on *Drawing 14* as the base of interior column B3. It is assumed that the bearing pressure on the ground is $200\,kN/m^2$. The area required for the base is determined by dividing the greatest service (i.e. unfactored) load, which is assumed to act concentrically, by the safe bearing pressure, as shown on *Calculation Sheets 17*.

If a vertical load acts concentrically on a base, the resulting bearing pressure beneath the base is normally assumed to be uniform. If, however, the base is loaded eccentrically or if the vertical load is combined with a horizontal force or bending moment or both, the distribution of pressure beneath the base will vary. The *Joint Institutions Design Manual* permits the assumption of concentric loading providing that the resulting eccentricity does not exceed 2% of the length of the base in the direction being considered.

From this point of view it is normally assumed that isolated bases behave in a rigid manner, so that the change in pressure is linear and the resulting pressure distribution is trapezoidal or triangular. In practice, of course, some deflection must actually occur since if there were no strain there would be no resulting stress and reinforcement would be unnecessary. Such deflection means that the resulting soil deformation is greatest in those areas most directly beneath the applied load. Thus the true bending moments and shearing forces will be somewhat less than those calculated on the assumption of a rigid base.

A major decision that must be made when designing spread foundations is whether uplift should be allowed (i.e. whether over part of the base the bearing pressure may fall to zero). One school of thought considers that this is extremely undesirable since it implies that over

this area there is no contact between the base and the soil. Thus the ingress of water is possible, softening the soil and reducing the bearing capacity. Uplift is not sanctioned by the *Joint Institutions Design Manual*.

One problem in determining the initial size of base required is in making allowance for the self-weight of the base itself which, on a soil having a low bearing capacity, may be considerable. If uplift is not allowed a simple solution is to assume a realistic base thickness, to calculate the resulting load per unit area due to the self-weight of such a base and to deduct this amount from the allowable bearing pressure before commencing the calculations. Then, providing the final calculated base thickness does not exceed that initially assumed, the design will be satisfactory. However, this strategy cannot be employed if uplift is allowed since the weight of the unsupported area of base must now be transferred to the area where the soil provides support, and it is thus necessary to assume initial base dimensions as well as thickness. Assuming a base self-weight of 5% to 10% of the vertical load supported is often satisfactory.

Although soil mechanics theory indicates that in a homogeneous soil the allowable bearing pressure increases with an increase in depth, the normal office design practice is to ignore such an increase when designing spread foundations, but to take it into account in the design of piers and piles. The usual procedure when designing spread foundations is to specify the depth to the underside of the base (which must be sufficiently deep to avoid the effects of water or frost) and the permissible bearing pressure at that depth. The calculated thickness of base is then added above this level. If this procedure brings the top of the base too close to the ground surface to satisfy considerations of appearance or to interfere with services, it may be necessary to specify the top level of the base instead and to thicken downwards.

The thickness of a base is determined principally by considering the shear resistance. The *Joint Institutions*

Design Manual provides (in Table 39) recommended minimum ratios of overall depth h to projection beyond column face a corresponding to various values of bearing pressure q. The relationship tabulated can be represented by the expression $h/a = (q/3300)^{0.34}$, where q is given in kN/m^2. In order to estimate an initial base thickness *Allen* suggests adopting a value of one-fifth of the minimum dimension in plan.

If wind forces act on the structure, the column bases may be subjected to overturning moments and horizontal forces inducing sliding. Resistance to sliding is provided by a combination of the friction between the underside of the base and the supporting soil (which is often taken as a proportion of the vertical load, e.g. $0.4V$) plus some assistance from cohesion (if such a soil is present) plus the passive resistance provided by the end of the base. It may be wise to adopt differing partial safety factors for these differing elements. A common factor for the frictional resistance is 1.5, with a similar or slightly greater value for cohesion, while a higher factor of 3 is normally used for passive resistance, since this is only invoked fully after some undesirable horizontal displacement has occurred. Note that, since the base may be founded on undisturbed ground but the soil at the sides and above the base may well have been excavated and replaced, it may be necessary to take account of different soil properties for these two areas, even if the soil material itself is the same.

17.2 REINFORCED CONCRETE BASE

According to BS8110, two separate conditions must be considered when determining the resistance of a reinforced concrete base to the shearing forces due to a concentrated load such as that produced by the reaction from a column. Firstly, the direct shear acting to one side of a vertical plane A–A extending across the full breadth of the base must be determined. Secondly, the effects of punching shear around the periphery of the concentrated load must be investigated.

The Code states that the critical position for the shear plane is at the face of the column. However, as in the case of slabs supporting concentrated loads described in section 6.5.1, where the section considered is within a distance of $1.5d$ of the concentrated load the shear resistance of the concrete v_c may be enhanced by multiplying by the factor $1.5d/a_v$, where a_v is the distance of the section considered from the concentrated load. Consequently, at this point the shear resistance v_c is at its maximum value of $0.8\sqrt{f_{cu}}$ or 5 N/mm^2, whichever is less, and the *Code Handbook* suggests that the critical section is more likely to occur at a distance d from the column face. However, investigations by the writer in the process of developing a computer program for the automatic design of isolated column bases have indicated that the critical plane tends to occur most frequently at a distance of $0.5d$ to $0.7d$ from the column. For important bases it is worth while checking the effects of both direct and punching shear at several successive planes if indications

are that the value might be critical. This procedure is adopted for the base for column B3 on *Calculation Sheets 17*. Of the two types of shear, punching is usually the more critical for normal bases where the punching shear perimeter falls wholly within the slab area (i.e. does not extend to the slab edge).

Although it is possible, if the applied shear force exceeds the resistance provided by the concrete alone, to provide inclined bars as may be necessary around the column heads with flat slabs, such a measure should be avoided by increasing the depth of the base if at all possible. The Code states that it is not advisable to use shear reinforcement in slabs less than 200 mm deep, and this must be borne in mind if bases are tapered as shown in *Figure 17.1*. In extreme circumstances an alternative strategy is to increase the shear resistance of the concrete v_c by increasing the proportion of main reinforcement passing through the critical plane, i.e. by employing it at less than its maximum design strength. This may be a less uneconomic tactic than it first appears, as the steel proportion rises as the base thickness decreases anyway. Furthermore, since the bars are not carrying the full tensile force a shorter bond length is needed, and they may thus be curtailed closer to the critical plane. Such a stratagem may be useful to achieve a slight reduction in base thickness where the depth available is limited.

The resistance to bending and the amount of reinforcement necessary are determined by applying the requirements of BS8110 that the critical plane for bending occurs at the face of the column (C–C in *Figure 17.1b*), and that the total bending moment acting across this plane is due to the moment of the forces over the entire area on one side of the plane, i.e. the area shown hatched on *Figure 17.1b*. The total ultimate bending moment on a square base is therefore $0.125F(l-c)^2/l$ kNm, in which F is the ultimate (i.e. factored) axial load on the column in kN, l is the length of the base in metres and c is the width of the square column in metres. If the base or column is not square, this formula may be adjusted accordingly.

The foregoing discussion assumes that the base supports a column carrying axial load only and that the base may be positioned symmetrically beneath the column. Thus the distribution of pressure beneath the base due to the column load may be assumed uniform. In practice a bending moment or moments about one or both principal axes may be transmitted from the column to the base. In such a situation it may be possible to so position the base that the centroid of action of the applied load and moments corresponds to the centroid of the base, so that the pressure beneath is again uniform. If this is not possible, however, and in those cases where the base may support various combinations of load and moment, the distribution of earth pressure under the base will vary accordingly and must be taken into account when calculating the moments and shears as described above.

The combinations of load and moment due to factored and unfactored values will almost invariably give rise to different positions of the centroid of action, and *Allen* recommends that in such circumstances it is preferable to position the base so that the pressure beneath is uniform when considering the action of the unfactored loads. While considering this point it should be noted that the particular loading combination that causes the maximum bearing pressure may not necessarily be the same as those inducing the maximum bending moment, direct shear and/or punching shear. The assumption usually made is that the load combination giving rise to the critical stress due to punching shear corresponds to that causing maximum moment due to bearing pressure.

In small bases the bars should be spaced uniformly across the full width of the base. However, if c is the column width, where the base dimension exceeds $1.5c + 4.5d$ BS8110 requires that two-thirds of the total reinforcement in that direction should be concentrated within a strip $c + 3d$ wide positioned symmetrically beneath the column. If bars of a uniform size are provided across the base, this requirement is equivalent to *reducing* the uniform *spacing* of the bars within the column strip by multiplying the normal spacing by the factor $1.5(c + 3d)/l$ (or alternatively by increasing the uniform amount per unit width by dividing by this factor). Similarly, the uniform spacing of the bars near the edges of the base must be *increased* by multiplying the normal *spacing* by the factor $3(l - c - 3d)/l$.

For example, with a base width l of 5 m, an effective depth d of 600 mm and a column width c of 400 mm, respacing is necessary. If the bars would otherwise have been uniformly spaced at 200 mm centres, over a central strip $(1.5 \times 400) + (4.5 \times 600) = 3.3$ m in width the spacing must be decreased to $200 \times 1.5 \times [400 + (3 \times 600)]/5000 = 132$ mm, and elsewhere the spacing should be increased to $200 \times 3 \times [5000 - 400 - (3 \times 600)]/5000 = 336$ mm. Although not explicitly stated in BS8110, the implication is that the spacing of bars in bases should never exceed 300 mm, and *Allen* works on this assumption. However, with steel proportions below 1% it is difficult to see why the Code rules regarding the maximum bar spacing in slabs should not apply. The *Joint Institutions Design Manual* specifies that maximum centre-to-centre bar spacings of 200 mm for high-yield bars and 300 mm for mild steel bars must be adopted. Note that even after the bars have been respaced the Code requirements regarding minimum nominal steel must be met.

Note also that respacing the reinforcement as described above modifies the values of the shear resistance v_c provided by the concrete; this is particularly advantageous in resisting punching shear, of course. It is interesting to note, however, that even the resistance in direct shear is usually increased by respacing the bars. For instance, in the example shown above, calculations indicate that making the theoretical adjustments to the bar spacings as shown increases the direct shear resistance of the section by about 4.4%. When calculating the punching shear on bases where the bars have been respaced, it is clearly necessary to check carefully those parts of the critical perimeter which fall within the central strip (and to which a higher value of v_c therefore applies) and those which fall outside.

If the bearing pressure beneath the base is not uniform, the appropriate bending moments must be determined from the resulting triangular or trapezoidal pressure distributions. For bending about the minor axis the resulting moments may vary (or even become zero) over part of the length, and in such circumstances it seems preferable to design reinforcement to resist the total force to one side of the plane being considered and then to respace this to satisfy the Code criteria as necessary.

Note that, if uplift occurs, bars will be required in the top of the slab to resist the moment due to the self-weight of the cantilevering part of the base plus the overburden to the level of the ground-floor slab, and also any imposed load on this floor slab if it is not suspended. However, such reinforcement is only needed over that length of base where uplift may occur.

It is also necessary to ensure that all bars have a sufficient anchorage-bond length from the section of maximum stress. In the example on *Calculation Sheets 17* the distance from the column face to the edge of the slab is about 1375 mm. According to *Data Sheet 18*, 25 mm mild steel straight bars in grade C30 concrete require a minimum bond length of 890 mm; if an end cover of say 75 mm is allowed, 25 mm bars would suffice. Reference 2 recommends that end anchorages should be avoided in bases. The *Joint Institutions Design Manual* states that it is usually possible to satisfy anchorage requirements while using straight bars, but if the bond length available is insufficient then the provision of bobs or hooks at the bar ends may be considered as an alternative to providing smaller bars at closer spacing. However, the provision of bobs on bar ends is not common current practice.

In larger bases it is often worth while terminating alternate bars at the minimum bond length by employing the staggered arrangement shown in *Figure 17.1c*, provided that the area of the continuing bars is sufficient to provide the necessary resistance to bending and (by influencing the value of v_c) shearing, and that the bars can develop the necessary anchorage bond beyond this point. In other words if, by curtailing 50% of the bars, the remainder are then stressed to their full design strength to provide the necessary resistance moment, they must also be provided with a full anchorage-bond length beyond this point. Such an arrangement complies fully with the requirements of BS8110 although, for some reason, the *Joint Institutions Design Manual* specifies that all reinforcement must extend for the full length of the base. This document also states that the minimum amounts of steel that should be provided at any point in each direction are $0.0025bh$ and $0.0013bh$ for mild steel and high-yield steel respectively; a similar requirement is also implied in BS8110.

The normal design procedure is thus first to check if the use of straight bars is possible and, if so, whether some of these can be stopped off. If not, then check whether bars with bobs can be employed and, if not, whether bars with hooks will be satisfactory.

Unlike previous codes, BS8110 does not require a check to be made on the local-bond stress (although it is interesting that such a requirement has been retained in the draft recommendations for permissible-stress design (ref. 3) currently available). Situations where the shearing force changes rapidly, as occurs in the case of small bases and high ground reactions, were those where local bond was a critical factor; although such a check is no longer required, the provision of a similar arrangement of relatively closely spaced small-diameter bars that used to be essential may still be prudent.

According to BS8110, the cover required for concrete in contact with non-aggressive soil corresponds to that of a moderate environment, i.e. ranging from 35 mm for grade C30 (the minimum permissible assuming a 'systematic checking regime' is employed; see *Data Sheet 42*) to 20 mm for grade C50. However, the Code also states that where concrete is cast directly against soil the nominal cover should not be less than 75 mm. In fact, to provide a clean surface on which to lay the reinforcement and to prevent the earth contaminating the structural concrete forming the foundation, it is usual to spread a layer of lean 'blinding' concrete over the bottom of an excavation. This layer may be 50 mm or more in thickness and of 1.8 concrete. Higher-grade concrete may be required if the permissible bearing pressure on the ground is high; for example, for foundations on rock this layer should be of concrete that is not weaker than that in the foundation. Since this blinding layer protects the reinforcement, BS8110 permits the minimum nominal cover to be reduced to 40 mm in such circumstances. The *Joint Institutions Design Manual* states that in foundations the minimum cover to all reinforcement should be 50 mm.

If the top of the base is splayed as shown on *Figure 17.1a*, it may be necessary to check the compressive stress in the concrete by calculating the moment of resistance of the central 400 mm width of base. If this central section does not provide sufficient resistance, the moments of resistance provided by the tapered sections may be added to that of the central section. If the resistance is still insufficient, the splays must be omitted, or the thickness of the base or the concrete grade increased.

BS8110 notes specifically (in clause 3.12.8.8) that the anchorage-bond stress in compression developed in the column starter bars in bases (or pile caps) need not be checked provided that the bars extend from the level of the bottom steel in the base, and that the base (or pile cap) has been designed to resist the full bending moments and shearing forces.

It is normal to ignore fire resistance and deflection requirements when designing foundations, and this is sanctioned by BS8110.

17.3 PLAIN CONCRETE BASE

An alternative type of concrete foundation comprises a plain concrete block which may sometimes be surmounted by a smaller block of higher-grade concrete. BS8110 does not give any guidance on the design of bases of plain concrete, so the following comments are based principally on the *Joint Institutions Design Manual*. Part 1 of BS8103 (which has replaced CP101 'Foundations and substructures for non-industrial buildings of not more than four storeys') states that the concrete for such bases should be an ordinary prescribed mix of at least grade C7.5P; alternatively, a designed mix with a maximum nominal aggregate size of 20 mm and containing at least 220 kg/m^3 of Portland cement may be used. In groundwater where sulphates may be present, this document specifies minimum concrete grades (made with sulphate-resisting cement to BS4027) of C20P, C25P and C30P for ground classes 2, 3 and 4 respectively, as classified in Table 6.2 in BS8110, Part 1 (which derive from the recommendations of *BRE Digest 250*). The depth of each base must be more than the horizontal distance from the column face to the base edge.

In addition to the latter requirement, the *Joint Institutions Design Manual* specifies that the minimum ratio of base depth h (which must not be less than 300 mm) to horizontal distance a must be not less than 0.15 $(q^2/f_{cu})^{0.25}$ (but never less than unity), where q is the unfactored bearing pressure in kN/m^2. The minimum strength of concrete that may be employed for plain bases is 20 N/mm^2, according to the manual, and the mix must contain at least 220 kg/m^3 of cement. For sulphate-bearing soils, the recommendations of *BRE Digest 250* (see above) must be followed.

If a mix having a strength of less than 20 N/mm^2 is used for the main base, it is suggested that this is surmounted by a block of higher-strength concrete to spread the concentrated load from the column on to the lower block, which distributes this load to the ground. The area of the upper block must be not less than the load on the column, divided by the permissible bearing pressure on the plain concrete. Reference 24 recommends that the local bearing capacity beneath a concentrated load be limited to 40% of the characteristic cube strength of the concrete. However, recent research by Cusens, Wang and Wong (ref. 25) proposes the relationship

$$f_b = 0.4f_{cu} + 0.33(f_{cu}a/a_1)^{0.75}$$

where a is the dimension of the area supporting the bearing and a_1 is the dimension of the bearing itself.

The thickness of the block should be sufficient to spread the load at an angle of not less than 45° from the column on to the lower block, noting that the dispersion should be reckoned diagonally from the corner of the column to the corresponding corner of the block. The splice bars for the column should be embedded for a length necessary to provide sufficient resistance to bond (see *Data Sheet 18*), which in this example is not less than 780 mm for the 25 mm bars in the octogonal column if grade C25 concrete is used. These requirements are met if the block is 1.050 m square and 550 mm thick.

The area of the lower concrete block should be sufficient to transmit the load on the column and the weight of the base to the ground at a pressure not exceeding 200 kN/m². The thickness of the block should be sufficient to disperse the load at an angle of not less than 45° to the horizontal, diagonally from one corner of the upper block to the corresponding corner of the lower block. The load producing shearing stress in planes at a distance from the periphery of the upper block equal to the depth of the lower block should be calculated. This stress should not exceed the normal permissible shearing stress. By ensuring that the thickness of the plain concrete base is such that the load is dispersed to the ground at an angle of not less than 45° as described, there are only small tensile stresses in the bottom of the base but, to guard against the possibility of cracks forming, large plain concrete bases may sometimes be strengthened by inserting reinforcement near the bottom. Resistance to horizontal forces is achieved by bonding the column to the plain concrete by bars, such as the splice bars for the column, extending from the bottom of the base into the shaft of the column.

17.4 COMBINED BASES

Where two columns are positioned so closely that it is impossible to locate separate isolated bases beneath each, it may be necessary to design a single footing to support both columns. A similar solution may be needed where, because of site boundary constraints for example, it is not possible to extend the edge of the base to the full distance required beyond the column face.

The design of such bases is not mentioned in BS8110 and is not dealt with in detail here, as the general principles to be observed are similar to those described for strip footings in section 17.5. It is most important to investigate carefully the effects of punching shear and the position of all alternative critical perimeters along which this may occur, especially where the columns are located close to the base edge. Hogging moments will almost always occur between the columns, and the *Joint Institutions Design Manual* recommends that the resulting top and bottom steel is tied together with links. In larger bases it may be necessary to respace the bars more closely beneath the columns as described for isolated bases in the previous section. Details of the criteria and requirements concerned are given in ref. 2, and are as follows with reference to *Figure 17.2a*:

1. Divide base as shown by line drawn midway between column centres.
2. (a) If l_{y1} exceeds $1.5(c_{y1} + 3d)$, then respace reinforcement so that two-thirds of that required over width l_{y1} is located within width $(c_{y1} + 3d)$ and arranged symmetrically beneath column.
   (b) If l_{y2} exceeds $1.5(c_{y2} + 3d)$, then respace reinforcement so that two-thirds of that required over width l_{y2} is located within width $(c_{y2} + 3d)$ and arranged symmetrically beneath column.

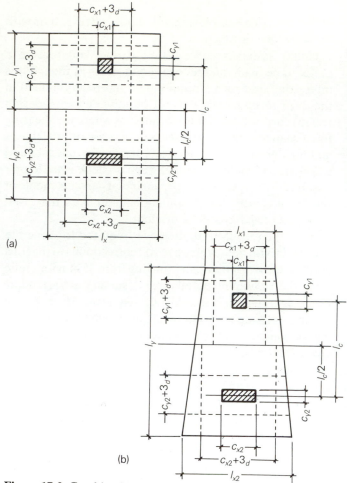

Figure 17.2 Combined bases.

(c) If c_{min} is the lesser of c_{x1} and c_{x2} and if l_x exceeds $1.5(c_{min} + 3d)$, then respace reinforcement so that two-thirds of that required over width l_x is located within width $(c_{min} + 3d)$ and arranged symmetrically beneath column.

17.5 STRIP FOOTINGS: WORKED EXAMPLE

When the permissible ground pressures are low, or when the columns are so close together that separate bases would overlap, a line of columns can be conveniently carried on a strip footing. Consider the footing shown in *Figure 17.3*, which is 20 m long, 2 m wide and carries five columns. The unfactored total loads on the footing are as shown. The first step in the design is to determine the position of the centre of gravity of the loads. Taking moments about the right-hand end of the footing:

$300 \times 2\,m =$	600
$350 \times 7.5\,m =$	$2\,625$
$400 \times 11.5\,m =$	$4\,600$
$450 \times 15\,m =$	$6\,750$
$500 \times 18\,m =$	$9\,000$
$2000\,kN$	$23\,575\,kN\,m$

The distance of the centre of gravity from the right-hand end is thus $23\,575/2000 = 11.788\,m$. The centroid of the footing is 10 m from each end; hence the eccentricity of

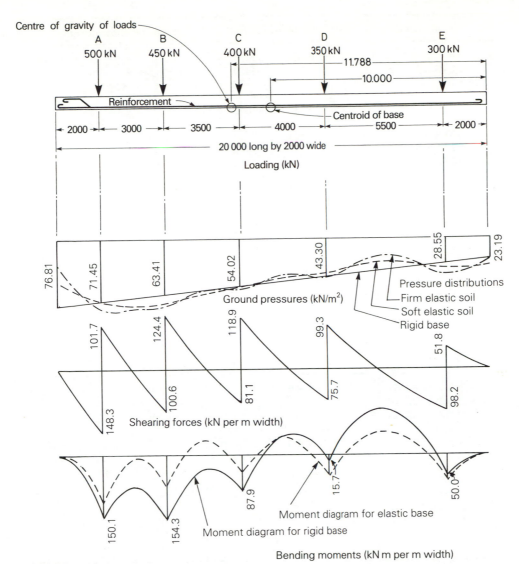

Figure 17.3 Strip footing.

the loading is 1.788 m. As this eccentricity is less than one-sixth (i.e. than 3.33 m) of the length of the footing, the maximum and minimum ground pressures are $(2000/20 \times 2)[1 \pm (6 \times 1.788/20)]$, that is 76.81 kN/m² maximum and 23.19 kN/m² minimum. These pressures are distributed as indicated by the full line in *Figure 17.3*. The self-weight of the footing must now be added and the total maximum pressure compared with the safe bearing pressure. If the maximum applied pressure is found to be excessive, the size of the footing must be increased. Alternatively it may be possible to alter the position of the footing so that the centroid coincides with the centre of gravity of the loads, thereby producing a uniform pressure throughout which is less than the maximum pressure under the same footing when eccentrically loaded with the same loads. In the present example, it is assumed that the calculated pressures are not excessive and that the dimensions or position of the footing cannot be altered. The above calculations also assume that it is sufficient to consider the effects of total loads only, but for very important bases it may be necessary to investigate the

effects of applying imposed loads to the columns at one end of the strip footing only.

The bending-moment and shearing-force diagrams due to the unfactored loads can be drawn by calculating the bending moments and shearing forces at a number of sections in accordance with the procedure shown in *Figure 17.4*; the pressures and loads are taken from *Figure 17.3*, and the calculations apply to a 1 m width of slab footing.

It is advisable at this stage to make a check calculation at C to ensure that no serious arithmetical error has occurred. Thus, considering the loads to the right of C,

$$
\begin{array}{llr}
-23.19 \times 11.5^2 \times \tfrac{1}{2} & = & -1533.4 \\
+(23.19 - 54.02) \times 11.5^2 \times \tfrac{1}{6} & = & -679.5 \\
+350 \times 4/2 & = \quad 700 & \\
+300 \times 9.5/2 & = \quad 1425 & \\
\cline{2-3}
& +2125 & -2212.9 \\
& & +2125.0 \\
\cline{3-3}
& & -87.9 \text{ kN m} \\
& & \text{per m width}
\end{array}
$$

Foundations

Section	Service bending moments (kN m per m width)			Service shearing forces (kN per m width)		
		Positive	Negative		Positive	Negative
A (loads to LHS)	$-71.45 \times 2^2 \times \frac{1}{2}$ $-5.36 \times 2^2 \times 0.333$		142.9 −7.2 ――― −150.1	-71.45×2 $-5.36 \times 2 \times \frac{1}{2}$		−142.9 −5.4 ――― −148.3
B (loads to LHS)	$+500 \times \frac{1}{2} \times 3$ $-63.41 \times 5^2 \times \frac{1}{2}$ $-13.40 \times 5^2 \times 0.333$	+750.0 +750.0 +750.0	−792.6 −111.7 ――― −904.3 ――― −154.3	$+500 \times \frac{1}{2}$ -63.41×5 $-13.40 \times 5 \times \frac{1}{2}$	+250.0	−317.1 −33.5 ――― −350.6 +250.0 ――― −100.6
C (loads to LHS)	$+500 \times \frac{1}{2} \times 6.5$ $+450 \times \frac{1}{2} \times 3.5$ $-54.02 \times 8.5^2 \times \frac{1}{2}$ $-22.79 \times 8.5^2 \times 0.333$	+1625.0 +787.5 +2412.5 +2412.5	−1951.5 −548.9 ――― −2500.4 ――― −87.9	$+500 \times \frac{1}{2}$ $+450 \times \frac{1}{2}$ -54.02×8.5 $-22.79 \times 8.5 \times \frac{1}{2}$	+250.0 +225.0 +475.0	−459.2 −96.9 ――― −556.1 +475.0 ――― −81.1
D (loads to RHS)	$+300 \times \frac{1}{2} \times 5.5$ $-23.19 \times 7.5^2 \times \frac{1}{2}$ $-20.11 \times 7.5^2 \times 0.167$	+825.0 +825.0 +825.0	−652.2 −188.5 ――― −840.7 ――― −15.7	$+300 \times \frac{1}{2}$ -23.19×7.5 $-20.11 \times 7.5 \times \frac{1}{2}$	+150.0 +150.0	−173.9 −75.4 ――― −249.3 +150.0 ――― −99.3
E (loads to RHS)	$-23.19 \times 2^2 \times \frac{1}{2}$ $-5.36 \times 2^2 \times 0.167$		−46.4 −3.6 ――― −50.0	-23.19×2 $-5.36 \times 2 \times \frac{1}{2}$		−46.4 −5.4 ――― −51.8

Figure 17.4

This corresponds to the value obtained by considering the loads on the left-hand side of C and confirms that the calculations are mathematically correct.

The thickness of the footing should be based on the maximum service moment of 154.3 kN m, the reinforcement being varied at the other sections. For simplicity, a partial safety factor for load γ_f of 1.6 may be adopted in order to obtain the required ultimate bending moments and shearing forces required to design the sections. This value of γ_f is that normally corresponding to imposed loads, and its use for total loads errs on the side of safety. It is probably quite in order to adopt a factor of 1.45 when considering dead and imposed loads together, and this is what the *Joint Institutions Design Manual* does; however, in this chapter a value of 1.6 has been assumed.

The maximum ultimate bending moment is thus $154.3 \times 1.6 = 247.0$ kN m per m and, with grade C35 concrete and mild steel reinforcement, the minimum effective depth that may be adopted is 215 mm. For grade C35 concrete it is advisable to have a minimum cover to all reinforcement of at least 40 mm, so that a suitable practical depth for the slab would be 350 mm. Then at sections A and B the reinforcement required would be 4310 mm²/m, which could be provided by using 32 mm bars at 175 mm centres. Throughout the remainder of the footing the maximum amount of reinforcement required is 2337 mm²/m: provide 25 mm bars at 200 mm centres. Owing to the length of the footing the bars would be provided in two or three lengths and lapped sufficiently to conform to the requirements for anchorage bond as described.

The foregoing bending moments have been calculated at the centre-lines of the columns, but in practice it is only necessary to consider the moments at the face of the columns. If this is done (assuming 400 mm square columns) it will be found that the critical values considered above reduce to ultimate moments of 216.8 and 116.6 kN m per m respectively, requiring only 3724 and 1912 mm²/m of reinforcement.

The maximum direct shearing force that must be considered is that at a distance of $1\frac{1}{2}d$ from the face of the column in question. Assuming that the effective depth d is

294 mm and the loads on the footing are applied through columns 400 mm square, the maximum shearing force, obtained from *Figure 17.3*, is about

$$1.6 \times 148.3 \left\{ \frac{2 - [0.2 + (1.5 \times 0.294)]}{2} \right\}$$
$$= 161.2 \, \text{kN per m width}$$

Then, since the proportion of main reinforcement provided is $4595/(10^3 \times 294) = 0.0156$, from *Data Sheet 15* the concrete alone will resist a total ultimate shearing force of $0.88 \times 294 \times 10^3 = 259 \, \text{kN/m}$, and thus no shearing reinforcement is necessary at this point. Where the smaller amount of reinforcement is provided, the maximum shearing force, as given on *Figure 17.3*, is 118.9 kN. Thus at a distance of $1\frac{1}{2}d$ from the face of this column the ultimate shearing force is about

$$1.6 \left[118.9 - \left(\frac{118.9 + 75.7}{4} \right) 0.641 \right] = 140.3 \, \text{kN}$$

Since $\varrho = 2454/(10^3 \times 294) = 0.0083$, the shearing resistance of the concrete alone is $0.71 \times 10^3 \times 294 = 208 \, \text{kN}$, which is again well in excess of the applied shearing force.

The punching shear around the most heavily loaded column supported by each designed area of main reinforcement must also be checked. The perimeter at a distance of $1.5d$ from the column face is first examined. With 400 mm square columns, the total length of perimeter here is $4 \times [0.4 + (2 \times 1.5 \times 0.294)] = 5.128 \, \text{m}$. With the larger of the two amounts of main reinforcement, the proportion of steel in this direction is 0.0156. With the minimum permissible proportion of $0.0025bh$ of mild steel provided transversely, the average value of ϱ for both directions (in terms of bd) is about 0.0090, and thus the shearing resistance of the concrete alone will be $0.73 \times 5128 \times 294 = 1100 \, \text{kN}$. Since the maximum shearing force cannot exceed $500 \times 1.6 = 800 \, \text{kN}$, the concrete alone will suffice. Checking the situation at the column face itself, the unit resistance of the concrete here is $0.8\sqrt{35} = 4.73 \, \text{N/mm}^2$, and so the resistance provided by the concrete is $4 \times 400 \times 294 \times 4.73 = 2226 \, \text{kN}$.

The smaller amount of reinforcement corresponds to an average proportion of about 0.005, and thus the shearing resistance of the concrete alone along a perimeter $1.5d$ from the column face is $0.6 \times 5128 \times 294 = 904 \, \text{kN}$, compared with the maximum ultimate axial load of $400 \times 1.6 = 640 \, \text{kN}$. Again the shearing resistance of the concrete is sufficient by itself.

It is advantageous not to have to provide shearing resistance in a strip footing if this can possibly be avoided, and it is well worth while increasing the thickness of the footing to that necessary to achieve this result. The actual shearing forces concerned are often so large that considerable amounts of steel are otherwise required to withstand the resulting difference between the applied shearing force and the resistance provided by the concrete.

Such a problem as that considered above involves, in the calculation of the bending moments, the determination of comparatively small differences between large quantities. It is therefore necessary for all calculations to be made with great care, avoiding approximations (especially for the value of the distributed ground pressures) that may give misleading results. Another factor is that the conventional method of calculation, as presented above, assumes that the base is sufficiently rigid to distribute the pressure so that it varies uniformly throughout its length. For the strains due to the calculated bending moments to be achieved, however, there must be some deformation of the base. This deflection will itself cause a redistribution of the ground pressures, which will tend to increase under the loaded points and consequently to decrease between these points. This redistribution reduces the resulting bending moments, and in extreme cases may even reverse the sign of the bending moments between the loaded points. For the extreme case the bending moment under any column, for example C, can be calculated from $\frac{1}{2} \times 400 \times 3.75 \times \frac{1}{10} = 75 \, \text{kN m}$ per metre width, where the unfactored load on column C is 400 kN and the mean of spans CB and DC is 3.75 m. If similar calculations are made for the other loaded points, a bending-moment diagram as shown dashed in *Figure 17.3* may result, from which it can be seen that generally the peak bending moments (i.e. the negative values) are smaller, and positive bending moments, which did not occur in the 'rigid base' analysis, now appear. These smaller bending moments, however, are accompanied by higher local ground pressures, the variations of which are somewhat as indicated on the pressure diagram in *Figure 17.3*. It is impossible to assess a close value for the maximum pressures but, if the ground can withstand these increased values without excessive settlement, the reduced moments will probably be realized. If these high local pressures were so much in excess of the safe pressure as to cause considerable settlement, a readjustment of the pressures would again occur, so that, in the case of poor soils with a symmetrical load, a uniformly distributed pressure would more nearly occur everywhere. With an eccentric load the pressure would vary nearly uniformly, in a similar manner to that indicated by the full line on *Figure 17.3*. This is the reason for employing the method of analysis adopted in the detailed design in this example for cases where the bearing capacity of the ground is small.

17.6 RAFT FOUNDATIONS: WORKED EXAMPLE

Consider the problem shown in *Figure 17.5*, where a total unfactored load of 13 600 kN from a group of twelve columns and other loads on the raft is to be distributed so that the ground pressure does not exceed $64 \, \text{kN/m}^2$. If it is assumed that, of the permissible ground pressure, $21 \, \text{kN/m}^2$ represents the weight of the raft and imposed load due to a depth of earth filling of 1 m, the pressure available for supporting the column loads is $43 \, \text{kN/m}^2$. It is

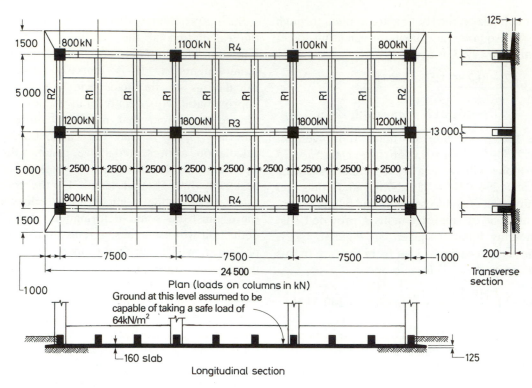

Figure 17.5 Raft foundation.

assumed that the distance from the centre-line of each end row of columns to the edge of the raft cannot exceed 1 m, and that there is no restriction on the width of the raft. Thus the maximum overall length of the raft is 24.5 m, and the minimum width is then

$$\frac{13\,600}{43 \times 24.5} = 12.91; \text{ say 13 metres.}$$

The raft comprises a slab over the whole site, stiffened by transverse ribs that, in conjunction with the longitudinal ribs, transmit the load from the columns to the ground. A suitable arrangement of ribs spaced at 2.5 m centres is shown in *Figure 17.5*. The raft is considered as an inverted floor carrying a uniform load of 43 kN/m², but with the difference that the column loads are also specified; this affects the bending moments in the ribs, as is shown later. Since the determination of the cross-sections of the concrete and the areas of reinforcement follow the methods described in Chapter 5 for slabs and T-beams, the present design does not proceed beyond determining the bending moments and shearing forces to be resisted by the slabs and ribs.

The ultimate bending moment on any interior panel of the slab, assuming the use of the arbitrary coefficients given in BS8110 (see section 3.2) is $43 \times 2.5^2 \times 0.063 \times 1.6 = 27.1$ kN m per metre width, which requires a 150 mm slab if grade 35 concrete is used and a minimum cover of 40 mm is to be provided to all bars. These bars should be arranged to resist tension in the top face of the slab between the transverse ribs, and in the bottom face under these ribs. The cantilever bending moment due to the 1 m projection beyond the end transver ribs is

$43 \times 1^2 \times \frac{1}{2} \times 1.6 = 34.4$ kN m per metre width. Hence the bending moment in the end panels spanning 2.5 m is about the same as that in the interior panels, and the same thickness of slab and amount of reinforcement can be used throughout.

Along the longitudinal edges of the raft, where the effective projection is 1.5 m, the ultimate bending moment is $1.6 \times 1.5^2 \times 43 \times \frac{1}{2} = 77.4$ kN m per m, which requires a 200 mm slab at the face of the rib. The projecting slabs can be tapered to, say, 150 mm at the outer edges, as shown in *Figure 17.5*.

The upward load on the interior transvere ribs R1 is, as shown in *Figure 17.6a*, a uniform pressure of $2.5 \times 43 = 107.5$ kN per m, neglecting any relief due to the weight of the rib. The total upward load on this rib is $107.5 \times 10 = 1075$ kN. The corresponding loads on the outer transverse ribs R2 are $[(2.5 \times \frac{1}{2}) + 1]43 = 96.8$ kN per m and 967.5 kN total. The sum of the load from ribs R1 and R2 on the central longitudinal rib R3 must be equal to the sum of the column loads on this rib, i.e. 6000 kN. Since the load on rib R2 is 90% of that on rib R1, the total number of ten ribs is equivalent to $8 + (2 \times 0.9) = 9.8$ ribs of R1. Therefore the load from each rib R1 is $6000/9.8 = 612.24$ kN, and that from each rib R2 is $0.9 \times 612.24 = 551.02$ kN. The load from R1 on each of the outer longitudinal ribs R4 is $\frac{1}{2}(1075 - 612.24) = 231.38$ kN. The corresponding load from rib R2 is $\frac{1}{2}(967.5 - 551.02) = 208.24$ kN. The loading and resulting *ultimate* bending-moment and shearing-force diagrams for the rib R1 (assuming a partial safety factor for loads of 1.6) are shown in *Figure 17.6a*; the net ultimate bending-moment diagram is the difference between a parabola of maximum

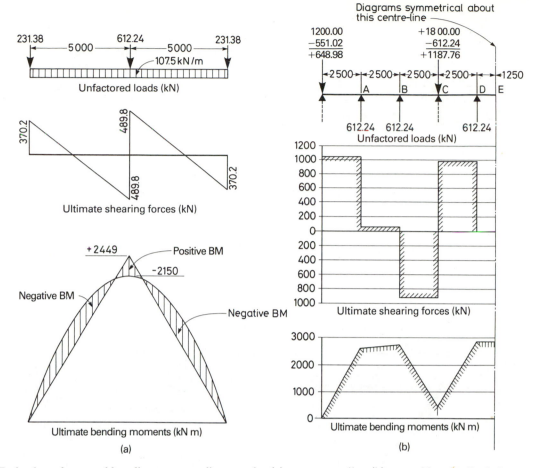

Figure 17.6 Raft: shear force and bending moment diagrams for (a) transverse ribs; (b) central longitudinal rib.

height $-\frac{1}{8} \times 1075 \times 10 \times 1.6 = -2150$ kN m and a triangle of maximum height $+\frac{1}{4} \times 612.24 \times 10 \times 1.6 = +2449$ kN m. Since the load on the outer ribs R2 is only 10% less than that on R1, the details adopted for R1 may be used for R2 without sacrificing much economy.

The load on the longitudinal ribs R3 and R4 consists of the column loads acting downwards, and the upward loads from the transverse ribs, the self-weight of the ribs being neglected. The diagrams of unfactored loads and ultimate shearing forces and bending moments are given in *Figure 17.6b* for the central longitudinal rib R3 and in *Figure 17.7* for the outer longitudinal rib R4. Since, in the example, the system is symmetrical, only one-half need be investigated. The calculation of the ultimate bending moments at a number of sections for the rib R3 is as follows:

Section A:
$$+648.98 \times \; 2.5 \; = +1622.4 \times 1.6 = \; +2595.9 \text{ kN m}$$

Section B:
$$
\begin{aligned}
+648.98 \times \; 5.0 \; &= +3244.9 \\
-612.24 \times \; 2.5 \; &= -1530.6 \\
\hline
+1714.3 \times 1.6 &= \; +2742.9 \text{ kN m}
\end{aligned}
$$

Section C:
$$
\begin{aligned}
+648.98 \times \; 7.5 \; &= +4867.3 \\
-612.24 \times \; 7.5 \; &= -4591.8 \\
\hline
+275.5 \times 1.6 &= \; + \; 440.9 \text{ kN m}
\end{aligned}
$$

Section D:
$$
\begin{aligned}
+648.98 \times 10.0 \; &= +6489.8 \\
+1187.76 \times \; 2.5 \; &= +2969.4 \\
-612.24 \times 12.5 \; &= -7653.0 \\
\hline
+1806.2 \times 1.6 &= \; +2889.9 \text{ kN m}
\end{aligned}
$$

Section E:
$$
\begin{aligned}
+648.98 \times 11.25 \; &= +7301.0 \\
+1187.76 \times \; 3.75 \; &= +4454.1 \\
-612.24 \times 16.25 \; &= -9948.9 \\
\hline
+1806.2 \times 1.6 &= \; +2889.9 \text{ kN m}
\end{aligned}
$$

In addition to the concentrated loads, the outer longitudinal ribs R4 carry a uniform load of $1.5 \times 43 = 64.5$ kN per metre. The resultant ultimate bending-moment diagram (*Figure 17.7*) can be derived either by a

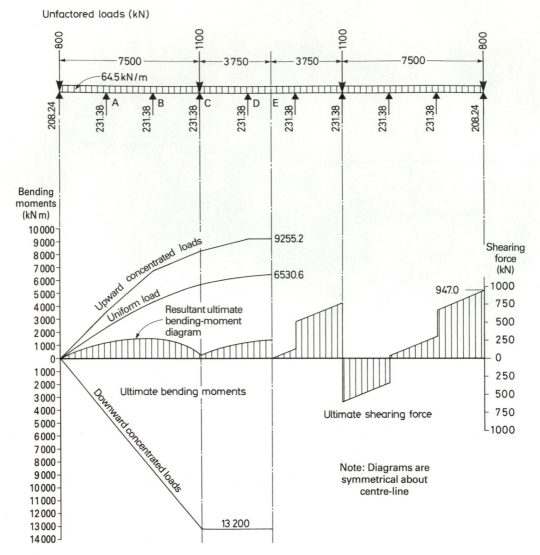

Unfactored loads (kN)

Bending moments (kN m)

9255.2

6530.6

Shearing force (kN)

947.0

Resultant ultimate bending-moment diagram

Ultimate bending moments

Downward concentrated loads

Ultimate shearing force

Note: Diagrams are symmetrical about centre-line

13 200

Figure 17.7 Diagrams for outer longitudinal ribs R4 of raft.

similar analysis or by algebraic addition of the bending-moment diagrams due to the following loads:

Column reactions: trapezoidal diagram having a maximum ordinate of

$$-1100 \times 7.5 \times 1.6 = -13\,200 \text{ kN m}$$

Distributed load: parabolic diagram having a maximum ordinate of

$$+\tfrac{1}{8} \times 64.5 \times 22.5^2 \times 1.6 = +6530.6 \text{ kN m}$$

Reactions from transverse ribs: diagram having the following ordinates:

Section A:
$$+925.52 \times 2.5 = +2313.8 \times 1.6 = +3702.1 \text{ kN m}$$

Section B:
$$+925.52 \times 5.0 = +4627.6$$
$$-231.38 \times 2.5 = -578.4$$
$$\overline{}$$
$$+4049.2 \times 1.6 = +6478.6 \text{ kN m}$$

Section C:
$$+925.52 \times 7.5 = +6941.4$$
$$-231.38 \times 7.5 = -1735.3$$
$$\overline{}$$
$$+5206.1 \times 1.6 = +8327.7 \text{ kN m}$$

Section D:
$$+925.52 \times 10.0 = +9255.2$$
$$-231.38 \times 15.0 = -3470.7$$
$$\overline{}$$
$$+5784.5 \times 1.6 = +9255.2 \text{ kN m}$$

Section E:
$$+925.52 \times 11.25 = +10412.1$$
$$-231.38 \times 20.0 = -4627.6$$
$$\overline{}$$
$$+5784.5 \times 1.6 = +9225.2 \text{ kN m}$$

At this stage an arithmetical check can be made on the whole of the preceding calculations by equating the upward and downward loads on one of the outer ribs, as given in *Figure 17.7*:

Downward load:

$$
\begin{array}{llll}
2 \times & 800 & = & 1600 \\
2 \times & 1100 & = & 2200 \\
\hline
& & & 3800 \text{ kN}
\end{array}
$$

Upward load:

$$
\begin{array}{lll}
2 \times & 208.24 = & 416.5 \\
8 \times & 231.38 = & 1851.0 \\
24.5 \times & 64.5 \;\; = & 1580.3 \\
\hline
& & 3847 \text{ kN}
\end{array}
$$

The small difference of about 1.25% between these two sums can be accounted for by the difference between the assumption of a ground pressure of 43 kN/m² and the actual pressure obtained by dividing the total column load by the area of the raft.

The remarks made in section 17.5 on the need for accurate computations, and the possible rearrangement of bending moments due to local increases in ground pressures under strip footings, apply equally to raft foundations.

More-sophisticated methods of analysis similar to those described for strip footings have also been devised for the analysis of rafts. These take account of the deflection of the structure and of the soil beneath the raft due to the imposed loading, and of the resulting effects on the distribution of soil pressure. Almost all such procedures currently used involve the use of finite-element analysis, for which access to a computer is essential.

17.7 PILED FOUNDATIONS

If the bearing resistance near the surface is so small that a large foundation is necessary to distribute the load, it is often more conomical to use piles. Both BS8110 and the *Joint Institutions Design Manual* provide some general rules for pile-cap design. The latter states that normally piles should be spaced at three times their diameter ϕ_p and that they should be arranged symmetrically beneath the loads they support. Pile-caps should extend 150 mm or more beyond the theoretical perimeter of the piles and, where caps are provided for individual piles or pairs of piles, the piles or ground beams should be designed to resist a moment corresponding to a column displacement of 75 mm.

Two principal methods are commonly used to determine the principal forces in pile-caps, and both are sanctioned by BS8110. The cap may be considered to behave as a deep short beam and the resulting moments found by simple statics. Alternatively the pile-cap may be imagined to act as a space frame; the inclined lines of force linking the underside of the column to the tops of the piles are assumed to act as compression members, and the pile heads are tied together by reinforcement acting as horizontal tension members. This latter method, which is considered in some detail in *Allen*, is particularly suitable for analysing the more 'three-dimensional' pile-caps, such

as those required for three or four piles. For example, for the three-pile cap sketched in *Figure 17.8a*, resolving the forces gives thrusts of $F\sqrt{h^2 + \frac{1}{3}l^2}/3h$ in the inclined struts (these may normally be neglected) and tensile forces of $Fl/9h$ between each pile head, if the actual size of

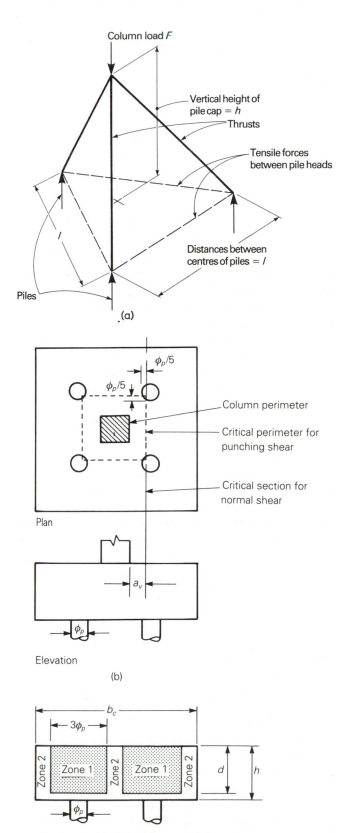

Figure 17.8

the column is ignored. The reinforcement is then designed to withstand these tensile forces by dividing the force by the design stress in the steel. Note that this method of analysis assumes that the tensile force remains constant to the centre-line of the pile, and the necessary anchorage-bond length corresponding to the stress in the steel must be provided beyond this point, normally by passing the bars involved outside the contained piles by means of large-radius bends. For further details, reference should be made to *Allen*.

With the beam method, however, only a nominal anchorage length beyond the pile is normally needed. In view of these widely differing requirements, as a guideline to assessing the actual length that should be provided the *Code Handbook* recommends that, if the distance between the column and the critical section for shear (see below) is at least twice the effective depth d, then only a nominal anchorage length of d or 12 times the bar size, whichever is greater, need be provided. However, if this distance is less than $0.6d$ then the design method for corbels described in clause 5.2.7 of BS8110 should be followed; this implies that an anchorage-bond length sufficient to develop the full tensile strength in the steel must be provided beyond a point one-fifth of the pile diameter within the outer face of the pile. For intermediate distances the *Code Handbook* suggests that linear interpolation be used to determine the bond length needed. Thus if l_{mx} and l_{mn} are the full anchorage-bond length and the minimum length respectively, and nd is the distance from the column face to the critical section, the actual length that should be provided is equal to $[5l_{mx}(2-n) - l_{mn}(3-5n)]/7$.

If the truss method of design is adopted and the spacing of the piles exceeds three times their diameter, only the reinforcing bars located within a strip $1.5\phi_p$ wide on each side of each pile centre-line contribute to the action of the truss. The *Joint Institutions Design Manual* recommends that all pile-caps should be reinforced near the top and bottom faces in each principal direction with not less than 0.25% of mild steel or 0.13% of high-yield steel, these percentages being in terms of the gross concrete section. The manual also states that circumferential steel consisting of at least 12 mm horizontal bars spaced vertically at not more than 250 mm centres must be provided.

According to BS8110, the critical section for shear occurs across a plane located at a distance of one-fifth of the size of the pile within the nearer edge of the pile to the supported column (see *Figure 17.8b*). If several rows of piles are provided, the shear at each row may require checking to determine which is the critical section. According to the Code the shear force to be resisted is that due to the action of all piles having centres lying beyond this plane. If the critical section is located at a distance of less than $2d$ from the column face, the shear resistance of the concrete is enhanced as described for beams in section 6.4. BS8110 specifies that if the pile spacing does not exceed three times the pile diameter, this enhancement

may be applied to the full width of the section, but with greater spacings the enhancement only applies to zones $1.5\phi_p$ wide on each side of each pile centre-line (see *Figure 17.8c*). The main tension steel must have a full anchorage-bond length l_{mx} beyond the critical section for shear: the *Code Handbook* states that this criterion is only critical where a shallow pile-cap surmounts small-diameter piles.

The Code also requires a check to be made that the shear along the column face does not exceed the maximum permissible value for the grade of concrete provided. In addition, if the piles are spaced at more than three times their diameter, the punching shear along the critical perimeter shown on *Figure 17.8b* must be checked. In determining the resistance of the concrete to punching shear, if the distance from the column face to the critical plane is less than $2d$ then the concrete shear resistance may be enhanced as described earlier. In cases where the distance to the critical section differs from one column face to another, the enhancement factor relating to each face must be determined separately and applied to the length of perimeter corresponding to that face only.

17.7.1 Pile-cap design: worked example of beam method

In the following, the beam method is used to design a suitable cap for six piles. A base supported on precast reinforced concrete piles, and suitable for a column similar to B3 in *Drawing 14*, is shown in *Figure 17.9a* and is designed as follows. The 450 mm square column is assumed to carry an unfactored load of 2680 kN, including the weight of the pile cap itself, and six 350 mm square piles are to be provided, each supporting 450 kN. The distance between adjacent piles must be not less than 1.1 m, and the distance between the face of a pile and the edge of the pile-cap not less than 125 mm. A suitable overall size for the pile-cap is therefore 2.75 m long and 1.5 m wide, as shown in *Figure 17.9a*. The thickness of the pile-cap is principally controlled by the shearing forces, either by direct shear across the cap or by punching shear around the column being supported. Since the shearing resistance of the concrete depends on the amount of main reinforcement provided, it is not possible to devise a direct design method to satisfy the Code requirements concerning shear. Instead it is necessary to guess a suitable thickness for the pile-cap, to calculate the main reinforcement needed, and then to check the design as regards shear.

Try a cap having a total thickness of 1 m. Then, with 50 mm cover to 12 mm links and 32 mm bars, the effective depth will be 922 mm. The effects of transverse bending may be neglected since, with a pile-cap of the thickness shown, the load dispersing from the column at an angle of 45° will cover the central pair of piles. Resistance to the longitudinal ultimate bending moment due to the cantilever action of the resistance of two piles acting at a distance of 1.1 m from the centre of the column must be provided. This ultimate bending moment is $2 \times 450 \times$

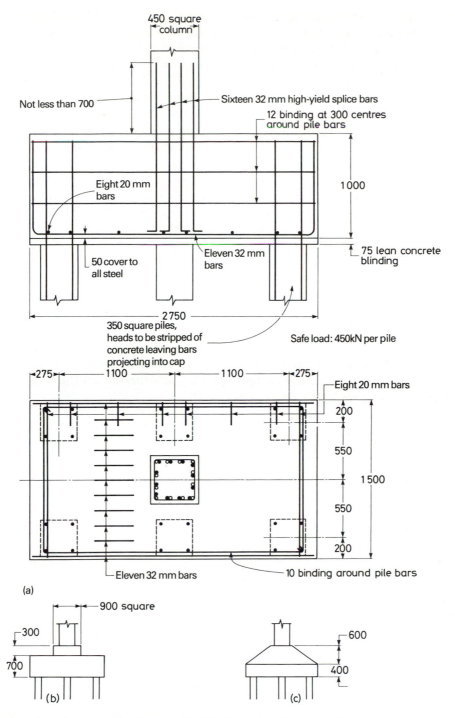

Figure 17.9 (a) Examples of reinforcement in pile-cap; (b), (c) alternative pile-cap designs.

$1.1 \times 1.6 = 1584 \, \text{kN m}$ (assuming a partial safety factor for loads of 1.6) and, with an effective depth of 922 mm, 8314 mm of mild steel reinforcement are required: provide eleven 32 mm bars. This gives a steel proportion of 0.0064.

According to BS8110, the critical section regarding shear is located at a distance of $350/5 = 70 \, \text{mm}$ from the inner face of the outer piles, i.e. $1100 - (450/2) - (350/2) + 70 = 770 \, \text{mm}$ or $0.835d$ from the face of the column. The *Code Handbook* states that, if this distance is $0.6d$ or less, the full tension anchorage-bond length of 1135 mm for 32 mm mild steel bars must be provided

beyond a plane located at a distance of 0.3 of the pile diameter beyond the centre-line of the pile, whereas if it is equal to $2d$ or more, only a nominal anchorage length of the greater of d or 12ϕ (922 mm in the present case) is required. For intermediate ratios the *Code Handbook* recommends linear interpolation between these two limits; so in the present case the anchorage-bond length required is $1135 - [(1135 - 922)(0.835 - 0.6)/(2 - 0.6)] = 1100 \, \text{mm}$. This can be achieved by turning the ends of the bars through a right angle and extending them to the top of the pile-cap. However, since the stress at the bend is considerable, the bearing stress at this point must be checked and a sufficiently large-radius bend employed to ensure that the limiting stress is not exceeded.

The shearing force must be checked at the same plane, 1100 mm from the column face. The ultimate shearing force is $2 \times 450 \times 1.6 = 1440$ kN. For a concentrated load, clause 3.4.5.8 of BS8110 permits the normal value of v_c corresponding to ϱ, as read from *Data Sheet 15*, to be increased to $2dv_c/a_v$, where a_v is the distance from the face of the support to the nearer edge of the load (see section 6.4), provided that the resulting value does not exceed the maximum permissible shearing stress. In the present example, $a_v = 770$ mm. Thus the normal value of v_c of 0.58 N/mm^2, corresponding to $\varrho = 0.0064$ and grade C30 concrete, can be increased to $2 \times 922 \times 0.58/770 = 1.38$ N/mm^2. Then the shearing resistance of the concrete alone is $1.38 \times 1500 \times 922 = 1916$ kN, which is well in excess of that required. Nominal 12 mm and 20 mm bars should be provided in addition to the main steel, as shown in *Figure 17.9a*. The ultimate resistance moment of the concrete in compression is 5977 kN m, which is nearly four times that required.

Since the spacing of the piles is less than $3\phi_p$, the only check regarding punching that need by made is around the column perimeter. Here $v_c = (2680 \times 10^3 \times 1.6)/(4 \times 450 \times 922) = 2.58$ N/mm^2. Since the maximum permissible shearing stress here is $0.8\sqrt{f_{cu}} = 4.38$ N/mm, this is satisfactory.

In view of the ample compressive and shearing strengths of the cap, other designs may be considered. If the thinner pile-cap shown in *Figure 17.9b* is adopted, reinforcement to resist a considerable shearing force from the end pairs of piles would be necessary, and it may also be necessary to introduce a block of reinforced concrete immediately beneath the column to cater for punching shear and to give sufficient embedment for the column steel. The main reinforcement would also be increased by about 50%. In the design shown in *Figure 17.9c*, the main longitudinal reinforcement would remain unaltered but shearing reinforcement would become essential. In both designs the cost of the extra reinforcement would be partly offset by saving in concrete and formwork.

The set to which the piles should be driven depends primarily on the load to be supported; the weight, type and fall of the hammer used; the pile length; and the nature of the ground. For detailed guidance on such matters reference should be made to specialized books dealing with this subject (for example ref. 29).

Chapter 18
Robustness and structural stability

The inclusion of specific requirements regarding structural stability marked yet another change of attitude between the appearance of the original version of CP114 in 1965 and that of CP110 in 1972 (although an amendment regarding stability was added to CP114 in 1969 when the metric version appeared). The need for a more extensive consideration of structural stability became evident in the late 1960s, when the possible consequences of the progressive collapse of multistorey buildings, such as happened at Ronan Point in 1968, became fully appreciated. As a result, additional stability requirements for buildings more than four storeys high were quickly added to the then current (1965) edition of the Building Regulations. Because the amendment containing these requirements was the fifth made to the 1965 document, these requirements themselves are still occasionally referred to as the 'Fifth amendment', although they now form regulation A1 of schedule 1 'Structural stability' of the 1985 Building Regulations. Similar requirements to those in the Building Regulations were also produced by the Institution of Structural Engineers.

When producing BS8110 it was considered that structural integrity must be given greater importance than had been the case in its predecessor. As a result the term 'robustness' replaced 'structural stability' and the relevant Code clauses were relocated in a more significant position in the document, although the majority of the requirements remain unchanged. The requirements regarding robustness given in BS8110 are 'deemed to satisfy' the 1985 Building Regulations.

18.1 CODE REQUIREMENTS

BS8110 states generally that the layout and behaviour of the structure should be robust and stable. While it is only necessary to design the building to resist the normal loads and forces, care should be taken to ensure that catastrophic failure will not occur if a foreseeable accident takes place, for example the effects of a minor explosion or, in the case of an elevated structure on columns, the possibility of a column or succession of columns being damaged by the collision of a vehicle. Of course, it is clearly impracticable to design every structure to withstand the effects of a major disaster. Thus, even in areas of the world where tornadoes are expected, structures are not normally designed to resist such phenomena unless they are vitally important or the resulting failure would be disastrous. For this reason the Code requires that any damage occurring to a structure should not be disproportionate to the original cause. BS8110 also states that, in the case of certain types of plant (for example a chemical factory), the effects of likely hazards should be considered to ensure that, if an accident should occur, the structure would remain stable even if it were damaged.

An important addition to the BS8110 requirements regarding robustness from its predecessor is the stipulation that, in structures of more than four storeys, 'key elements', whose failure would cause the collapse of more than a limited area close to the member concerned, should be identified. If possible the structural layout should then be amended to eliminate such key members, but if this cannot be done they must be specially designed to take account of their importance according to the recommendations given in Part 2 of the Code. The restriction of damage caused by the removal of any vertical load-bearing wall or column (apart from the 'key elements' already discussed) to the area close to the member concerned can normally be achieved by providing suitable vertical and horizontal ties, as described in sections 18.2 and 18.3. However, where the structural arrangement adopted makes this impossible or unsuitable, each member concerned should be assumed to be removed from the structure, and the structural elements that it supports should be so designed that they 'bridge' the resulting space as required in Part 2 of BS8110.

To ensure that the general requirements concerning robustness are met, the Code sets out in clauses 3.1.4 various specific recommendations. The first of these is that any structure should be capable of resisting at any level a notional horizontal ultimate load of not less than

0.015 times the total characteristic dead load above that level, the resulting horizontal force being shared among the structural members in proportion to their stiffnesses and strengths. This requirement is apparently necessary because the combinations of load including wind that are permitted by BS8110 may result in the consideration of very low lateral loads for certain types of structure. The second Code requirement is the provision of suitable protective earthworks, barriers etc. to prevent damage to, or failure of, vital load-bearing structural members at road level.

The remaining recommendations in the Code are concerned with the provision of vertical and horizontal ties. These ties are designed to withstand sufficient forces to ensure that the structure formed by the component members is robust and stable and is capable of withstanding some overloading, or the effects of accidental damage or limited mususe. Details of the Code requirements are given below.

When determining the reinforcement in the ties, BS8110 states that it may be assumed that no other forces are acting simultaneously and that the full characteristic strength of the steel may be utilized (i.e. $\gamma_m = 1.0$ for the reinforcement in this case). Thus the reinforcement that has already been determined to meet the structural requirements may also be assumed to provide part or all of each area of reinforcement required. As a result the *Code Handbook* suggests that, since the amounts of steel provided for normal design purposes will normally meet the stability requirements with little or no modification, it is perhaps best to first follow normal structural design procedures and to then check the resulting design to ensure that the stability requirements have also been met, any adjustments being made as necessary. This procedure is followed in the worked example that forms Part Two

of this book, the stability calculations being given on *Calculation Sheets 5, 8, 11, 12, 16* and *34*. It will be seen that, in the present example, the resulting areas of steel required are insignificant and are easily accommodated by careful detailing of the reinforcement required to meet normal design requirements.

18.2 VERTICAL TIES

In all structures that are more than four storeys in height, effective continuous ties must be provided from foundation to roof level in all columns and walls. For reinforced concrete members the Code specifically requires that each tie must be capable of withstanding a tensile force equal to the maximum total ultimate load transferred to the member in question from the roof or any single floor, although the resulting minimum permissible area of reinforcement required will almost always meet the requirement regarding vertical ties. For precast concrete construction the somewhat complex requirements incorporated into CP110 have been replaced by the foregoing conditions plus stringent provisions concerning the continuity of the ties and the reinforcement forming them. For more detailed information reference should be made directly to clauses 5.1.8.2 and 5.3.4 in Part 1 of the Code.

18.3 HORIZONTAL TIES

All structures of any height must be provided with effective horizontal ties, both peripherally and internally, as shown in *Figure 18.1*. In addition, column and wall ties must be provided at each intersection between the column or wall and the floor. In the words of the Code, the ties should be so positioned as to minimize, by the effects of

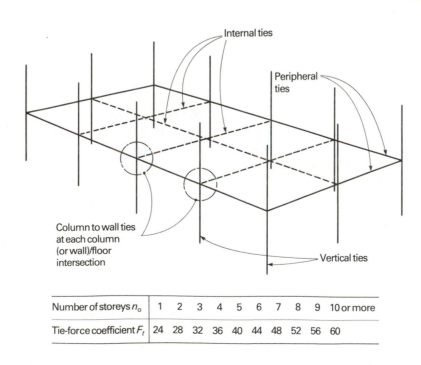

Number of storeys n_o	1	2	3	4	5	6	7	8	9	10 or more
Tie-force coefficient F_t	24	28	32	36	40	44	48	52	56	60

Figure 18.1

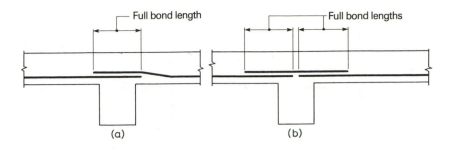

Figure 18.2

cantilever, catenary or any other appropriate action, 'the results of extreme damage by accidental causes'. The actual forces to be resisted are related to a factor F_t which depends on the number of storeys in the building, as shown in the table on *Figure 18.1*. The basis for this relationship is the philosophy that the taller the building, the more serious the consequences of such a failure. The tie requirements are then as follows.

18.3.1 Peripheral ties

At each floor and roof level in the peripheral wall or within 1200 mm of the edge of the building, effectively uninterrupted peripheral ties capable of withstanding a tension of F_t kN must be provided.

18.3.2 Internal ties

At each floor level, either spaced evenly through the slab or concentrated at beams or in walls within 500 mm of the top or bottom of the floor slab, effectively uninterrupted internal ties must be provided in two directions at right angles. Each tie must be effectively anchored to the peripheral ties at both ends, unless they continue as column/wall ties. If $(g_k + q_k) l$ is less than 37.5 kN/m, the tie must be capable of resisting a tension of F_t kN per metre of width; for greater values of $(g_k + q_k)l$, the tie must be able to resist a force of $0.0267(g_k + q_l) lF_t$ kN per metre of width. In the foregoing expressions, l (in metres) is the lesser of either five times the clear storey height (below beams if provided) or the greatest distance between the centres of the vertical load-bearing elements in the direction concerned. If the vertical load-bearing elements are walls occurring in plan in one direction only, then the ties parallel to the walls must be capable of resisting a tension of F_t kN per metre of width.

18.3.3 Column/wall ties

Each external column or wall element must be effectively anchored to the floor structure at each floor level by providing resistance to the following tensile forces. If l_0, the floor to ceiling height in metres, is less than 5 m, the tie must be able to resist a tension of at least $0.4l_0 F_t$ kN; for greater values of l_0, the corresponding minimum tension is $2F_t$ kN. In addition, the tie must be capable of resisting a tension of 0.03 ties the total ultimate vertical load for which the member being considered has been designed, at the intersection in question. The ties provided for corner columns must be sufficient to resist these forces in both directions. The reinforcement required for the column/wall ties may be formed partly or wholly by that provided for the internal or peripheral ties.

18.4 TIE REINFORCEMENT

The following notes are based mainly on the detailed and helpful discussion in *Allen*, to which the designer is referred for further information.

Since the internal ties must be effectively uninterrupted throughout their length, the laps employed between the successive lengths of bar provided to meet stability requirements must be sufficient to cater for the full characteristic strength of the bar. Thus, if some of the bottom bars in the beams or floor slab are also considered as stability steel, the laps provided to these particular bars over their supports must be increased beyond the nominal lengths that usually suffice (*Figure 18.2a*), or separate splice bars must be provided (*Figure 18.2b*). Alternatively, it may presumably be assumed that all the bottom reinforcement in the floor slab and beams acts as an internal tie. Then the resulting tensile stress in the bars, and hence the bond (i.e. lap) length required, will be

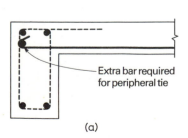

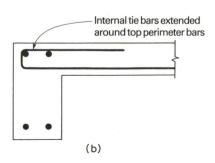

Figure 18.3

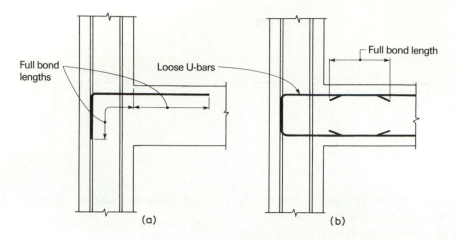

Figure 18.4

reduced accordingly, although if this assumption is made it must be remembered that *all* this steel must be effectively anchored around the peripheral ties, as described below. It should be noted that lapping bars must be in contact; thus the steel in the top of a beam or slab over the supports cannot normally be considered to contribute to an effectively uninterrupted tie.

BS8110 also requires that the reinforcement in the internal ties is effectively anchored to that in the peripheral ties. Since the level of the internal tie reinforcement normally differs from that of the both top and bottom bars in the perimeter beam, if one is provided, it is necessary either to provide an extra bar for the peripheral tie around which the internal tie bars can pass, as shown in *Figure 18.3a*, or to extend the internal tie bars, as indicated in *Figure 18.3b*, to pass around the top perimeter bars which can then be made continuous.

If the peripheral beam spans between external columns, it seems most sensible to use this beam as the peripheral tie, and the laps provided for the reinforcement therein should be increased accordingly. Again, all the appropriate steel in the beams and slab within 1200 mm of the edge of the building can be utilized if desired, thus reducing the stress in the bars and the bond lengths needed. However, if the edge of the structure cantilevers beyond the columns by more than 1200 mm, the peripheral tie must be formed in the slab and the contribution of any edge beam between the columns cannot be considered.

The column/wall ties must provide adequate anchorage to withstand the forces for which they are designed. Thus a full anchorage length on both sides of the internal face of the column or wall is necessary (*Figure 18.4a*). If the stress in the bars is high, the radius of the bend of the bar within the column or wall may need to be greater than that normally provided. To avoid this complication, it is often better to provide extra steel, thus reducing the stress in the bars. The use of loose U-bars, as shown in *Figure 18.4b*, often provides a convenient way of detailing such junctions.

When considering stability, BS8110 permits the anchorage-bond stresses to be increased by 15% to cater for the difference between the partial safety factors for materials when stressed normally and when stressed by the effects of excessive loading or localized damage. This may at first appear to indicate that the bond lengths required may be reduced accordingly, but since the steel is now assumed to be working at its full characteristic strength f_y instead of at a maximum of $0.87f_y$, as is the case in normal design, the appropriate values cancel out. Thus tables for bond length and bearing radius, such as those on *Data Sheets 18 and 19*, may be used as they stand to determine bond lengths and bearing radii for stability requirements.

Chapter 19
Fire resistance

Detailed recommendations are given in section 4 of BS8110, Part 2, relating to the resistance to fire of various types of reinforced and prestressed concrete members; simpler but safe rules for minimum member sizes and cover are provided in Part 1 of the Code. This resistance, which depends primarily on the type of aggregate, the thickness of the member and the cover of concrete over the reinforcement, is expressed by the number of hours of effective resistance as established by tests which have been made in accordance with BS476 'Fire tests on building materials and structures'. Some of the values specified in the Code for normal-weight concrete made with siliceous aggregates are set out on *Data Sheet 62*; for full details of all the relevant requirements reference must be made to the Code itself.

The degree of fire resistance required depends on the size of the building and the use to which it is to be put. Part B of the Building Regulations 1985 deals with fire spread, although these general requirements may be modified by the local authority by-laws applicable to the location of the building under consideration.

The Building Regulations 1985 recognize the following nine types of use, designated as 'purpose groups' (unfortunately, unlike its predecessor, the 1985 document does not give reference numbers to these groups):

Residential group:
　Dwellinghouse: private dwellings other than flats etc.
　Flat: including maisonettes
　Institutional: hospitals, homes, schools etc. used as
　　living accommodation
　Other residential: hotels, boarding houses, hostels etc.

Non-residential group:
　Assembly: public buildings or places for social or
　　recreational activities
　Offices
　Shop
　Industrial: factories etc.
　Other non-residential: stores, garages etc.

For buildings intended to be used for various purposes, the appropriate purpose group is that relating to the principal use. However, in the following cases the section concerned may be considered as having an ancillary purpose in its own right: a maisonette or flat; a shop storage area exceeding more than one-third of the total shop floor area; and any other situation where the floor area exceeds 20% of the total building floor area. The *Guide to the Building Regulations* assumes this to mean that separation or compartmentation (see below) is not required here.

For each purpose group, certain limits to the height, the floor area of each storey and the cubic capacity are specified. If the overall sizes of the building exceed any of these limits the building must be divided into a number of individual *compartments*, each of which is separated from its neighbours by walls and/or floors having the required fire-resistance periods. Then the actual period of fire resistance required depends on the purpose group and the height, floor area and cubic capacity of the building (or compartment of the building) concerned.

Since the main area of the building considered in Part Two is intended for office use, the general requirements for offices apply throughout since, although the ground floor is intended for shops and for garaging vehicles, the floor area concerned does not exceed one-fifth of the total floor area of the building. Thus according to the Building Regulations, for a building 29.25 m long, 15.45 m wide and 25.75 m high (floor area per storey = 452 m² and cubic capacity < 11 600 m³), a minimum fire-resistance period of one hour is required for the ground and upper storeys, and of one and a half hours for the basement (including the ground-floor slab over). It may be of interest to note that if the principal use of the same building was as a shopping structure, the fire period requirement would rise to two hours for the basement but otherwise remain unchanged; while the required periods for a multistorey garage of the same size would be four hours for the basement and two hours elsewhere.

The separate elements of the buildings in Part Two are now considered.

19.1 SLABS

The 100 mm solid slabs (*Drawings 3 and 4*) in the upper beam-and-slab floors are sufficiently thick to give a resistance of one hour, and the cover of 20 mm adopted also corresponds to the minimum value required. The

Fire resistance

200 mm flat-slab floors (*Drawings 12 and 13*) and the 150 mm two-way roof slabs (*Drawings 8 and 9*) also exceed the minimum requirement. For the alternative designs for the floors of the shops on the ground floor (Chapter 12), the minimum thickness of the precast slabs (*Figure 12.4*) is 40 mm, compared with the 110 mm required for a one and a half hour period; this deficiency must be made up by the thickness of the topping and finishes, although in such circumstances it would clearly be more sensible to redesign the slabs. The rib width of 90 mm is insufficient for a one and a half hour period, for which a minimum of 125 mm is required, and both the side and bottom cover to the main bars would require to be increased unless additional resistance were provided (as indicated on *Figure 12.4*) by employing a ceiling formed of a suitable thickness of plaster, lightweight insulation or vermiculite. To achieve a four hour fire resistance, were this to be required for the basement ceiling of a multistorey structure used for garaging vehicles, the overall thickness of the floor section would have to be increased to 170 mm, and thus the use of precast sections would be impracticable. However, such sections would be unsuitable anyway to act as garage flooring.

The total thickness of 175 mm of the hollow clay-block floor with its topping exceeds the 110 mm required for a one and a half hour period. However, the bottom cover is only 20 mm which, while satisfactory for a continuous span, must be increased to 25 mm if the floor is designed as simply supported bays with anti-crack steel over the supports. For the solid slab design for the ground floor over the basement, BS8110 again requires a minimum thickness of 110 mm; since only 100 mm is provided, this additional thickness must be made up by the finish or suitable treatment must be applied to the soffit of the slab. The minimum cover of 20 mm provided to the reinforcement here to meet exposure conditions is sufficient for continuous construction.

19.2 BEAMS

The continuous beams of the upper floors shown on *Drawings 5, 6 and 7* and of the roof on *Drawing 9* provide the required resistance of one hour for the upper storeys,

since the concrete cover is more than 20 mm to the main reinforcement resisting tension. The cover provided to the continuous beams supporting the ground floor is also sufficient to meet the one and a half hour requirement for the structure over the basement, and the minimum needs regarding rib width are amply met by the web dimensions provided.

19.3 COLUMNS

In all storeys except the top, the columns (*Drawings 14 to 16*) are not less than 300 mm square, and therefore either meet or exceed the minimum sizes of 200 mm and 250 mm needed for fully exposed columns with resistance periods of one and of one and a half hours respectively. In the topmost storey the minimum thickness of the columns is only 130 mm against a minimum of 200 mm specified, but a layer of sprayed lightweight insulation could be applied to all exposed column faces to raise the period of fire resistance to the necessary value.

19.4 WALL AND STAIRS

The transverse walls are either of brickwork or of reinforced or plain concrete 225 mm in thickness. The requirements for brickwork fall outside the scope of BS8110 and of this book. For concrete walls with more than 1% of reinforcement, a thickness of 225 mm is well in excess of the 100 mm required for a period of one and a half hours. The reinforced concrete end walls considered in the building in Part Two, which contain between 0.4% and 1% of steel, should have a minimum thickness of 140 mm to provide a one and a half hour fire period. With less than 0.4% of reinforcement, a minimum wall thickness of 175 mm is required to give a one and a half hour fire period. Consequently both the transverse walls and the 200 mm basement wall would be quite satisfactory in this respect.

The Code gives no special recommendations regarding the fire resistance of stairs, but presumably they may be considered as floor slabs. The 125 mm landings then provide a resistance of two hours, which is well in excess of that required.

Part Two

Design of a Typical Building

Typical calculations and details, which have been prepared as described in the previous chapters, are given on the following pages. The drawings and calculations provided are not intended to be sufficient to construct, or to obtain the necessary authority to construct, a building of the type and size considered. Instead, the specimen calculations provided for the building shown have been selected to illustrate as many as possible of the types of design for which calculations may have to be prepared. For instance, examples are included of freely supported and continuous slabs spanning both one way and two ways, as well as of a two-way slab supporting a central concentrated load, and of flat-slab construction. As far as convenient, the calculations are presented in accordance with the Concrete Society technical report entitled *Model Procedure for the Presentation of Calculations*, 1981 (ref. 30). However, where it has been thought that rigorous adherence to these requirements would result in unnecessary repetition, the presentation has been shortened somewhat in order to save space. Also, to permit rapid reference to the data sheets used to facilitate the calculations, the left-hand column on each sheet (which normally gives the reference number of the member being considered) is here used to list the number of the data sheet concerned. Typical calculation sheets prepared strictly in accordance with the foregoing Concrete Society technical report are presented in *Designed and Detailed* by J. B. Higgins and M. R. Hollington (ref. 31).

Note that, in normal design office practice, each individual A4 Calculation Sheet would be numbered separately. However, in order to permit cross-reference between Parts One and Two of this book, such an arrangement was found unworkable, and so here the same calculation page number has been given to all the sheets relating to a particular structural element. Thus,

for example, all the pages dealing with the design of the flat-slab floor are designated "Calculation sheet 16/1, 16/2", and so on.

It should be noted that the dimensions used in the following calculations have been taken from preliminary drawings and in a few cases may differ slightly from those shown on the full-page details. Any variations in the resulting moments and forces are insignificant, however. In addition, differences may occasionally be observed between the bar system needed to meet the minimum requirements according to the calculations (shown on the sketches on the calculation pages) and that adopted on the working drawings. For the main floor beams, for example, it may appear that unnecessary extra bars are provided at certain sections. Such bars are not considered to contribute to the bending resistance at these points, however, but the arrangement adopted is thought particularly suitable for detailing purposes.

It is important to realize that the drawings provided are not intended to represent typical general arrangement drawings or working details. The direct reduction of a large working drawing to a size sufficient to fit on a page of this book would be unsatisfactory since it would be almost unreadable and the important details would be lost in the general mass of information. Instead, therefore, particular working details have been included because they illustrate specific points. Again, only certain slab panels or beam spans may be detailed if they are also representative of similar adjoining members. However, the method of detailing adopted conforms in most respects to that recommended in the *Standard Method of Detailing Structural Concrete* (ref. 2). The production of working drawings is dealt with in detail in a splendid publication entitled *Drawing for the Structural Concrete Engineer* by K. K. McKelvey (ref. 32).

Schedule of Drawings and Calculation sheets

Drawing	Components	Type of construction	Calc. sheet	Chapter
1, 2	General arrangement drawings			
	Beam-and-slab construction: one-way slabs			
3, 4	Upper-floor slabs	Continuous one-way slab	1	9
	Canopy	Simple cantilever	2	9
	Parapet	Vertical cantilever	3	9
	Roof slab over rear stair-well	Freely supported one-way slab	4	9
5	Secondary beams supporting upper-floor slabs (51 to 55)	Continuous beam supporting uniform load	5	9
5	Secondary beams supporting upper-floor slabs (51A to 55A)	Continuous beam supporting uniform load	6	9
6	Fascia beams (loading)	—	7	9
7	Main beams for upper floors (M1 to M3)	Continuous beam supporting uniform and central concentrated loads	8	9
	Beam S6 at stair-well	Freely supported beam	9	9
	Main roof: beam-and-slab construction			
8, 9	Roof slabs	Continuous two-way slabs	10	10
	Typical longitudinal beams	Continuous beams supporting two-way slabs	11	10
	Typical transverse beam	Continuous beams supporting two-way slabs	12	10
10, 11	Floor of tank room	Continuous two-way slab supporting central concentrated load	13	10
	Roof of tank room	Single two-way slab with restrained corners	14	10
	Roof of motor room	Single two-way slab with unrestrained corners	15	10
	Flat-slab construction			
12, 13	Upper floor slab	Flat slab	16	12
	Columns: axial load			
14	Interior column B3	Column supporting axial load only	17	13
	Base to column B3	Isolated foundation	18	17
	Columns: axial load and bending			
15	Exterior column D3	Column subjected to combined axial load and bending in one direction	19	14
16	Exterior column C1	Column subjected to combined axial load and bending in two directions	20	14

Drawing 1 General arrangement: sections

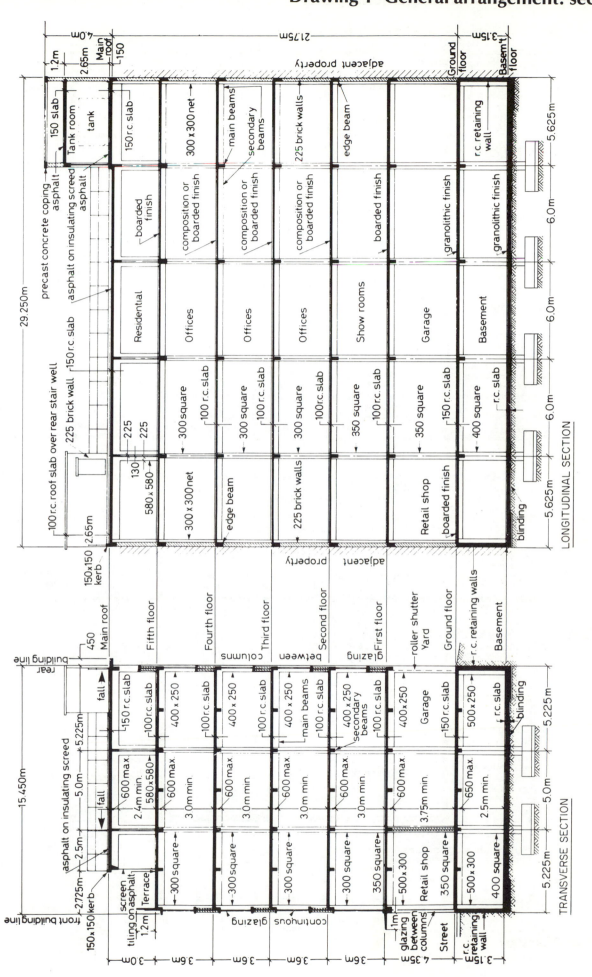

167

Drawing 2 General arrangement: plans

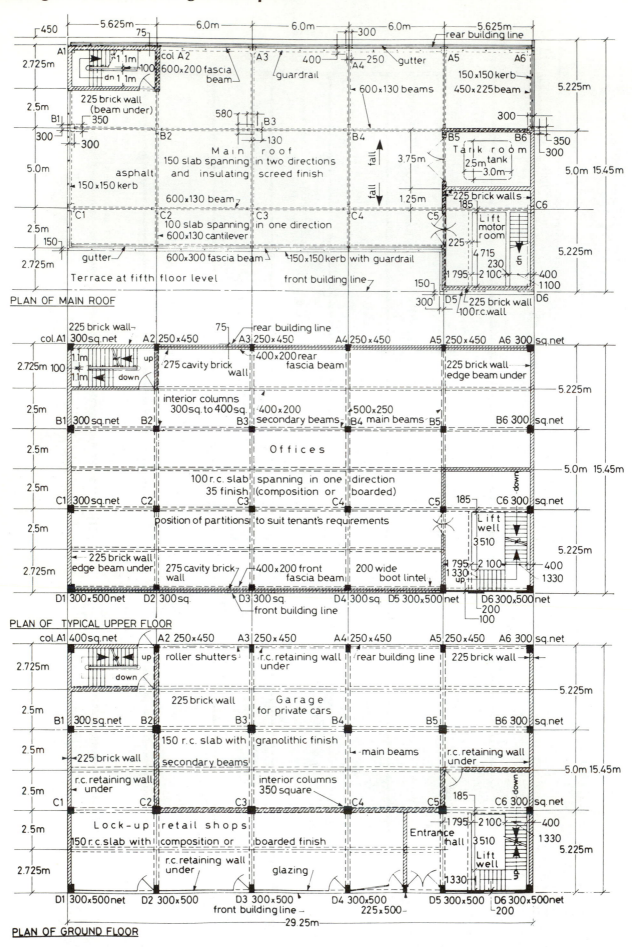

PLAN OF MAIN ROOF

PLAN OF TYPICAL UPPER FLOOR

PLAN OF GROUND FLOOR

Notes: abr = alternate bars reversed
Bar layer notation:
top outer layer   T1   bottom second layer B2
top second layer   T2   bottom outer layer   B1

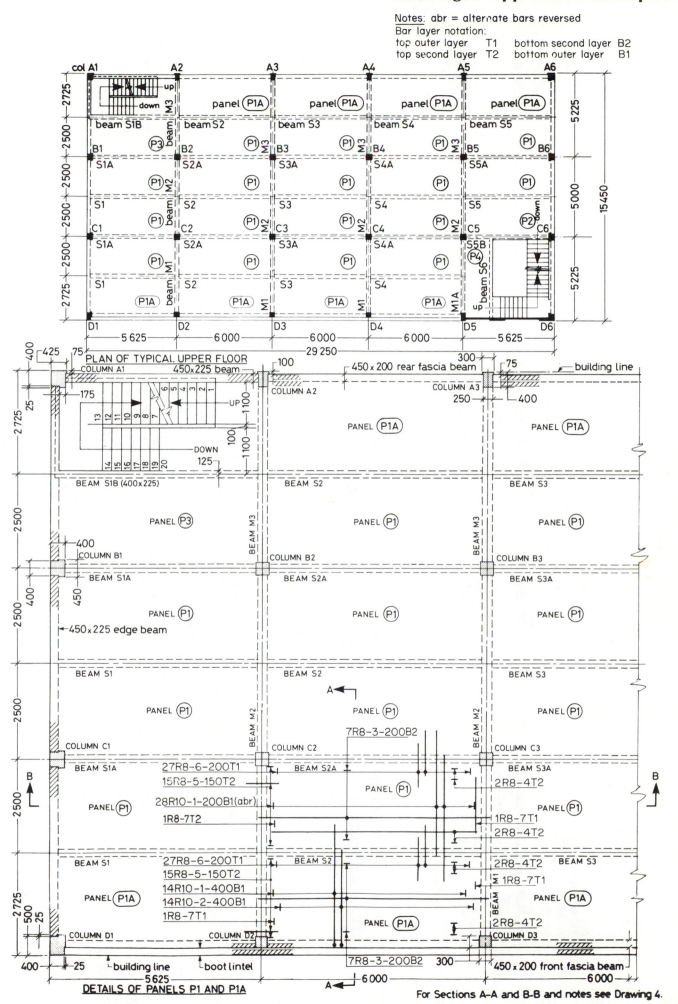

PLAN OF TYPICAL UPPER FLOOR

DETAILS OF PANELS P1 AND P1A

For Sections A-A and B-B and notes see Drawing 4.

169

Drawing 4 Upper floor slabs: sections

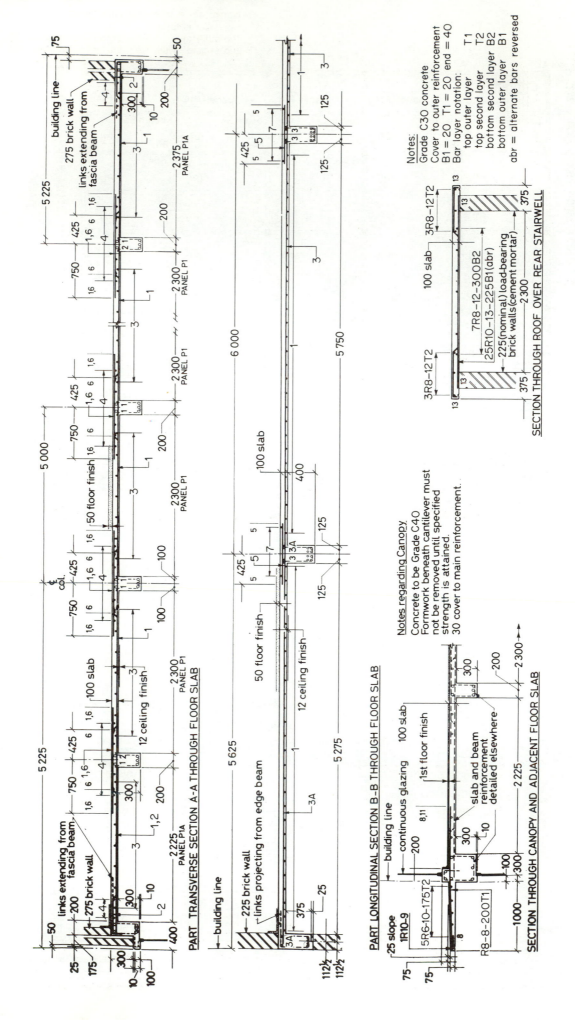

PART TRANSVERSE SECTION A-A THROUGH FLOOR SLAB

PART LONGITUDINAL SECTION B-B THROUGH FLOOR SLAB

SECTION THROUGH CANOPY AND ADJACENT FLOOR SLAB

SECTION THROUGH ROOF OVER REAR STAIRWELL

Notes:
Grade C30 concrete
Cover to outer reinforcement
B1 = 20 T1 = 20 end = 40
Bar layer notation:
top outer layer   T1
top second layer   T2
bottom second layer   B2
bottom outer layer   B1
abr = alternate bars reversed

Notes regarding Canopy.
Concrete to be Grade C40
Formwork beneath cantilever must
not be removed until specified
strength is attained.
30 cover to main reinforcement.

170

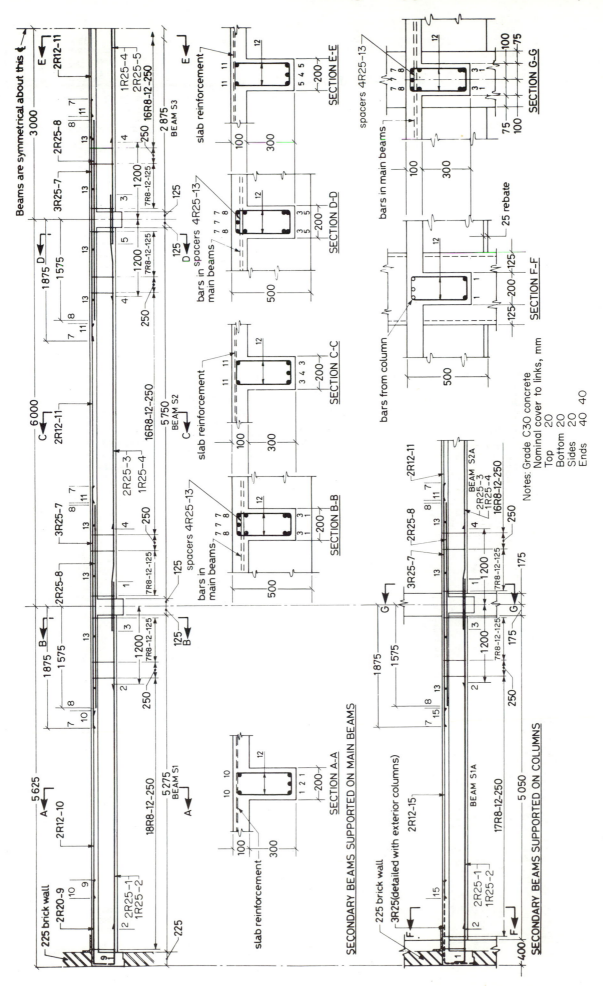

SECTION E-E

SECTION D-D

slab reinforcement

bars in main beams   spacers 4R25-13

SECTION G-G

spacers 4R25-13

bars in main beams

25 rebate

bars from column

SECTION F-F

SECTION C-C

slab reinforcement

SECTION B-B

bars in main beams   spacers 4R25-13

SECTION A-A

slab reinforcement

SECONDARY BEAMS SUPPORTED ON MAIN BEAMS

225 brick wall

3R25(detailed with exterior columns)

2R12-15

BEAM S1A

17R8-12-250

SECONDARY BEAMS SUPPORTED ON COLUMNS

Notes: Grade C30 concrete
Nominal cover to links, mm
Top   20
Bottom   20
Sides   20
Ends   40   40

171

Drawing 6 Upper floors: fascia beams, etc.

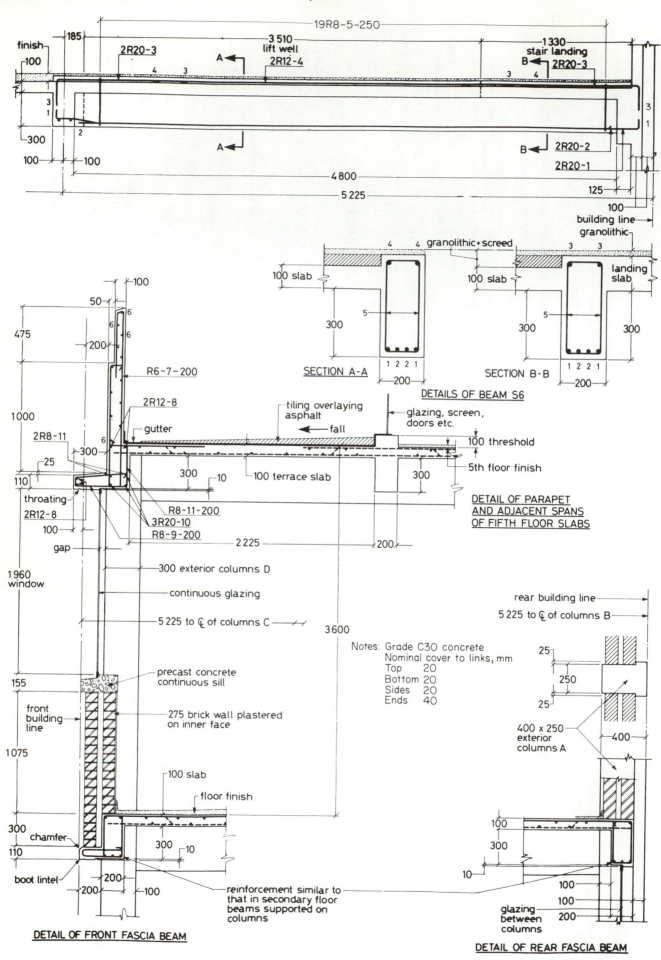

19R8-5-250

finish
185
2R20-3
3 510
lift well
2R12-4
A
4   3
1 330
stair landing
B
2R20-3
3   4
100
300
3
1
2
A
B
2R20-2
3
1
2R20-1
100   100
4 800
5 225
125
100
building line
granolithic

granolithic + screed

4   4
100 slab
300
5
SECTION A-A
1 2 2 1
200

100 slab
300
100 slab
300
5
landing slab
300
SECTION B-B
1 2 2 1
200
DETAILS OF BEAM S6

100
50
475
6
6 6
6 6
200
R6-7-200
2R12-8
6
gutter
1000
2R8-11
300
25
110
throating
2R12-8
100
gap
1960
window
155
front
building
line
1075
300 chamfer
110
boot lintel
200   200   100

R8-11-200
3R20-10
R8-9-200
2 225
300 exterior columns D
continuous glazing
5 225 to ℄ of columns C
precast concrete
continuous sill
275 brick wall plastered
on inner face
100 slab
floor finish
reinforcement similar to
that in secondary floor
beams supported on
columns
DETAIL OF FRONT FASCIA BEAM

tiling overlaying
asphalt
fall
glazing, screen,
doors etc.
100 threshold
5th floor finish
100 terrace slab
300
300
10
300
200
DETAIL OF PARAPET
AND ADJACENT SPANS
OF FIFTH FLOOR SLABS

3600

Notes: Grade C30 concrete
Nominal cover to links, mm
Top   20
Bottom 20
Sides   20
Ends   40

rear building line
5 225 to ℄ of columns B
25
250
25
400 x 250
exterior
columns A
400
100
300
10
DETAIL OF REAR FASCIA BEAM
400 x 250
glazing
between
columns
100
100
200

172

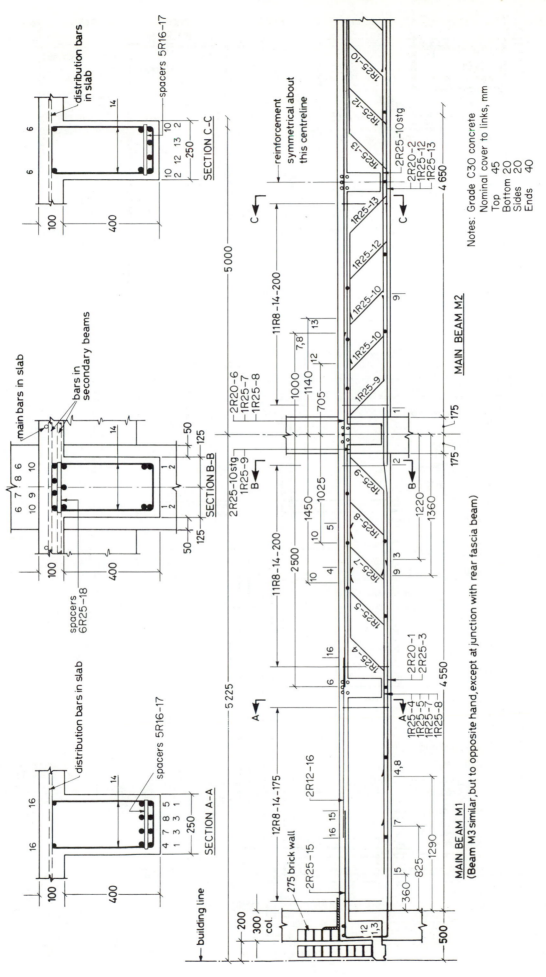

SECTION C-C

distribution bars in slab

spacers 5R16-17

SECTION B-B

main bars in slab

bars in secondary beams

spacers 6R25-18

SECTION A-A

distribution bars in slab

spacers 5R16-17

building line

reinforcement symmetrical about this centreline

2R20-6
1R25-7
1R25-8

2R25-10stg
1R25-9

11R8-14-200

11R8-14-200

12R8-14-175

2R12-16

275 brick wall

2R25-15

col.

1R25-10
1R25-12
1R25-13

2R20-2
1R25-12
1R25-13

2R25-10stg

1R25-13

1R25-9

1R25-8

1R25-7

1R25-5

1R25-4

2R20-1
2R25-3

1R25-4
1R25-5
1R25-7
1R25-8

MAIN BEAM M2

MAIN BEAM M1
(Beam M3 similar, but to opposite hand, except at junction with rear fascia beam)

Notes: Grade C30 concrete
Nominal cover to links, mm
Top   45
Bottom 20
Sides 20
Ends   40

173

Drawing 8 Main roof: plans

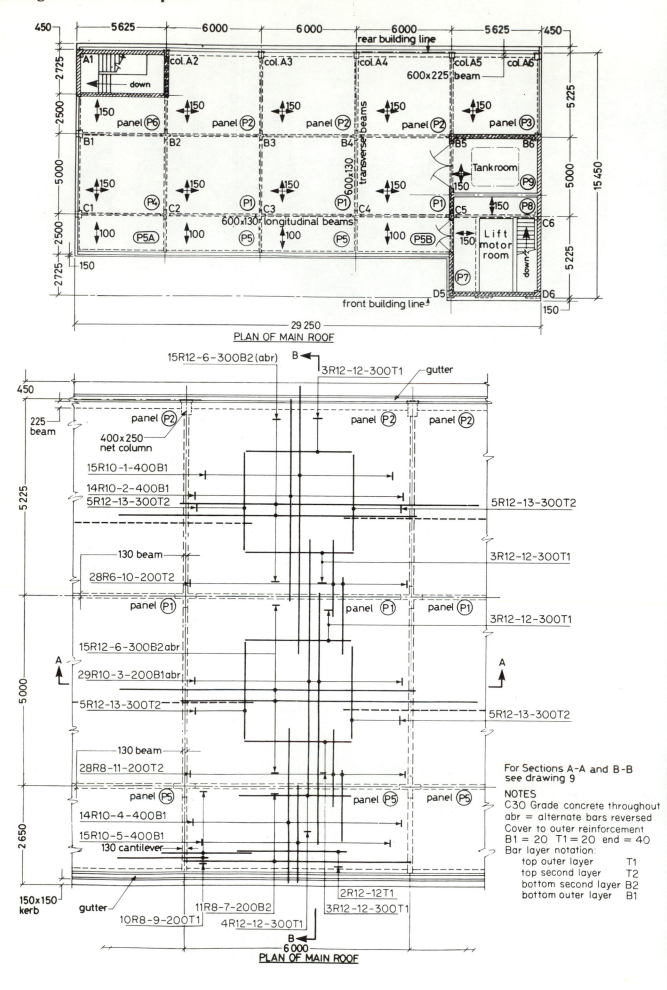

PLAN OF MAIN ROOF

15R12-6-300B2(abr)   B ⟵   3R12-12-300T1   gutter

225 beam

panel (P2)   panel (P2)   panel (P2)

400×250 net column

15R10-1-400B1
14R10-2-400B1
5R12-13-300T2   5R12-13-300T2

130 beam

28R6-10-200T2   3R12-12-300T1

panel (P1)   panel (P1)   panel (P1)

3R12-12-300T1

A ⟵
15R12-6-300B2abr
29R10-3-200B1abr
5R12-13-300T2   5R12-13-300T2

A

130 beam
28R8-11-200T2

For Sections A-A and B-B see drawing 9

NOTES
C30 Grade concrete throughout
abr = alternate bars reversed
Cover to outer reinforcement
B1 = 20 T1 = 20 end = 40
Bar layer notation:
    top outer layer   T1
    top second layer   T2
    bottom second layer   B2
    bottom outer layer   B1

panel (P5)   panel (P5)   panel (P5)
14R10-4-400B1
15R10-5-400B1
130 cantilever

150×150 kerb   gutter

2R12-12T1
3R12-12-300T1

11R8-7-200B2
10R8-9-200T1   4R12-12-300T1

B ⟵
6 000
PLAN OF MAIN ROOF

174

Drawing 9 Main roof: slab sections and beam details

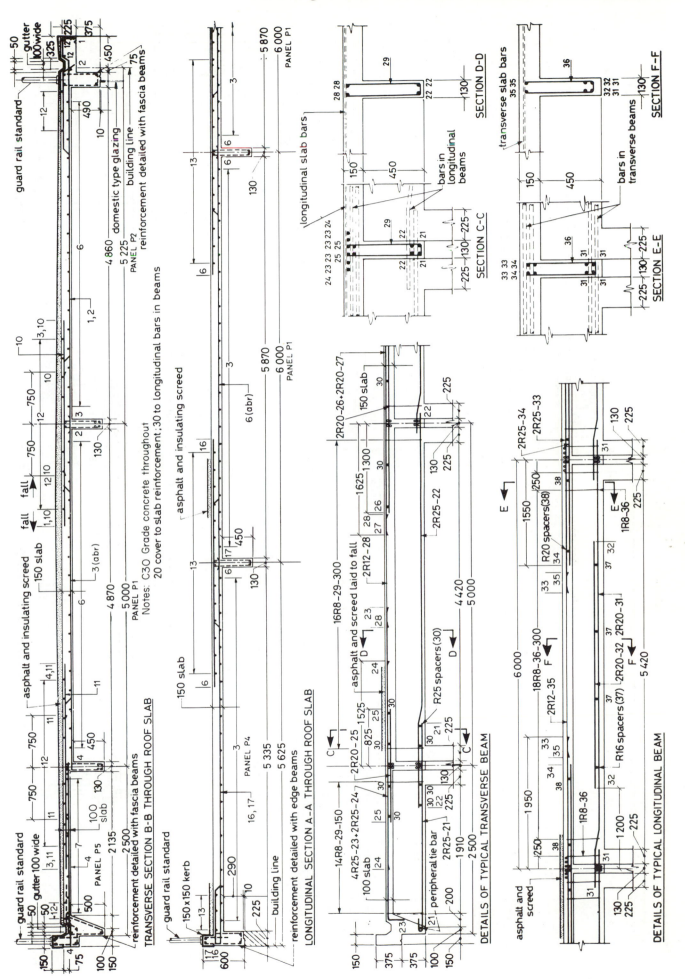

Drawing 10 Tank room and motor room—1

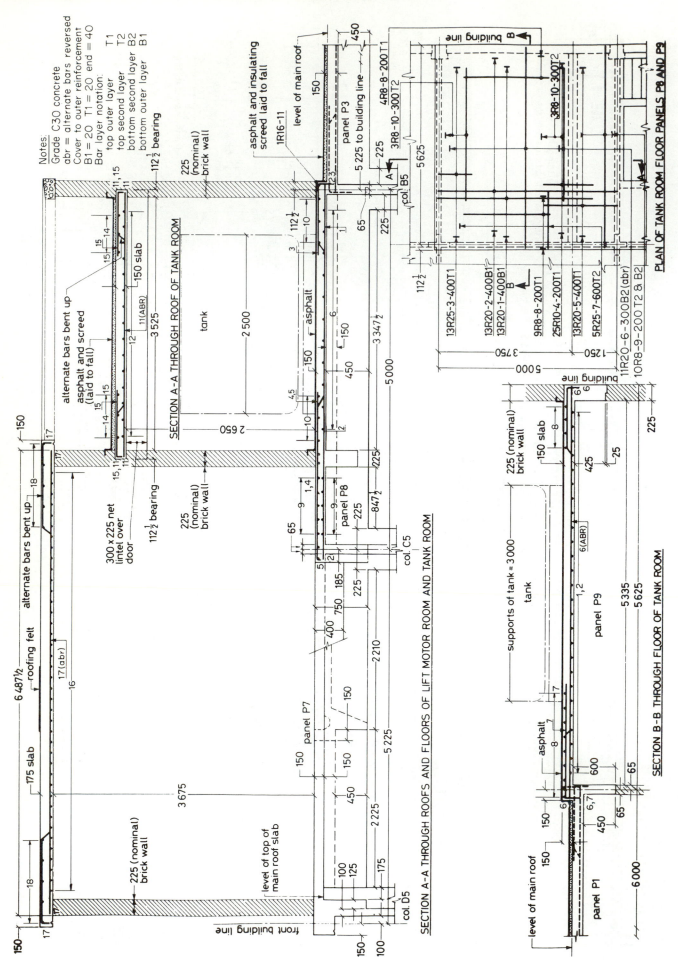

Notes:
Grade C30 concrete
abr = alternate bars reversed
Cover to outer reinforcement
B1 = 20   T1 = 20   end = 40
Bar layer notation:
top outer layer   T1
top second layer   T2
bottom second layer   B2
bottom outer layer   B1

SECTION A-A THROUGH ROOF OF TANK ROOM

SECTION A-A THROUGH ROOFS AND FLOORS OF LIFT MOTOR ROOM AND TANK ROOM

PLAN OF TANK ROOM FLOOR PANELS P9 AND P9

SECTION B-B THROUGH FLOOR OF TANK ROOM

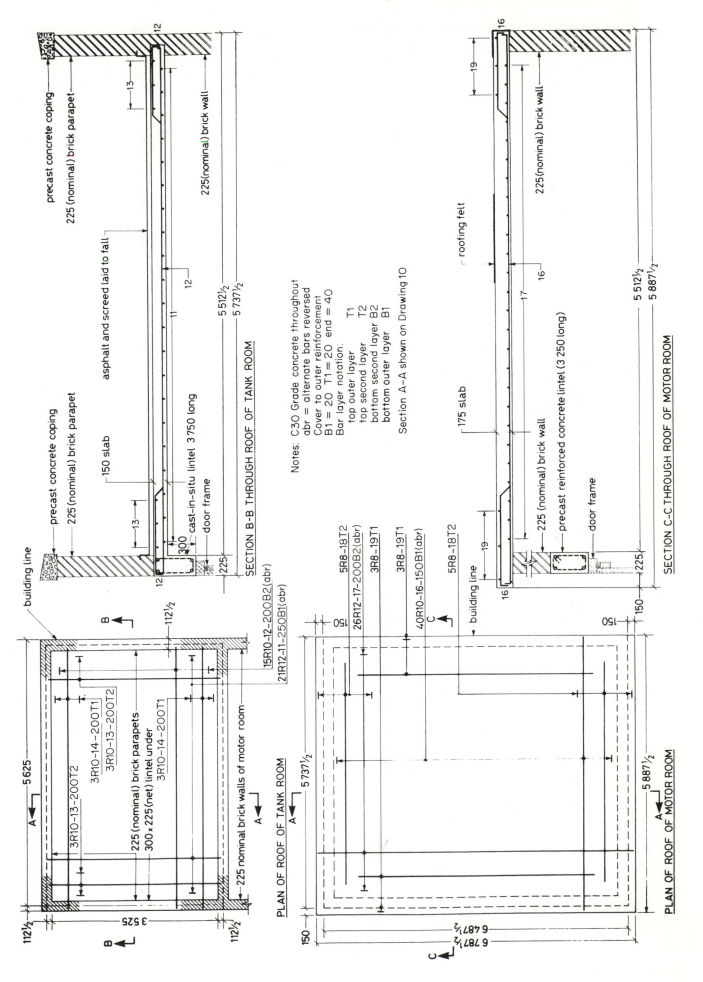

Notes: C30 Grade concrete throughout
abr = alternate bars reversed
Cover to outer reinforcement
B1 = 20 T1 = 20 end = 40
Bar layer notation:
top outer layer   T1
top second layer   T2
bottom second layer   B2
bottom outer layer   B1

Section A–A shown on Drawing 10

precast concrete coping

225 (nominal) brick parapet

asphalt and screed laid to fall

150 slab

cast-in-situ lintel 3 750 long

door frame

225 (nominal) brick wall

SECTION B–B THROUGH ROOF OF TANK ROOM

precast concrete coping

225 (nominal) brick parapet

roofing felt

175 slab

225 (nominal) brick wall

precast reinforced concrete lintel (3 250 long)

door frame

225 (nominal) brick wall

SECTION C–C THROUGH ROOF OF MOTOR ROOM

building line

15R10-12-200B2(abr)
21R12-11-250B1(abr)

3R10-13-200T2
3R10-14-200T1
3R10-13-200T2
3R10-14-200T1

225 (nominal) brick parapets
300 x 225 (net) lintel under
225 nominal brick walls of motor room

PLAN OF ROOF OF TANK ROOM

5R8-18T2
26R12-17-200B2(abr)
3R8-19T1
3R8-19T1
40R10-16-150B1(abr)
5R8-18T2

building line

PLAN OF ROOF OF MOTOR ROOM

177

Drawing 12 Flat-slab panels: plans

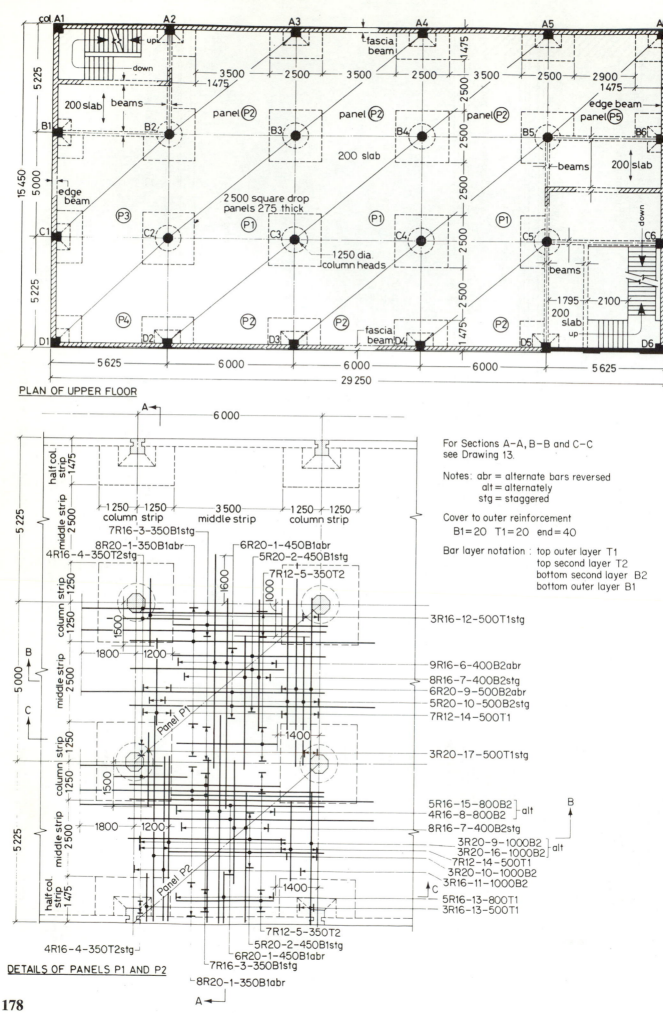

PLAN OF UPPER FLOOR

For Sections A–A, B–B and C–C
see Drawing 13.

Notes: abr = alternate bars reversed
        alt = alternately
        stg = staggered

Cover to outer reinforcement
        B1 = 20   T1 = 20   end = 40

Bar layer notation : top outer layer T1
                      top second layer T2
                      bottom second layer B2
                      bottom outer layer B1

DETAILS OF PANELS P1 AND P2

Drawing 13 Flat-slab panels: sections

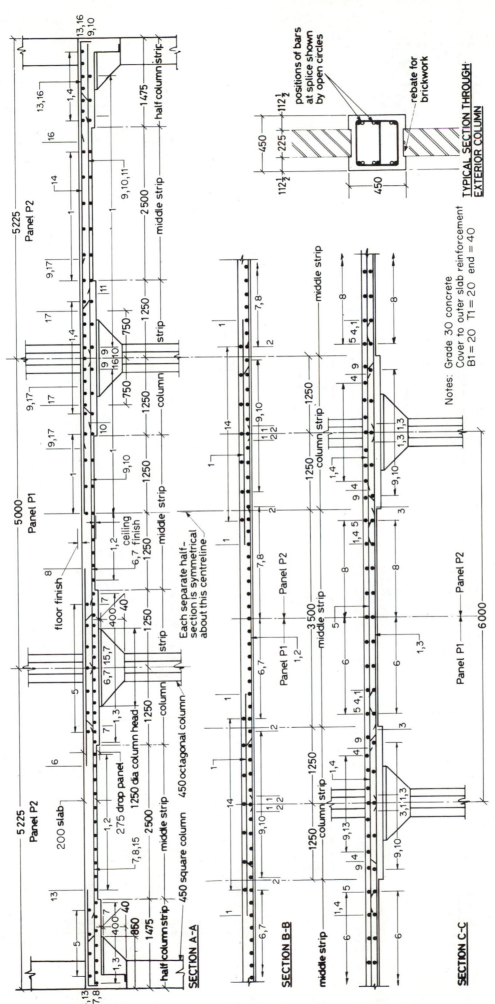

179

Drawing 14 Interior column B3

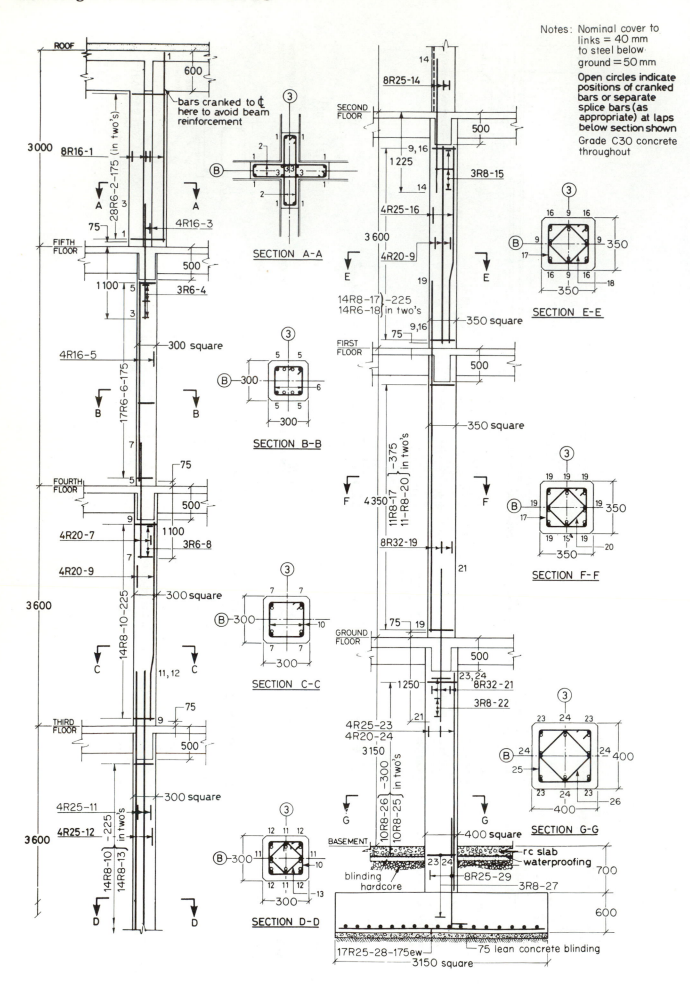

Notes: Nominal cover to links = 40 mm to steel below ground = 50 mm

Open circles indicate positions of cranked bars or separate splice bars (as appropriate) at laps below section shown

Grade C30 concrete throughout

ROOF

600

bars cranked to ₵ here to avoid beam reinforcement

3000   8R16-1

28R6-2-175 (in two's)

4R16-3

75

FIFTH FLOOR

500

1100

3R6-4

4R16-5   300 square

17R6-6-175

FOURTH FLOOR

500

4R20-7   1100

3R6-8

4R20-9

3600   14R8-10-225   300 square

11, 12

THIRD FLOOR

500

4R25-11   300 square

3600   4R25-12   -225 in two's

14R8-10
14R8-13

SECTION A-A

SECTION B-B

SECTION C-C

SECTION D-D

14   8R25-14

SECOND FLOOR   500

9, 16   1225   3R8-15

14

4R25-16

3600   4R20-9

19

14R8-17)-225
14R6-18) in two's   350 square

75   9,16

FIRST FLOOR   500

350 square

11R8-17)-375
11-R8-20) in two's

4 350

8R32-19

21

75   19

GROUND FLOOR   500

1250   23,24   8R32-21

3R8-22

21

4R25-23
4R20-24

3150

10R8-26)-300
10R8-25) in two's

BASEMENT   400 square

rc slab
waterproofing

23,24

blinding
hardcore   8R25-29   3R8-27

17R25-28-175ew   75 lean concrete blinding

3150 square

SECTION E-E   350

SECTION F-F   350

SECTION G-G   400

700

600

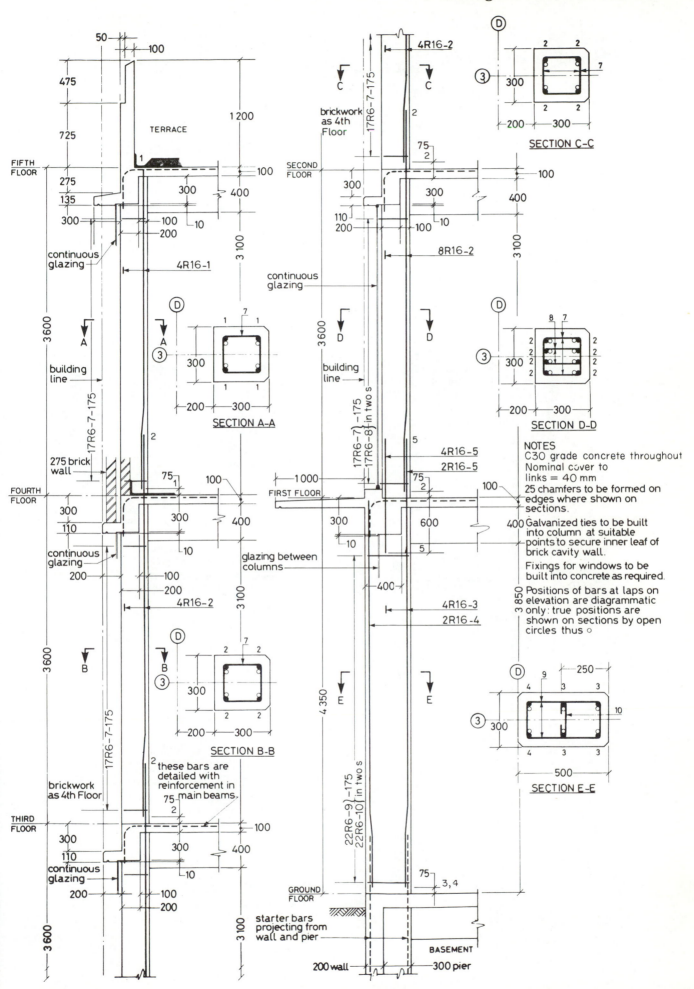

SECTION C-C

SECTION A-A

SECTION D-D

SECTION B-B

SECTION E-E

TERRACE

FIFTH FLOOR

continuous glazing

4R16-1

building line

275 brick wall

FOURTH FLOOR

continuous glazing

4R16-2

brickwork as 4th Floor

THIRD FLOOR

continuous glazing

SECOND FLOOR

4R16-2

8R16-2

continuous glazing

building line

FIRST FLOOR

4R16-5
2R16-5

glazing between columns

4R16-3
2R16-4

GROUND FLOOR

starter bars projecting from wall and pier

200 wall

300 pier

BASEMENT

brickwork as 4th Floor

these bars are detailed with reinforcement in main beams

NOTES
C30 grade concrete throughout
Nominal cover to
links = 40 mm
25 chamfers to be formed on
edges where shown on
sections.
Galvanized ties to be built
into column at suitable
points to secure inner leaf of
brick cavity wall.

Fixings for windows to be
built into concrete as required.

Positions of bars at laps on
elevation are diagrammatic
only: true positions are
shown on sections by open
circles thus ∘

Drawing 16 Exterior column C1

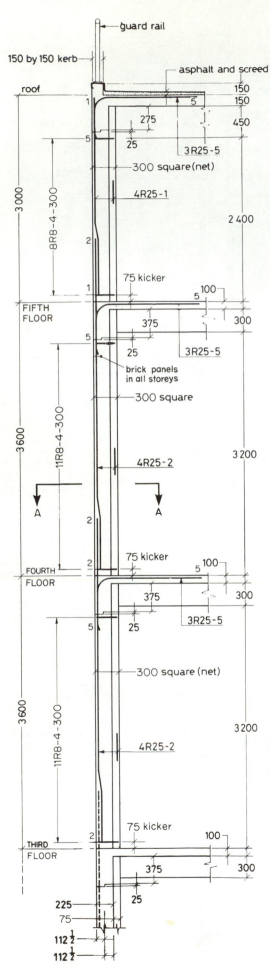

guard rail

150 by 150 kerb

asphalt and screed

roof

150
150
275
5
450
25
3R25-5

3 000

300 square (net)

8R8-4-300

4R25-1

2 400

75 kicker

1
5
100
FIFTH FLOOR
375
300
5
25
3R25-5

3 600

brick panels in all storeys

11R8-4-300

300 square

4R25-2

3 200

A

A

75 kicker

2
5
100
FOURTH FLOOR
375
300
5
25
3R25-5

3 600

300 square (net)

11R8-4-300

4R25-2

3 200

75 kicker

2
100
THIRD FLOOR
375
300
25

225
75
112 ½
112 ½

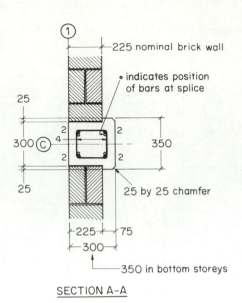

1

225 nominal brick wall

• indicates position of bars at splice

25

2   2
300 (C) 4
2   2

350

25

25 by 25 chamfer

225   75
300

350 in bottom storeys

SECTION A-A

Notes:
For position of bars at overlap
see Section A–A
Concrete: grade C30 in each storey
Nominal cover to links = 40 mm
Galvanised metal ties to be built
into column at suitable spacings
to secure back panels

182

DATA SHEET	CALCULATIONS	OUTPUT
	Upper floors: beam-and-slab construction The upper floors are designed as monolithic beam-and-slab construction with solid slabs spanning one way and supported on secondary beams, which in turn are carried by main beams framing into the columns	
	Continuous solid one-way slab (see Drawings 3 and 4) Effective span $l = 2.50$ m; $h = 100$ mm; $f_{cu} = 30$ N/mm^2; $f_y = 250$ N/mm^2	
42 62	**Durability and fire resistance** Minimum cover for 'mild' exposure $= 20$ mm Maximum fire resistance of 100 mm slab with 20 mm cover $= 1$ hour	Cover $= 20$ mm $\therefore$ Fire resistance OK
2	**Loading** Self-weight of 100 mm slab: $0.1 \times 24.0 = 2.40$ Finishes (say) $= 0.72$ Partitions (minimum) $= 1.00$ Characteristic dead load $\quad = 4.12$ kN/m^2 Characteristic imposed load $\quad = 5.00$ kN/m^2 Minimum design load $= g_k = 4.12$ kN/m^2 Maximum design load $= 1.4g_k + 1.6q_k = 5.77 + 8.00 = 13.77$ kN/m^2	$g_k = 4.12$ kN/m^2 $q_k = 5.00$ kN/m^2 $n = 13.77$ kN/m^2
7 7	**Approximate maximum ultimate bending moments** *Over supports* $M = nl^2/12 = 13.77 \times 2.5^2/12 = 7.17$ kNm/m *At midspan* $M = \frac{1}{24}(2n - g_k)\,l^2 = \frac{1}{24}(2 \times 13.77 - 4.12)\,2.5^2 = 6.10$ kNm/m	
13 41 13 41 41	**Main reinforcement** *Over supports* $\dfrac{M}{bd^2} = \dfrac{7.17 \times 10^6}{75^2 \times 10^3} = 1.28$ N/mm^2; hence $\varrho f_y = 1.54$ $A_{s\,req} = \dfrac{1.54 \times 75 \times 10^3}{250} = 463$ mm^2/m *At midspan* $\dfrac{M}{bd^2} = \dfrac{6.10 \times 10^6}{75^2 \times 10^3} = 1.08$ N/mm^2; hence $\varrho f_y = 1.31$ $A_{s\,req} = \dfrac{1.31 \times 10^3 \times 75}{250} = 393$ mm^2/m Minimum permissible reinforcement $\qquad = 0.0024bn = 0.0024 \times 10^3 \times 100 = 240$ mm^2/m	$d = 100 - 20 - (\frac{1}{2} \times 10)$ $= 75$ mm TOP R10 @ 200 + R8 @ 200 (644 mm^2/m) BOTTOM R10 @ 200 (393 mm^2/m) $\therefore$ OK

DATA SHEET	CALCULATIONS	OUTPUT
	Secondary reinforcement	
41	Generally: $0.0024bh$ at maximum clear spacing of $3d$ $A_{s\,req} = 0.0024 \times 10^3 \times 100 = 240 \text{ mm}^2/\text{m}$	R8 @ 200 $(251 \text{ mm}^2/\text{m})$
41	Over effective flange width of main beams: $0.0015bh$ $A_{s\,req} = 0.0015 \times 10^3 \times 100 = 150 \text{ mm}^2/\text{m}$	R8 @ 200 $(251 \text{ mm}^2/\text{m})$
	Deflection	
20	Basic minimum effective depth for 2.5 m continuous span $= 96 \text{ mm}$ Multiplier for $f_y = 250 \text{ N/mm}^2$ and $M/bd^2 = 1.08$ is 0.52 $\therefore$ Minimum allowable effective depth $= 96 \times 0.52 = 50 \text{ mm}$	$\therefore$ Deflection OK
	Cracking	
37	Since $h < 250 \text{ mm}$, criterion is that clear spacing between bars must not exceed $3d$, i.e. 225 mm	$\therefore$ Crack widths OK

184

DATA SHEET	CALCULATIONS	OUTPUT
	Canopy (see Drawing 4)	
	Clear cantilever span = 1.0 m; effective span = 1.025 m; h = 100 mm tapering to 75 mm; f_{cu} = 40 N/mm²; f_y = 250 N/mm²	
	Durability and fire resistance	
42	Minimum cover for severe exposure (with f_{cu} = 40 N/mm²) = 30 mm	Cover to main bars = 30 mm
62	Maximum fire resistance of 100 mm slab with 30 mm cover = 1 hour	$\therefore$ Fire resistance OK
	Loading	
2	Self-weight of 75 mm to 100 mm slab (no finishes): $0.088 \times 24.0 = 2.10$ kN/m²	$g_k = 2.10$ kN/m²
	Characteristic imposed load $= 1.50$ kN/m²	$q_k = 1.50$ kN/m²
	Design load $= 1.4g_k + 1.6q_k = 2.94 + 2.40 = 5.34$ kN/m²	$n = 5.34$ kN/m²
	Ultimate bending moment	
	$M = \frac{1}{2}nl^2 = \frac{1}{2} \times 5.34 \times 1.025^2 = 2.81$ kNm/m	
	Reinforcement	
13	$\dfrac{M}{bd^2} = \dfrac{2.81 \times 10^6}{10^3 \times 65^2} = 0.67$ N/mm²; hence $\varrho f_y = 0.80$	$d = 100 - 30 - (\frac{1}{2} \times 10)$ $= 65$ mm
41	$A_{s\,req} = \dfrac{0.80 \times 10^3 \times 65}{250} = 208$ mm²/m	
41	Minimum permissible reinforcement $= 0.0024 \times 100 \times 10^3 = 240$ mm²/m	TOP R8 @ 200 (251 mm²/m)
	Deflection	
20	Basic minimum effective depth for 1.025 m cantilever = 146 mm	Deflection does not meet simplified rules. $\therefore$ Rigorous investigation required (see section 9.3.1)
	Multiplier for f_y = 250 N/mm² and M/bd^2 = 0.67 is 0.50	
	$\therefore$ Minimum allowable effective depth = $146 \times 0.50 = 73$ mm	
	Cracking	
37	Since $h < 250$ mm, criterion is that clear spacing must not exceed $3d$, i.e. $3 \times 65 = 195$ mm; thus maximum centre-to-centre spacing = $195 + 8 = 203$ mm	$\therefore$ Crack widths OK

DATA SHEET	CALCULATIONS	OUTPUT
	Parapet (see Drawing 6)	
	Effective span $= 1.2 + (\frac{1}{2} \times 0.097) = 1.249$ m; $h = 150$ mm; $f_{cu} = 35$ N/mm^2; $f_y = 250$ N/mm^2	
	Durability and fire resistance	
42	Minimum cover for 'severe' exposure $= 40$ mm	Cover $= 40$ mm
62	Maximum fire resistance of 150 mm slab with 40 mm cover $= 3$ hours	$\therefore$ Fire resistance OK
	Loading, bending moment and reinforcement	
2	Horizontal force at top $= 740$ N/m	
	Since $l = 1.249$ m, UBM $= 740 \times 1.6 \times 1.249 = 1.48$ kNm/m	
13	$\dfrac{M}{bd^2} = \dfrac{1.48 \times 10^6}{10^3 \times 107^2} = 0.13$ N/mm^2; hence $\varrho f_y = 0.16$	$d = 150 - 40 - (\frac{1}{2} \times 6)$ $= 107$ mm
41	$A_{s\,req} = \dfrac{0.16 \times 10^3 \times 107}{250} = 68$ mm^2/m	
41	Minimum permissible main reinforcement: $\quad 0.0024bh = 0.0024 \times 150 \times 10^3 = 360$ mm^2/m	R8 @ 200 (251 mm^2/m) each way near each face
	Deflection	
	Requirements do not apply	
	Cracking	
37	Since $h < 250$ mm, criterion is that clear spacing must not exceed $3d$, i.e. $3 \times 107 = 321$ mm	$\therefore$ Crack widths OK

DATA SHEET	CALCULATIONS	OUTPUT
	Roof slab over rear stair-well (see Drawing 4)	
	Effective span $= 2.39$ m; freely supported slab; $h = 100$ mm; $f_{cu} = 30$ N/mm^2; $f_y = 250$ N/mm^2	
	Durability and fire resistance	
42	Minimum cover for 'mild' exposure $= 20$ mm	Cover $= 20$ mm
62	Maximum resistance of 100 mm slab with 20 mm cover $= 1$ hour	$\therefore$ Fire resistance OK
	Loading, bending moment and reinforcement	
	Self-weight of 100 mm slab: $0.10 \times 24.0 = 2.40$ Asphalt and screed: (say) $= 0.96$	
	Characteristic dead load $= 3.36$ kN/m^2	$g_k = 3.36$ kN/m^2
2	Characteristic imposed load (access) $= 1.50$ kN/m^2	$q_k = 1.50$ kN/m^2
	$\therefore$ Design load $= 1.4g_k + 1.6q_k = 7.10$ kN/m^2	$n = 7.10$ kN/m^2
	Effective span $=$ clear span $+$ effective depth $= 2.31 + 0.08 = 2.39$ m	
	$\therefore$ UBM $= \frac{1}{8}nl^2 = \frac{1}{8} \times 7.10 \times 2.39^2 = 5.07$ kNm/m	$d = 100 - 20 - (\frac{1}{2} \times 10)$ $= 75$ mm
13	$\dfrac{M}{bd^2} = \dfrac{5.07 \times 10^6}{10^3 \times 75^2} = 0.90$ N/mm^2; hence $\varrho f_y = 1.09$	
41	$A_{s\,req} = \dfrac{1.09 \times 75 \times 10^3}{250} = 327$ mm^2/m	BOTTOM R10 @ 255 (349 mm^2/m)
	Deflection	
20	Basic effective depth required for 2.39 m freely supported span $= 120$ mm	
	Multiplier for $f_y = 250$ N/mm^2 and $M/bd^2 = 0.90$ is 0.5	
	$\therefore$ Minimum permissible effective depth $= 120 \times 0.5 = 60$ mm	$\therefore$ Deflection OK
	Cracking	
37	Since $h < 250$ mm, criterion is that clear spacing must not exceed $3d$, i.e. $3 \times 75 = 225$ mm	$\therefore$ Crack widths OK

DATA SHEET	CALCULATIONS	OUTPUT
	Secondary beams supporting upper-floor slabs (S1 to S5) (see Drawing 5)	
	Supported on main beams; $f_{cu} = 30$ N/mm^2; $f_y = 250$ N/mm^2	
	Beam assumed continuous over supports and capable of free rotation at them Ends assumed freely supported where supported by main beams	
	Durability and fire resistance	
42	Minimum cover to all steel for 'mild' exposure = 20 mm	Cover to links = 20 mm
62	Fire resistance of 200 mm wide flanged beam with 30 mm cover to main steel = 1½ hours	∴ Fire resistance OK
	Loading	
	From slab, including finishes etc.: $4.12 \times 2.5 = 10.30$ Beam rib and finishes: $(0.3 \times 0.2 \times 24) + 5\% = 1.50$	
2	Characteristic dead load $= 11.80$ kN/m Characteristic imposed load: $5.0 \times 2.5 = 12.50$ kN/m	$g_k = 11.80$ kN/m $q_k = 12.50$ kN/m
	Minimum design load: g_k $= 11.80$ kN/m Maximum design load: $1.4g_k + 1.6q_k = 36.52$ kN/m	$n = 36.52$ kN/m
	Maximum bending moments and shearing forces (approximate) *At support A (and F)*	
	Bending moment: $M = 0$	
14	Shearing force: $V_{max} = [(0.393 \times 11.80 \times 1.4) + (0.446 \times 12.50 \times 1.6)]5.425 = 83.6$ kN	
	Midspan AB (and EF)	
7	Bending moment: $M = (0.3162n - 0.0374g_k)^2 l^2/n$ $= [(0.3162 \times 36.52) - (0.0374 \times 11.80)]^2 5.425^2/36.52 = 99.4$ kN m	
	At support B (and E)	
	Considering span of length $= \frac{1}{2}(5.425 + 6.000)$ Bending moment:	
7	$M = 0.1056nl^2 = 0.1056 \times 36.52 \times 5.713^2 = 125.9$ kN m	
14	Shearing force to left: $V_{max} = [(0.607 \times 11.80 \times 1.4) + (0.621 \times 12.50 \times 1.6)]5.425 = 121.8$ kN	
14	Shearing force to right (and each side of C and D): $V_{max} = [(0.536 \times 11.80 \times 1.4) + (0.603 \times 12.50 \times 1.6)]6.000 = 125.5$ kN	
	Midspan BC (and CD and DE)	
7	Bending moment: $M = (2n - g_k)l^2/24 = [(2 \times 36.52) - 11.80]6.0^2/24 = 91.9$ kN m	
	At support C (and D)	
7	Bending moment: $M = nl^2/12 = 36.52 \times 6.0^2/12 = 109.6$ kN m	

DATA SHEET	CALCULATIONS	OUTPUT

Longitudinal reinforcement

Midspan AB (and EF)

Minimum flange width required

$$= \frac{5M}{f_{cu}h_f(2d - h_f)} = \frac{5 \times 99.4 \times 10^6}{30 \times 100[(357.5 \times 2) - 100]} = 270 \text{ mm}$$

Flange width provided $= b + (\frac{1}{5} \times 0.85l) = 200 + 922 = 1122 \text{ mm}$

13    $\dfrac{M}{bd^2} = \dfrac{99.4 \times 10^6}{1122 \times 357.5^2} = 0.69 \text{ N/mm}^2$; hence $\varrho f_y = 0.84$

41    $A_{s\,req} = 0.84 \times 1122 \times 357.5/250 = 1348 \text{ mm}^2$

∴ Flange width OK

BOTTOM
3R25
(1473 mm²)

Support B (and E)

13    $\dfrac{M}{bd^2} = \dfrac{125.9 \times 10^6}{200 \times 337.5^2} = 5.53 \text{ N/mm}^2$; hence $\varrho' f_y = 1.2$ and $\varrho f_y = 8.15$

41   Thus $A'_{s\,req} = 1.2 \times 200 \times 337.5/250 = 324 \text{ mm}^2$

41   and $A_{s\,req} = 8.15 \times 200 \times 337.5/250 = 2200 \text{ mm}^2$

TOP
5R25
(2454 mm²)

BOTTOM
2R25
(982 mm²)

Midspan BC (and CD and DE)

Minimum flange width required $= \dfrac{5 \times 91.9 \times 10^6}{30 \times 100[(357.5 \times 2) - 100]} = 249 \text{ mm}$

Flange width provided $= b + (\frac{1}{5} \times 0.7l) = 1040 \text{ mm}$

13    $\dfrac{M}{bd^2} = \dfrac{91.9 \times 10^6}{1040 \times 357.5^2} = 0.69 \text{ N/mm}^2$; hence $\varrho f_y = 0.84$

41    $A_{s\,req} = 0.84 \times 1040 \times 357.5/250 = 1250 \text{ mm}^2$

∴ Flange width OK

BOTTOM
3R25
(1473 mm²)

Support C (and D)

13    $\dfrac{M}{bd^2} = \dfrac{109.6 \times 10^6}{200 \times 337.5^2} = 4.81 \text{ N/mm}^2$; hence $\varrho' f_y = 0.19$ and $\varrho f_y = 7.1$

41   Thus $A'_{s\,req} = 0.19 \times 200 \times 337.5/250 = 52 \text{ mm}^2$

A minimum area of $0.002 \times 200 \times 400 = 160 \text{ mm}^2$ must be provided

and $A_{s\,req} = 7.1 \times 200 \times 337.5/250 = 1925 \text{ mm}^2$

TOP
5R25
(2454 mm²)

BOTTOM
2R25
(983 mm²)

DATA SHEET	CALCULATIONS	OUTPUT
	Shearing reinforcement	
	Support A (and F)	
15	Shearing resistance of concrete (with 2R25 bars longitudinal reinforcement): $\varrho = 982/(357.5 \times 200) = 0.014$, so that $v_c = 0.77$ N/mm^2	
16	Thus concrete alone carries $0.77 \times 200 \times 357.5 = 55.2$ kN Shearing force to be carried by links = $83.6 - 55.2 = 28.4$ kN $K_u = 28.4 \times 10^3/357.5 = 80$ N/mm depth	R8 @ 250 (87 N/mm)
	Support B (and E)	
15	To left: shearing resistance of concrete alone: Since $\varrho = 2454/(337.5 \times 200) = 0.036$, $v_c = 1.01$ N/mm^2 Concrete alone carries $1.01 \times 200 \times 337.5 = 68.2$ kN	
16	Shearing force to be carried by links = $121.8 - 68.2 = 53.6$ kN $K_u = 53.6 \times 10^3/337.5 = 159$ N/mm depth	
	To right: shearing resistance of concrete alone = 68.2 kN (as above) Shearing force to be carried by links = $125.5 - 68.2 = 57.3$ kN	R8 @ 125 (175 N/mm^2) Increase spacing to nominal 250 mm centres where concrete alone suffices
16	$K_u = 57.3 \times 10^3/337.5 = 170$ N/mm depth	
	Support C (and D)	
	As for support B (right)	
	Deflection	
20	For 6.0 m continuous span, basic minimum effective depth = 230 mm Multiplier for $f_y = 250$ N/mm^2 and $M/bd^2 = 0.69$ is 0.5 Multiplier for compression reinforcement = 1 Multiplier for flanged beam (with $b_w/b = 200/1040 = 0.19$) = 1.25 $\therefore$ Minimum permissible effective depth = $230 \times 0.5 \times 1.25 = 144$ mm	$\therefore$ Deflection OK
	Cracking	
37	Crack widths controlled by limiting bar spacing Maximum spacing allowed (no redistribution) = 300 mm	$\therefore$ Crack widths OK
	Robustness	
	Since $n_s = 6$, $F_t = 44$:	
	Internal ties	
	Greatest distance between centres of vertical load-bearing elements = 6 m With $g_k = 4.12$ kN/m^2 and $q_k = 5.0$ kN/m^2, $(g_k + q_k)l = 9.12 \times 6.0 = 54.7$ kN/m Tie must be capable of resisting $0.0267 \times 54.7 \times 44 = 64.2$ kN/m	Minimum slab reinforcement provided (R8 @ 200) gives 251 mm^2/m, and main beam steel adds 393 mm^2/m
41	$A_{s\,req} = \dfrac{64.2 \times 10^3}{250} = 257$ mm^2/m	
	Column/floor ties	
	Since $l_o (= 3$ m$) < 5$ m, all ties must resist $0.4 \times 3.0 \times 44 = 52.8$ kN or $0.03 \times$ vertical load Maximum vertical load = 1262 kN and thus $0.03 \times 1262 = 37.9$ kN	
41	$A_{s\,req} = \dfrac{52.8 \times 10^3}{250} = 212$ mm^2/m	$\therefore$ OK

DATA SHEET	CALCULATIONS	OUTPUT

Secondary beams supporting upper-floor slabs (S1A to S5A) (see Drawing 5)

Supported on columns
Loads, basic ultimate bending moments, section dimensions etc. as for secondary beams S1 to S5 above

Assume ultimate BM at support A (and F) = 77 kN m

Longitudinal reinforcement
Support A (and F)

OUTPUT:
TOP
3R25
(1473 mm^2)

13

$$\frac{M}{bd^2} = \frac{77 \times 10^6}{200 \times 357.5^2} = 3.01 \text{ N/mm}^2; \text{ hence } \varrho f_y = 3.97$$

41

$$A_{s\,req} = \frac{3.97 \times 200 \times 357.5}{250} = 1138 \text{ mm}^2/\text{m}$$

Neglect reduction of moment of ¼ × 77.0 = 19.3 kN m at B (and E)
Increase of moment at C (and D) may safely be neglected

Shearing reinforcement
Support A (and F)

Shearing force calculated for beam S1 = 83.6
Additional shearing force due to moment = 77.0 × 0.105/5.425 = 1.5

_____

85.1 kN

Shearing resistance of concrete alone (with 3R25 bars longitudinal reinforcement):

15
$\varrho = 1473/(357.5 \times 200) = 0.021$, so that $v_c = 0.88$ N/mm^2
Thus concrete alone carries $0.88 \times 200 \times 357.5 = 63.0$ kN
Shearing force to be carried by links = 85.1 − 63.0 = 22.1 kN

16
$K_u = 22.1 \times 10^3/357.5 = 62$ N/mm depth

OUTPUT: R8 @ 250 (87 N/mm)

DATA SHEET	CALCULATIONS	OUTPUT
	Fascia beams (loading) (see Drawing 6)	
	Spans etc. as secondary beams S1 to S5 Compare loadings	
2	**Front fascia beam: loading**	
	Self-weight of 100 mm slab with finishes: $1.265 \times 4.12 = 5.21$ Window: 1.96×0.24 $= 0.47$ Wall (inner leaf): 1.08×2.42 $= 2.61$ (outer leaf): 1.38×2.18 $= 3.01$ Sill: $0.34 \times 0.15 \times 24.0$ $= 1.22$ Beam (including boot lintel): 0.1×24.0 $= 2.40$	
	Characteristic dead load $= 14.92$ kN/m	$g_k = 14.92$ kN/m
	Characteristic imposed load: 5.0×1.265 $= 6.33$ kN/m	$g_k = 6.33$ kN/m
	Design load $= (14.92 \times 1.4) + (6.33 \times 1.6) = 31.0$ kN/m, compared with 36.52 kN/m on main secondary beams	$n = 31.00$ kN/m
2	**Rear fascia beam: loading**	
	Self-weight of 100 mm slab, finishes etc.: $1.403 \times 4.12 = 5.78$ Window: 1.96×0.24 $= 0.47$ 275 mm cavity wall: 1.08×4.60 $= 4.97$ Sill: $0.34 \times 0.15 \times 24.0$ $= 1.22$ Beam: $0.20 \times 0.35 \times 24.0$ $= 1.68$	
	Characteristic dead load $= 14.12$ kN/m	$g_k = 14.12$ kN/m
	Characteristic imposed load: 5.0×1.403 $= 7.02$ kN/m	$g_k = 7.02$ kN/m
	Design load $= (14.12 \times 1.4) + (7.02 \times 1.6) = 31.0$ kN/m, compared with 36.52 kN/m on main secondary beams	$n = 31.00$ kN/m
	Since loadings are less than on main secondary beams, provide similar reinforcement	

DATA SHEET	CALCULATIONS	OUTPUT
	Main beams for upper floors (M1 to M3) (see Drawing 7)	
	$f_{cu} = 30\ \text{N/mm}^2; f_y = 250\ \text{N/mm}^2$	
	Durability and fire resistance	
42	Minimum cover to all steel for 'mild' exposure = 20 mm	Cover to links = 20 mm
62	Fire resistance of 250 mm wide flanged beam with 30 mm cover to main steel = 1 hour	∴ Fire resistance OK

Loading

F = load from secondary beams

A — 5.000 — B — 5.000 — C — 5.000 — D

Uniform loads

Dead load from slab, finishes etc.: 4.12×0.25	= 1.03
Beam rib and finishes: $0.4 \times 0.25 \times 24.0$	= 2.40
Characteristic dead load	= 3.43 kN/m

$g_k = 3.43$ kN/m

2: Characteristic imposed load from slab: $5.0 \times 0.25 = 1.25$ kN/m — $q_k = 1.25$ kN/m

Equivalent uniform design dead load = $1.0g_k = 3.43$ kN/m
Equivalent uniform design imposed load = $0.4g_k + 1.6q_k = 3.37$ kN/m

Concentrated loads from secondary beams

14: Characteristic dead load: $[(0.607 \times 5.18) + (0.536 \times 5.75)]11.80 = 73.5$ kN — $G_k = 73.5$ kN

14: Characteristic imposed load: $[(0.621 \times 5.18) + (0.603 \times 5.75)]12.50 = 83.6$ kN — $Q_k = 83.6$ kN

Equivalent concentrated design dead load = $1.0G_k = 73.5$ kN
Equivalent concentrated design imposed load = $0.4G_k + 1.6Q_k = 163.2$ kN

Ultimate bending moments
Midspan AB (and CD)

9: UBM due to dead loads (10% reduction of support moments):
Uniform load: $0.084 \times 3.43 \times 5.0^2 = 7.2$
Concentrated load: $0.183 \times 73.5 \times 5.0 = 67.3$
74.5 kN m

9: UBM due to imposed loads (10% reduction of moments):
Uniform load: $0.091 \times 3.37 \times 5.0^2 = 7.7$
Concentrated load: $0.191 \times 163.2 \times 5.0 = 155.9$
163.6 kN m

∴ Total UBM = 74.5 + 163.6 = 238.1 kN/m

At support B (and C)

9: UBM due to dead loads (10% reduction of moments):
Uniform load: $0.090 \times 3.43 \times 5.0^2 = 7.7$
Concentrated load: $0.135 \times 73.5 \times 5.0 = 49.6$
57.3 kN m

9: UBM due to imposed loads (10% reduction of moments):
Uniform load: $0.090 \times 3.37 \times 5.0^2 = 7.6$
Concentrated load: $0.135 \times 163.2 \times 5.0 = 110.2$
117.8 kN m

∴ Total UBM = 57.3 + 117.8 = 175.1 kN/m

DATA SHEET	CALCULATIONS	OUTPUT
	Midspan BC	
9	UBM due to dead loads (10% reduction of support moments): Uniform load: $0.035 \times 3.43 \times 5.0^2$ $= 3.0$ Concentrated load: $0.115 \times 73.5 \times 5.0$ $= 42.3$ $\overline{}$ $45.3\,\text{kN m}$	
9	UBM due to imposed loads (10% reduction of moments): Uniform load: $0.068 \times 3.37 \times 5.0^2$ $= 5.7$ Concentrated load: $0.158 \times 163.2 \times 5.0 = 128.9$ $\overline{}$ $134.6\,\text{kN m}$ $\therefore$ Total UBM $= 45.3 + 134.6 = 179.9\,\text{kN/m}$ *At support A (and D)* Assumed UBM $=$ (say) $65.0\,\text{kN/m}$	
	Ultimate shearing forces *At support A (and D)*	
14	Uniform dead load: $0.400 \times 3.43 \times 5.0$ $= 6.9$ Uniform imposed load: $0.450 \times 3.37 \times 5.0 = 7.6$ Concentrated dead load: 0.350×73.5 $= 25.7$ Concentrated imposed load: $0.425 \times 163.2 = 69.4$ Add, due to fixity: $(65 \times 10^6/5 \times 10^3) \times 1.2 = 15.6$ $\overline{}$ USF $= 125.2\,\text{kN}$	
	At support B (towards A)	
14	Uniform dead load: $0.600 \times 3.43 \times 5.0$ $= 10.3$ Uniform imposed load: $0.617 \times 3.37 \times 5.0 = 10.4$ Concentrated dead load: 0.650×73.5 $= 47.8$ Concentrated imposed load: $0.675 \times 163.2 = 110.2$ $\overline{}$ USF $= 178.7\,\text{kN}$ Neglect any reduction due to fixity at A	
	At support B (towards C)	
14	Uniform dead load: $0.500 \times 3.43 \times 5.0$ $= 8.6$ Uniform imposed load: $0.583 \times 3.37 \times 5.0 = 9.8$ Concentrated dead load: 0.500×73.5 $= 36.8$ Concentrated imposed load: $0.625 \times 163.2 = 102.0$ $\overline{}$ USF $= 157.2\,\text{kN}$	
	Main reinforcement *At support A*	
13	$\dfrac{M}{bd^2} = \dfrac{65 \times 10^6}{250 \times 457.5^2} = 1.24\,\text{N/mm}^2$; hence $\varrho f_y = 1.51$	$d = 500 - 20 - 10 - (25 \times \frac{1}{2})$ $= 457.5\,\text{mm}$
41	$A_{s\,req} = 1.51 \times 250 \times 457.5/250 = 687\,\text{mm}^2$	TOP 2R25 $(982\,\text{mm}^2)$

DATA SHEET	CALCULATIONS	OUTPUT
		$\therefore$ Flange width OK

Midspan AB

Minimum flange width required

$$= \frac{5M}{f_{cu}h_f(2d - h_f)} = \frac{5 \times 238.1 \times 10^6}{30 \times 100(2 \times 437.5 - 100)} = 512 \text{ mm}$$

Flange width provided $= \frac{1}{5} \times 5.0 \times 0.85 + 0.25 = 1100 \text{ mm}$

13
$$\frac{M}{bd^2} = \frac{238.1 \times 10^6}{1100 \times 437.5^2} = 1.13 \text{ N/mm}^2; \text{ hence } \varrho f_y = 1.37$$

41
$$A_{s\,req} = 1.37 \times 1100 \times 437.5/250 = 2638 \text{ m}^2$$

Output: BOTTOM 6R20 + 2R25 (2866 mm²)

At support B

13
$$\frac{M}{bd^2} = \frac{175.1 \times 10^6}{250 \times 412.5^2} = 4.12 \text{ N/mm}^2; \text{ hence } \varrho f_y = 5.84$$

Output: TOP 6R20 + 2R25 (2866 mm²)

41
$$A_{s\,req} = 5.84 \times 250 \times 412.5/250 = 2409 \text{ mm}^2$$

Output: BOTTOM 2R25 (982 mm²)

Midspan BC

Minimum flange width required

$$= \frac{5M}{f_{cu}h_f(2d - h_f)} = \frac{5 \times 179.9 \times 10^6}{30 \times 100(2 \times 437.5 - 100)} = 387 \text{ mm}$$

Flange width provided $= \frac{1}{5} \times 5.0 \times 0.7 + 0.25 = 950 \text{ mm}$

13
$$\frac{M}{bd^2} = \frac{179.9 \times 10^6}{950 \times 437.5^2} = 0.99 \text{ N/mm}^2; \text{ hence } \varrho f_y = 1.20$$

41
$$A_{s\,req} = 1.20 \times 950 \times 437.5/250 = 1996 \text{ m}^2$$

Output: $\therefore$ Flange width OK; BOTTOM 2R20 + 3R25 (2101 mm²)

Shearing reinforcement
At support A

Since $A_{s\,prov} = 981 \text{ mm}^2$, $\varrho = \dfrac{981}{(457.5 \times 250)} = 0.0086$

15
Hence v_c of concrete alone $= 0.63 \text{ N/mm}^2$ and concrete alone will support $0.63 \times 457.5 \times 250/10^3 = 72.0 \text{ kN}$

16
$\therefore$ Resistance required from reinforcement $= 125.2 - 72.0 = 53.2 \text{ kN}$, i.e. $K_u = 117 \text{ N/mm depth}$

Output: R8 @ 175 (125 N/mm)

195

DATA SHEET	CALCULATIONS	OUTPUT
	At support B (towards A)	
15	Since $A_{s\,prov} = 2.866\,\text{mm}^2$, $\varrho = \dfrac{2866}{(412.5 \times 250)} = 0.028$ $\therefore v_c$ of concrete alone $= 0.94\,\text{N/mm}^2$, and thus concrete will support $0.94 \times 412.5 \times 250/10^3 = 96.9\,\text{kN}$ $\therefore$ Resistance required from reinforcement $= 178.7 - 96.9 = 81.8\,\text{kN}$	
16	Minimum resistance required from links $= \frac{1}{2} \times 81.8 = 40.9\,\text{kN}$, i.e. $K_u = 40.9 \times 10^3/412.5 = 99\,\text{N/mm}$	R8 @ 200 (109 N/mm)
17	$\therefore$ Resistance required from inclined bars $= 81.8 - (109 \times 412.5/10^3) = 36.9\,\text{kN}$ One 25 mm bar at 45° will provide $250 \times 0.87 \times 491 \times \sin 45° = 75.5\,\text{kN}$	IR25 @ 45°
	At support B (towards C)	
	As above, resistance of concrete alone $= 96.9\,\text{kN}$ Resistance required from reinforcement $= 157.2 - 96.9 = 60.3\,\text{kN}$ Minimum resistance required from links $= \frac{1}{2} \times 60.3 = 30.2\,\text{kN}$	
16	i.e. $K_u = 30.2 \times 10^3/412.5 = 74\,\text{N/mm}^2$	R8 @ 200 (109 N/mm)
17	Resistance required from inclined bars $= 60.3 - (109 \times 412.5/10^3) = 15.4\,\text{kN}$ One 25 mm bar at 45° will provide $75.5\,\text{kN}$	IR25 @ 45°
20	**Deflection** For 5.0 m continuous span, basic minimum effective depth $= 192\,\text{mm}$ Multiplier for $f_y = 250\,\text{N/mm}^2$ and $M/bd^2 = 1.13$ is 0.54 Multiplier for flanged beam (with $b_w/b = 250/1100 = 0.23$) = 1.25 $\therefore$ Minimum permissible effective depth $= 192 \times 0.54 \times 1.25 = 130\,\text{mm}$	$\therefore$ Deflection OK
37	**Cracking** Crack widths controlled by limiting bars spacing to 300 mm	$\therefore$ Crack widths OK
	Robustness As for secondary beams on *Calculation Sheet 5* For internal ties, $A_{s\,req} = 257\,\text{mm}^2$ For column/floor ties, $A_{s\,req} = 212\,\text{mm}^2$	

DATA SHEET	CALCULATIONS	OUTPUT
	Beam S6 (adjoining stair-well)	
	Simply supported beam spanning 6.0 m; $f_{cu} = 30$ N/mm^2; $f_y = 250$ N/mm^2	
	Durability and fire resistance	
42	Minimum cover for 'mild' exposure = 20 mm	Cover to main bars = 30 mm
62	Maximum fire resistance of 200 mm wide flanged beam with 30 mm cover to main reinforcement = 1 hour	$\therefore$ Fire resistance OK
	Loading	
	Uniform loads	
	Self-weight of 100 mm slab, finishes etc.: 0.9×4.12 = 3.71	
	Beam (including finishes): $0.3 \times 0.2 \times 24.0$ (say) = 1.50	
2	Characteristic dead load = 5.21 kN/m	$g_k = 5.21$ kN/m
	Characteristic imposed load: 5.0×0.9 = 4.50 kN/m	$q_k = 4.50$ kN/m
	Design load = $(5.21 \times 1.4) + (4.50 \times 1.6)$ = 14.49 kN/m	$n = 14.49$ kN/m
	Loads from stairs	
	Characteristic dead load:	
	$\dfrac{1}{3.70}[(0.62 \times 3.39 \times 3.57) + (1.53 \times 2.315 \times 6.55) + (\frac{1}{2} \times 1.55^2 \times 1.79)]$	
	= 8.88 kN/m	$g_k = 8.88$ kN/m
	Characteristic imposed load:	
	$\dfrac{1}{3.70}[(0.62 \times 3.39 \times 4.25) + (1.53 \times 2.315 \times 4.25) + (\frac{1}{2} \times 1.55^2 \times 2.13)]$	
	= 7.17 kN/m	$q_k = 7.17$ kN/m
	Design load = $(8.88 \times 1.4) + (7.17 \times 1.6) = 23.91$ kN/m	$n = 23.91$ kN/m

DATA SHEET	CALCULATIONS	OUTPUT
	Ultimate bending moments and shearing forces	
	$R_1 = (\frac{1}{2} \times 6.0 \times 14.49) + \dfrac{(23.91 \times 1.23 \times 0.715)}{6.0} = 47.0$ kN	
	$R_2 = (\frac{1}{2} \times 6.0 \times 14.49) + \dfrac{(23.91 \times 1.23 \times 5.285)}{6.0} = 69.4$ kN	
	Zero shearing force occurs at $\dfrac{47.0}{14.49} = 3.24$ m from R_1	
	Maximum ultimate bending moment $= 47.0 \times 3.24 - \frac{1}{2} \times 3.24^2 \times 14.49 = 76.1$ kN m	

DATA SHEET	CALCULATIONS	OUTPUT
	Main reinforcement	
	Maximum flange width required $= \dfrac{5 \times 76.1 \times 10^6}{30 \times 100(2 \times 357.5 - 100)} = 207 \text{ mm}$	
	Flange width provided $= b_w + \frac{1}{10}l_e = 200 + 600 = 800 \text{ mm}$	$\therefore$ Flange width OK
13	$\dfrac{M}{bd^2} = \dfrac{76.1 \times 10^6}{800 \times 357.5^2} = 0.74 \text{ N/mm}^2$; hence $\varrho f_y = 0.90$	
41	$A_{s\,req} = \dfrac{0.90 \times 800 \times 357.5}{250} = 1030 \text{ mm}^2/\text{m}$	
		BOTTOM 4R20 (1256 mm²/m)
	Shearing reinforcement	
15	Since $\varrho = 1256/(200 \times 357.5) = 1.76\%$, v_c for concrete alone $= 0.83 \text{ N/mm}^2$ Thus concrete alone will support $0.83 \times 200 \times 357.5 = 59.3 \text{ kN}$ Maximum support required from links $= 69.4 - 59.3 = 10.1 \text{ kN}$	
16	i.e. $K_u = 10.1 \times 10^3/357.5 = 29 \text{ N/mm}^2$	Provide nominal R8 @ 250 (87 N/mm²) throughout
	Deflection	
20	Basic minimum effective depth for 6.0 m freely supported beam $= 300 \text{ mm}$	
	Multiplier for $f_y = 250 \text{ N/mm}^2$ and $M/bd^2 = 0.75$ is 0.5	
	Multiplier for flanged beam (with $b_w/b = 200/800 = 0.25$) $= 1.25$	
	$\therefore$ Minimum permissible effective depth $= 300 \times 0.5 \times 1.25 = 188 \text{ mm}$	$\therefore$ Deflection OK
	Cracking	
37	With $f_y = 250 \text{ N/mm}^2$ and no redistribution of moments, criterion is that spacing must not exceed 300 mm	$\therefore$ Crack widths OK

DATA SHEET	CALCULATIONS	OUTPUT
	Main roof: beam-and-slab construction	
	The main roof is designed as monolithic beam-and-slab construction with solid slabs spanning two ways and supported on transverse and longitudinal beams framing directly into the columns	
	Roof-slab: two-way slab (panels P1 to P4) (see Drawings 8 and 9)	
	$f_{cu} = 30\,\text{N/mm}^2; f_y = 250\,\text{N/mm}^2; h = 150\,\text{mm}$	
	Durability and fire resistance	
42	Minimum cover for mild exposure = 20 mm	Cover = 20 mm
62	Maximum fire resistance of 150 mm slab with 20 mm cover = 1½ hours	∴ Fire resistance OK
	Loading	
	Self-weight of 150 mm slab: 0.15×24.0 = 3.60	
	20 mm asphalt (0.48), 50 mm insulating screed (0.72) = 1.20	
	Ceiling finish = 0.24	
2	Characteristic dead load = 5.04 kN/m²	$g_k = 5.04\,\text{kN/m}$
	Characteristic imposed load (roof with access) = 1.50 kN/m²	$g_k = 1.50\,\text{kN/m}$
	Design load $n = 1.4g_k + 1.6q_k = 7.06 + 2.40$ = 9.46 kN/m²	$n = 9.46\,\text{kN/m}$
	Ultimate bending moments	
	Panels P1 (continuous at all edges)	
	$l_x = 5.0\,\text{m}; l_y = 6.0\,\text{m}; l_y/l_x = 1.2$	
	Shorter span:	
44	UBM at edges $= -0.042 \times 5.0^2 \times 9.46 = -10.0$ kNm/m	
	UBM at midspan $= +0.032 \times 5.0^2 \times 9.46 = +7.6$ kNm/m	
	Longer span:	
44	UBM at edges $= -0.032 \times 5.0^2 \times 9.46 = -7.6$ kNm/m	
	UBM at midspan $= +0.024 \times 5.0^2 \times 9.46 = +5.7$ kNm/m	
	Panels P2 (freely supported along one long side)	
	$l_x = 5.0\,\text{m}; l_y = 6.0\,\text{m}; l_y/l_x = 1.2$	
	Shorter span:	
44	UBM at continuous edge $= -0.056 \times 5.0^2 \times 9.46 = -13.3$ kNm/m	
	UBM at midspan $= +0.042 \times 5.0^2 \times 9.46 = +10.0$ kNm/m	
	Longer span:	
44	UBM at edges $= -0.037 \times 5.0^2 \times 9.46 = -8.8$ kNm/m	
	UBM at midspan $= +0.028 \times 5.0^2 \times 9.46 = +6.6$ kNm/m	
	Panels P3 (freely supported along two adjacent sides)	
	$l_x = 5.0\,\text{m}; l_y = 5.425\,\text{m}; l_y/l_x = 1.085$	
	Shorter span:	
44	UBM at continuous edge $= -0.063 \times 5.0^2 \times 9.46 = -14.9$ kNm/m	
	UBM at midspan $= +0.047 \times 5.0^2 \times 9.46 = +11.2$ kNm/m	
	Longer span:	
44	UBM at continuous edge $= -0.045 \times 5.0^2 \times 9.46 = -10.7$ kNm/m	
	UBM at midspan $= +0.034 \times 5.0^2 \times 9.46 = +8.1$ kNm/m	
	Panels P4 (freely supported along one short side)	
	$l_x = 5.0\,\text{m}; l_y = 5.425\,\text{m}; l_y/l_x = 1.085$	
	Shorter span:	
44	UBM at edges $= -0.048 \times 5.0^2 \times 9.46 = -11.4$ kNm/m	
	UBM at midspan $= +0.036 \times 5.0^2 \times 9.46 = +8.6$ kNm/m	

DATA SHEET	CALCULATIONS	OUTPUT
44	**Longer span:** UBM at continuous edge $= -0.037 \times 5.0^2 \times 9.46 = -8.8$ kNm/m UBM at midspan $= +0.028 \times 5.0^2 \times 9.46 = +6.6$ kNm/m	

Main reinforcement
Panels P1

Assuming the use of maximum 12 mm bars
In direction of short span: $d = 150 - 20 - 6 = 124$ mm
In direction of long span: $d = 150 - 20 - 12 - 6 = 112$ mm

$\therefore$Minimum reinforcement in both directions
$= 0.0024 \times 150 \times 10^3$
$= 360$ mm^2/m
Maximum spacing $= 300$ mm

Shorter span, at edges:

| 13 | $\dfrac{M}{bd^2} = \dfrac{10.0 \times 10^6}{10^3 \times 124^2} = 0.65 \text{ N/mm}^2$; hence $\varrho f_y = 0.79$ | |
| 41 | $A_{s\,req} = \dfrac{0.79 \times 10^3 \times 124}{250} = 390 \text{ mm}^2/\text{m}$ | TOP
R10 @ 200
(392 mm^2/m) |

Shorter span, at midspan:

| 13 | $\dfrac{M}{bd^2} = \dfrac{7.6 \times 10^6}{10^3 \times 124^2} = 0.49 \text{ N/mm}^2$; hence $\varrho f_y = 0.60$ | |
| 41 | $A_{s\,req} = \dfrac{0.60 \times 10^3 \times 124}{250} = 297 \text{ mm}^2/\text{m}$ | BOTTOM
R10 @ 200
(392 mm^2/m) |

Longer span, at edges:

| 13 | $\dfrac{M}{bd^2} = \dfrac{7.6 \times 10^6}{10^3 \times 112^2} = 0.61 \text{ N/mm}^2$; hence $\varrho f_y = 0.73$ | |
| 41 | $A_{s\,req} = \dfrac{0.73 \times 10^3 \times 112}{250} = 328 \text{ mm}^2/\text{m}$ | TOP
R12 @ 300
(376 mm^2/m) |

Longer span, at midspan:

| 13 | $\dfrac{M}{bd^2} = \dfrac{5.7 \times 10^6}{10^3 \times 112^2} = 0.45 \text{ N/mm}^2$; hence $\varrho f_y = 0.55$ | |
| 41 | $A_{s\,req} = \dfrac{0.55 \times 10^3 \times 112}{250} = 246 \text{ mm}^2/\text{m}$ | BOTTOM
R12 @ 300
(376 mm^2/m) |

Minimum permissible reinforcement $= 0.0024 \times 10^3 \times 150 = 360$ mm^2/m

Panels P2, P3 and P4

Similar calculations to the foregoing give the following areas of steel for panels P2, P3 and P4 (mm^2/m):

		P2	P3	P4	
13	Shorter span, at supports	519	582	445	
	Shorter span, at midspan	390	437	336	
	Longer span, at supports	380	463	380	
	Longer span, at midspan	285	350	285	

DATA SHEET	CALCULATIONS	OUTPUT
41	Thus the following systems of reinforcement are provided: *Panels P2:* Shorter span, at supports	TOP R6 @ 200 + R10 @ 200 (534 mm^2/m)
	Shorter span, at midspan	BOTTOM R10 @ 200 (392 mm^2/m)
	Longer span, at supports	TOP R12 @ 300 (381 mm^2/m)
	Longer span, at midspan	BOTTOM R12 @ 300 (381 mm^2/m)
	Panels P3: Shorter span, at supports	TOP R8 @ 200 + R10 @ 200 (644 mm^2/m)
	Shorter span, at midspan	BOTTOM R10 @ 400 + R12 @ 400 (479 mm^2/m)
	Longer span, at supports	TOP R6 @ 300 + R12 @ 300 (470 mm^2/m)
	Longer span, at midspan	BOTTOM R12 @ 300 (376 mm^2/m)
	Panels P4: Shorter span, at supports	TOP R10 @ 200 + R12 @ 200 (479 mm^2/m)
	Shorter span, at midspan	BOTTOM R10 @ 200 (392 mm^2/m)
	Longer span, at supports	TOP R12 @ 300 (376 mm^2/m)
	Longer span, at midspan	BOTTOM R12 @ 300 (376 mm^2/m)
41	**Secondary reinforcement** $A_{s\,req} = 0.0024 \times 10^3 \times 150 = 360$ mm^2/m Maximum spacing $= 3d = 3 \times 112 = 336$ mm	Minimum (where required) R12 @ 300 (376 mm^2/m) ∴ Spacing OK

DATA SHEET	CALCULATIONS	OUTPUT
44	**Torsion reinforcement** At corner of panel P3: $A_{s\ req} = \tfrac{3}{4} \times 428 = 321\ mm^2/m$	TOP AND BOTTOM R12 @ 300 ($376\ mm^2/m$)
	At corners between panels P2 and P3, and P3 and P4: $A_{s\ req} = \tfrac{3}{8} \times 428 = 161\ mm^2/m$	As secondary reinforcement above
	At corners between adjoining pairs of panels P2: $A_{s\ req} = \tfrac{3}{8} \times 381 = 143\ mm^2/m$	As secondary reinforcement above
	At corners between adjoining pairs of panels P4: $A_{s\ req} = \tfrac{3}{8} \times 328 = 123\ mm^2/m$	As secondary reinforcement above
20	**Deflection** Basic minimum effective depth for continuous 5 m span $= 192\ mm$ Tension reinforcement multiplier for $f_y = 250\ N/mm^2$ and $M/bd^2 = 0.49$ is 0.5 $\therefore$ Minimum permissible effective depth $= 192 \times 0.5 = 96\ mm$	$\therefore$ Deflection OK
37	**Cracking** Since $h\ (= 150\ mm) < 250\ mm$, criterion is that bar spacing must not exceed $3d$, i.e. 372 mm for shorter span and 336 mm for longer span. These requirements are met	$\therefore$ Crack widths OK

DATA SHEET	CALCULATIONS	OUTPUT
	Typical longitudinal beam supporting two-way roof slabs (see Drawings 8 and 9)	
	Interior span of theoretically infinite system. $f_{cu} = 30$ N/mm^2; $f_y = 250$ N/mm^2; $l = 6.0$ m	
42 62	**Durability and fire resistance** Minimum cover to all steel for 'mild' exposure $= 20$ mm Fire resistance of 130 mm wide flanged beam with 30 mm cover to main reinforcement $= 2$ hours	Cover to main bars $= 30$ mm ∴ Fire resistance OK
44 44	**Loading** For panel P1, since $l_y/l_x = 6.0/5.0 = 1.2$ and all edges are continuous, $\beta_{vx} = 0.39$ For panel P2 with one long edge discontinuous, β_{vx} $\qquad = 0.44$	
	From panel P1: dead load $= 5.04 \times 0.39 \times 5.0 \qquad = 9.83$ From panel P2: dead load $= 5.04 \times 0.44 \times 5.0 \qquad = 11.09$	
	Characteristic dead load from two-way slabs $\qquad = 20.92$ kN/m	$g_k = 20.92$ kN/m
	Self-weight of beam rib and finishes (say) $\qquad = 1.73$ kN/m	$g_k = 1.73$ kN/m
	From panel P1: imposed load $= 1.5 \times 0.39 \times 5.0 \qquad = 2.93$ From panel P2: imposed load $= 1.5 \times 0.44 \times 5.0 \qquad = 3.30$	
	Characteristic imposed load from two-way slabs $\qquad = 6.23$ kN/m	$q_k = 6.23$ kN/m
	Maximum design load from slabs $= 1.4G_k + 1.6Q_k$ $\qquad = 20.92 \times 1.4 + 6.23 \times 1.6 = 39.26$ kN/m Maximum design load due to self-weight $= 1.4 \times 1.73 \qquad = 2.42$ kN/m	$n = 39.26$ kN/m $n = 2.42$ kN/m
	Ultimate bending moments Considering penultimate support of theoretically infinate system:	
8	Due to load from slabs: $\quad$ UBM $= 0.0965nl^2 = 0.0965 \times 39.26 \times 6.0^2 \qquad = 136.4$	
7	Due to self-weight of rib etc.: $\quad$ UBM $= 0.1056nl^2 = 0.1056 \times 2.42 \times 6.0^2 \qquad = 9.2$ $\qquad\qquad\qquad\qquad\qquad\qquad\qquad$ 145.6 kN m	
	Considering end span of theoretically infinite system:	
8	Due to load from slabs: $\quad$ UBM $= (0.0942n^2 - 0.0218ng + 0.0012g^2)l^2/n$ $\quad = (0.0942 \times 39.26^2 - 0.0218 \times 39.26$ $\qquad \times 20.92 + 0.0012 \times 20.92^2) \times 6^2/39.26 \qquad = 117.2$	
7	Due to self-weight of beam rib etc.: $\quad$ UBM $= (0.3162n - 0.0374g)^2 l^2/n$ $\quad = (0.3162 \times 2.42 - 0.0374 \times 1.73)^2 \times 6^2/2.42 \qquad = 7.3$ $\qquad\qquad\qquad\qquad\qquad\qquad\qquad$ 124.5 kN m	

DATA SHEET	CALCULATIONS	OUTPUT

Longitudinal reinforcement

At support

13

$$\frac{M}{bd^2} = \frac{145.6 \times 10^6}{130 \times 509.5^2} = 4.31 \text{ N/mm}^2; \text{ hence } \varrho f_y = 6.21$$

41

$$\therefore A_{s\,req} = \frac{6.21 \times 130 \times 509.5}{250} = 1646 \text{ mm}^2/\text{m}$$

At midspan:

Minimum required flange width

$$= \frac{5M}{f_{cu} h_f (2d - h_f)} = \frac{5 \times 124.5 \times 10^6}{30 \times 150 (2 \times 542.5 - 150)} = 148 \text{ mm}$$

Effective flange width provided $= (0.2 \times 5.425 \times 0.85) + 0.13 = 1.052$ m

13

$$\frac{M}{bd^2} = \frac{124.5 \times 10^6}{1052 \times 542.5^2} = 0.40 \text{ N/mm}^2; \text{ hence } \varrho f_y = 0.48$$

41

$$\therefore A_{s\,req} = \frac{0.48 \times 1052 \times 542.5}{250} = 1107 \text{ mm}^2/\text{m}$$

OUTPUT:

TOP
4R25
(1963 mm²)

150 — 90.5

450 — 424

85.5

130

∴ Flange width OK

1052

150

450 — 542.5

130

BOTTOM
4R20
(1256 mm²/m)

Shearing force and reinforcement

USF at support:

From panel P1:
due to dead load: $9.83 \times 5.425 \times 0.75 \times 0.6 \times 1.4$ = 33.6
due to imposed load: $2.93 \times 5.425 \times 0.75 \times 0.6 \times 1.6$ = 11.5
From panel P2:
due to lead load: $11.09 \times 5.425 \times 0.75 \times 0.6 \times 1.4$ = 37.9
due to imposed load: $3.30 \times 5.425 \times 0.75 \times 0.6 \times 1.6$ = 12.9
Due to self-weight of rib etc.: $1.73 \times 5.425 \times 0.6 \times 1.4$ = 7.9

Design shearing force = 103.8 kN

Since $\varrho = 1963/(509.5 \times 130) = 0.030$, $v_c = 0.97$ N/mm², and thus concrete alone carries $0.97 \times 130 \times 509.5 = 64.2$ kN. Reinforcement required to carry $103.8 - 64.2 = 39.6$ kN,

i.e. $K_u = \dfrac{39.6 \times 10^3}{509.5} = 78$ N/mm depth

OUTPUT: R8 @ 150 (82 N/mm)

Deflection

20

Basic effective depth for 6 m continuous span = 231 mm

Modifier for $f_y = 250$ N/mm² and $M/bd^2 = 0.40$ is 0.5

Since $b_w/b = 130/1052 = 0.12$, modifier for flanged beam is 1.25

Thus minimum permissible effective depth = $231 \times 0.5 \times 1.25 = 145$ mm

OUTPUT: ∴ Deflection OK

Cracking

37

For $f_y = 250$ N/mm² and no redistribution, limiting spacing = 300 mm

OUTPUT: ∴ Crack widths OK

Robustness

As for secondary beams on *Calculation Sheet 7*
For internal ties, $A_{s\,req} = 257$ mm²/m
For column/floor ties, $A_{s\,req} = 212$ mm

DATA SHEET	CALCULATIONS	OUTPUT
	Typical transverse beam supporting two-way roof-slabs (see Drawings 8 and 9)	
	Two-span beam with cantilever at one end $f_{cu} = 30 \text{ N/mm}^2; f_y = 250 \text{ N/mm}^2$	

DATA SHEET	CALCULATIONS	OUTPUT
42 62	**Durability and fire resistance** Minimum cover to all steel for 'mild' exposure = 20 mm Fire resistance of 130 mm wide flanged beam with 30 mm cover to main reinforcement = 2 hours	Cover to main bars = 30 mm ∴ Fire resistance OK

Loading

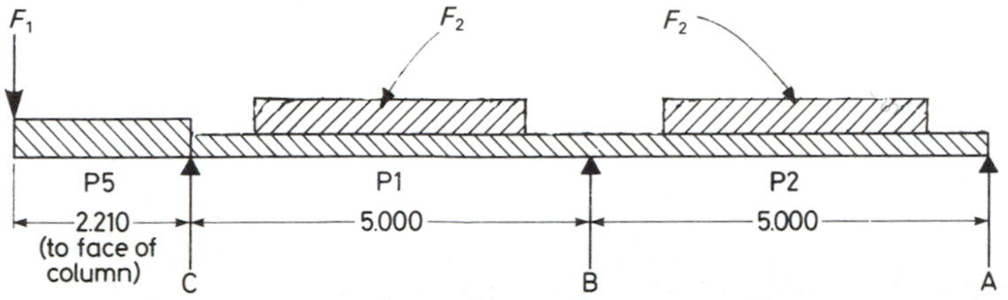

Concentrated load from edge beam of cantilever (F_1):
Dead loads from roof slab = $4.12 \times 1.25 = 5.15$ kN/m $\times 6.0 = 30.90$
Self-weight of edge beam (say) 3.0×6.0 = 18.00

Characteristic dead load   = 48.90 kN $G_k = 48.90$ kN

Characteristic imposed load = $5.0 \times 1.25 = 6.25 \times 6.0$ = 37.50 kN $Q_k = 37.50$ kN

Design dead load = $1.4 G_k = 48.90 \times 1.4 = 68.49$ kN
Design imposed load = $1.6 Q_k = 37.50 \times 1.6 = 60.00$ kN

Dead load due to self-weight of cantilever rib etc. (say) = 1.90 kN/m
Design dead load = $1.9 \times 1.4 = 2.66$ kN/m $n = 2.66$ kN/m

From *Data Sheet 44*, reaction coefficient at short edge where l_y/l_x is 1.2 is 0.33.
Thus from panel P1:

Characteristic dead load = $5.04 \times 6.0 \times 0.33 \times 2 = 19.96$ kN/m $g_k = 19.96$ kN/m
Characteristic imposed load = $1.50 \times 6.0 \times 0.33 \times 2 = 5.94$ kN/m $q_k = 5.94$ kN/m

Maximum design load = $1.4 g_k + 1.6 q_k = 27.95 + 9.51 = 37.46$ kN/m $n = 37.46$ kN/m

Dead load due to self-weight (say) = 1.73 kN/m
Design dead load = $1.73 \times 1.4 = 2.42$ kN/m $n = 2.42$ kN/m

Ultimate bending moments
At support C (at face of column)

> UBM = $\frac{1}{2}nl^2 + Fl = (\frac{1}{2} \times 2.66 \times 2.21^2) + (68.49 + 60.00)2.21 = 290.5$ kN m

At midspan BC and AB (neglecting relief due to cantilever)

From two-way slab:
UBM = $(0.0903n^2 - 0.0253ng + 0.0016g^2)l^2/n$
    = $[(0.0903 \times 37.46^2) - (0.0253 \times 37.46 \times 19.96) + (0.0016 \times 19.96^2)]$
    $\times 5.0^2/37.46 = 72.4$ kN m

From self-weight:
UBM = $(7n - g_k)^2 l^2/512n$
    = $[(7 \times 2.42) - 1.73]^2 5.0^2/(512 \times 2.42) = 4.7$ kN m

Additional moment on AB due to cantilever $\simeq \frac{1}{8} \times 290.5 = 36.3$ kN m

Total UBM = $72.4 + 4.7 + 36.3 = 113.4$ kN m

At support B (neglecting relief due to cantilever)

> UBM = $(0.1143 \times 37.46 \times 5.0^2) + (\frac{1}{8} \times 2.42 \times 5.0^2) = 114.6$ kN m

(left margin data sheet references: 8, 7)

DATA SHEET	CALCULATIONS	OUTPUT
	Longitudinal reinforcement	
	At midspan AB	
	Minimum breadth of flange required	
	$$= \frac{5M}{f_{cu}h_f(2d - h_f)} = \frac{5 \times 113.4 \times 10^6}{30 \times 150(2 \times 542.5 - 150)} = 135 \text{ mm}$$	
	Effective flange width provided $= (0.2 \times 5.0 \times 0.7) + 0.13 = 830$ mm	$\therefore$ Flange width OK
13	$$\frac{M}{bd^2} = \frac{113.4 \times 10^6}{830 \times 559.5^2} = 0.44 \text{ N/mm}^2; \text{ hence } \varrho f_y = 0.53$$	
41	Thus $A_{s\,req} = \dfrac{0.53 \times 559.5 \times 830}{250} = 981 \text{ mm}^2$	
		BOTTOM 4R25 (982 mm²/m) $\therefore$ OK
	Minimum permissible reinforcement $= 0.0024 \times 130 \times 600 = 188$ mm²	
	At support B	
13	$$\frac{M}{bd^2} = \frac{114.6 \times 10^6}{130 \times 542.5^2} = 3.00 \text{ N/mm}^2; \text{ hence } \varrho f_y = 3.95$$	
41	$A_{s\,req} = \dfrac{3.95 \times 130 \times 542.5}{250} = 1114 \text{ mm}^2/\text{m}$	TOP 4R20 (1256 mm²/m)
	At support C	
	Slenderness: clear distance from cantilever tip to support face must not exceed $25b_c = 25 \times 0.13 = 3.25$ m or $100b_c^2/d = 100 \times 0.13^2/0.537 = 3.15$ m	$\therefore$ Slenderness OK
13	$$\frac{M}{bd^2} = \frac{290.5 \times 10^6}{130 \times 537.5^2} = 7.73$$	
	Since section cannot be designed by means of design chart, use formulae in clause 3.4.4.4 of BS8110	
	Resistance provided by concrete $= 0.156bd^2 f_{cu}$ $= 0.156 \times 130 \times 537.5^2 \times 30 = 175.8 \times 10^6$ N mm $z = 537.5[0.5 + \sqrt{(0.25 - 0.156/0.9)}] = 417.6$ mm	TOP 4R25 + 4R20 (3220 mm²)
41	Thus resistance required from compression steel $= 290.5 \times 10^6 - 175.8 \times 10^6 = 114.7 \times 10^6$ N mm	
	$A_{s\,req} = \dfrac{114.7 \times 10^6}{0.87 \times 250(537.5 - 62.5)} = 1111 \text{ mm}^2/\text{m}$	
	$A_{s\,req} = \dfrac{175.8 \times 10^6}{(0.87 \times 250 \times 417.6)} + 1111 = 3047 \text{ mm}^2/\text{m}$	BOTTOM 4R25 (1963 mm²)
	Percentage of tension steel in terms of gross section $= \dfrac{3047}{130 \times 600} = 3.9\%$	$\therefore$ Reinforcement within allowable limit of 4%

DATA SHEET	CALCULATIONS	OUTPUT
	Ultimate shearing force and reinforcement	
	At support C	
	USF from $F_1 = 68.5 + 60.0$ $\qquad = 128.5$	
	USF due to cantilever rib etc. $= 2.66 \times 2.21 = \quad 5.9$	
	$\overline{}$	
	134.4 kN	
15	Since $d > 400$ and $\varrho > 0.03$, v_c of concrete $= 0.97$ N/mm^2 and shearing resistance of concrete $= 0.97 \times 130 \times 537.5 = 67.8$ kN	
	Shearing reinforcement is required to withstand $134.4 - 67.8 = 66.6$ kN	
16	i.e. $K_u = 66.6 \times 10^3/537.5 = 124$ N/mm depth	R8 @ 150 (146 N/mm)
	At support B	
15	USF $\simeq (37.46 \times 2.5 \times 0.75) + (2.42 \times 2.5) + (114.6/5.0) = 99.2$ kN	
	Since $\varrho = 1256/(130 \times 542.5) = 0.018$, v_c of concrete alone $= 0.82$ N/mm^2, and concrete alone will support $0.82 \times 130 \times 542.5 = 57.8$ kN	
	Shearing reinforcement is required to withstand $99.2 - 57.8 = 41.4$ kN	
16	i.e. $K_u = 41.4 \times 10^3/542.5 = 76$ N/mm depth	R8 @ 275 (79 N/mm)
	Deflection	
20	For 5.0 m continuous span, basic effective depth $= 192$ mm	
	Multiplier for $f_y = 250$ N/mm^2 and $M/bd^2 = 0.44$ is 0.5	
	Multiplier for flanged beam ($b_w/b = 130/830 = 0.16$) $= 1.25$	
	$\therefore$ Minimum permissible effective depth $= 192 \times 0.5 \times 1.25 = 120$ mm	$\therefore$ Deflection OK
	Cracking	
37	Crack widths controlled by limiting bar spacing	
	Maximum spacing allowed (no redistribution) $= 300$ mm	$\therefore$ Crack widths OK
	Robustness	
	As for secondary beams on *Calculation Sheet 7*	
	For internal ties, $A_{s\,req} = 257$ mm^2/m	
	For column/floor ties, $A_{s\,req} = 212$ mm^2	

DATA SHEET	CALCULATIONS	OUTPUT
	Floor of tank room (see Drawings 10 and 11)	
	Two-way slab continuous over three edges and supporting concentrated load $f_{cu} = 30 \text{ N/mm}^2$; $f_y = 250 \text{ N/mm}^2$; $h = 150 \text{ mm}$	
42 62	**Durability and fire resistance** Minimum cover for 'mild' exposure = 20 mm Maximum fire resistance of 150 mm slab with 20 mm cover = 2 hours	Cover = 20 mm ∴ Fire resistance OK

2	**Loading**
	Uniform dead loads:
	  From slab: 0.15×24.0 = 3.60
	  Asphalt and ceiling: (say) = 0.72
	  ───────
	  4.32 kN/m²
	
	Uniform imposed load: = 2.50 kN/m²
	
	Uniform design load = $(4.32 \times 1.4) + (2.50 \times 1.6) = 10.05 \text{ kN/m}^2$
	
	Characteristic concentrated load = 230 kN spread over area 3.0 m × 2.5 m
	
	Net concentrated design load = $[230 - (2.50 \times 3.0 \times 2.5)]1.6 = 338 \text{ kN}$
		$g_k = 4.32 \text{ kN/m}^2$
		
		$q_k = 2.50 \text{ kN/m}^2$
		
		$n = 10.05 \text{ kN/m}^2$

45 44	**Ultimate bending moments** Since $a_x/l_x = 2.50/3.75 = 0.67$ and $a_y/l_x = 3.0/5.625 = 0.55$, Pigeaud's coefficients are $m_x = 0.104$ and $m_y = 0.067$. Continuity coefficients are 0.95 at edges and 0.75 at centre for short span (l_x) and 0.90 at edges and 0.70 at centre for long span (l_y) For uniform load, $a_x = -0.057$ and $+0.044$, and $a_y = -0.037$ and $+0.028$ *Shorter span: at edges* UBM = $-(0.057 \times 10.05 \times 3.75^2) - (0.104 \times 0.95 \times 338) = -41.5 \text{ kN m/m}$ *Shorter span: at midspan* UBM = $+(0.044 \times 10.05 \times 3.75^2) + (0.104 \times 0.75 \times 338) = +32.6 \text{ kN m/m}$ *Longer span: at edge* UBM = $-(0.037 \times 10.05 \times 3.75^2) - (0.067 \times 0.90 \times 338) = -25.6 \text{ kN m}$ *Longer span: at midspan* UBM = $+(0.028 \times 10.05 \times 3.75^2) + (0.067 \times 0.70 \times 338) = +19.8 \text{ kN m}$	
13 41	**Main reinforcement** *Shorter span: at edges* $\dfrac{M}{bd^2} = \dfrac{41.5 \times 10^6}{10^3 \times 120^2} = 2.88$; hence $\varrho f_y = 3.78$ $A_{s\,req} = \dfrac{3.78 \times 10^3 \times 120}{250} = 1813 \text{ mm}^2/\text{m}$	TOP R20 @ 200 + R10 @ 200 (1964 mm²/m)

(Drawing dimensions: $l_x = 3.750$; $l_y = 5.625$; 3.000; 2.500)

DATA SHEET	CALCULATIONS	OUTPUT
	Shorter span: at midspan	
13	$\dfrac{M}{bd^2} = \dfrac{32.6 \times 10^6}{10^3 \times 120^2} = 2.26$; hence $\varrho f_y = 2.87$	
41	$A_{s\,req} = \dfrac{2.87 \times 10^3 \times 120}{250} = 1378\ \text{mm}^2/\text{m}$	BOTTOM R20 @ 200 (1571 mm²/m)
	Longer span: at edges	
13	$\dfrac{M}{bd^2} = \dfrac{25.6 \times 10^6}{10^3 \times 100^2} = 2.56$; hence $\varrho f_y = 3.29$	
41	$A_{s\,req} = \dfrac{3.29 \times 10^3 \times 100}{250} = 1318\ \text{mm}^2/\text{m}$	TOP R20 @ 300 + R12 @ 300 (1424 mm²/m)
	Longer span: at midspan	
13	$\dfrac{M}{bd^2} = \dfrac{19.8 \times 10^6}{10^3 \times 100^2} = 1.98$; hence $\varrho f_y = 2.47$	
41	$A_{s\,req} = \dfrac{2.47 \times 10^3 \times 100}{250} = 990\ \text{mm}^2/\text{m}$	BOTTOM R20 @ 300 (1047 mm²/m)
	Minimum permissible reinforcement: $0.0024 \times 120 \times 10^3 = 288\ \text{mm}^2/\text{m}$	

Ultimate shearing force

Total design load N within critical perimeter:

Due to uniform loads:
$\qquad 3.300 \times 2.860 \times 10.05 = \quad 94.9$
Due to net concentrated load $\qquad = 338.0$
$\qquad\qquad\qquad\qquad\qquad\qquad\overline{\quad 432.9\ \text{kN}}$

Length of critical perimeter = $2(3.30 + 2.86) = 12\,320$ mm

Since average $d = 110$ mm, $v = 432.9 \times 10^3/(12\,320 \times 110) = 0.32\ \text{N/mm}^2$

Since average ϱ provided in both directions = $\frac{1}{2}(1571 + 1047)/(110 \times 10^3)$
$\qquad\qquad\qquad\qquad = 0.012$,
with $f_{cu} = 30\ \text{N/mm}^2$ and $d = 110$ mm, $v_c = 0.98\ \text{N/mm}^2$

∴ Shearing resistance OK

Deflection

20   Basic effective depth for 3.75 m continuous span = 144 mm
Multiplier for $f_y = 250$ and $M/bd^2 = 2.9 = 0.80$
∴ Minimum allowable effective depth = $144 \times 0.80 = 116$ mm

∴ Deflection OK

Cracking

37   Since $h < 250$ mm, criterion is that bar spacing must not exceed $3d$, i.e. 300 mm for longer span

∴ Crack widths OK

DATA SHEET	CALCULATIONS	OUTPUT
	Roof of tank room (see Drawings 10 and 11) Two-way freely supported slab restrained at corners $f_{cu} = 30 \text{ N/mm}^2; f_y = 250 \text{ N/mm}^2; h = 150 \text{ mm}$	
2	**Loading** Self-weight of 150 mm slab: $0.15 \times 24.0 = 3.60$ Asphalt and screed: (say) $= 0.96$ Characteristic dead load $= 4.56 \text{ kN/m}^2$ Characteristic imposed load $= 1.50 \text{ kN/m}^2$ Design load $= 1.4g_k + 1.6q_k = 6.40 + 2.40 = 8.80 \text{ kN/m}^2$	$g_k = 4.56 \text{ kN/m}^2$ $q_k = 1.50 \text{ kN/m}^2$ $n = 8.80 \text{ kN/m}^2$
44	**Ultimate bending moments** Since $l_x = 3.64$ m and $l_y = 5.41$ m, $l_y/l_x = 1.48$ For slab freely supported on all sides and with restrained corners, $a_x = 0.091$ and $a_y = 0.056$ *Shorter span: at midspan* $\quad$ UBM $= a_x n l_x^2 = 0.091 \times 8.8 \times 3.64^2 = 10.6 \text{ kNm/m}$ *Longer span: at midspan* $\quad$ UBM $= a_y n l_x^2 = 0.056 \times 8.8 \times 3.64^2 = 6.5 \text{ kNm/m}$	
41	**Reinforcement** Minimum permissible reinforcement: $0.0024 \times 150 \times 10^3 = 360 \text{ mm}^2/\text{m}$	
13	*Shorter span: at midspan* $\quad \dfrac{M}{bd^2} = \dfrac{10.6 \times 10^6}{10^3 \times 124^2} = 0.69$; hence $\varrho f_y = 0.83$	With 20 mm cover, $d = 150 - 20 - (\frac{1}{2} \times 12)$ $= 124 \text{ mm}$
41	$\quad A_{s\,req} = \dfrac{0.83 \times 124 \times 10^3}{250} = 412 \text{ mm}^2/\text{m}$	BOTTOM R12 @ 250 $(452 \text{ mm}^2/\text{m})$
13	*Longer span: at midspan* $\quad \dfrac{M}{bd^2} = \dfrac{6.5 \times 10^6}{10^3 \times 112^2} = 0.52$; hence $\varrho f_y = 0.63$	$d = 150 - 12 - (\frac{1}{2} \times 12)$ $= 112 \text{ mm}$
41	$\quad A_{s\,req} = \dfrac{0.63 \times 112 \times 10^3}{250} = 283 \text{ mm}^2/\text{m}$ Provide minimum area of main reinforcement of 360 mm^2/m *Torsional reinforcement in each direction at each corner* $\quad A_{s\,req} = \frac{3}{4} \times 452 = 339 \text{ mm}^2/\text{m}$	BOTTOM R10 @ 200 $(392 \text{ mm}^2/\text{m})$ TOP R10 @ 200 $(392 \text{ mm}^2/\text{m})$

DATA SHEET	CALCULATIONS	OUTPUT
20	**Deflection** Basic minimum effective depth of 3.64 m freely supported span = 182 mm Multiplier for f_y = 250 N/mm^2 and M/bd^2 = 0.69 is 0.5 ∴ Minimum allowable effective depth = 182×0.5 = 91 mm	∴ Deflection OK
37	**Cracking** Since $h < 250$ mm, criterion is that bar spacing must not exceed 300 mm	∴ Crack widths OK

DATA SHEET	CALCULATIONS	OUTPUT
	Roof of motor room (see Drawings 10 and 11) Two-way freely supported slab with corners not restrained $f_{cu} = 30$ N/mm^2; $f_y = 250$ N/mm^2; $h = 175$ mm	

DATA SHEET	CALCULATIONS	OUTPUT
2	**Loading** Self-weight of 175 mm slab: 0.175×24.0 = 4.20 Screed, finish etc.: (say) = 0.96 Characteristic dead load = 5.16 kN/m^2 Characteristic imposed load = 0.75 kN/m^2 Total design load $= (5.16 \times 1.4) + (0.75 \times 1.6) = 8.42$ kN/m^2	$g_k = 5.16$ kN/m^2 $q_k = 0.75$ kN/m^2 $n = 8.42$ kN/m^2
44	**Ultimate bending moments** Since $l_x = 5.42$ m and $l_y = 6.16$ m, $l_y/l_x = 6.16/5.42 = 1.14$ and thus, for a slab freely supported on all sides with unrestrained corners, $a_x = 0.078$ and $a_y = 0.061$. Thus: For shorter span: UBM $= 0.078 \times 5.42^2 \times 8.42 = 19.2$ kNm/m For longer span: UBM $= 0.061 \times 5.42^2 \times 8.42 = 15.0$ kNm/m	
13 41	**Reinforcement** *For shorter span* $\dfrac{M}{bd^2} = \dfrac{19.2 \times 10^6}{10^3 \times 149^2} = 0.86$; hence $\varrho f_y = 1.05$ $A_{s\,req} = \dfrac{1.05 \times 10^3 \times 149}{250} = 626$ mm^2/m	$d = 175 - 20 - (\frac{1}{2} \times 12)$ $= 149$ mm BOTTOM R12 @ 150 (754 mm^2/m)
13 41 41	*For longer span* $\dfrac{M}{bd^2} = \dfrac{15.0 \times 10^6}{10^3 \times 137^2} = 0.80$; hence $\varrho f_y = 0.97$ $A_{s\,req} = \dfrac{0.97 \times 10^3 \times 137}{250} = 532$ mm^2/m Minimum permissible reinforcement $= 0.0024 \times 175 \times 10^3 = 420$ mm^2/m	$d = 175 - 20 - 12 -$ $(\frac{1}{2} \times 12) = 137$ mm BOTTOM R12 @ 200 (565 mm^2/m) $\therefore$ OK
20	**Deflection** Basic minimum effective depth for 5.42 m freely supported span = 271 mm Multiplier for $f_y = 250$ N/mm^2 and $M/bd^2 = 0.86$ is 0.5 $\therefore$ Minimum permissible effective depth $= 271 \times 0.5 = 136$ mm^2	$\therefore$ Deflection OK
37	**Cracking** Since $h < 250$ mm, criterion is that maximum spacing must not exceed $3d$	$\therefore$ Crack widths OK

DATA SHEET	CALCULATIONS	OUTPUT
	Upper-floor slab: flat-slab construction (see Drawings 12 and 13)	

Alternative design for upper floors as flat slabs using Code coefficients
$f_{cu} = 30$ N/mm^2; $f_y = 250$ N/mm^2; interior panels P1 and P2; 6.0 m $\times$ 5.0 m;
$h = 200$ mm; 2.5 m square drop 75 mm deep; column head $= 1.25$ m in diameter

Durability and fire resistance

DATA SHEET	CALCULATIONS	OUTPUT
42	Minimum cover for 'mild' exposure $= 20$ mm	Cover $= 20$ mm
62	Maximum fire resistance of 200 mm slab with 20 mm cover $= 2$ hours	$\therefore$ Fire resistance OK

Loading

Self-weight of 200 mm slab: 0.2×24 $= 4.80$
Allow for 75 mm drop panels: $= 0.38$
Finishes: floor (0.48) and ceiling (0.24) $= 0.72$
Partitions: 1 (minimum) plus 1 (additional) $= 2.00$

$\therefore$ Characteristic dead load $= 7.90$ kN/m^2 $g_k = 7.90$ kN/m^2

| 2 | Characteristic imposed load $= 5.00$ kN/m^2 | $q_k = 5.00$ kN/m^2 |

Thus design load $n = 1.4g_k + 1.6q_k = 11.06 + 8.0 = 19.06$ kN/m^2 $n = 19.06$ kN/m^2

Ultimate bending moments in interior panels P1
Longer span ($l_1 = 6.0$ m and $l_2 = 5.0$ m)

46
Total positive UBM on middle strip 2.5 m wide
$= 0.03195nl_1 l_2(l_1 - h_c/1.5)$
$= 0.03195 \times 19.06 \times 5.0 \times 6.0(6.0 - 1.25/1.5) = 94.4$ kN m
i.e. $94.4/2.5 = 37.8$ kN m/m width

46
Total negative UBM on middle strip 2.5 m wide
$= 0.01375nl_1 l_2(l_1 - h_c/1.5)$
$= 0.01375 \times 19.06 \times 5.0 \times 6.0(6.0 - 1.25/1.5) = 40.6$ kN m
i.e. $40.6/2.5 = 16.3$ kN m/m width

46
Total positive UBM on column strip 2.5 m wide
$= 0.03905nl_1 l_2(l_1 - h_c/1.5)$
$= 0.03905 \times 19.06 \times 5.0 \times 6.0(6.0 - 1.25/1.5) = 115.4$ kN m
i.e. $115.4/2.5 = 46.2$ kN m/m width

46
Total negative UBM on column strip 2.5 m wide
$= 0.04125nl_1 l_2(l_1 - h_c/1.5)$
$= 0.04125 \times 19.06 \times 5.0 \times 6.0(6.0 - 1.25/1.5) = 121.9$ kN m
i.e. $121.9/2.5 = 48.8$ kN m/m width

Shorter span ($l_1 = 5.0$ m and $l_2 = 6.0$ m)

46
Total positive UBM on middle strip 3.0 m wide
$= 0.03195nl_1 l_2(l_1 - h_c/1.5)$
$= 0.03195 \times 19.06 \times 5.0 \times 6.0(5.0 - 1.25/1.5) = 76.1$ kN m
However, since middle strip is 3.5 m wide, this must be increased by
$76.1 \times 0.5/3.0 = 12.7$ kN m; so
Total UBM $= 76.1 + 12.7 = 88.8$ kN m, i.e. $= 88.8/3.5 = 25.4$ kN m/m

46
Total negative UBM on middle strip 3.0 m wide
$= 0.01375nl_1 l_2(l_1 - h_c/1.5)$
$= 0.01375 \times 19.06 \times 5.0 \times 6.0(5.0 - 1.25/1.5) = 32.8$ kN m
However, since middle strip is 3.5 m wide, this must be increased by
$32.8 \times 0.5/3.0 = 5.5$ kN m; so
Total UBM $= 32.8 + 5.5 = 38.3$ kN m, i.e. $= 38.3/3.5 = 10.9$ kN m/m

DATA SHEET	CALCULATIONS	OUTPUT

46

Total positive UBM on column strip 3.0 m wide
$$= 0.03905 n l_1 l_2 (l_1 - h_c/1.5)$$
$$= 0.03905 \times 19.06 \times 5.0 \times 6.0(5.0 - 1.25/1.5) = 93.0 \text{ kN m}$$
However, since column strip is only 2.5 m wide, decrease this total by similar amount by which middle strip is increased; so
Total UBM = 93.0 − 12.7 = 80.3 kN m, i.e. = 80.3/2.5 = 32.1 kN m/m

46

Total negative UBM on column strip 3.0 m wide
$$= 0.04125 n l_1 l_2 (l_1 - h_c/1.5)$$
$$= 0.04125 \times 19.06 \times 5.0 \times 6.0(5.0 - 1.25/1.5) = 98.3 \text{ kN m}$$
However, since column strip is only 2.5 m wide, decrease this total by similar amount by which middle strip is increased; so
Total UBM = 98.3 − 5.5 = 92.8 kN m, i.e. = 92.8/2.5 = 37.1 kN m/m

UBM Summary

By making similar calculations for the remaining panels, the following table can be prepared for UBMs in kN m per metre width:

	Direction of short span		Direction of long span	
	Middle strip	Column strip	Middle strip	Column strip
At interior support	−10.9	−37.1	−16.3	−48.8
In interior span	25.4	32.1	37.8	46.2
At penultimate support	−12.5	−42.5	−18.6	−55.9
In end span	29.7	37.6	44.2	54.0
At edge of slab	−7.9	−27.0	−11.8	−35.5

Reinforcement in interior panels P1

Longer span

Middle strip at midspan:

d provided = 200 − 20 − 10 = 170 mm

Output: $d = 170$ mm

13

$$\frac{M}{bd^2} = \frac{37.8 \times 10^6}{10^3 \times 170^2} = 1.31 \text{ N/mm}^2; \text{ hence } \varrho f_y = 1.58$$

41

$$A_{s\,req} = \frac{1.58 \times 10^3 \times 170}{250} = 1078 \text{ mm}^2/\text{m width}$$

Output: BOTTOM R20 @ 225 (1396 mm²/m)

Middle strip over supports:

d provided = 200 − 20 − 20 − 10 = 150 mm

Output: $d = 150$ mm

13

$$\frac{M}{bd^2} = \frac{16.3 \times 10^6}{10^3 \times 150^2} = 0.72 \text{ N/mm}^2; \text{ hence } \varrho f_y = 0.88$$

41

$$A_{s\,req} = \frac{0.88 \times 10^3 \times 150}{250} = 526 \text{ mm}^2/\text{m width}$$

Output: TOP R20 @ 450 (698 mm²/m)

Column strip at midspan:

d provided = 200 − 20 − 10 = 170 mm

Output: $d = 170$ mm

13

$$\frac{M}{bd^2} = \frac{46.2 \times 10^6}{10^3 \times 170^2} = 1.60 \text{ N/mm}^2; \text{ hence } \varrho f_y = 1.96$$

41

$$A_{s\,req} = \frac{1.96 \times 10^3 \times 170}{250} = 1335 \text{ mm}^2/\text{m width}$$

Output: BOTTOM R16 and R20 alternately @ 175 (1472 mm²/m)

DATA SHEET	CALCULATIONS	OUTPUT
	Column strip over supports:	
	d provided $= 275 - 20 - 20 - 10 = 225$ mm	$d = 225$ mm
13	$\dfrac{M}{bd^2} = \dfrac{48.8 \times 10^6}{10^3 \times 225^2} = 0.96$ N/mm^2; hence $\varrho f_y = 1.17$	
41	$A_{s\,req} = \dfrac{1.17 \times 10^3 \times 225}{250} = 1050$ mm^2/m width	TOP Centre: R16 and R20 alternately @ 175 (1472 mm^2/m) Edges: R20 @ 350 (898 mm^2/m)
	Shorter span	
	Middle strip at midspan:	
	d provided $= 200 - 20 - 20 - 10 = 150$ mm	$d = 150$ mm
13	$\dfrac{M}{bd^2} = \dfrac{25.4 \times 10^6}{10^3 \times 150^2} = 1.13$ N/mm^2; hence $\varrho f_y = 1.37$	
41	$A_{s\,req} = \dfrac{1.37 \times 10^3 \times 150}{250} = 820$ mm^2/m width	BOTTOM R16 @ 200 (1005 mm^2/m)
	Middle strip over supports:	
	d provided $= 200 - 20 - 20 - 10 = 170$ mm	$d = 170$ mm
13	$\dfrac{M}{bd^2} = \dfrac{10.9 \times 10^6}{10^3 \times 170^2} = 0.38$ N/mm^2; hence $\varrho f_y = 0.46$	
41	$A_{s\,req} = \dfrac{0.46 \times 10^3 \times 170}{250} = 311$ mm^2/m width	TOP R16 @ 400 (503 mm^2/m)
	A minimum area of $0.0024 \times 1000 \times 200 = 480$ mm^2/m must be provided	
	Column strip at midspan:	
	d provided $= 200 - 20 - 20 - 10 = 150$ mm	$d = 150$ mm
13	$\dfrac{M}{bd^2} = \dfrac{32.1 \times 10^6}{10^3 \times 150^2} = 1.43$ N/mm^2; hence $\varrho f_y = 1.74$	
41	$A_{s\,req} = \dfrac{1.74 \times 10^3 \times 150}{250} = 1043$ mm^2/m width	BOTTOM R20 @ 250 (1257 mm^2/m)
	Column strip over supports:	
	d provided $= 275 - 20 - 10 = 245$ mm	$d = 245$ mm
13	$\dfrac{M}{bd^2} = \dfrac{37.1 \times 10^6}{10^3 \times 245^2} = 0.62$ N/mm^2; hence $\varrho f_y = 0.75$	
41	$A_{s\,req} = \dfrac{0.75 \times 10^3 \times 245}{250} = 733$ mm^2/m width	TOP Centre: R16 and R20 alternately @ 250 (1030 mm^2/m) Edges: R20 @ 500 (628 mm^2/m)

DATA SHEET	CALCULATIONS	OUTPUT

Reinforcement summary

Similar calculations for the remaining panels lead to the following areas of reinforcement (in mm²/m) and bar arrangements being required:

In direction of short span:

		In middle strip			In column strip	
At internal support	480	Provide 16 @ 400	(503)	480	Provide alternate 16 and 20 @ 250 over central 1.25 m width and 20 @ 500 elsewhere	(1030) (628)
In internal span	820	Provide 16 @ 200	(1005)	820	Provide 20 @ 250	(1257)
At penultimate support	480	Provide 16 @ 400	(503)	480	Provide 20 @ 250 over central 1.25 m width and 20 @ 500 elsewhere	(1257) (628)
In end span	820	Provide 16 @ 200	(1005)	820	Provide 20 @ 250	(1257)
At end support	480	Provide 16 @ 400	(503)	480	Provide 20 @ 250 over central 1.25 m width and 16 @ 500 elsewhere	(1257) (402)

In direction of long span:

		In middle strip			In column strip	
At internal support	526	Provide 20 @ 450	(698)	1050	Provide alternate 16 and 20 @ 175 over central 1.25 m width and 20 @ 350 elsewhere	(1472) (898)
In internal span	1078	Provide 20 @ 225	(1396)	1335	Provide alternate 16 and 20 @ 175	(1472)
At penultimate support	600	Provide 20 @ 450	(698)	1211	Provide 20 @ 175 over central 1.25 m width and 20 @ 350 elsewhere	(1795) (898)
In end span	1273	Provide 20 @ 225	(1396)	1580	Provide 20 @ 175	(1795)
At end support	480	Provide 20 @ 450	(698)	764	Provide alternate 16 and 20 @ 175 over central 1.25 m width and 16 @ 350 elsewhere	(1472) (574)

Punching shear

Around head of internal column

$V_{eff} = 1.15\, V_t = 1.15 \times 5.0 \times 6.0 \times 19.06 = 657.6$ kN

At critical plane 1.5d from edge of column head,
  perimeter length $= 4 \times (1250 + 3 \times 235) = 7820$ mm

∴ Maximum shear stress $= 657.6 \times 10^3/(7820 \times 235) = 0.36$ N/mm²

Average A_s in each direction over column head $= (1030 + 628 + 1472 + 898)/4$
  $= 1007$ mm²/m

Thus $\varrho = 1007/(235 \times 10^3) = 0.0042$, and with $f_{cu} = 30$ N/mm², $v_c = 0.58$ N/mm²

Around edge of drop panel

Panel area within perimeter $= (2.5 + 3 \times 0.16)^2 = 8.88$ m²

∴ Loaded area supported by column $= 6.0 \times 5.0 - 8.88 = 21.12$ m²

$V_{eff} = 1.15\, V_t = 1.15 \times 21.12 \times 19.06 = 463$ kN

At critical plane 1.5d from edge of drop panel,
  perimeter length $= 4 \times (2500 + 3 \times 160) = 11\,920$ mm

∴ Maximum shear stress $= 463 \times 10^3/(11\,920 \times 160) = 0.24$ N/mm²

15 *(data sheet)*

∴ No shearing reinforcement required round column head

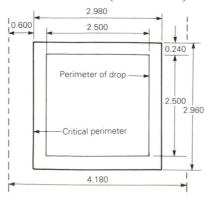

DATA SHEET	CALCULATIONS	OUTPUT
15	Slab width over which proportion of reinforcement controls shearing resistance $= 2.5 + (3 \times 0.16) + 6h = 4.18$ m Average A_s in each direction in column strips $= 1007$ mm^2/m (as above) Average A_s in each direction in middle strips $= (503 + 698)/2 = 600$ mm^2/m Average A_s over width concerned $= [(2.5 \times 1007) + (1.68 \times 600)]/4.18$ $= 843$ mm^2/m Thus $\varrho = 843/(160 \times 10^3) = 0.0053$, and with $f_{cu} = 30$ N/mm^2, $v_c = 0.68$ N/mm^2 *At edge column* Since $V_{eff} = 1.4\ V_t$ here, column load will be approximately $0.5 \times 657.6 \times 1.4/ 1.15 = 400.3$ kN. If the length of the critical perimeter is one-half of that at an internal column head, the maximum punching shear stress will be approximately $400.3/(3910 \times 235) = 0.44$ N/mm^2. Since the proportion of main steel passing over the column head is similar to that at an internal column, V_c will be about 0.58 N/mm^2 and thus no shearing reinforcement is needed. By inspection it is clear that special shearing reinforcement is also unnecessary around the edge of the drop panel, and a similar situation applies to both critical planes adjoining the corner columns	∴ No shearing reinforcement required round drop panel ∴ Shear OK
20	**Deflection** Critical situation occurs in penultimate longer span Moment on total panel width is $(44.2 \times 2.5) + (54.0 \times 2.5) = 245.5$ kN With basic span/depth ratio for 6 m continuous span $= 26$, modifier for $M/bd^2 = 245.5 \times 10^6/(5 \times 10^3 \times 170^2) = 1.70$ and $f_y = 250$ N/mm^2 is 1.58 Thus maximum span/effective-depth ratio is $26 \times 1.58 = 41.0$ ∴ Minimum effective depth $= 6000/41 = 147$ mm	∴ Deflection OK
	Cracking Since h is less than 250 mm, except where drop panels occur, criterion is that clear spacing must not exceed $3d = 450$ mm. Where drop panels occur, maximum spacing $= 3 \times 225 = 675$ mm, provided $A_s/bd < 0.003$; i.e. $A_s < 0.003 \times 10^3 \times 225 = 675$ mm^2/m. The only situations where this condition is not met is at the edges of the column strip over the internal and penultimate supports in the direction of the longer span. Here, since $\varrho_1 = 898/(1000 \times 225) = 0.004$ and $f_y = 250$ N/mm^2, maximum clear spacing $= 300/0.4 = 750$ mm	∴ Cracking OK
	Robustness Lateral and longitudinal internal ties. Since $5 \times$ clear storey height exceeds greatest distance between centres of vertical load-bearing members, $l = 6.0$ m. Then, since $(g_k + q_k)l = 12.9 \times 6.0 = 77.4$ kN/m exceeds 37.5 kN/m, the tie force to be considered, where $F_e = 44$ kN, $= 0.0267(g_k + q_k)\ lF_t = 0.0267 \times 77.4 \times 44 = 90.8$ kN/m width $A_{s\,req} = \dfrac{90.8 \times 10^3}{250} = 364$ mm^2/m	∴ Reinforcement for column strip in bottom of slab sufficient

DATA SHEET	CALCULATIONS	OUTPUT
	Interior column B3	
	Design as axially loaded column supporting approximately symmetrical beam arrangement $f_{cu} = 30$ N/mm^2; $f_y = 250$ N/mm^2	
42 62	**Durability and fire resistance** Minimum cover to all steel for 'moderate' exposure $= 35$ mm Fire resistance of 300×300 column $= 2$ hours	Cover to main bars $= 50$ mm $\therefore$ Fire resistance OK

Loading

	Unfactored load (kN)	Design load (kN)
Fifth floor to roof		
Roof slab, finishes etc.:	$30.0 \times 5.04 = 151.2 \times 1.4 =$	211.7
Roof beams etc.:	$11.0 \times 1.73 = 19.0 \times 1.4 =$	26.6
Self-weight of column:	$2.4 \times 3.29 = 7.9 \times 1.4 =$	11.1
Imposed load on roof:	$30.0 \times 1.50 = 45.0 \times 1.6 =$	72.0
Loads above fifth floor	223.1	321.4
Fourth to fifth floor		
Fifth-floor slab, finishes:	$30.0 \times 4.12 = 123.6 \times 1.4 =$	173.1
Secondary beams:	$12.0 \times 1.50 = 18.0 \times 1.4 =$	25.2
Main beams:	$5.0 \times 2.40 = 12.0 \times 1.4 =$	16.8
Self-weight of column:	$2.16 \times 3.10 = 6.7 \times 1.4 =$	9.4
Additional imposed load: $30.0[(5.0 \times 0.9) - (1.5 \times 0.1)]$	$= 130.5 \times 1.6 =$	208.8
Loads above fourth floor	513.9	754.7
Third to fourth floor		
Fourth-floor dead load (as fifth floor)	$= 153.6 \times 1.4 =$	215.1
Self-weight of column:	$2.16 \times 3.10 = 6.7 \times 1.4 =$	9.4
Additional imposed load: $30.0[(5.0 \times 0.7) - (1.5 \times 0.1)]$	$= 100.5 \times 1.6 =$	160.8
Loads above third floor	774.7	1140.0
Second to third floor		
Third-floor dead load (as fourth floor)	$= 153.6 \times 1.4 =$	215.1
Self-weight of column:	$2.16 \times 3.10 = 6.7 \times 1.4 =$	9.4
Additional imposed load: $30.0[(5.0 \times 0.5) - (1.5 \times 0.1)]$	$= 70.5 \times 1.6 =$	112.8
Loads above second floor	1005.5	1477.3
First to second floor		
Second-floor dead load (as third floor)	$= 153.6 \times 1.4 =$	215.1
Self-weight of column:	$2.94 \times 3.10 = 9.1 \times 1.4 =$	12.8
Additional imposed load: $30.0[(5.0 \times 0.3) - (1.5 \times 0.1)]$	$= 40.5 \times 1.6 =$	64.8
Loads above first floor	1208.7	1770.0

DATA SHEET	CALCULATIONS	OUTPUT
	Ground to first floor	

Ground to first floor

First-floor dead load (as second floor) $= 153.6 \times 1.4 =$ 215.1
Self-weight of column: $2.94 \times 3.85 = 11.3 \times 1.4 =$ 15.9
Additional imposed load: $30.0 \times 5.0 \times 0.6 = 90.0 \times 1.6 =$ 144.0

Loads above ground floor   1463.6   2145.0

Basement to ground floor

Ground-floor dead load: $30.0 \times 6.8 = 204.0 \times 1.4 =$ 285.6
Self-weight of column: $3.84 \times 2.55 = 9.8 \times 1.4 =$ 13.7
Imposed load (no reduction): $30.0 \times 5.0 = 150.0 \times 1.6 =$ 240.0

Total loads above basement   1827.4   2684.3

Reinforcement

Fifth floor to roof

Total area of concrete $= (4 \times 225 \times 130) + 130^2 = 133\,900$ mm^2
Minimum reinforcement necessary (0.4%) $= 536$ mm^2
Provide 8R16 as shown on sketch (1.2%)

47   Load-carrying capacity of section $= 133\,900 \times 12.39 = 1659$ kN

41   Actual design load to be supported $= 321.4$ kN

←225→ 130 ←225→
8R16
(1608 mm^2)

Fourth to fifth floor

Design load to be supported $= 754.7$ kN
300 mm square column reinforced with 0.4% (i.e. minimum) steel will support
$90\,000 \times 11.13 = 1002$ kN

41 $A_{sc\,req} = \dfrac{300^2}{250} = 360$ mm^2/m

Assuming $l_e = l_0$, slenderness $= \dfrac{3100}{300} = 10.3$

300 / 300
4R16
(804 mm^2/m)

∴ Slenderness OK

Third to fourth floor

Design load to be supported $= 1140.0$ kN

47   Load per unit area $= \dfrac{1140.0 \times 10^3}{300^2} = 12.67$ N/mm^2

41   Provide 1.4% steel: $A_{sc\,req} = \dfrac{1.4 \times 300^2}{100} = 1260$ mm^2

Slenderness $= \dfrac{3100}{300} = 10.3$, assuming $l_e = l_0$

300 / 300
4R20
(1257 mm^2)

∴ Slenderness OK

Second to third floor

Design load to be supported $= 1477.3$ kN

47   Load per unit area $= \dfrac{1477.3 \times 10^3}{300^2} = 16.41$ N/mm^2

41   Provide 3.8% steel: $A_{sc\,req} = \dfrac{3.8 \times 300^2}{100} = 3420$ mm^2

Slenderness $= \dfrac{3100}{300} = 10.3$

Design load to be supported $= 1770.0$ kN

300 / 300
8R25
(3927 mm^2)

∴ Slenderness OK

219

DATA SHEET	CALCULATIONS	OUTPUT
	First to second floor	
47	Load per unit area $= \dfrac{1770.0 \times 10^3}{350^2} = 14.45$ N/mm^2	
41	Provide 2.6% steel: $A_{sc\,req} = \dfrac{2.6 \times 350^2}{100} = 3185$ mm^2	4R20 + 4R25 (3220 mm^2)
	Slenderness $= \dfrac{3100}{350} = 8.9$	$\therefore$ Slenderness OK
	Ground to first floor	
	Design load to be supported $= 2145.0$ kN	
47	Load per unit area $= \dfrac{2145.0 \times 10^3}{350^2} = 17.51$ N/mm^2	
41	Provide 4.5% steel: $A_{sc\,req} = \dfrac{4.5 \times 350^2}{100} = 5513$ mm^2	8R32 (6434 mm^2)
	Slenderness $= \dfrac{3850}{350} = 11.0$, assuming $l_e = l_0$	$\therefore$ Slenderness OK
	Basement to ground floor	
	Design load to be supported $= 2692.0$ kN	
47	Load per unit area $= \dfrac{2692.0 \times 10^3}{400^2} = 16.83$ N/mm^2	
41	Provide 4.1% steel: $A_{sc\,req} = \dfrac{4.1 \times 400^2}{100} = 6560$ mm^2	4R40 + 4R25 (6990 mm^2)
	Slenderness $= \dfrac{2650}{400} = 6.6$, assuming $l_e = l_0$	$\therefore$ Slenderness OK

220

DATA SHEET	CALCULATIONS	OUTPUT
	Base to column B3 (see Drawing 14) $f_{cu} = 30$ N/mm^2; $f_y = 250$ N/mm^2	
42	**Durability** Nominal minimum cover for concrete in contact with non-aggressive soil (BS8110) = 35 mm Nominal minimum cover for bases (*Joint Institutions Design Manual*) = 50 mm	Cover = 50 mm

	Loading
	
	                        Unfactored Factored
	                        (i.e. 'service') (i.e. 'design')
	From column 1827.4 2684.3
	Self-weight of base: $3.15^2 \times 0.6 \times 24 = $ 142.9 $\times$ 1.4 = 200.1
	
	                          1970.3 kN 2884.4 kN
	
	Permissible bearing pressure on ground = 200 kN/m^2
	
	Size of base required = $\sqrt{[(1970.3 \times 10^3)/(200 \times 10^3)]}$ = 3.14 m $\times$ 3.14 m
	
	Provide base 3.15 m square and 600 mm thick
	
	'Design' pressure = $\dfrac{2884.4}{3.15^2}$ = 290.7 kN/m^2
		Square base
		3.15 m $\times$ 3.15 m
		$\times$ 0.6 m deep

13
 	
41
	Ultimate bending moment
	
	With 50 mm cover to 25 mm bars,
	   minimum $d = 600 - 50 - 25 - 12.5 = 512.5$ mm
	Maximum ultimate applied moment = $290.7 \times 3.15 \times (3.15 - 0.4)^2/8$
	   = 866 kN m
	Resistance moment provided by concrete
	   $0.156bd^2 f_{cu} = 0.156 \times 3150 \times 512.5^2 \times 30 = 3872$ kN m
	   which is more than 4 times that required
	
	$M/bd^2 = 866 \times 10^6/(3150 \times 512.5^2) = 1.05$ and $f_y = 250$ N/mm^2;
	   hence $\varrho f_y = 1.265$
	
	$A_{s\,req} = 1.265 \times 3150 \times 512.5/250 = 8168$ mm^2
	
	Since the base dimension (= 3.15 m) exceeds 1.5$(b + 3d)$(= 2.91 m),
	   two-thirds of this reinforcement must be concentrated within a strip
	   $b + 3d = 0.4 + (3 \times 0.5125) = 1.938$ m wide centred beneath column
	
	Thus within central strip 1.938 m in width provide $0.67 \times 8168/1.938$
	   = 2810 mm^2 per m width, and at edges provide $0.33 \times 8168/(3.15 - 1.938)$
	   = 2247 mm^2 per m
	
	In practice, rather than respace the (only 2) bars in each edge strip, it would be sensible to adopt the closer spacing throughout the entire base width
		$d = 512.5$ mm
		
		∴ Resistance of concrete OK
		
		R25 @ 175
		R25 @ 215
		
		17R25 each way
		(8345 mm^2)

15	**Punching shear** Since $\varrho = 8345/(3150 \times 512.5) = 0.0052$ and $f_{cu} = 30$ N/mm^2, $v_c = 0.539$ N/mm^2 *At distance 1.5d from column face* Length of one side of punching shear plane = $0.4 + (2 \times 1.5 \times 0.5125)$ = 1.9375 m ∴ Force causing punching = $290.7 \times (3.15^2 - 1.9375^2) = 1793$ kN ∴ Punching shear stress = $1793 \times 10^3/(4 \times 1937.5 \times 512.5) = 0.451$ N/mm^2 Ratio of applied stress to resistance provided by concrete = $0.451/0.539 = 0.837$	∴ Punching shear OK

DATA SHEET	CALCULATIONS	OUTPUT
	At distance d from column face	
	Length of one side of punching shear plane = 0.4 + (2 × 0.5125) = 1.425 m ∴ Force causing punching = 290.7 × (3.15² − 1.425²) = 2294 kN ∴ Punching shear stress = 2294 × 10³/(4 × 1425 × 512.5) = 0.785 N/mm² Resistance provided by concrete (with enhancement) = 0.539 × 1.5 = 0.808 N/mm² Ratio of applied stress to resistance provided by concrete = 0.785/0.808 = 0.972	∴ Punching shear OK
	At distance 0.9d from column face	

	Length of one side of punching shear plane = 0.4 + (2 × 0.9 × 0.5125)
	     = 1.3225 m
	 ∴ Force causing punching = 290.7 × (3.15² − 1.3225²) = 2376 kN
	 ∴ Punching shear stress = 2376 × 10³/(4 × 1322.5 × 512.5) = 0.876 N/mm²
	 Resistance provided by concrete (with enhancement) = 0.539 × 1.5/0.9
	     = 0.898 N/mm²
	 Ratio of applied stress to resistance provided by concrete = 0.876/0.898 = 0.976
		∴ Punching shear OK

	At distance 0.8d from column face	
	Length of one side of punching shear plane = 0.4 + (2 × 0.8 × 0.5125) = 1.22 m ∴ Force causing punching = 290.7 × (3.15² − 1.22²) = 2451.8 kN ∴ Punching shear stress = 2451.8 × 10³/(4 × 1220 × 512.5) = 0.98 N/mm² Resistance provided by concrete (with enhancement) = 0.539 × 1.5/0.8 = 1.01 N/mm² Ratio of applied stress to resistance provided by concrete = 0.981/1.01 = 0.970	∴ Punching shear OK
	Thus the critical plane for punching is at 0.9d from the column face	
	Punching shear at column face = 2884.4 × 10³/(4 × 400 × 512.5) = 3.52 N/mm² Maximum permissible value = 0.8 √f_cu = 0.8 √30 = 4.38 N/mm²	∴ Punching shear OK
	Direct shear	
	At plane 1.5d from column face	
	Direct shearing force = 290.7 × 3.15 × [(0.5 × 3.15) − 0.2 − (1.5 × 0.5125)] = 555.2 kN Applied shearing stress = 555.2 × 10³/(3150 × 512.5) = 0.344 N/mm² Shear resistance provided by concrete alone = 0.539 N/mm²	∴ Direct shear OK
	Anchorage bond	

	Minimum bond length required for bending with 25 mm bars = 890 mm from
	     column face, i.e. 890 + 200 = 1090 mm from centre-line of column
	
	 Minimum bond length required from critical punching-shear plane is
	     d = 512.5 mm,
	     i.e. 512.5 + (0.9 × 512.5) + 200 = 1174 mm from centre-line of column
	 Maximum bond length available = (0.5 × 3.15) − 0.05 = 1525 mm
		∴ Anchorage bond OK

	Cracking	
	Clear distance between bars = 175 − 25 = 150 mm Maximum permissible value = 300 mm	∴ Crack widths OK

DATA SHEET	CALCULATIONS	OUTPUT
	Exterior column D3 (see Drawing 15)	
	Design as column subjected to uniaxial bending and direct thrust	
	$f_{cu} = 30$ N/mm^2; $f_y = 250$ N/mm^2	

Durability and fire resistance

42	Minimum cover to all steel for moderate exposure = 35 mm	Cover to main bars = 50 mm
62	Fire resistance of 300 × 300 column = 2 hours	∴ Fire resistance OK

Loading

	Unfactored load (kN)	Design load (kN)

Fourth to fifth floor

Fifth-floor slab and finishes:	6.0 × 2.5 × 4.12 =	61.8 × 1.4 =	86.5
From parapet: 6.0[(0.75 × 0.15) + (0.45 × 0.1)]24.0 =		22.7 × 1.4 =	31.8
Secondary beam:	½ × 5.75 × 1.50 =	4.3 × 1.4 =	6.0
Main beam:	½ × 4.56 × 2.40 =	5.5 × 1.4 =	7.7
Fascia beam:	5.70 × 2.40 =	13.7 × 1.4 =	19.2
Imposed load on fifth floor:	6.0 × 2.5 × 5.0 =	75.0 × 1.6 =	120.0

Loads at soffit of fifth floor = 183.0   271.2

Maximum design load = 271.2 kN
Minimum design load = 183.0 − 75.0 = 108.0 kN

Third to fourth floor

Fourth-floor dead loads (as fifth floor):

	61.8+4.3+5.5+13.7 =	85.3 × 1.4 =	119.4
Wall, sill etc. (say)	=	41.0 × 1.4 =	57.4
Window:	0.47 × 6.0 =	2.8 × 1.4 =	4.0
Self-weight of column:	3.6 × 0.3^2 × 24.0 =	7.8 × 1.4 =	10.9
Additional imposed load:	(1.8 − 1.0) × 75.0 =	60.0 × 1.6 =	96.0

Loads at soffit of fourth floor = 379.9   558.9

Maximum design load = 558.9 kN
Minimum design load = 379.9 − 135.0 = 244.9 kN

Second to third floor

Third-floor dead loads (as fourth floor):

	85.3+41.0+2.8+7.8 =	136.9 × 1.4 =	191.7
Additional imposed load:	(2.4 − 1.8) × 75.0 =	45.0 × 1.6 =	72.0

Loads at soffit of third floor = 561.8   822.6

Maximum design load = 822.6 kN
Minimum design load = 561.8 − 180.0 = 381.8 kN

First to second floor

Second-floor dead loads (as fourth floor)	=	136.9 × 1.4 =	191.7
Additional imposed load:	(2.8 − 2.4) × 75.0 =	30.0 × 1.6 =	48.0

Loads at soffit of second floor = 728.7   1062.3

Maximum design load = 1062.3 kN
Minimum design load = 728.7 − 210.0 = 518.7 kN

DATA SHEET	CALCULATIONS	OUTPUT
	Ground to first floor	

Ground to first floor

First-floor dead loads (as fifth floor)	$= 85.3 \times 1.4 =$	119.4
Canopy dead load:	$6.0 \times 1.0 \times 2.1 = 12.6 \times 1.4 =$	17.6
Canopy imposed load:	$6.0 \times 1.0 \times 1.5 = 9.0 \times 1.6 =$	14.4
Windows:	$6.0 \times 0.76 = 4.6 \times 1.4 =$	6.4
Self-weight of column:	$3.6 \times 0.3 \times 0.5 \times 24.0 = 13.0 \times 1.4 =$	18.2
Additional imposed load:	$(3.0 - 2.8) \times 75.0 = 15.0 \times 1.6 =$	24.0

Loads at soffit of first floor   868.2   1262.3

Maximum design load = 1262.3 kN
Minimum design load = 868.2 − 225.0 = 643.2 kN

Ultimate bending moments
At fifth-floor level

Maximum fixing moment at support of fifth-floor beam due to:

48
Uniform load: $\frac{1}{12} \times 5.0 \times 5.0[(3.43 \times 1.4) + (1.25 \times 1.6)] =$ 14.2 kN m
Concentrated load: $\frac{1}{8} \times 5.0[(73.5 \times 1.4) + (83.6 \times 1.6)] =$ 148.0 kN m

162.2 kN m

49
Moment of inertia of column section $= \frac{1}{12}h^4 = \frac{1}{12} \times 300^4 = 0.68 \times 10^9 \text{mm}^4$
Length of column $= 3.6 \times 10^3 \text{mm}$

∴ Stiffness of column $= 0.68 \times 10^9/3.6 \times 10^3 = 0.188 \times 10^6 \text{ mm}^3$

49
With $h_c/h = 100/500 = 0.2$ and
$b_w/b = 250/1100 = 0.23$, $K = 0.034$

∴ Moment of inertia of beam section
$= 0.034 \times 1100 \times 500^3 = 4.68 \times 10^9 \text{ mm}^4$

Span of beam $= 5.0 \times 10^3$ mm

∴ Stiffness of beam $= 4.68 \times 10^9/5.0 \times 10^3$
$= 0.935 \times 10^6 \text{ mm}^3$
$b = (\frac{1}{5} \times 5.0 \times 0.85) + 0.25 = 1.1$ m

Thus final moment transferred to column

$$= 162.2 \times \frac{0.188}{(\frac{1}{2} \times 0.935) + 0.188} = 46.5 \text{ kN m}$$

∴ Moment at soffit of beam $= 46.5 - (46.5 + 36.2)\frac{0.25}{3.60} = 40.8 \text{ kN m}$

At fourth-floor level

Maximum fixing moment at support of fourth-floor beam (as fifth-floor beam) = 162.2 kN m
Stiffness of each column $= 0.188 \times 10^6 \text{ mm}^3$ (as fifth floor)
Stiffness of beam $= 0.935 \times 10^6 \text{ mm}^3$ (as fifth floor)
Thus final moment transferred to each column

$$= 162.2 \times \frac{0.188}{(2 \times 0.188) + (\frac{1}{2} \times 0.935)} = 36.2 \text{ kN m}$$

∴ Final moment in beam at support $= 2 \times 36.2 = 72.4$ kN m

∴ Moment at soffit of beam $= 36.2 \times \frac{1.80 - 0.25}{1.80} = 31.1$ kN m

At third-floor level

UBM in column at soffit of beam (as fourth floor) = 31.1 kN m

At second-floor level

UBM in column at soffit of beam (as fourth floor) = 31.1 kN m

224

DATA SHEET	CALCULATIONS	OUTPUT

At first-floor level

Check slenderness (transverse to plane of bending) of lower column

49

For flanged edge beam to one side of column, with $h_f/h = 100/400 = 0.25$ and $b_w/b = 200/620 = 0.32$, $K = 0.043$; hence moment of inertia $= 0.043 \times 620 \times 400^3 = 1.71 \times 10^9$ mm^4 and stiffness $= 0.284 \times 10^6$ mm^3

Stiffness of upper column $= 300^4/12 \times 3600 = 0.188 \times 10^6$ mm^3
Stiffness of lower column $= 500 \times 300^3/12 \times 4350 = 0.259 \times 10^6$ mm^3
Thus at first floor, $\alpha_{c1} = (0.188 + 0.259)/(0.284 \times 2) = 0.78 = \alpha_{c\,min}$
At ground-floor level, $\alpha_{c2} \not< 1$
Hence $l_{e\,min} = l_0(0.85 + 0.05\alpha_{c\,min}) = 3.85[0.85 + (0.05 \times 0.78)] = 3.42$ m
Since $l_e/b = 3.42/0.3 = 11.4$, member acts as 'short' column

49

Moment of inertia of lower column (in plane of bending)
$= \frac{1}{12}bh^3 = \frac{1}{12} \times 300 \times 500^3 = 3.13 \times 10^9$ mm^4
Stiffness $= 3.13 \times 10^9/4.35 \times 10^3 = 0.718 \times 10^6$ mm^3
Stiffness of upper column $= 0.188 \times 10^6$ mm^3 (as fifth floor)
Stiffness of beam $= 0.935 \times 10^6$ mm^3 (as fifth floor)
Maximum fixing moment at support of first-floor beam $= 162.2$ kN m
Thus final moment transferred to upper column

$$= 162.2 \times \frac{0.188}{0.188 + 0.718 + (\frac{1}{2} \times 0.935)} = 22.2 \text{ kN m}$$

and final moment transferred to lower column

$$= 162.2 \times \frac{0.718}{0.188 + 0.718 + 0.468} = 84.8 \text{ kN m}$$

$\therefore$ Final moment in beam at support $= 84.8 + 22.2 = 107.0$ kN m
$\therefore$ Moment at soffit of beam $= 84.8 \times (2.90 - 0.25)/2.90 = 77.5$ kN m

Summary diagram of ultimate bending moments due to frame action on column D3

DATA SHEET	CALCULATIONS	OUTPUT
	Vertical reinforcement	
	Fifth-floor soffit level	
51	Since $\dfrac{M}{bh^2 f_{cu}} = \dfrac{40.8 \times 10^6}{300^3 \times 30} = 0.050$ and $\dfrac{N}{bhf_{cu}} = \dfrac{271.2 \times 10^3}{300^2 \times 30} = 0.100$,	240 · 300 · 300
41	$\varrho_1/f_{cu} = 0.0002$, and hence $A_{sc\ req} = 0.0002 \times 30 \times 300^2 = 540$ mm²	4R16
	Minimum permissible proportion of 0.4% $= 0.004 \times 300^2 = 360$ mm²	(804 mm²)
	Fourth-floor soffit level	
51	Since $\dfrac{M}{bh^2 f_{cu}} = \dfrac{31.1 \times 10^6}{300^3 \times 30} = 0.038$ and $\dfrac{N}{bhf_{cu}} = \dfrac{558.9 \times 10^3}{300^2 \times 30} = 0.207$,	
41	$\varrho_1/f_{cu} = $ nil	4R16
	Minimum permissible proportion of 0.4% $= 360$ mm²	(804 mm²) arranged as above
	Third-floor soffit level	
51	Since $\dfrac{M}{bh^2 f_{cu}} = \dfrac{31.1 \times 10^6}{300^3 \times 30} = 0.038$ and $\dfrac{N}{bhf_{cu}} = \dfrac{822.6 \times 10^3}{300^2 \times 30} = 0.305$,	
41	$\varrho_1/f_{cu} = $ nil	4R16
	Minimum permissible proportion of 0.4% $= 360$ mm²	(804 mm²) arranged as above
	Second-floor soffit level	
	Since $\dfrac{M}{bh^2 f_{cu}} = \dfrac{31.1 \times 10^6}{300^3 \times 30} = 0.038$ and $\dfrac{N}{bhf_{cu}} = \dfrac{1062.3 \times 10^3}{300^2 \times 30} = 0.393$,	4R16
41	$\varrho_1/f_{cu} = 0.00025$, and hence $A_{sc\ req} = 0.00025 \times 30 \times 300^2 = 675$ mm²	(804 mm²) arranged as above
	First-floor soffit level	440 · 500 · 300
51	Since $\dfrac{M}{bh^2 f_{cu}} = \dfrac{77.5 \times 10^6}{500^2 \times 300 \times 30} = 0.034$ and $\dfrac{N}{bhf_{cu}} = \dfrac{1262.3 \times 10^3}{500 \times 300 \times 30} = 0.281$,	
	$\varrho_1/f_{cu} = $ nil	6R16
41	Minimum permissible proportion of 0.4% $= 0.004 \times 500 \times 300 = 600$ mm²	(1206 mm²) arranged as shown

DATA SHEET	CALCULATIONS	OUTPUT
	Exterior column C1 (see Drawing 16)	
	Design as column subjected to biaxial bending and direct thrust $f_{cu} = 30$ N/mm^2; $f_y = 250$ N/mm^2	
	Durability and fire resistance	
42	Minimum cover for 'moderate' exposure = 35 mm to all steel	∴ Cover to main bars = 63 mm
62	Fire resistance of 300×300 column = 2 hours	∴ Fire resistance OK

Loading

	Char. dead loads (kN)	Char. imposed loads (kN)
Fifth floor to roof		
100 mm roof slab, finishes etc.: $2.5 \times 2.8 \times 3.36$ =	23.5	
150 mm roof slab, finishes etc.: $2.5 \times 2.8 \times 5.04$ =	35.3	
Guard rail, kerb etc.: 8.0×1.05 (say) =	8.4	
Fascia beam, eaves etc.: $2.67 \times 0.156 \times 24.0$ =	10.0	
Roof beam: $2.67 \times 0.13 \times 0.60 \times 24.0$ =	5.0	
Edge beam: $4.70 \times 0.45 \times 0.225 \times 24.0$ =	11.4	
Imposed load on roof: $5.0 \times 2.8 \times 1.50$ =		21.0
Characteristic loads at soffit of roof =	93.6	21.0

Minimum design load = 93.6 kN
Maximum design load (with wind) = $(93.6 + 21.0) \times 1.2 = 137.5$ kN
Maximum design load (without wind) = $(93.6 \times 1.4) + (21.0 \times 1.6) = 164.6$ kN

Fourth to fifth floor		
Fifth-floor slab, finishes etc.: $5.0 \times 2.8 \times 4.12$ =	57.7	
Secondary beam: $\frac{1}{2}[5.05 + (5.28 \times 2 \times \frac{1}{2})] \times 1.50$ =	7.7	
Edge beam (as at roof) =	11.4	
225 mm brick wall: $2.55 \times 4.60 \times 4.30$ =	50.4	
Self-weight of column: $3.0 \times 0.3 \times 0.3 \times 24.0$ =	6.5	
Imposed load on fifth floor: $(5.0 \times 2.8 \times 5.0 \times 0.9) - (5.0 \times 2.8 \times 1.5 \times 0.1)$ =		60.9
Characteristic loads at soffit of fifth floor	227.3	81.9

Minimum design load = 227.3 kN
Maximum design load (with wind) = $(227.3 + 81.9) \times 1.2 = 371.1$ kN
Maximum design load (without wind) = $(227.3 \times 1.4) + (81.9 \times 1.6) = 449.3$ kN

Third to fourth floor		
Fourth-floor dead loads (as fifth floor): $57.7 + 7.7 + 11.4$ =	76.8	
225 mm brick wall: $3.15 \times 4.60 \times 4.30$ =	62.3	
Self-weight of column: $3.6 \times 0.3 \times 0.3 \times 24.0$ =	7.8	
Additional imposed load: $5.0 \times 2.8 \times [(5.0 \times 0.7) - (1.5 \times 0.1)]$ =		46.9
Characteristic loads at soffit of fourth floor =	374.2	128.8

Minimum design load = 374.2 kN
Maximum design load (with wind) = $(374.2 + 128.8) \times 1.2 = 603.6$ kN
Maximum design load (without wind) = $(374.2 \times 1.4) + (128.8 \times 1.6) = 730.0$ kN

DATA SHEET	CALCULATIONS	OUTPUT
	Second to third floor	

Second to third floor

Third-floor dead loads (as fourth floor): $76.8 + 62.3 + 7.8$ $= 146.9$
Additional imposed load:
$5.0 \times 2.8 \times [(5.0 \times 0.5) - (1.5 \times 0.1)]$ $=$ 32.9

$\overline{}$
$521.1 \quad 161.7$

Minimum design load $= 521.1$ kN
Maximum design load (with wind) $= (521.1 + 161.7) \times 1.2 = 819.4$ kN
Maximum design load (without wind) $= (521.1 \times 1.4) + (161.7 \times 1.6) = 988.3$ kN

Ultimate bending moments
At roof level: in longitudinal plane

48

Maximum fixing moment at support of roof beam
Dead load:
Trapezoidal: $\frac{1}{10} \times \frac{1}{2}(5.60 + 0.60)2.5 \times 5.04 \times 5.60 \times 1.2$ $= 26.3$
Uniform: $\frac{1}{12} \times 5.60^2(1.25 \times 3.36 + 0.60 \times 0.13 \times 24.0) \times 1.2 = 19.1$
Imposed load:
Trapezoidal: $\frac{1}{10} \times \frac{1}{2}(5.60 + 0.60)2.5 \times 1.50 \times 5.60 \times 1.2$ $= 7.8$
Uniform: $\frac{1}{12} \times 5.60^2 \times 1.25 \times 1.50 \times 1.2$ $= 5.9$

$\overline{}$

Total fixing moment $= 59.1$ kN m

49

Moment of inertia of column section
$= \frac{1}{12}h^4 = \frac{1}{12} \times 300^4 = 0.675 \times 10^9$ mm^4

Length of column $= 3.0 \times 10^3$ mm
$\therefore$ Stiffness of column $= 0.675 \times 10^9/(3.0 \times 10^3) = 0.225 \times 10^6$ mm^3

$b = (\frac{1}{5} \times 5.425 \times 0.85) + 0.13 = 1.052$ m

49

With $h_f/h = 150/600 = 0.25$ and
$b_w/b = 130/1052 = 0.12$, $K = 0.0237$

$\therefore$ Moment of inertia of beam section
$= 0.0237 \times 1052 \times 600^3$
$= 5.38 \times 10^9$ mm^4

Span of beam $= 5425$ mm

$\therefore$ Stiffness of beam $= 538 \times 10^9/5425$
$= 0.992 \times 10^6$ mm^3

Thus final moment transferred to column

$$= 59.1 \times \frac{0.225}{0.225 + (\frac{1}{2} \times 0.992)} = 18.5 \text{ kN m}$$

$\therefore$ Moment at soffit of beam $= 18.5 - (18.5 + 26.6)\dfrac{0.30}{3.00} = 14.0$ kN m

At roof level: in lateral plane

From analysis outlined in section 14.11:

Moment at soffit of beam $= 41.96 - (41.96 + 7.70)\dfrac{0.30}{3.00} = 37.0$ kN m

At fifth-floor level: in longitudinal plane

Maximum fixing moment at support of fifth-floor beam:

$\frac{1}{12} \times (11.80 + 12.50) \times 5.425^2 \times 1.2 = 71.5$ kN m

Stiffness of upper column (as floor above) $= 0.225 \times 10^6$ mm^3

Moment of inertia of lower column (as floor above) $= 0.675 \times 10^9$ mm^4
Length of lower column $= 3600$ mm
$\therefore$ Stiffness of lower column $= 0.675 \times 10^9/3600 = 0.188 \times 10^6$ mm^3

DATA SHEET	CALCULATIONS	OUTPUT
49		

With $h_f/h = 100/400 = 0.25$ and $b_w/b = 0.200/1.122 = 0.18$, $K = 0.029$
∴ Moment of inertia of beam $= 0.029 \times 1122 \times 400^3 = 2.08 \times 10^9$ mm^4
Span of beam $= 5425$ mm
So stiffness of beam $= 2.08 \times 10^9/5425 = 0.384 \times 10^6$ mm^3

Final moment transferred to upper column:

$$71.5 \times \frac{0.225}{0.225 + 0.188 + (\frac{1}{2} \times 0.384)} = 26.6 \text{ kN m}$$

Final moment transferred to lower column:

$$71.5 \times \frac{0.188}{0.225 + 0.188 + 0.192} = 22.2 \text{ kN m}$$

∴ Moment at soffit of fifth-floor beam $= 22.2 \times \dfrac{3.20}{3.60} = 19.7$ kN m

Final moment in beam $= 26.6 + 22.2 = 48.8$ kN m

At fifth-floor level: in lateral plane

From analysis in section 14.11:

∴ Moment at soffit of fifth-floor beam $= 13.40 \times \dfrac{3.20}{3.60} = 11.9$ kN m

At fourth-floor level: in longitudinal plane

Maximum fixing moment at support of fourth-floor beam (as fifth floor)
   $= 71.5$ kN m

Stiffness of upper and lower columns (as lower column of fifth floor)
   $= 0.188 \times 10^6$ mm^3

Stiffness of beam (as fifth floor) $= 0.384 \times 10^6$ mm^3

Final moment transferred to upper and to lower column:

$$71.5 \times \frac{0.188}{0.188 + 0.188 + (\frac{1}{2} \times 0.384)} = 23.7 \text{ kN m}$$

Final moment in beam $= 2 \times 23.7 = 47.4$ kN m

∴ Moment at soffit of fourth-floor beam $= 23.7 \times \dfrac{3.20}{3.60} = 21.1$ kN m

At fourth-floor level: in lateral plane

From analysis in section 14.11:

∴ Moment at soffit of fourth-floor beam $= 19.56 \times \dfrac{3.20}{3.60} = 17.4$ kN m

At third-floor level: in longitudinal plane

Moment at soffit of third-floor beam (as fourth floor) $= 21.1$ kN m

At third-floor level: in lateral plane

From analysis in section 14.11:

Moment at soffit of third-floor beam $= 26.76 \times \dfrac{3.20}{3.60} = 23.8$ kN m

DATA SHEET	CALCULATIONS	OUTPUT

Reinforcement

At soffit of roof construction

$N = 137.5$ kN, $M_x = 37.0$ kN m and $M_y = 14.0$ kN m

Since $M_x/h' (= 37.0 \times 10^6/220) > M_y/b' (= 14.0 \times 10^6/220)$ and
$\beta = 1 - [7 \times 137.5 \times 10^3/(6 \times 300 \times 300 \times 30)] = 0.94$, design section
for $M_x' = M_x + \beta M_y = (37 \times 10^6) + (0.94 \times 14 \times 10^6) = 50.17 \times 10^6$

If $d/h = 0.733$, $M/bh^2 f_{cu} = 50.17 \times 10^6/(300 \times 300^2 \times 30) = 0.062$ and
$N/bhf_{cu} = 137.5 \times 10^3/(300 \times 300 \times 30) = 0.051$, extrapolating from the
chart for $d/h = 0.75$, $\varrho_1/f_{cu} = 0.0007$, so that $\varrho_1 = 0.021$ and $A_{sc} = 1890$ mm^2

Output: ≈220, 300, 300
4R25
(1963 mm^2)

At soffit of fifth floor

$N = 371.1$ kN, $M_x = 11.9$ kN m and $M_y = 19.7$ kN m

Since $M_x/h' (= 11.9 \times 10^6/220) < M_y/b' (= 19.7 \times 10^6/220)$ and
$\beta = 1 - [7 \times 371.1 \times 10^3/(6 \times 300 \times 300 \times 30)] = 0.84$, design section
for $M_y' = M_y + \beta M_x = (19.7 \times 10^6) + (0.84 \times 11.9 \times 10^6) = 29.7 \times 10^6$

54

If $d/h = 0.733$, $M/bh^2 f_{cu} = 29.7 \times 10^6/(300 \times 300^2 \times 30) = 0.037$ and
$N/bhf_{cu} = 371.1 \times 10^3/(300 \times 300 \times 30) = 0.138$, extrapolating from the
chart for $d/h = 0.75$, $\varrho_1/f_{cu} < 0$, so that only nominal reinforcement is required

Output: ≈220, 300, 300
4R25
(1963 mm^2)

At soffit of fourth floor

$N = 603.6$ kN, $M_x = 17.4$ kN m and $M_y = 21.1$ kN m

Since $M_x/h' (= 17.4 \times 10^6/220) < M_y/b' (= 21.1 \times 10^6/220)$ and
$\beta = 1 - [7 \times 603.6 \times 10^3/(6 \times 300 \times 300 \times 30)] = 0.74$, design section
for $M_y' = M_y + \beta M_x = (21.1 \times 10^6) + (0.74 \times 17.4 \times 10^6) = 34 \times 10^6$

If $d/h = 0.733$, $M/bh^2 f_{cu} = 34 \times 10^6/(300 \times 300^2 \times 30) = 0.042$ and
$N/bhf_{cu} = 603.6 \times 10^3/(300 \times 300 \times 30) = 0.224$, extrapolating from the
chart for $d/h = 0.75$, $\varrho_1/f_{cu} < 0$, so that only nominal reinforcement is required

Output: ≈220, 300, 300
4R25
(1963 mm^2)

At soffit of third floor

$N = 819.4$ kN, $M_x = 23.8$ kN m and $M_y = 21.1$ kN m

Since $M_x/h' (= 23.8 \times 10^6/220) > M_y/b' (= 21.1 \times 10^6/220)$ and
$\beta = 1 - [7 \times 819.4 \times 10^3/(6 \times 300 \times 300 \times 30)] = 0.65$, design section
for $M_x' = M_x + \beta M_y = (23.8 \times 10^6) + (0.65 \times 21.1 \times 10^6) = 37.4 \times 10^6$

54

If $d/h = 0.733$, $M/bh^2 f_{cu} = 37.4 \times 10^6/(300 \times 300^2 \times 30) = 0.046$ and
$N/bhf_{cu} = 819.4 \times 10^3/(300 \times 300 \times 30) = 0.304$, extrapolating from the
chart for $d/h = 0.75$, $\varrho_1/f_{cu} = 0$, so that only nominal reinforcement is required

Output: ≈220, 300, 300
4R25
(1963 mm^2)

Part Three

Design Data

Schedule of
Data sheets

233

Design loads and strengths, and partial safety factors
(*see section 1.2*)

	Condition	Ultimate limit-state	Serviceability limit-state	Notes
Design loads	Dead + imposed load (+ earth pressure)	Maximum loads of $1.4G_k + 1.6Q_k + 1.4E_n$ and minimum loads of $1.0G_k$ so arranged as to achieve the most unfavourable condition*	$1.0(G_k + Q_k + E_n)$	G_k: characteristic dead load Q_k: characteristic imposed load W_k: characteristic wind load E_n: nominal earth and/or water load (if applicable) Design load = characteristic load × partial safety factor for loads γ_f To consider probable effects of (a) excessive loading or (b) localized damage, take $\gamma_f = 1.05$ and consider only loads likely to occur simultaneously for (a) and loads likely to occur before remedial measures are taken for (b)
	Dead + wind load (+ earth pressure)	Maximum loads of $1.4(G_k + W_k + E_n)$ and minimum loads of $1.0G_k$ so arranged as to achieve the most unfavourable condition*	$1.0(G_k + W_k + E_n)$	
	Dead + imposed + wind load (+ earth pressure)	$1.2(G_k + Q_k + W_k + E_n)$	$1.0(G_k + Q_k + W_k + E_n)$	

* The *Joint Institutions Design Manual* states that where water pressure results from an accidental head of water at ground level, a value of $1.2E_n$ may be employed.

	Material	Fore effects of excessive load or damage	Otherwise	Notes
Design strengths	Concrete	$\dfrac{f_{cu}}{1.3}$	$\dfrac{f_{cu}}{1.5}$	f_{cu}: characteristic strength of concrete f_y: characteristic strength of reinforcement $\text{Design strength} = \dfrac{\text{characteristic strength}}{\text{partial safety factor for materials } \gamma_m}$ To calculate crack widths or service stresses, take $\gamma_m = 1.3$ for concrete and $\gamma_m = 1.0$ for reinforcement To calculate deflections, take $\gamma_m = 1$ for both concrete and reinforcement For shear without shear reinforcement, $\gamma_m = 1.25$ For bond, $\gamma_m = 1.4$ For other situations (e.g. bearing stress), $\gamma_m = 1.5$
	Reinforcement	$\dfrac{f_y}{1.0}$	$\dfrac{f_{cu}}{1.15}$	

Imposed loads on floors (*see section 2.1*)

Uniform load in (kN/m²)	Alternative concentrated load (kN) on 300 mm × 300 mm	Type of floor
1.5	1.4	Bungalows, dwellings, flats, houses and maisonettes
	1.8	Bedrooms and dormitories in institutional and residential buildings, colleges and clubs
2.0	nil	Toilet rooms
	1.8	Bedrooms in hotels and hospitals; wards; dressing rooms; equipment areas
	2.7	Dining and billiard rooms; lounges
	4.5	Utility and X-ray rooms, operating theatres etc. in hospitals
2.5	nil	Grids
	1.8	Light workshops without storage facilities
	2.7	General offices
	4.5	Library reading rooms without storage facilities
	9.0	Parking for vehicles not exceeding 2.5 tonnes in gross weight
3.0	nil	Banking halls
	2.7	Places of worship; classrooms
	4.5	Laundries; kitchens; laboratories
3.5	4.5	Offices containing fixed computing equipment and the like
4.0	nil	Areas for public assembly with fixed seating
	3.6	Shop floors
	4.5	Library reading rooms with storage; exhibition floors in museums and art galleries; machinery halls
5.0	nil	Bars; projection rooms
	3.6	Areas for public assembly without fixed seating; dance halls; grandstands; gymnasia
	4.5	Workshops, factories and the like; filing rooms
	9.0	Drill halls and rooms; vehicle repair workshops; parking for vehicles exceeding 2.5 tonnes in gross weight
7.5	4.5	Theatre and studio stages; motor, boiler and fan rooms (inc. machinery)
12.5	9.0	Printers' type stores

Uniform load (kN/m² per metre of height)	Specified minimum (if any)	Alternative concentrated load	Type of floor
2.4	—	7.0	Storage (other than defined elsewhere)
2.4	6.5	7.0	Book stacking rooms
4.0	—	9.0	Paper and stationery stores
4.8	9.6	7.0	Dense book stacks on mobile trucks
5.0	15.0	9.0	Cold stores

Minimum equivalent imposed loads on garage floors
(*see section 2.2*)

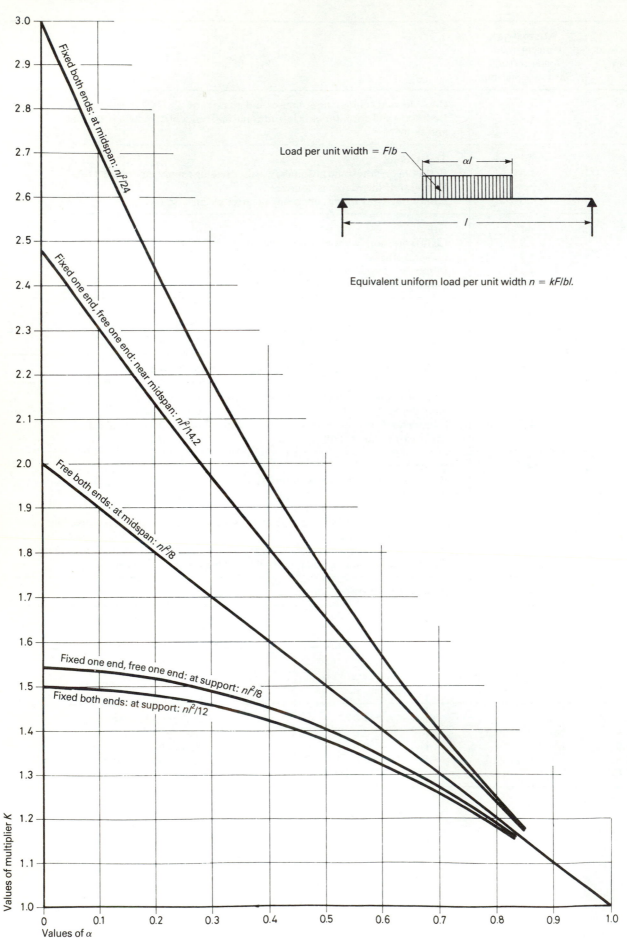

Load per unit width = *F/b*

Equivalent uniform load per unit width *n* = *kF/bl*.

Fixed both ends: at midspan: $nl^2/24$

Fixed one end, free one end: near midspan: $nl^2/14.2$

Free both ends: at midspan: $nl^2/8$

Fixed one end, free one end: at support: $nl^2/8$

Fixed both ends: at support: $nl^2/12$

Values of multiplier *K*

Values of α

Imposed loads on roofs and permissible reductions
(*see sections 2.4 and 2.5*)

Roof loading data

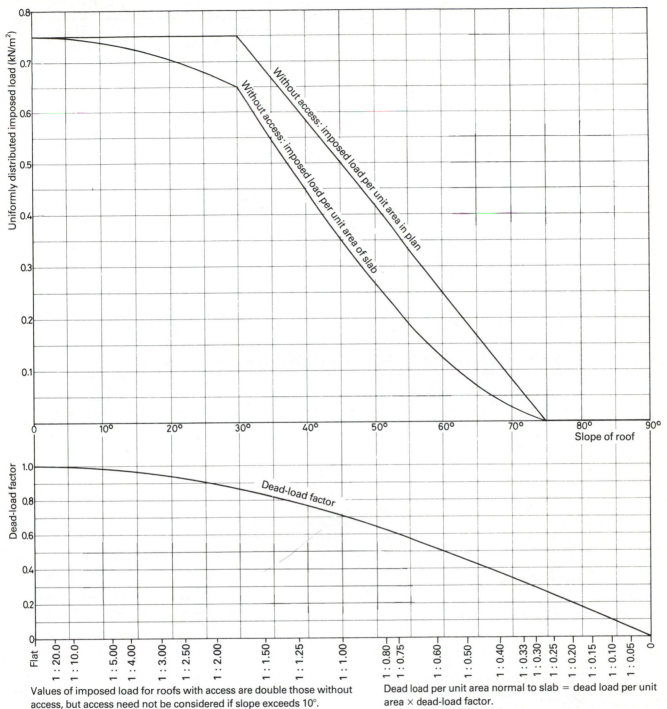

Values of imposed load for roofs with access are double those without access, but access need not be considered if slope exceeds 10°.

Dead load per unit area normal to slab = dead load per unit area × dead-load factor.

Reduction of imposed loads on columns, walls, piers, foundations etc.

Number of floors supported		1	2	3	4	5	6	7	8	9	10	More than 10
Values factor K	Non-storage floors etc.	1.0	1.8	2.4	2.8	3.0	3.6	4.2	4.8	5.4	6.0	0.5n
	Factories and workshops designed for 5 kN/m² or more	5	10	15	20	25	30	35	40	45	50	5n

(In the Factories row: K is as above but Kq_k not less than (kN/m²))

The foregoing reductions apply to all types of building except warehouses and other stores, garages, and office areas used for storage or filing. No reduction should be made for machinery or similar particular loads.

Total imposed load on column supporting two or more floors = $Kq_k A$, if area A of each floor is the same.

237

Wind loading–1 (see section 2.6)

Values of external pressure coefficient C_{pe} on rectangular structure with flat roof

	$h \leqslant \frac{1}{2}b$	$\frac{1}{2}b < h \leqslant 1\frac{1}{2}b$	$1\frac{1}{2}b < h \leqslant 6b$
$b < l \leqslant 1\frac{1}{2}b$	↓Wind +0.70 −0.50 −0.50 −0.20	↓Wind +0.70 −0.60 −0.60 −0.25	↓Wind +0.80 −0.80 −0.80 −0.25
$1\frac{1}{2}b < l \leqslant 4b$	↓Wind +0.70 −0.60 −0.60 −0.25	↓Wind +0.70 −0.70 −0.70 −0.30	↓Wind +0.70 −0.70 −0.70 −0.40
$b < l \leqslant 1\frac{1}{2}b$	Wind→ −0.50 +0.70 −0.20 −0.50	Wind→ −0.60 +0.70 −0.25 −0.60	Wind→ −0.80 +0.80 −0.25 −0.80
$1\frac{1}{2}b < l \leqslant 4b$	Wind→ −0.50 +0.70 −0.10 −0.50	Wind→ −0.50 +0.70 −0.10 −0.50	Wind→ −0.50 +0.80 −0.10 −0.50
Cladding: $b < l \leqslant 1\frac{1}{2}b$	−0.8	−1.1	−1.2
Cladding: $1\frac{1}{2}b < l \leqslant 4b$	−1.0	−1.1	−1.2

h: height of structure l: greater length in plan b: lesser length in plan

Typical values of internal pressure coefficients C_{pi} on cladding

	All four faces impermeable	Two opposite faces permeable Two opposite faces impermeable
Wind on impermeable face	−0.3	−0.3
Wind on permeable face	—	+0.2

Outline of basic procedure for determining wind force

It is assumed that $S_1 = S_3 = 1$. If this is not so, see section 2.6 where procedure is described in more detail.

1. Read off basic wind speed V corresponding to location, from adjoining map.

2. Read off appropriate value of factor S_2 for given topographical factor, overall dimension and height above terrain from graphs on *Data Sheet 6*.

3. Multiply V by S_2 to obtain characteristic wind speed V_s.

4. Read off characteristic wind pressure w_k corresponding to V_s from scale on *Data Sheet 6*.

5. Determine appropriate external and internal pressure coefficients from above table.

6. Total wind force F on area A of structure as a whole $= w_k A(C_{pe1} - C_{pe2})$, where C_{pe1} and C_{pe2} are external pressure coefficients on windward and leeward faces respectively.
   Total wind force F on area A of particular face of structure $= w_k A C_{pe}$.
   Total wind force F on cladding element $= w_k A(C_{pe} - C_{pi})$, where C_{pe} and C_{pi} are external and internal pressure coefficients respectively.

To obtain total force on entire structure, divide structure into parts, determine force on each part by steps 1–6 and then sum results vectorially. Consider appropriate value of h for each individual part (but for approx. analysis, use of single value of w_k corresponding to height to top of building errs on side of safety).

Alternatively, first follow steps 1–4. Next, obtain value of force coefficient C_f from graph on left. Then: total wind force on area $A = w_k A C_f$. For greater accuracy, subdivide structure and sum individual results vectorially as before.

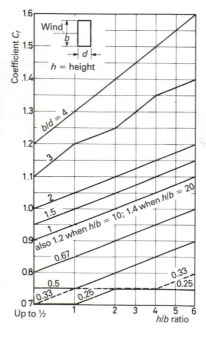

Basic wind speed (m/s)

56 54 52 50 48 46 46 48 50 44 42 40

Coefficient C_f

Wind h = height

$b/d = 4$

3

2

1.5

1

0.67

0.5

0.33

0.25

0.33

also 1.2 when h/b = 10; 1.4 when h/b = 20

Up to ½   1   2   3   4   5   6 h/b ratio

Wind loading–2 (see section 2.6)

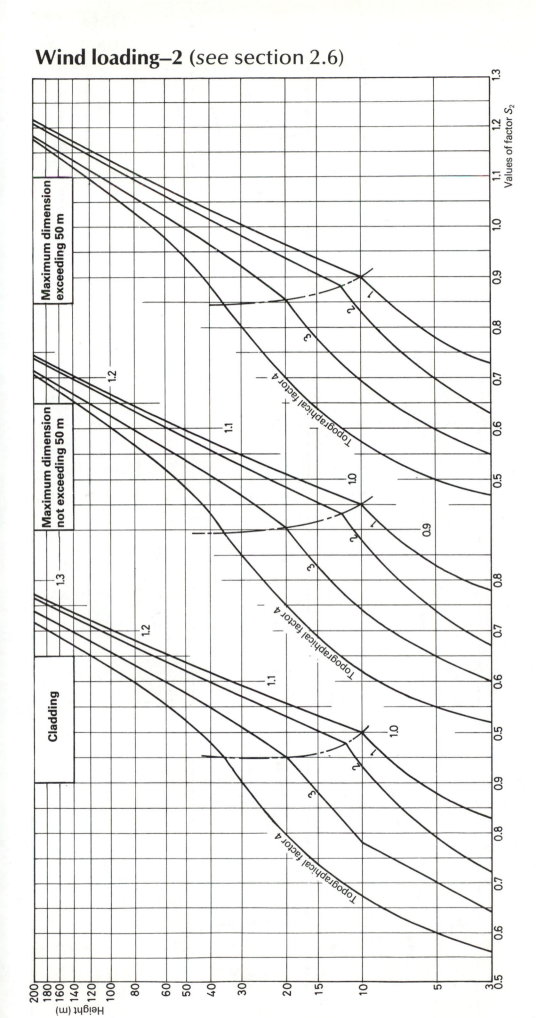

This data sheet should be read in conjunction with the notes on *Data Sheet 5*.

Continuous beams: bending-moment formulae for uniform loads on equal spans (see section 3.4)

Number of spans	Key diagrams	Imposed load on span	Critical moment conditions	Maximum bending moment		
				In span 1	At support	In span 2
Two		1	Span 1 max. Span 2 min.	$\dfrac{1}{512n}(7n-g)^2 l^2$	$\dfrac{1}{16}(n+g)l^2$	$\dfrac{1}{512g}(7g-n)^2 l^2$
		1+2	Support max.	$\dfrac{9}{128}nl^2$	$\dfrac{1}{8}nl^2$	$\dfrac{9}{128}nl^2$
Three		1+3	Span 1 max. Span 2 min.	$\dfrac{1}{800n}(9n-g)^2 l^2$	$\dfrac{1}{20}(n+g)l^2$	$\dfrac{1}{40}(3g-2n)l^2$
		1+2+3	Support max.	$\dfrac{2}{25}nl^2$	$\dfrac{1}{10}nl^2$	$\dfrac{1}{40}nl^2$
		2	Span 2 max. Span 1 min.	$\dfrac{1}{800g}(9g-n)^2 l^2$	$\dfrac{1}{20}(n+g)l^2$	$\dfrac{1}{40}(3n-2g)l^2$
Infinite (internal span)		Alternate	Span 1 max. Span 2 min.	$\dfrac{1}{24}(2n-g)l^2$	$\dfrac{1}{24}(n+g)l^2$	$\dfrac{1}{24}(2g-n)l^2$
		All	Support max.	$\dfrac{1}{24}nl^2$	$\dfrac{1}{12}nl^2$	$\dfrac{1}{24}nl^2$
Infinite (end span)		1, 3 etc.	Span 1 max. Span 2 min.	$\dfrac{1}{n}(0.3162n-0.0374g)^2 l^2$	$0.0528(n+g)l^2$	$\left[\dfrac{1}{g}(0.0100n+0.3636g)^2-0.0528(n+g)\right]l^2$
		All	Support max.	$0.0778nl^2$	$0.1056nl^2$	$0.0340nl^2$
		2, 4 etc.	Span 2 max. Span 1 min.	$\dfrac{1}{g}(0.3162g-0.0374n)^2 l^2$	$0.0528(n+g)l^2$	$\left[\dfrac{1}{n}(0.0100g+0.3636n)^2-0.0528(n+g)\right]l^2$

Continuous beams: bending-moment formulae for loads on equal spans from two-way slabs (see section 3.4)

Number of spans	Key diagrams	Imposed load on span	Critical moment conditions	Maximum bending moment		
				In span 1	At support	In span 2
Two		1	Span 1 max. Span 2 min.	$\frac{1}{n}(0.0903n^2 - 0.0253ng + 0.0016g^2)\,l^2$	$0.0571(n+g)\,l^2$	$\frac{1}{g}(0.0903g^2 - 0.0253ng + 0.0016n^2)\,l^2$
		1 + 2	Support max.	$0.0666nl^2$	$0.1143nl^2$	$0.0666nl^2$
Three		1 + 3	Span 1 max. Span 2 min.	$\frac{1}{n}(0.0954n^2 - 0.0208ng + 0.0011g^2)\,l^2$	$0.0457(n+g)\,l^2$	$(0.0715g - 0.0457n)\,l^2$
		1 + 2 + 3	Support max.	$0.0757nl^2$	$0.0914nl^2$	$0.0258nl^2$
		2	Span 2 max. Span 1 min.	$\frac{1}{g}(0.0954g^2 - 0.0208ng + 0.0011n^2)\,l^2$	$0.0457(n+g)\,l^2$	$(0.0715n - 0.0457g)\,l^2$
Infinite (internal span)		Alternate	Span 1 max. Span 2 min.	$(0.0791n - 0.0381g)\,l^2$		$(0.0791g - 0.0381n)\,l^2$
		All	Support max.	$0.0410nl^2$	$0.0762nl^2$	$0.0410nl^2$
Infinite (end span)		1, 3 etc.	Span 1 max. Span 2 min.	$\frac{1}{n}(0.0942n^2 - 0.0218ng + 0.0012g^2)\,l^2$	$0.0483(n+g)\,l^2$	$\simeq(0.0754n - 0.0418g)\,l^2$
		All	Support max.	$0.0736nl^2$	$0.0965nl^2$	$0.0335nl^2$
		2, 4 etc.	Span 2 max. Span 1 min.	$\frac{1}{g}(0.0942g^2 - 0.0218ng + 0.0012n^2)\,l^2$	$0.0483(n+g)\,l^2$	$\simeq(0.0754n - 0.0418g)\,l^2$

Continuous beams: bending-moment coefficients for equal spans and uniform moment of inertia (*see section 3.4*)

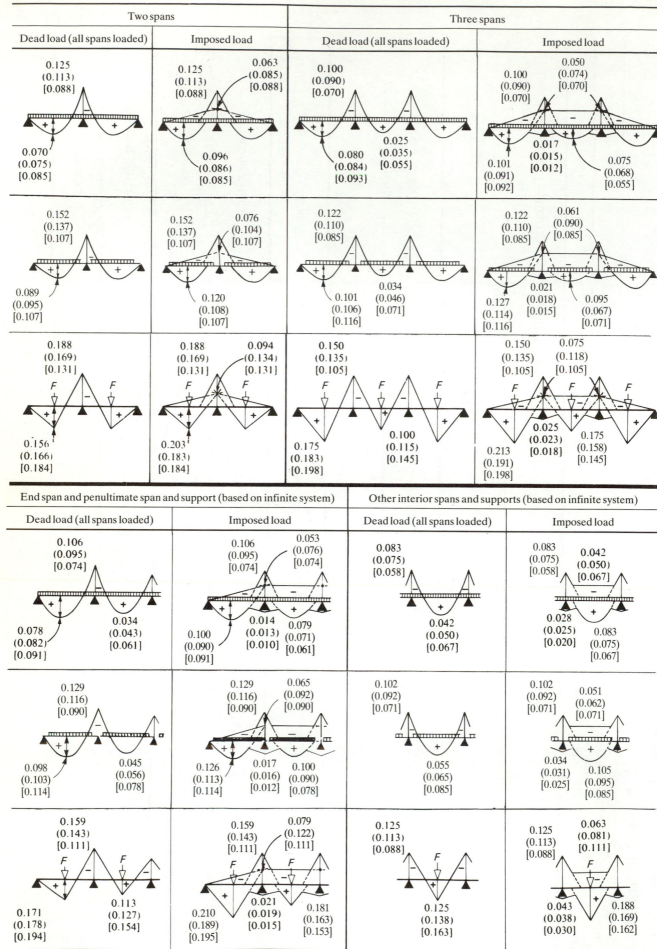

Diagrams are symmetrical but not drawn to scale.
Equal load *F* (total load) on each loaded span.
Bending moment = coefficient × *F* × span.
Coefficients thus: 0.125 : theoretical bending moments
(0.106): bending moments with 10% redistribution at supports and in spans
[0.191]: bending moments with 30% redistribution at supports and maximum corresponding redistribution in spans (see text).
Positive moments at supports do not result from loading arrangement prescribed in BS8110, which gives zero positive moment at all supports.
To meet BS8110 requirements, consider dead load of $1.0G_k$ and 'imposed load' of $0.4G_k + 1.6Q_k$.

Precise moment distribution: continuity and distribution factors (see section 3.7)

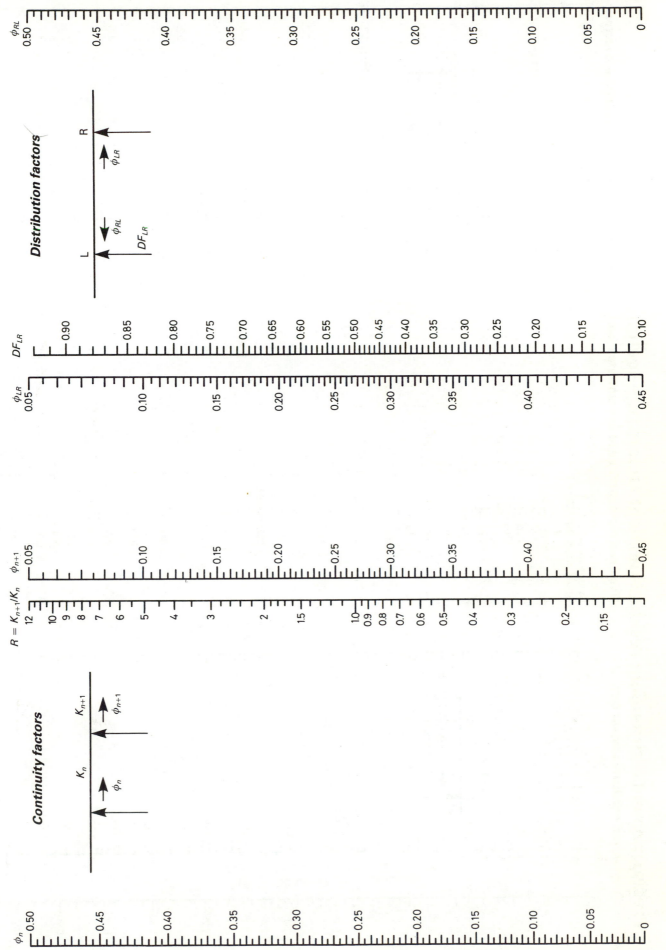

Precise moment distribution: residual-moment factors (see section 3.7)

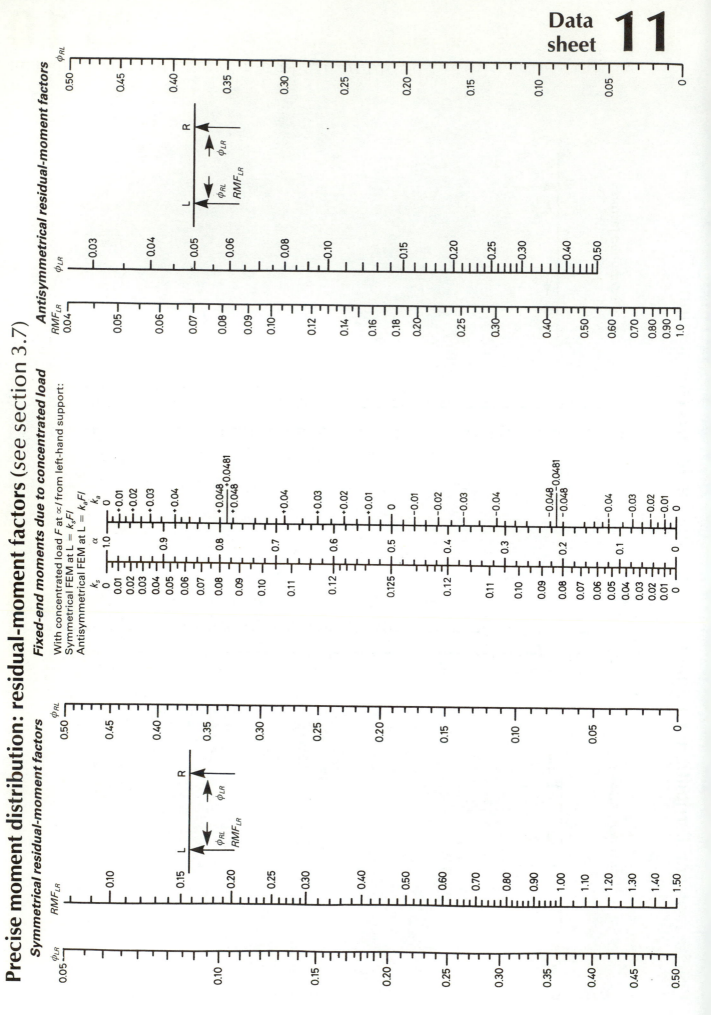

Symmetrical residual-moment factors

Antisymmetrical residual-moment factors

Fixed-end moments due to concentrated load

With concentrated load F at $\propto l$ from left-hand support:
Symmetrical FEM at $L = k_s Fl$
Antisymmetrical FEM at $L = k_a Fl$

244

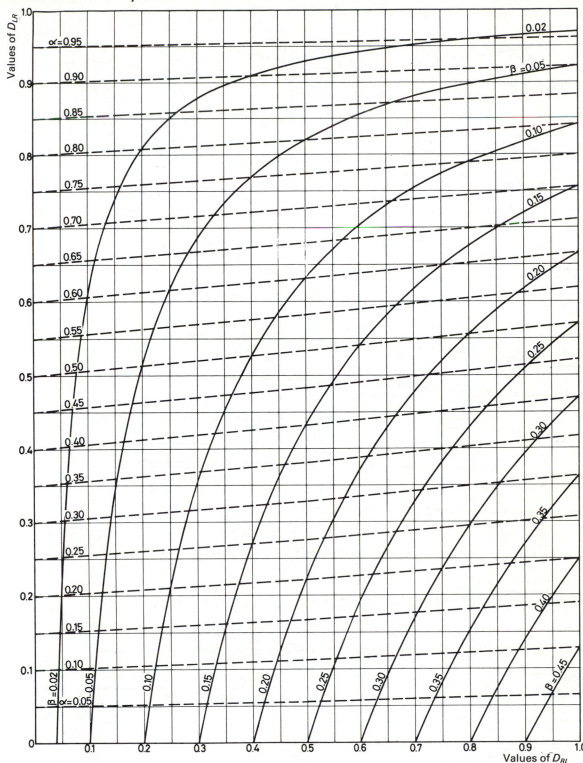

Values of D_{LR} (vertical axis)

Values of D_{RL} (horizontal axis)

Final support moment in beam

$$M_{LR} = FEM_{LR} - \alpha F'_L + \beta F'_R$$

where FEM_{LR} is fixed-end moment at L on span LR,

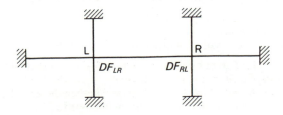

F'_L *and* F'_R are out-of-balance fixed-end moments at L and R respectively, and α and β are coefficients read from adjoining graph corresponding to given distribution factors D_{LR} and D_{RL} at supports L and R of span LR.

To obtain M_{RL}, transpose values of D_{LR} and D_{RL} to determine α' and β'. Then

$$M_{RL} = FEM_{RL} - \alpha' F'_R + \beta F'_L$$

Final moment in column meeting at L and having no initial fixed-end moment

$$M = \text{distribution factor for column} \times \frac{2D_{RL}F'_R + 4F'_L}{4 - D_{LR}D_{RL}}$$

where notation is as given previously.

Design chart for rectangular members (see section 5.1)

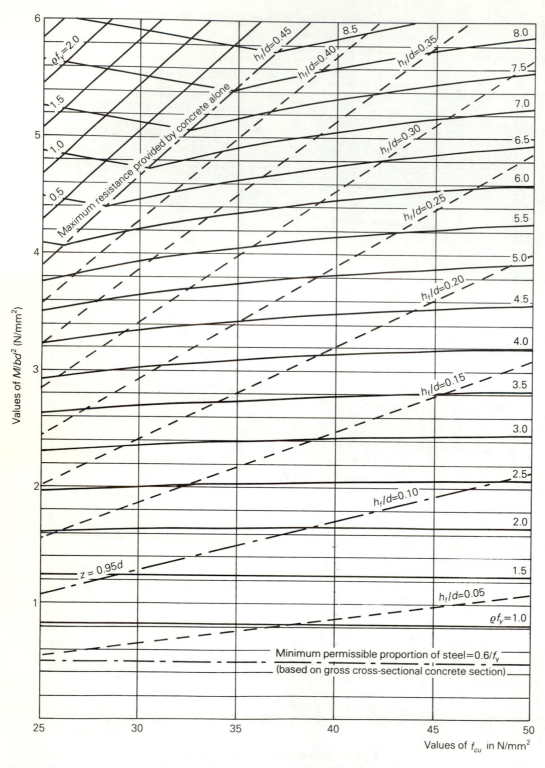
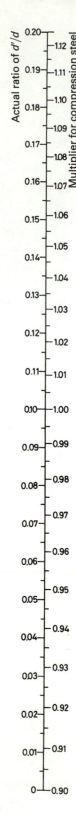

$A_s = \varrho bd$ $A_s' = \varrho' bd$

Curves are based on rigorous analysis using BS8110 uniform rectangular concrete stress-block (clause 3.4.4.1 of Code).

Minimum permissible proportion of main steel = 0.0013 (high-yield reinforcement) or 0.0024 (mild steel reinforcement) based on gross cross-sectional area of section.

Outline of design procedure

Slabs and singly reinforced rectangular beams
Select value of ϱf_y corresponding to intersection of given values of M_u/bd^2 and f_{cu}.

Doubly reinforced rectangular beams
If $d' = 0.1d$, read appropriate values of ϱf_y and $\varrho' f_y$ directly from chart.

If $d' \neq 0.1d$, determine ϱf_y and $\varrho' f_y$ as before. Then multiply $\varrho' f_y$ by coefficient relating to actual cover ratio of d'/d, as read from the scale on the right of the data sheet. Finally, add or subtract to ϱf_y (depending on whether the actual ratio of d'/d is greater or less than 0.1) the same adjustment made to $\varrho' f_y$.

Continuous beams: shearing-force coefficients
(*see section 6.1*)

Coefficients for maximum shearing forces

Type of load			V_{AR}	V_{BL}	V_{BR}	V_{CL}	V_{AR}	V_{BL}	V_{BR}	V_{CL}	V_{AR}	V_{BL}	V_{BR}	V_{CL}
2 spans	Dead load		0.375	0.625			0.313	0.688			0.333	0.667		
	Imposed load		0.438	0.625			0.406	0.688			0.417	0.667		
	imposed load / dead load	0.5	0.600	0.917			0.533	1.008			0.556	0.978		
		1	0.625	0.938			0.563	1.031			0.583	1.000		
		2	0.650	0.958			0.592	1.054			0.611	1.022		
		3	0.663	0.969			0.606	1.066			0.625	1.033		
3 spans	Dead load		0.400	0.600	0.500		0.350	0.650	0.500		0.367	0.633	0.500	
	Imposed load		0.450	0.617	0.583		0.425	0.675	0.625		0.433	0.656	0.611	
	imposed load / dead load	0.5	0.627	0.893	0.800		0.573	0.973	0.833		0.591	0.947	0.822	
		1	0.650	0.917	0.833		0.600	1.000	0.875		0.617	0.972	0.861	
		2	0.673	0.940	0.867		0.627	1.027	0.917		0.642	0.998	0.900	
		3	0.685	0.952	0.883		0.640	1.040	0.938		0.655	1.011	0.919	
4 spans	Dead load		0.393	0.607	0.536	0.464	0.339	0.661	0.554	0.446	0.357	0.643	0.548	0.452
	Imposed load		0.446	0.621	0.603	0.571	0.420	0.681	0.654	0.607	0.429	0.661	0.637	0.595
	imposed load / dead load	0.5	0.619	0.901	0.839	0.767	0.562	0.985	0.892	0.783	0.581	0.957	0.875	0.778
		1	0.643	0.924	0.871	0.804	0.589	1.011	0.931	0.830	0.607	0.982	0.911	0.821
		2	0.667	0.947	0.902	0.840	0.617	1.037	0.969	0.877	0.633	1.007	0.947	0.865
		3	0.679	0.958	0.917	0.859	0.630	1.050	0.989	0.901	0.646	1.020	0.967	0.887

Single span (imposed or dead load)

Freely supported: 0.500 — 0.500

Fixed at both ends: 0.500 — 0.500 — For any symmetrical load

Fixed at one support and free at the other: 0.625 — 0.375 | 0.688 — 0.313 | 0.667 — 0.333

Equal spans and uniform moment of inertia throughout.

Shearing force = coefficient for dead or imposed load × *ultimate* load.

Coefficients for various ratios of dead to imposed load *take into account* partial safety factors of 1.4 and 1.0 for dead load and 1.6 for imposed load, so shearing force = coefficient × total *characteristic* load.

Adjustment factors due to bending moments at end supports

		Unit bending moment applied at A only				Unit bending moment applied at A and K simultaneously			
Number of spans		2	3	4	5	2	3	4	5
Bending moment	M_A	−1.000	−1.000	−1.000	−1.000	−1.000	−1.000	−1.000	−1.000
	M_B	+0.250	+0.267	+0.268	+0.268	+0.500	+0.200	+0.286	+0.263
	M_C	—	—	−0.071	−0.072	—	—	−0.143	−0.053
	M_H	—	—	—	+0.019	—	—	—	−0.053
	M_J	—	−0.067	+0.018	−0.005	—	+0.200	+0.286	+0.263
	M_K	0	0	0	0	−1.000	−1.000	−1.000	−1.000
Shearing force	V_{AR}	+1.250	+1.267	+1.268	+1.268	+1.500	+1.200	+1.286	+1.263
	V_{BL}	−1.250	−1.267	−1.268	−1.268	−1.500	−1.200	−1.286	−1.263
	V_{BR}	−0.250	−0.333	−0.340	−0.340	—	0	−0.429	−0.316
	V_{CL}	—	—	+0.340	+0.340	—	—	+0.429	+0.316
	V_{CR}	—	—	+0.089	+0.091	—	—	—	0
	V_{HL}	—	—	—	−0.091	—	—	—	0
	V_{HR}	—	—	—	−0.024	—	—	+0.429	+0.316
	V_{JL}	—	+0.333	−0.089	+0.024	—	0	−0.429	−0.316
	V_{JR}	—	+0.067	−0.018	+0.005	−1.500	−1.200	−1.286	−1.263
	V_{KL}	+0.250	−0.067	+0.018	−0.005	+1.500	+1.200	+1.286	+1.263

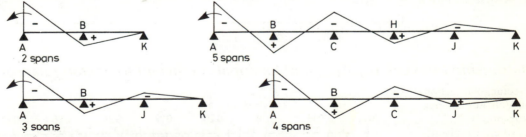

Key showing typical moment diagrams with unit moment applied at one end only.

2 spans | 5 spans | 3 spans | 4 spans

Adjustment to bending moment = *M*-coefficient × applied bending moment.

Adjustment to shearing force = (*V*-coefficient × applied bending moment)/span.

Shearing and torsional resistance of concrete
(*see section 6.2*)

Shearing resistance of concrete without shearing reinforcement

First value of resistance relates to normal-weight concrete. Second value (i.e. that in brackets) relates to lightweight-aggregate concrete. These values only apply directly for grade 25 concrete. For other grades multiply the value read from the graph by the modifying factor from the scale below.

Percentage of longitudinal reinforcement to be considered is that which extends a distance at least equal to effective depth beyond point being considered, except at supports where total area of tension steel may be taken into account providing that Code anchorage requirements are met.

Values are calculated from

$$v_c = 13.12 \varrho^{1/3}/d^{1/4}$$

where $\varrho = A_s/b_v\,d$ and does not exceed 0.03, and d does not exceed 400 mm. For lightweight concrete, multiply these values by 0.8.

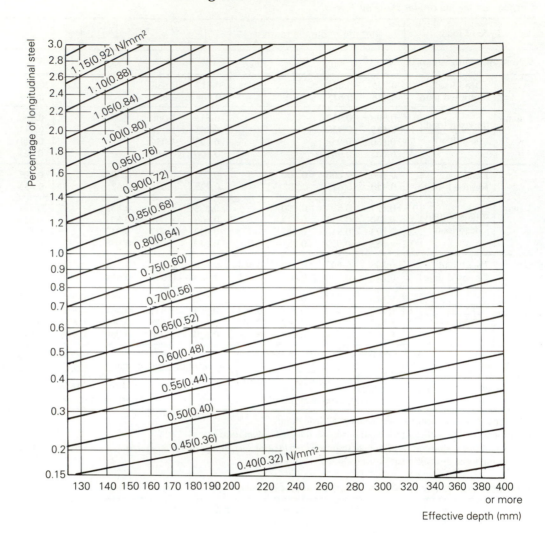

Modifying factor for different concrete grades

Multiply the value of v_c, obtained from the above chart, by the multiplier read from the adjoining scale corresponding to the concrete strength.

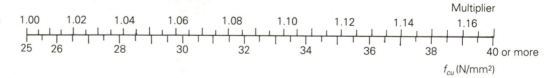

Limiting shearing resistance v_{max} and torsional resistances $v_{t\,min}$ and v_{tu} in beams

For lightweight-aggregate concrete, multiply values of v read from these scales by 0.8.

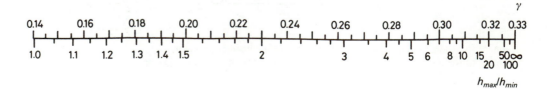

Relationship between h_{min}/h_{max} and torsional coefficient for rectangular sections

For rectangular sections, Saint Venant torsional coefficient $J = \gamma h_{min}^3 h_{max}$, where γ is given by adjoining scale. To comply with BS8110, take torsional constant $C = \frac{1}{2}J$.

Values of K_u and b_t for given link system

Type of reinforcement	Bar (mm)	Factor	Spacing of links in single system (mm)											Shearing resistance (N) when $s = 0.75d$
			50	75	100	125	150	175	200	225	250	275	300	
Mild steel $f_y = 250 \, \text{N/mm}^2$	6	K_u	246	164	123	98	82	70	61	55	49	45	41	9225
		b_t	615	410	307	246	205	176	154	137	123	112	102	
	8	K_u	437	292	219	175	146	125	109	97	87	80	73	16 399
		b_t	1093	729	547	437	364	312	273	243	219	199	182	
	10	K_u	683	456	342	273	228	195	171	152	137	124	114	25 624
		b_t	1708	1139	854	683	569	488	427	380	342	311	285	
	12	K_u	984	656	492	394	328	281	246	219	197	179	164	36 898
		b_t	2460	1640	1230	984	820	703	615	547	492	447	410	
	16	K_u	1749	1166	875	700	583	500	437	389	350	318	292	65 596
		b_t	4373	2915	2187	1749	1458	1249	1093	972	875	795	729	
High-yield steel $f_y = 460 \, \text{N/mm}^2$	6	K_u	453	302	226	181	151	129	113	101	91	82	75	16 973
		b_t	1132	754	566	453	377	323	283	251	226	206	189	
	8	K_u	805	536	402	322	268	230	201	179	161	146	134	30 174
		b_t	2012	1341	1006	805	671	575	503	447	402	366	335	
	10	K_u	1257	838	629	503	419	359	314	279	251	229	210	47 147
		b_t	3143	2095	1572	1257	1048	898	786	698	629	571	524	
	12	K_u	1810	1207	905	724	603	517	453	402	362	329	302	67 892
		b_t	4526	3017	2263	1810	1509	1293	1132	1006	905	823	754	
	16	K_u	3219	2146	1609	1287	1073	920	805	715	644	585	536	120 697
		b_t	8046	5364	4023	3219	2682	2299	2012	1788	1609	1463	1341	
Minimum size of compression bar (mm) for given link spacing			6	8	10	12	16	16	20	20	25	25	25	$\not> 12\phi$

K_u = link-reinforcement factor = ultimate shearing resistance provided by system in N per mm of effective depth.
b_t = maximum permissible width of section with nominal links unless $v < \frac{1}{2}v_c$.

Values of K_u provided by system in N per mm of effective depth

Bar arrangement	f_y (N/mm^2)		Bar arrangement	f_y (N/mm^2)		Bar arrangement	f_y (N/mm^2)		Bar arrangement	f_y (N/mm^2)	
	250	460		250	460		250	460		250	460
6 @ 300	41	75	6 @ 100	123	226	12 @ 225	219	402	8 @ 50	437	805
6 @ 275	45	82	10 @ 275	124	229	10 @ 150	228	419	16 @ 200	437	805
6 @ 250	49	91	8 @ 175	125	230	6 @ 50	246	453	10 @ 75	456	838
6 @ 225	55	101	10 @ 250	137	251	12 @ 200	246	453	12 @ 100	492	905
6 @ 200	61	113	8 @ 150	146	268	10 @ 125	273	503	16 @ 175	500	920
6 @ 175	70	129	10 @ 225	152	279	12 @ 175	281	517	16 @ 150	583	1073
8 @ 300	73	134	6 @ 75	164	302	8 @ 75	292	536	12 @ 75	656	1207
8 @ 275	80	146	12 @ 300	164	302	16 @ 300	292	536	10 @ 50	683	1257
6 @ 150	82	151	10 @ 200	171	314	16 @ 275	318	585	16 @ 125	700	1287
8 @ 250	87	161	8 @ 125	175	322	12 @ 150	328	603	16 @ 100	875	1609
8 @ 225	97	179	12 @ 275	179	329	10 @ 100	342	629	12 @ 50	984	1810
6 @ 125	98	181	10 @ 175	195	359	16 @ 250	350	644	16 @ 75	1166	2146
8 @ 200	109	201	12 @ 250	197	362	16 @ 225	389	715	16 @ 50	1749	3219
10 @ 300	114	210	8 @ 100	219	402	12 @ 125	394	724			

Arrangement for maximum design stress

Inclined bars arranged for design stress of $0.87f_y$ in bending reinforcement (horizontal part of bar) to be equal to tensile stress of $0.87f_{yv}$ in shearing reinforcement (inclined part of bar) assuming $f_{yv} = f_y$.

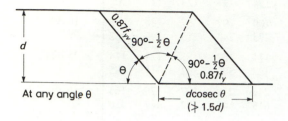

At any angle θ

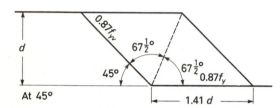

At 45°

Spacing and resistance of bars inclined at 45°

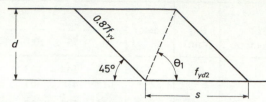

Tensile design stress in shearing reinforcement = $0.87f_{yv}$. Any design stress f_{yd2} in bending reinforcement.

f_{yd2}	θ_1	s max.	Equivalent number of single systems
$0.615f_{yv}$	90.0°	1.000d	1.414
$0.669f_{yv}$	85.0°	1.087d	1.300
$0.700f_{yv}$	82.2°	1.138d	1.243
$0.724f_{yv}$	80.0°	1.176d	1.202
$0.750f_{yv}$	77.7°	1.219d	1.160
$0.780f_{yv}$	75.0°	1.268d	1.115
$0.800f_{yv}$	73.3°	1.301d	1.087
$0.839f_{yv}$	70.0°	1.364d	1.037
$0.850f_{yv}$	69.1°	1.382d	1.023
$0.870f_{yv}$	67.5°	1.414d	1.000
$0.900f_{yv}$	65.2°	1.463d	0.967
$0.902f_{yv}$	65.0°	1.466d	0.964
$0.923f_{yv}$	63.4°	1.500d	0.943

Ultimate shearing resistance V_i of inclined bars in kN

Size of bar (mm)	f_{yv} (N/mm²)	Single system	Double system	Treble system	Quadruple system
			$\theta = 45°$		
16	250	30.92	61.84	92.77	123.69
	460	56.90	113.79	170.69	227.59
20	250	48.32	96.63	144.95	193.27
	460	88.90	177.80	266.71	355.61
25	250	75.49	150.99	226.48	301.98
	460	138.91	277.82	416.73	555.64
32	250	123.69	247.38	371.07	494.76
	460	227.59	455.18	682.77	910.36
40	250	193.27	386.53	579.80	773.06
	460	355.61	711.22	1066.83	1422.43
Spacing s (maximum)		1.41d	0.71d	0.47d	0.35d

This maximum spacing applies only if design stress in horizontal part of bar is $0.87f_y$.

Single system

$V_i = 0.87f_{yv} A_{sv} \sin\theta$:

if $f_{yv} = 250$ N/mm², then $V_i = 153.8A_{sv}$
if $f_{yv} = 460$ N/mm², then $V_i = 283.0A_{sv}$

Double system

$V_i = 1.74f_{yv} A_{sv} \sin\theta$:

if $f_{yv} = 250$ N/mm², then $V_i = 307.6A_{sv}$
if $f_{yv} = 460$ N/mm², then $V_i = 566.0A_{sv}$

Treble system

$V_i = 2.61f_{yv} A_{sv} \sin\theta$:

if $f_{yv} = 250$ N/mm², then $V_i = 461.4A_{sv}$
if $f_{yv} = 460$ N/mm², then $V_i = 849.0A_{sv}$

Quadruple system

$V_i = 3.48f_{yv} A_{sv} \sin\theta$:

if $f_{yv} = 250$ N/mm², then $V_i = 615.2A_{sv}$
if $f_{yv} = 460$ N/mm², then $V_i = 1131.9A_{sv}$

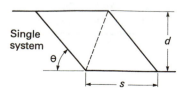

Single system

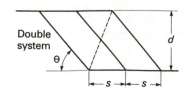

Double system

Anchorage bond: minimum lengths in millimetres (see section 7.1)

Diam. of bar (mm)	Min. 12φ (mm)	Min. lap 15φ or 300mm	Type of anchorage	$f_{cu}=25\,N/mm^2$ Tension 0°	90°	180°	1.4×	2×	Comp.	$f_{cu}=30\,N/mm^2$ 0°	90°	Tension 180°	1.4×	2×	Comp.	$f_{cu}=40\,N/mm^2$ 0°	90°	Tension 180°	1.4×	2×	Comp.
6	75	300	250	235	185	140	330	470	190	215	165	120	300	430	175	185	140	90	260	370	150
			460-0	430	360	285	600	860	345	395	320	250	550	785	315	340	270	195	475	680	275
			460-1	300	230	160	420	600	240	275	205	130	385	550	220	240	170	95	335	475	190
			460-2	240	170	100	340	480	195	220	150	80	310	440	175	190	120	50	270	380	155
8	100	300	250	315	250	185	435	625	250	285	220	160	400	570	230	250	185	120	345	495	200
			460-0	575	480	380	800	1145	460	525	430	330	735	1045	420	455	360	260	635	905	365
			460-1	400	305	210	560	800	320	370	270	175	515	735	295	320	225	125	445	635	255
			460-2	320	225	130	450	640	255	295	200	105	410	585	235	255	160	65	355	510	205
10	120	300	250	390	310	230	545	780	315	355	275	195	500	710	285	310	230	150	430	615	250
			460-0	715	595	475	1000	1430	575	655	535	415	915	1305	525	565	445	325	795	1130	455
			460-1	500	380	260	700	1000	400	460	340	220	640	915	370	400	280	160	555	795	320
			460-2	400	280	160	560	800	320	370	250	130	515	735	290	320	200	80	445	635	255
12	145	300	250	470	370	275	655	935	375	430	330	235	600	855	345	370	275	180	520	740	295
			460-0	860	715	570	1200	1715	690	785	640	495	1100	1565	630	680	535	390	950	1360	545
			460-1	600	455	310	840	1200	480	550	405	260	770	1100	440	475	335	190	665	950	380
			460-2	480	335	190	675	960	385	440	295	155	615	880	350	380	240	95	535	760	305
16	195	300	250	625	495	370	870	1245	500	570	440	315	795	1135	455	495	365	240	690	985	395
			460-0	1145	955	760	1600	2290	915	1045	855	660	1465	2090	835	905	715	520	1265	1810	725
			460-1	800	605	415	1120	1600	640	735	540	350	1025	1465	585	635	445	250	890	1265	510
			460-2	640	445	255	900	1280	510	585	395	205	820	1170	465	510	315	125	710	1015	405
20	240	300	250	780	620	460	1090	1555	625	710	550	390	995	1420	570	615	455	295	860	1230	495
			460-0	1430	1190	950	2000	2860	1145	1305	1065	825	1830	2610	1045	1130	890	650	1585	2260	905
			460-1	1000	760	520	1400	2000	800	915	675	435	1280	1830	735	795	555	315	1110	1585	635
			460-2	800	560	320	1120	1600	635	735	495	255	1025	1465	580	635	395	155	890	1265	505
25	300	375	250	975	775	575	1360	1945	780	890	690	490	1245	1775	710	770	570	370	1075	1535	615
			460-0	1790	1490	1190	2500	3575	1430	1635	1335	1035	2285	3265	1305	1415	1115	815	1980	2825	1130
			460-1	1250	950	650	1750	2500	1000	1145	845	545	1600	2285	915	990	690	390	1385	1980	795
			460-2	1000	700	400	1400	2000	795	915	615	315	1280	1830	725	795	495	195	1110	1585	630
32	385	480	250	1245	990	735	1740	2485	995	1135	880	625	1590	2270	910	985	730	475	1375	1965	790
			460-0	2290	1905	1520	3200	4575	1830	2090	1705	1320	2925	4175	1670	1810	1425	1040	2530	3615	1450
			460-1	1600	1215	830	2240	3200	1280	1465	1080	695	2045	2925	1170	1265	885	500	1775	2530	1015
			460-2	1280	895	510	1795	2560	1020	1170	785	405	1640	2340	930	1015	630	245	1425	2025	805
40	480	600	250	1555	1235	915	2175	3110	1245	1420	1100	780	1985	2840	1135	1230	910	590	1720	2460	985
			460-0	2860	2380	1900	4000	5715	2290	2610	2130	1650	3655	5220	2090	2260	1780	1300	3165	4520	1810
			460-1	2000	1520	1040	2800	4000	1600	1830	1350	870	2560	3655	1465	1585	1105	625	2215	3165	1265
			460-2	1600	1120	640	2240	3200	1270	1465	985	505	2045	2925	1160	1265	785	305	1775	2530	1005

Notes to Data sheet 18

1. All lengths are rounded to 5 mm value above exact figure.
2. Minimum stopping-off length = 12φ or d, whichever is greater.
3. Minimum lap in tension: The greater of either 15φ or 300mm, whichever is greater, or anchorage length of smaller bar.
   (a) Where lap occurs at top of section as cast, and size of lapped bar exceeds half the minimum cover, multiply lap length by 1.4.
   (b) Where lap occurs near section corner and size of lapped bars exceeds half the minimum cover to either face, or where clear distance between adjacent bars is less than 75 mm or six times size of lapped bars, whichever is greater, multiply lap lengths given by 1.4.

Where conditions (a) and (b) both apply, multiply lap lengths by 2.0.

4. Minimum lap in compression: The greater of either 15φ or 300mm, which is greater, or 1.25 times anchorage length of smaller bar.
5. 250 indicates mild steel.
   460-0 indicates plain high-yield bars.
   460-1 indicates high-yield bars of Type 1.
   460-2 indicates high-yield bars of Type 2.
6. 0°, 90° and 180° indicate no-hook, right-angled hook or bob, and standard hook, respectively.
7. Values for hooks correspond to internal radius of 2φ for mild steel bars and 3φ for high-yield bars.
8. Bars must extend a minimum distance of 4φ beyond any bend.
9. Lengths given correspond to maximum design stresses in steel of $0.87f_y$, in tension and compression. For lower design stresses at point beyond which anchorage is to be provided, determine length required from no-hook value on pro-rata basis. Then if hook is provided, subtract length equal to difference between appropriate values given in table.
10. In beams only, where sufficient links to meet nominal requirements are not provided employ anchorage-bond length corresponding to plain bars, irrespective of actual type of bars provided.
11. For lightweight-aggregate concrete, multiply no-hook length by 1.25. Then, if hook is provided, subtract length equal to difference between appropriate values given on table.

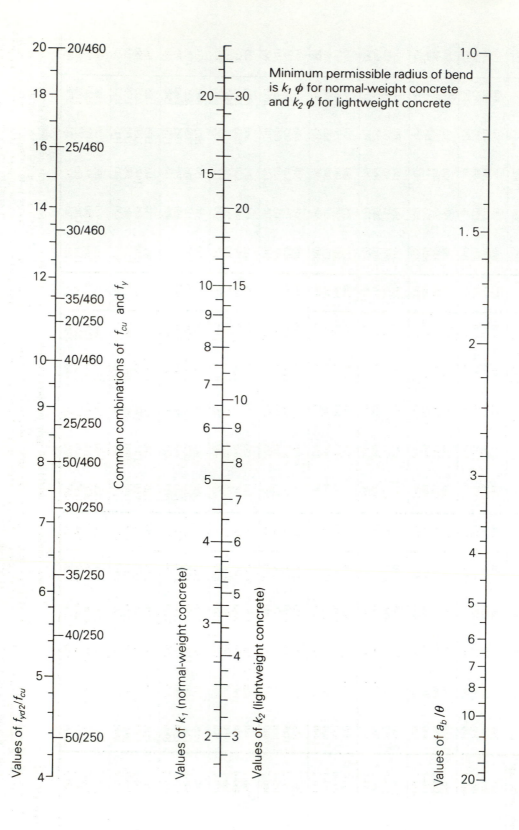

Minimum permissible radius of bend
is $k_1\phi$ for normal-weight concrete
and $k_2\phi$ for lightweight concrete

Values of f_{yd2}/f_{cu}

Common combinations of f_{cu} and f_y

Values of k_1 (normal-weight concrete)

Values of k_2 (lightweight concrete)

Values of a_b/θ

Deflection: BS8110 simplified requirements
(see section 8.2)

Design procedure

1. From the appropriate scale below, read the basic effective depth corresponding to the given length of span and fixity. For cantilevers spanning more than 10m, the minimum effective depth must be justified by rigorous analysis.

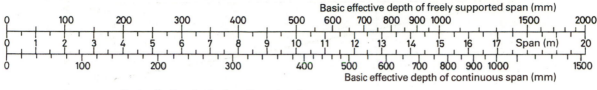

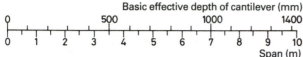

2. Multiply the basic effective depth by the coefficient from the graph below relating to the ultimate applied moment and the service stress in the tension reinforcement.

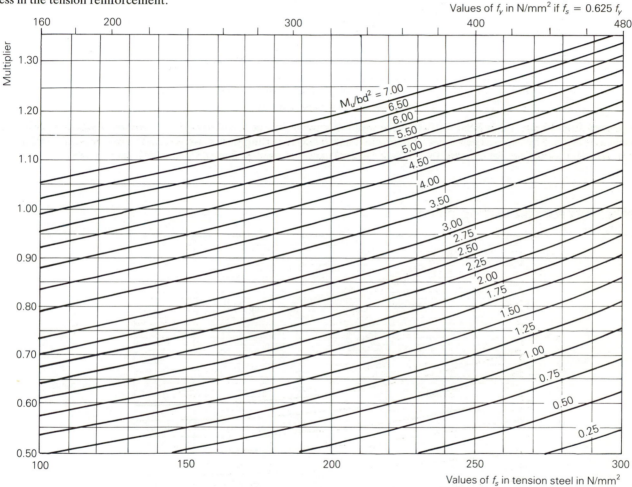

3. Multiply the modified effective depth by the coefficient from the scale below relating to the proportion of compression reinforcement (if any).

4. For flanged sections (including hollow-block and similar floors), multiply the modified effective depth by the coefficient from the scale below, relating to the web width divided by the effective flange width.

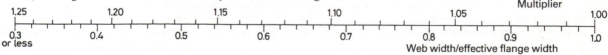

5. For flat slabs not having a width of drop panel in both directions of at least one-third of the total width or length in that direction, multiply the modified effective depth by a further factor of 1.11.

6. Lightweight-aggregate concrete only: for slabs supporting a characteristic imposed load of more than $4\,kN/m^2$ and for all beams of lightweight-aggregate concrete, multiply the modified effective depth by a further factor of 1.18.

Data for rigorous deflection analysis

Coefficients for calculating shrinkage curvature of plain concrete

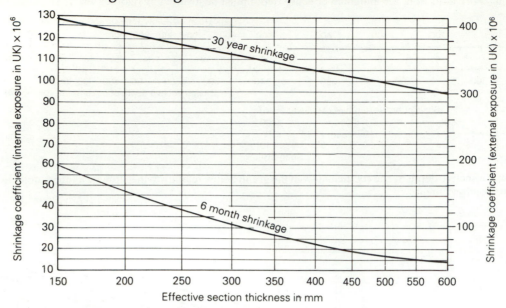

Data derived from Figure 7.2 in BS8110, Part 2. For uniform sections, the code defines *effective section thickness* as twice cross-sectional area divided by exposed perimeter. If permanent immersion or sealing prevents drying, take effective section thickness of 600 mm. Relative humidities for external (in UK) and internal exposure employed are 85% and 45% respectively.

Coefficients for calculating creep

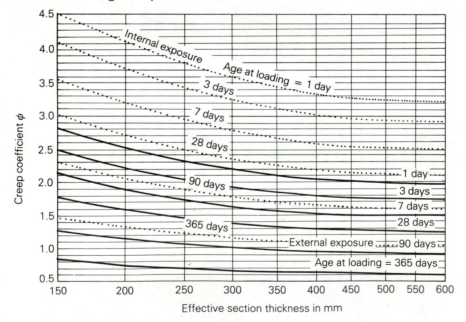

Data derived from Figure 7.1 in BS8110, Part 2. Effective section thickness is as defined above. Thirty-year creep strain in concrete = creep coefficient read from graph above, multiplied by the stress, and divided by the elastic modulus of the concrete at the age at which it is loaded.

According to BS8110 the mean value of the static modulus of normal-weight concrete at 28 days $E_{c28} = 20 + 0.2f_{cu28}$, with a typical range of 6 N/mm² above and below this mean value. The elastic modulus for concrete of any age t from 3 days upwards can be assessed from the expression $E_{c28}(0.4 + 0.6f_{cut}/f_{cu28})$, where a value of f_{cut} can be read from *Figure 4.1*.

For lightweight-aggregate concrete with a density of w kg/m³, multiply the value of static modulus obtained above by $(w/2400)^2$.

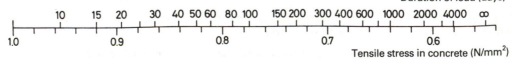

Neutral-axis, lever-arm, and moment-of-inertia factors: $d' = 0.05d$ (see section 8.3)

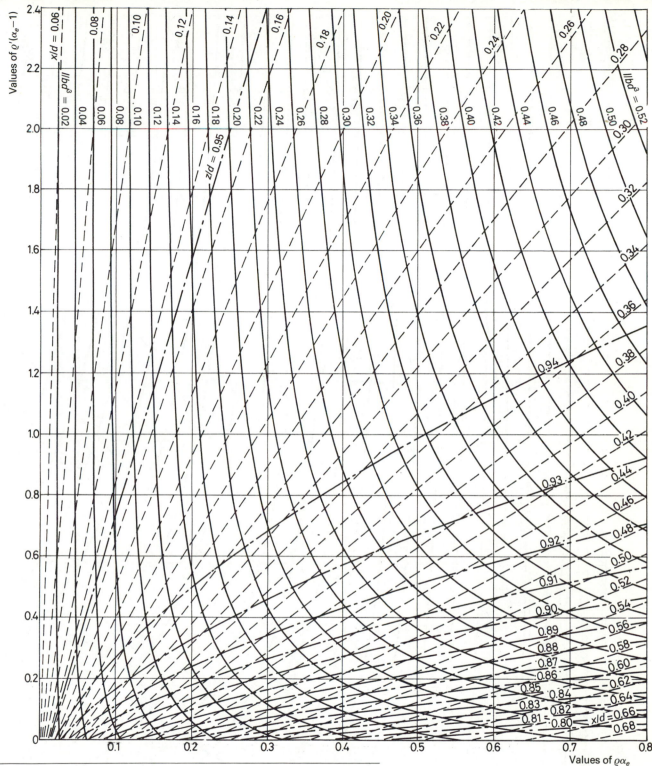

Flanged beams: equivalent factors

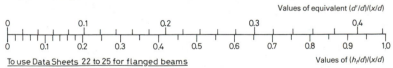

Values of equivalent $(d'/d)/(x/d)$

Values of $(h_f/d)/(x/d)$

To use Data Sheets 22 to 25 for flanged beams

1) Calculate $\rho'(\alpha_e-1) = \frac{h_f}{d}\left(\frac{b}{b_w}-1\right)$.

2) With this value of $\rho'(\alpha_e-1)$ and taking $d'/d = \frac{1}{2}h_f/d$, find the corresponding value of x/d from the appropriate chart, interpolating between charts if necessary.

3) If x/d is less than h_f/d, consider the section as a rectangular beam of breadth b.

If x/d exceeds h_f/d, calculate $\left(\frac{h_f/d}{x/d}\right)$ and read equivalent value of $\left(\frac{d'/d}{x/d}\right)$ from above scale, and thus determine the equivalent ratio of d'/d. Then with the equivalent value of $\rho'(\alpha_e-1)$ obtained from step 1 and with the value of d'/d just obtained, use the appropriate chart or charts to determine I/bd^3.

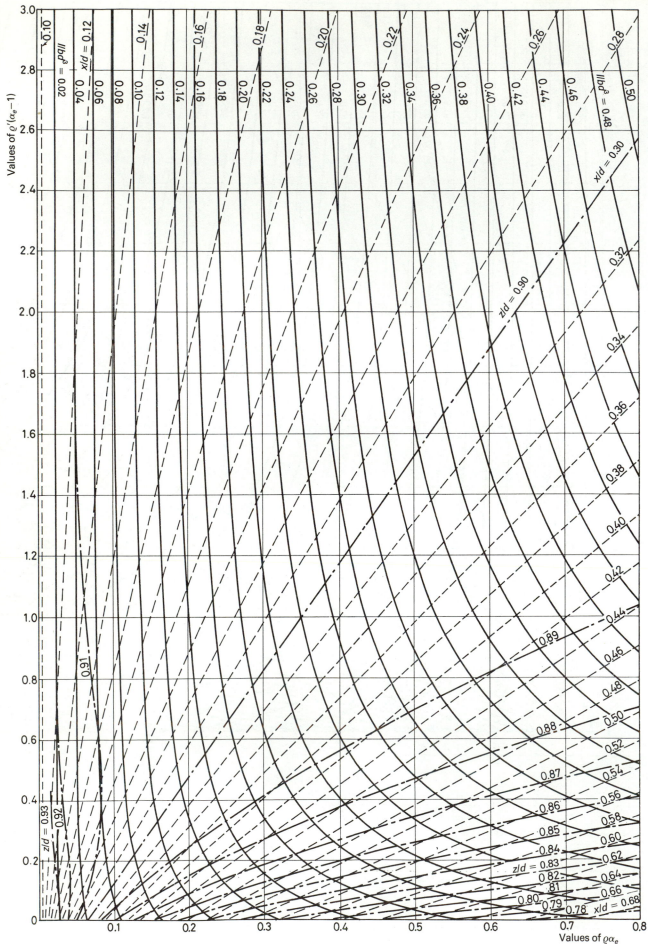

Neutral-axis, lever-arm, and moment-of-inertia factors: $d' = 0.15d$ (*see section 8.3*)

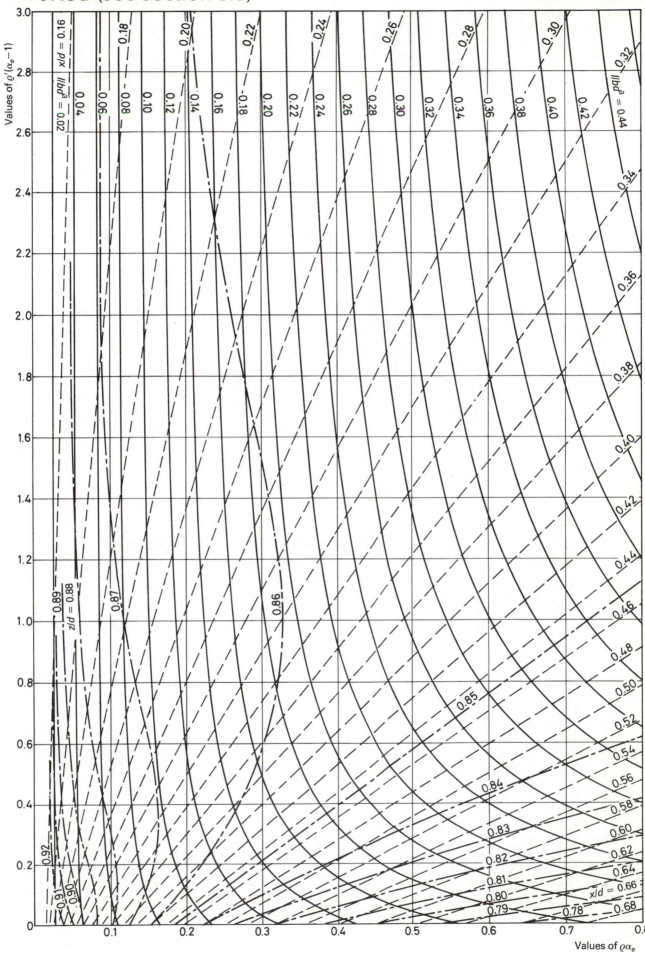

Values of $\varrho'(\alpha_e - 1)$ (vertical axis)

Values of $\varrho\alpha_e$ (horizontal axis)

$\varrho'/\varrho = 0.16$

$I/bd^3 = 0.02$

$z/d = p$

$x/d = 0.66$

Neutral-axis, lever-arm, and moment-of-inertia factors: *d*′ = 0.20*d* (see section 8.3)

Data sheet 25

Values of $\varrho'(\alpha_e - 1)$
Values of $\varrho\alpha_e$
x/d = 0.20
I/bd³ = 0.04
I/bd³ = 0.38
z/d = 0.89
z/d = 0.80 : x/d = 0.60
0.90
0.88
0.87
0.86
0.85
0.84
0.83
0.82
0.81
0.79
0.78

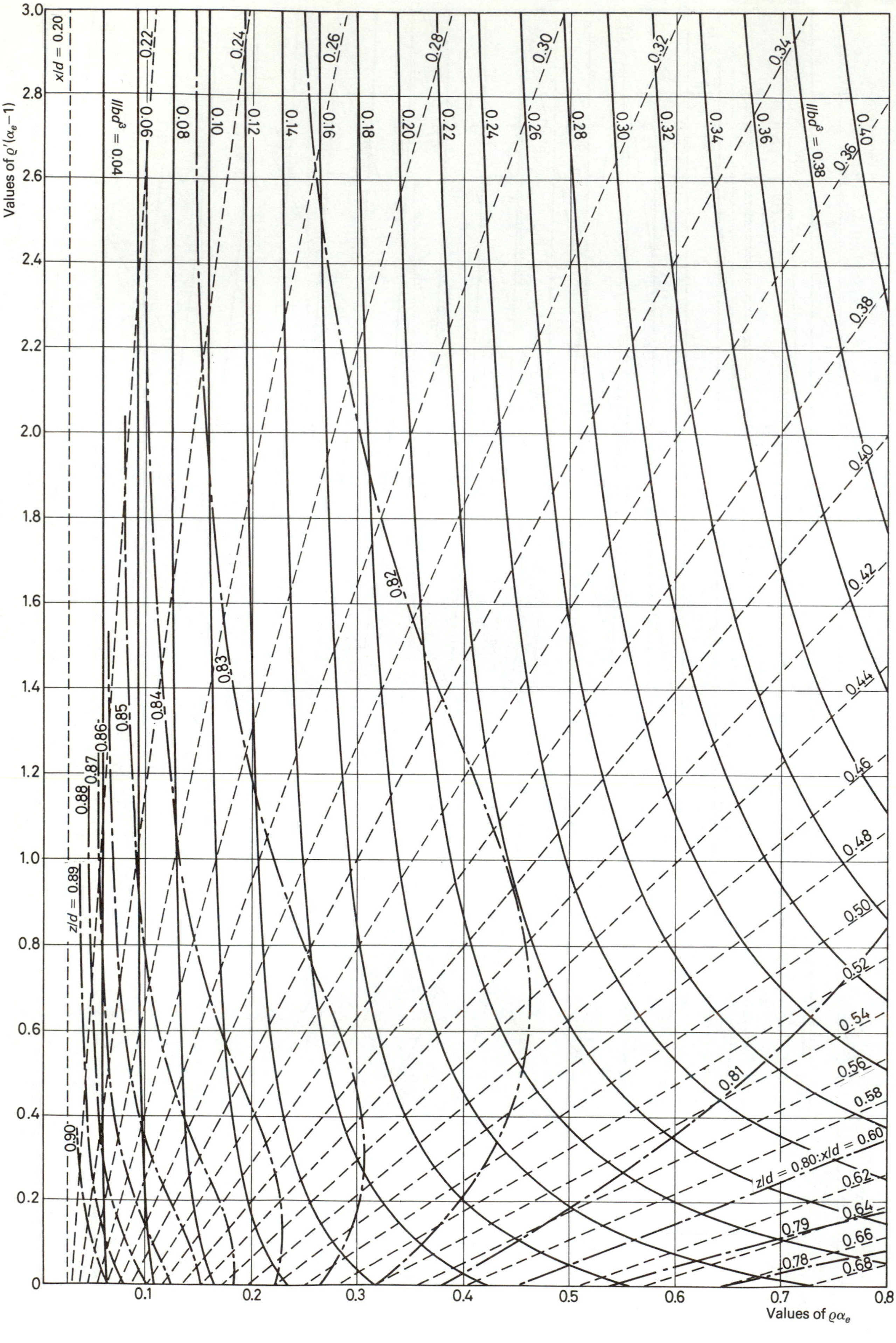

Deflection and bending-moment coefficients: simple cantilever with partial uniform load (*see section 8.3*)

Deflection coefficients

Deflection = coefficient $\times (Fl^3/EI)$.

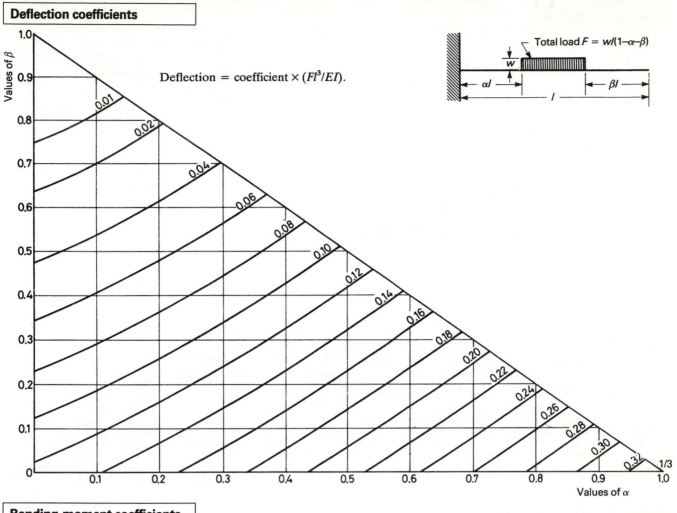

Total load $F = wl(1-\alpha-\beta)$

Values of β

Values of α

Bending-moment coefficients

Bending moment = coefficient $\times Fl$.

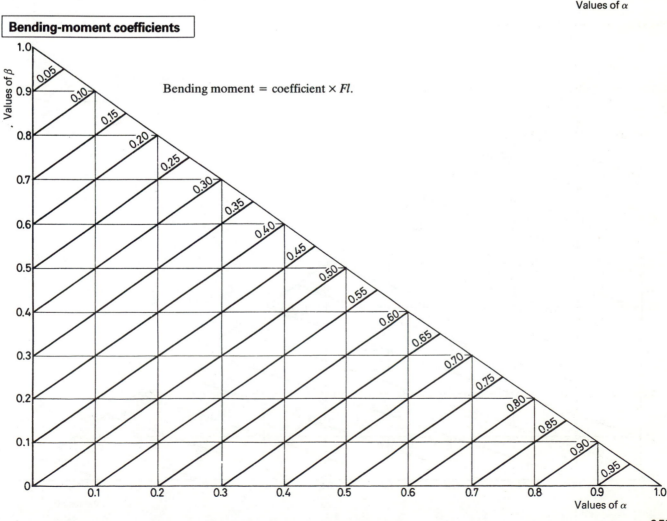

Values of β

Values of α

Deflection and bending-moment coefficients: simple cantilever with partial triangular load 1 (*see* section 8.3)

Data sheet **27**

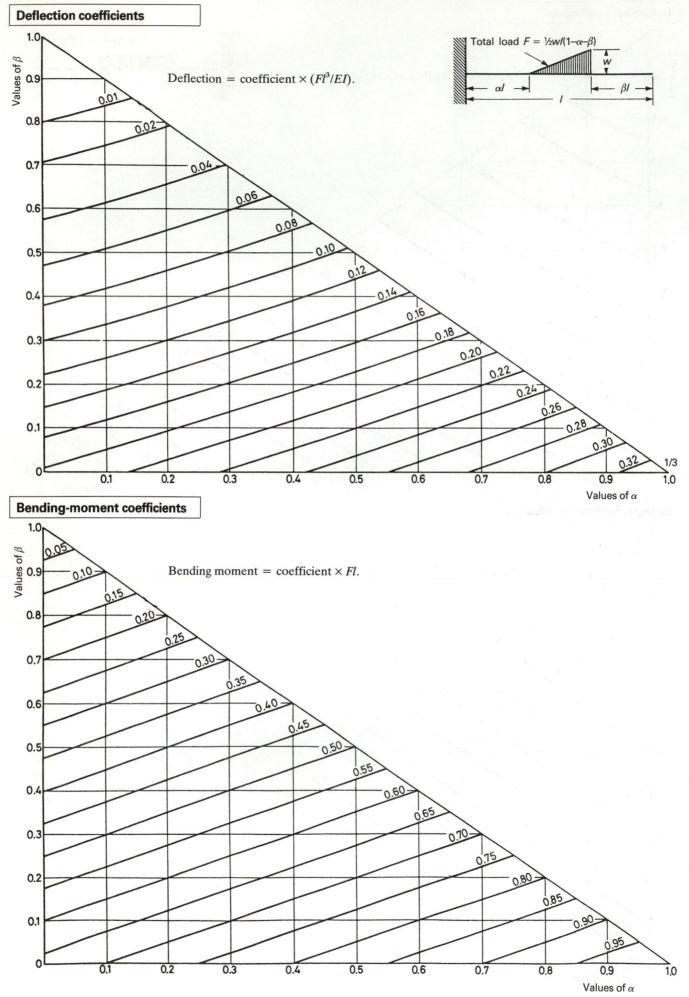

Deflection coefficients

Deflection = coefficient $\times (Fl^3/EI)$.

Total load $F = \frac{1}{2}wl(1-\alpha-\beta)$

Values of β

Values of α

Bending-moment coefficients

Bending moment = coefficient $\times Fl$.

Values of β

Values of α

260

Deflection coefficients

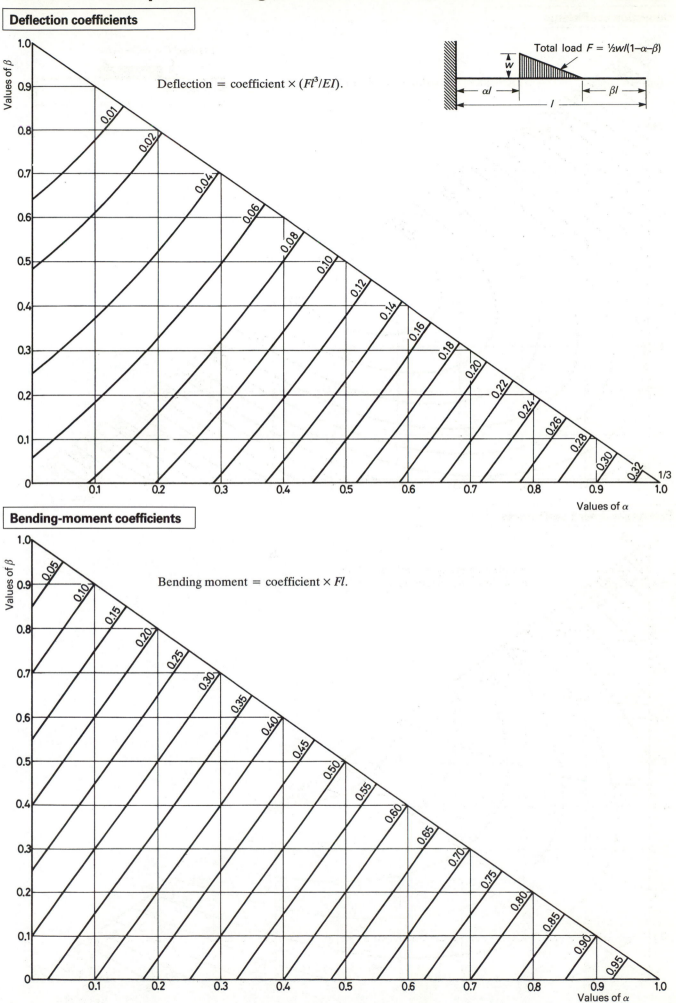

Deflection = coefficient × (Fl^3/EI).

Total load $F = \frac{1}{2}wl(1-\alpha-\beta)$

Values of β

Values of α

Bending-moment coefficients

Bending moment = coefficient × Fl.

Values of β

Values of α

Deflection and bending-moment coefficients: freely supported span with partial uniform load (*see* section 8.3)

Deflection coefficients

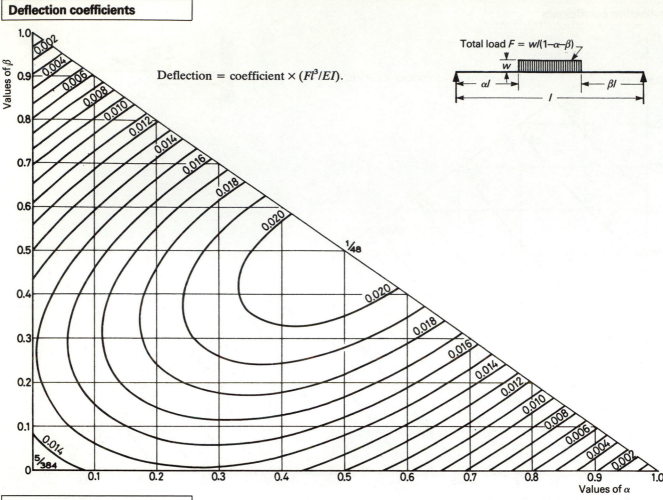

Deflection = coefficient × (Fl^3/EI).

Total load $F = wl(1-\alpha-\beta)$

Bending-moment coefficients

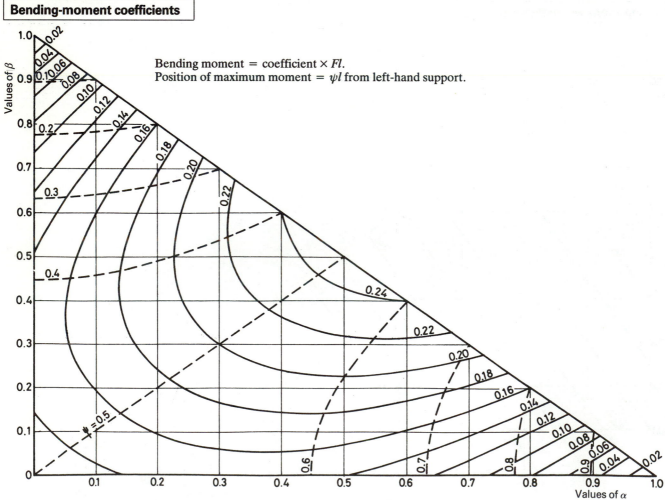

Bending moment = coefficient × Fl.
Position of maximum moment = ψl from left-hand support.

Deflection and bending-moment coefficients: freely supported span with partial triangular load (*see* section 8.3)

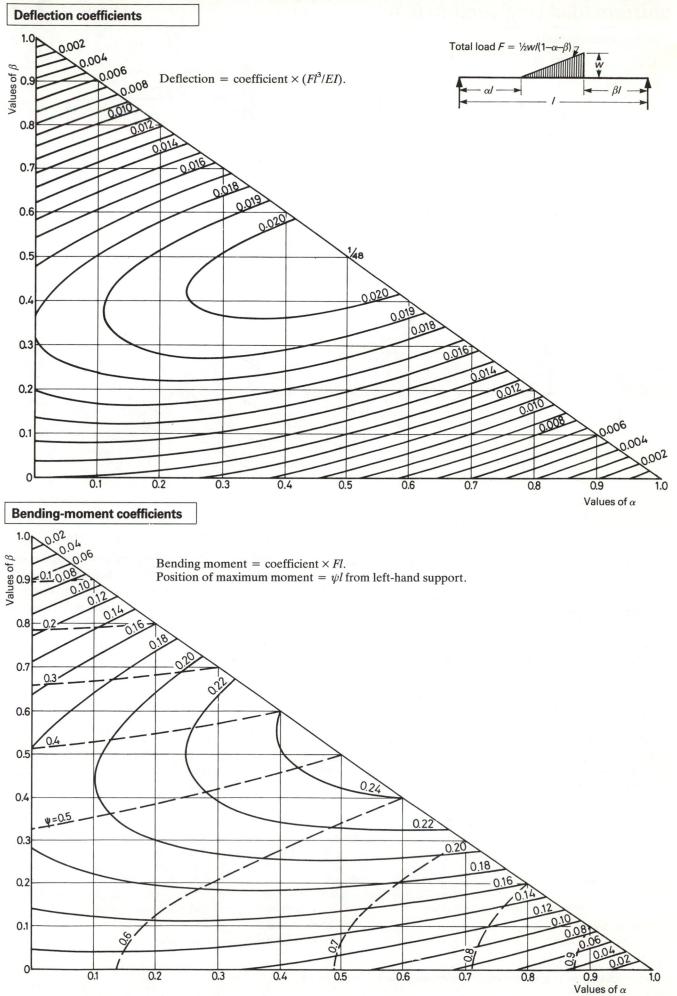

Deflection coefficients

Deflection = coefficient × (Fl^3/EI).

Total load $F = \frac{1}{2}wl(1-\alpha-\beta)$

Bending-moment coefficients

Bending moment = coefficient × Fl.
Position of maximum moment = ψl from left-hand support.

Deflection coefficients: propped cantilever with partial uniform load (*see* section 8.3)

Deflection coefficients

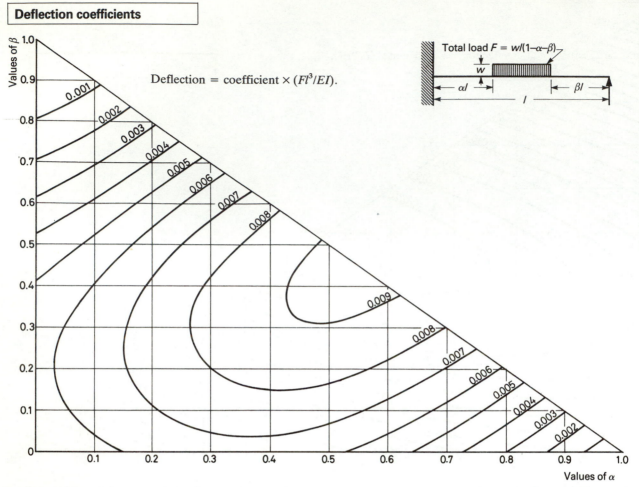

Deflection = coefficient × (Fl^3/EI).

Total load $F = wl(1-\alpha-\beta)$

To determine maximum positive moment in span if one or both supports are fixed:

1. sketch bending-moment diagram due to given loading on freely supported span;
2. calculate corresponding fixed-end moments from information given on *Data Sheet 48* (or Tables 29 to 31 of *RCDH*);

3. then using these fixed-end moments as new base line read off resulting positive bending moment.

If the support moments actually calculated for the system indicate less than complete fixity, determine the central deflection and the maximum positive moment with both fully fixed and free supports, and then interpolate linearly between these limiting values.

Position and value of maximum deflection due to concentrated load

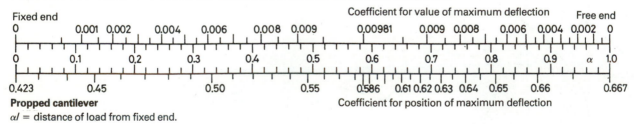

Propped cantilever

αl = distance of load from fixed end.

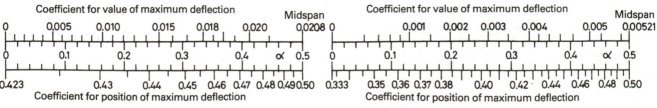

Freely supported span

αl = distance of load from nearer support.

Fully fixed span

αl = distance of load from nearer support.

Deflection = coefficient × (Fl^3/EI).
Distance from support (or from fixed end of propped cantilever) = coefficient × l.

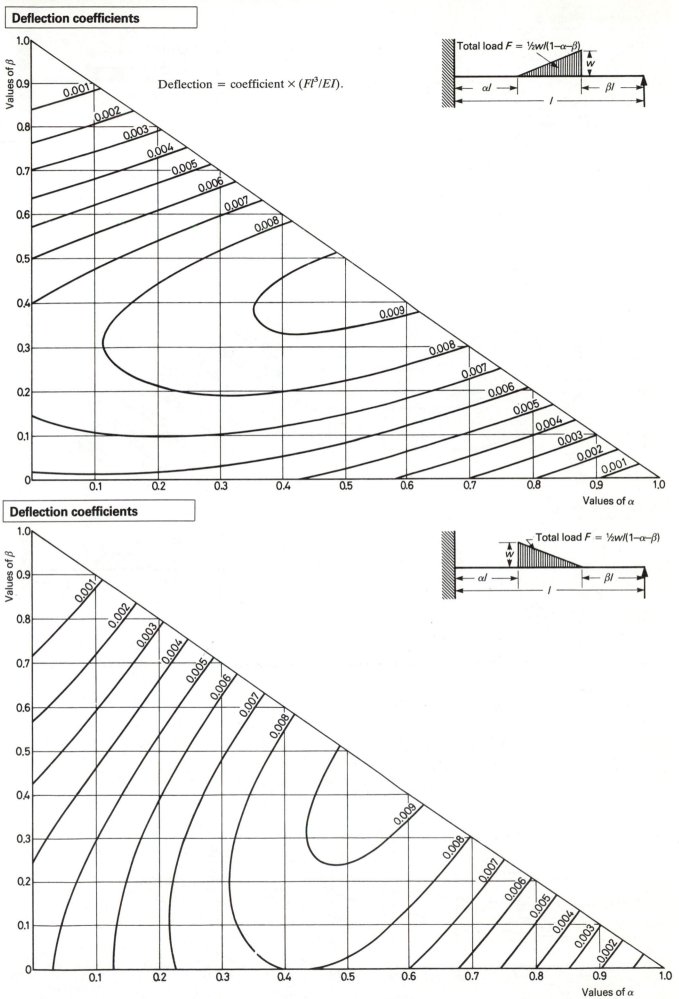

Deflection coefficients

Deflection = coefficient $\times (Fl^3/EI)$.

Total load $F = \tfrac{1}{2}wl(1-\alpha-\beta)$

Deflection coefficients

Total load $F = \tfrac{1}{2}wl(1-\alpha-\beta)$

Deflection coefficients: fully fixed span with partial uniform and triangular loads (*see* section 8.3)

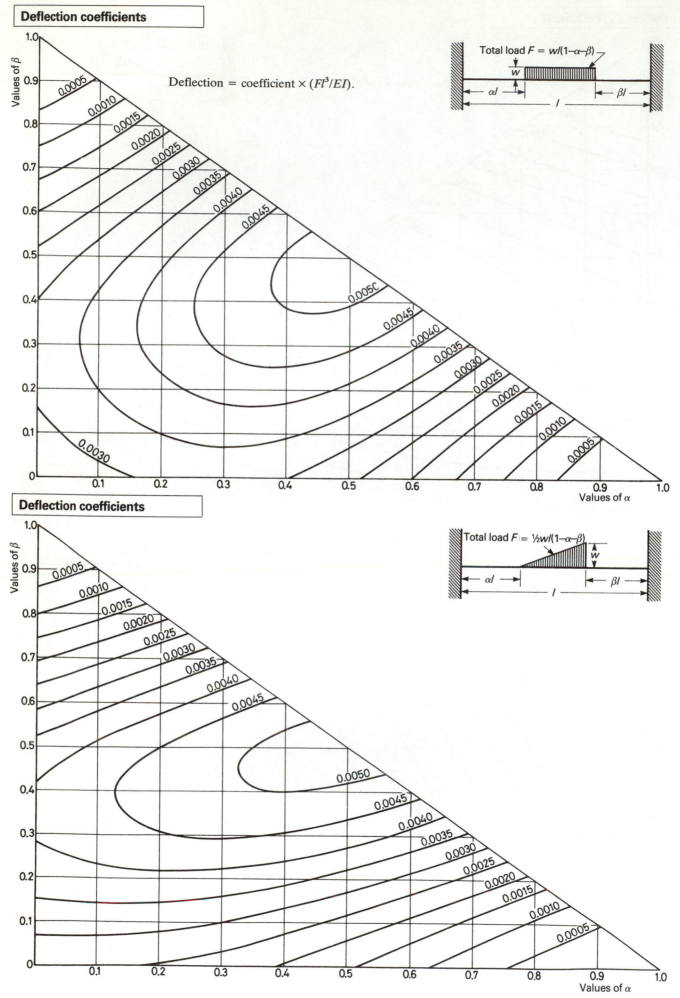

Deflection coefficients

Deflection = coefficient × (Fl^3/EI).

Total load $F = wl(1-\alpha-\beta)$

Deflection coefficients

Total load $F = \frac{1}{2}wl(1-\alpha-\beta)$

Area-moment coefficients: partial uniform load
(see section 8.3)

Area-moment coefficients

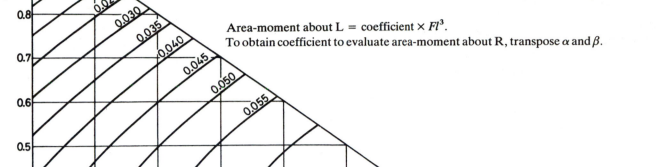

Area-moment about L = coefficient $\times Fl^3$.
To obtain coefficient to evaluate area-moment about R, transpose α and β.

Total load $F = wl(1-\alpha-\beta)$

Values of β

Values of α

Rotation of cantilever support

In order to calculate the rotation of a cantilever support, an appropriate value of K_1 for a particular partial load may be obtained by multiplying the area-moment coefficient (read from the above chart or those on *Data Sheet 35*) by the total partial load times the span, and dividing by the free bending moment due to this partial load.

To consider *any* arrangement of loading, subdivide the loading into a combination of partial uniform and triangular loads. Sketch the free bending-moment diagram due to the total load (by superimposing maximum values and positions for the individual partial loads, obtained with the aid of the charts on *Data Sheets 26* to *28* if required) and hence determine the maximum free moment. Next, determine the area-moment value for each partial load, sum the values obtained, and divide this result by the square of the span and by the value of the maximum free moment to obtain K_1.

Values of K_1 for common arrangements of load

Uniform load across span:

$$K_1 = \tfrac{1}{3}$$

Central concentrated load:

$$K_1 = \tfrac{1}{4}$$

Any *even* number of equally spaced loads:

$$K_1 = \tfrac{1}{3}$$

Any *odd* number j of equally spaced loads:

$$K_1 = \frac{j(j+2)}{3(j+1)^2}$$

Single concentrated load at distance αl from L:

About L:	$K_1 = \tfrac{1}{6}(1+\alpha)$
About R:	$K_1 = \tfrac{1}{6}(2-\alpha)$

For combinations of the above loadings, the appropriate values of K_1 may not be summed. Instead, the procedure outlined on the left must be followed.

Area-moment coefficients: partial triangular load
(*see* section 8.3)

Area-moment about L

Area-moment about L = coefficient $\times Fl^3$.

Total load $F = \frac{1}{2}wl(1-\alpha-\beta)$

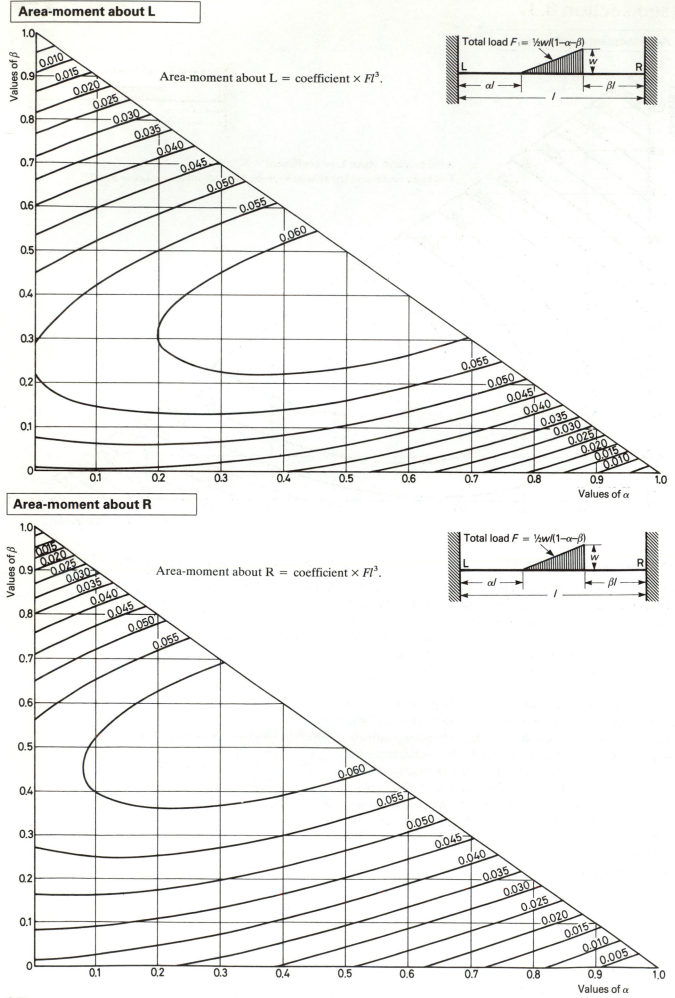

Values of β

0.010
0.015
0.020
0.025
0.030
0.035
0.040
0.045
0.050
0.055
0.060

0.055
0.050
0.045
0.040
0.035
0.030
0.025
0.020
0.015
0.010

Values of α

Area-moment about R

Area-moment about R = coefficient $\times Fl^3$.

Total load $F = \frac{1}{2}wl(1-\alpha-\beta)$

0.015
0.020
0.025
0.030
0.035
0.040
0.045
0.050
0.055

0.060
0.055
0.050
0.045
0.040
0.035
0.030
0.025
0.020
0.015
0.010
0.005

Values of β

Values of α

Maximum bar spacings in sides of deep beams
(*see* section 8.6)

Breadth of beam (mm)	$f_y = 250\,\text{N/mm}^2$ Size of anti-cracking bars (mm)					$f_y = 460\,\text{N/mm}^2$ Size of anti-cracking bars (mm)				
	8	10	12	16	20	8	10	12	16	20
75	213	250	250	250	250	250	250	250	250	250
100	160	250	250	250	250	250	250	250	250	250
125	128	200	250	250	250	236	250	250	250	250
150	107	167	240	250	250	196	250	250	250	250
175	91	143	206	250	250	168	250	250	250	250
200	80	125	180	250	250	147	230	250	250	250
225	*	111	160	250	250	131	204	250	250	250
250	*	100	144	250	250	118	184	250	250	250
275	*	91	131	233	250	107	167	241	250	250
300	*	83	120	213	250	98	153	221	250	250
325	*	77	111	197	250	91	142	204	250	250
350	*	*	103	183	250	84	131	189	250	250
375	*	*	96	171	250	79	123	177	250	250
400	*	*	90	160	250	*	115	166	250	250
425	*	*	85	151	235	*	108	156	250	250
450	*	*	80	142	222	*	102	147	250	250
475	*	*	76	135	211	*	97	139	248	250
500 or more	*	*	*	128	200	*	92	132	236	250

Maximum bar spacings in millimetres in sides of beams exceeding 750 mm deep.
* Indicates larger bar should be used, as calculated maximum spacing is less than 75 mm.
In no case may spacing exceed 250 mm.
Bars may also be assumed to contribute to resistance moment of section.

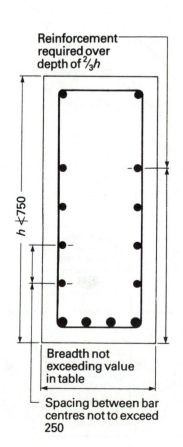

Reinforcement required over depth of $\frac{2}{3}h$

$h \not< 750$

Breadth not exceeding value in table

Spacing between bar centres not to exceed 250

Bold lines indicate maximum spacing given by expression in Code from which numerical values presented in BS8110 (indicated by heavy dots) have been obtained.

In particularly aggressive conditions, chart is only valid when $f_y \not> 300\,\text{N/mm}^2$.

If desired, spacing may be increased to $s_{b\,max}(A_{s\,prov}/A_{s\,req})$.

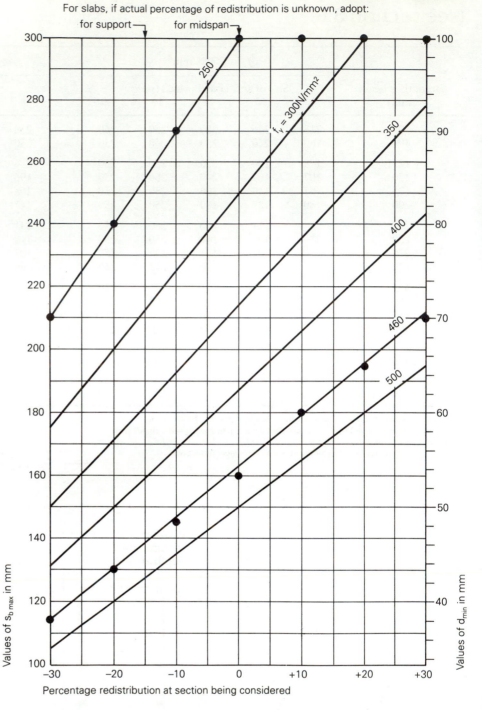

For slabs, if actual percentage of redistribution is unknown, adopt:
for support →   for midspan →

Values of $s_{b\,max}$ in mm

Values of d_{min} in mm

Percentage redistribution at section being considered

Flow chart for slabs subjected to normal internal and external environments

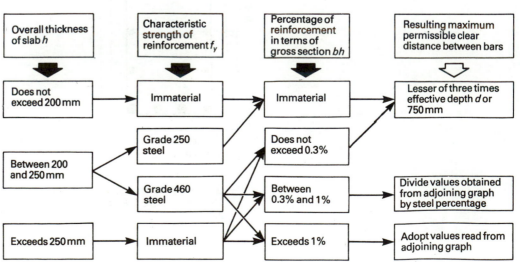

Overall thickness of slab h	Characteristic strength of reinforcement f_y	Percentage of reinforcement in terms of gross section bh	Resulting maximum permissible clear distance between bars
Does not exceed 200 mm	Immaterial	Immaterial	Lesser of three times effective depth *d* or 750 mm
Between 200 and 250 mm	Grade 250 steel	Does not exceed 0.3%	
	Grade 460 steel	Between 0.3% and 1%	Divide values obtained from adjoining graph by steel percentage
Exceeds 250 mm	Immaterial	Exceeds 1%	Adopt values read from adjoining graph

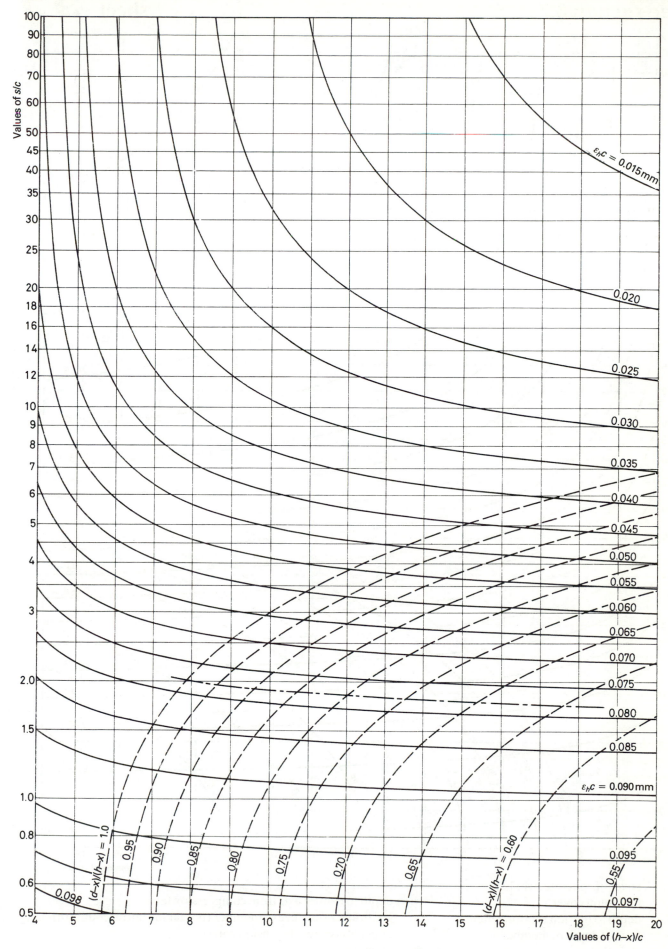

For notation see *Data Sheet 39*.

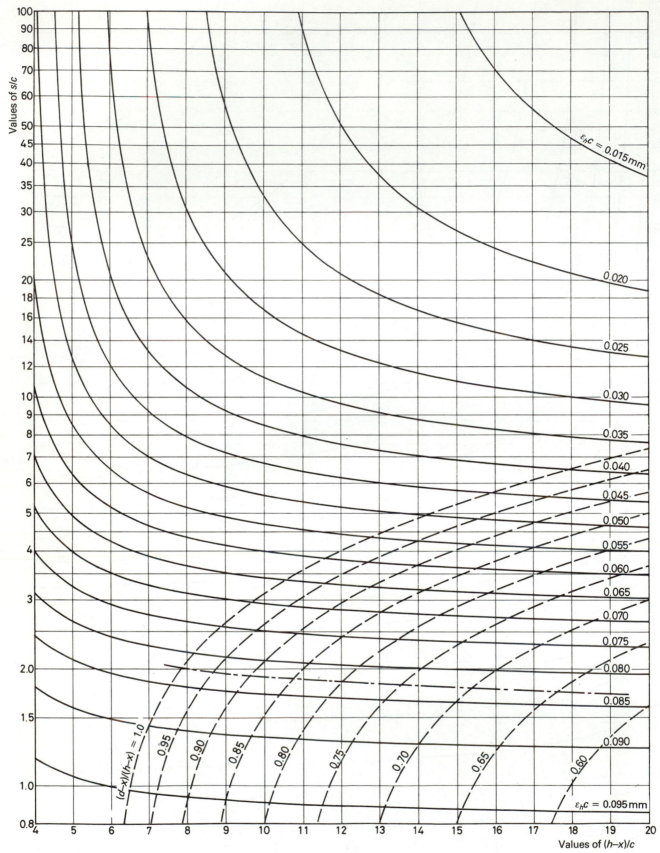

c: cover of concrete over bar

c_{min}:   minimum cover

d: depth to tension reinforcement

h: overall depth of member

s: clear distance between bars

x: depth to neutral axis from compression face

ε_h: strain at tension face of member (not taking into account stiffening effect of concrete in tension zone)

ϕ: diameter of bar

Solid slabs: resistance and reinforcement
(*see section 9.1*)

Characteristic strengths (N/mm²)		Basic factors	Thickness h of slab (mm)									
f_y	f_{cu}		75	100	125	150	175	200	225	250	275	300
250	25	Eff. depth d	44	67	90	112	137	149	174	199	224	240
		$M_u = 3.89d^2$	7 523	17 440	31 470	49 180	73 470	86 280	117 600	153 900	195 000	223 800
		$A_s = 23.14d$	1 018	1 550	2 083	2 604	3 182	3 448	4 027	4 606	5 184	5 555
		Arrangement	12 @ 100	16 @ 125	20 @ 150	25 @ 175	25 @ 150	32 @ 225	32 @ 175	32 @ 150	32 @ 150	40 @ 225
	30	Eff. depth d	49	70	87	112	124	149	165	190	215	240
		$M_u = 4.63d^2$	11 110	22 680	35 440	58 580	71 170	102 700	126 000	167 100	213 900	266 600
		$A_s = 27.50d$	1 347	1 925	2 406	3 093	3 410	4 097	4 537	5 225	5 912	6 600
		Arrangement	12 @75	20 @ 150	25 @ 200	25 @ 150	32 @ 225	32 @ 175	40 @ 275	40 @ 225	40 @ 200	40 @ 175
	35	Eff. depth d	47	70	87	99	124	140	165	190	215	240
		$M_u = 5.36d^2$	11 840	26 280	41 060	52 560	82 460	105 100	146 000	193 600	247 900	308 900
		$A_s = 31.79d$	1 494	2 225	2 781	3 146	3 941	4 450	5 244	6 039	6 834	7 629
		Arrangement	16 @ 125	20 @ 125	25 @ 175	32 @ 250	32 @ 200	40 @ 275	40 @ 225	40 @ 200	40 @ 175	40 @ 150
	40	Eff. depth d	47	70	87	99	124	140	165	190	215	240
		$M_u = 6.09d^2$	13 450	29 830	46 620	59 680	93 630	119 300	165 700	219 800	281 400	350 700
		$A_s = 36.01d$	1 692	2 520	3 151	3 565	4 465	5 041	5 942	6 842	7 742	8 643
		Arrangement	16 @ 100	20 @ 100	25 @ 150	32 @ 225	32 @ 175	40 @ 225	40 @ 200	40 @ 175	40 @ 150	40 @ 125
460	25	Eff. depth d	45	69	92	115	137	162	187	199	224	249
		$M_u = 3.89d^2$	7 860	18 500	32 890	51 390	73 470	102 600	136 600	153 900	195 000	240 900
		$A_s = 12.58d$	566	868	1 157	1 446	1 729	2 044	2 358	2 503	2 817	3 132
		Arrangement	10 @ 125	12 @ 125	16 @ 150	20 @ 200	25 @ 275	25 @ 225	25 @ 200	32 @ 300	32 @ 275	32 @ 250
	30	Eff. depth d	50	72	95	112	137	162	174	199	224	249
		$M_u = 4.63d^2$	11 570	23 990	41 770	58 580	87 520	122 200	140 100	183 300	232 200	287 000
		$A_s = 14.94d$	747	1 076	1 419	1 681	2 055	2 428	2 600	2 974	3 347	3 721
		Arrangement	10 @ 100	16 @ 175	20 @ 200	25 @ 275	25 @ 225	25 @ 200	32 @ 300	32 @ 250	32 @ 225	32 @ 200
	35	Eff. depth d	49	72	95	112	137	149	174	199	224	240
		$M_u = 5.36d^2$	12 870	27 800	48 400	67 870	101 400	119 000	162 300	212 300	269 100	308 900
		$A_s = 17.27d$	846	1 243	1 641	1 943	2 375	2 574	3 006	3 437	3 869	4 146
		Arrangement	12 @ 125	16 @ 150	20 @ 175	25 @ 250	25 @ 200	32 @ 300	32 @ 250	32 @ 225	32 @ 200	40 @ 300
	40	Eff. depth d	54	72	95	112	137	149	174	199	215	240
		$M_u = 6.09d^2$	17 750	31 560	54 950	77 060	115 100	135 100	184 300	241 100	281 400	350 700
		$A_s = 19.57d$	1 056	1 409	1 859	2 201	2 691	2 916	3 405	3 894	4 208	4 697
		Arrangement	12 @ 100	16 @ 125	20 @ 150	25 @ 200	25 @ 175	32 @ 275	32 @ 225	32 @ 200	40 @ 275	40 @ 250
Distribution steel		Mild steel $A_s = 2.4h$	180	240	300	360	420	480	540	600	660	720
		Arrangement	6 @ 125*	8 @ 200	10 @ 250	12 @ 300	12 @ 250	12 @ 225	12 @ 200	12 @ 175	12 @ 150	16 @ 275
		High-yield steel $A_s = 1.3h$	98	130	163	195	228	260	293	325	358	390
		Arrangement	6 @ 125*	6 @ 200	6 @ 150	8 @ 250	8 @ 200	10 @ 300	10 @ 250	10 @ 225	10 @ 300	12 @ 275

M_u ultimate moment of resistance in N m or kN mm) per m width of slab reinforced in tension only

A_s area of tension reinforcement in mm² per m width of slab

Diameter and spacing given (e.g. 12 @ 150 denotes 12 mm bars at 150 mm centres) are possible arrangements of tension reinforcement. If combination of actual cover and bar size adopted gives a different effective depth, calculate M_u and A_s from basic factors given. Cover assumed: 25 mm or bar diameter for $f_u = 25$ N/mm²; 20 mm or bar diameter for $f_{cu} = 30$ N/mm² and 35 N/mm²; 15 mm or bar diameter for $f_{cu} = 40$ N/mm² (rounded to 5 mm diameter dimension above).

* Indicates spacing determined by restriction that it must not exceed 3d.

Reinforcement data (see section 9.1)

Cross-sectional areas (mm²) of specific numbers of bars

Size (mm)	1	2	3	4	5	6	7	8	9	10	11	12	Weight (kg/m)
						Number of bars							
6	28	56	84	113	141	169	197	226	254	282	311	339	0.222
8	50	100	150	201	251	301	351	402	452	502	552	603	0.395
10	78	157	235	314	392	471	549	628	706	785	863	942	0.617
12	113	226	339	452	565	678	791	904	1 017	1 130	1 244	1 357	0.888
16	201	402	603	804	1 005	1 206	1 407	1 608	1 809	2 010	2 211	2 412	1.578
20	314	628	942	1 256	1 570	1 884	2 199	2 513	2 827	3 141	3 455	3 769	2.466
25	490	981	1 472	1 963	2 454	2 945	3 436	3 926	4 417	4 908	5 399	5 890	3.853
32	804	1 608	2 412	3 216	4 021	4 825	5 629	6 433	7 238	8 042	8 846	9 650	6.313
40	1 256	2 513	3 769	5 026	6 283	7 539	8 796	10 053	11 309	12 566	13 823	15 079	9.864

Cross-sectional areas (mm²) of bars at specific spacings

Size (mm)	75	100	125	150	175	200	225	250	300	350	400	450	500
						Bar spacing (mm)							
6	376	282	226	188	161	141	125	113	94	80	70	62	56
8	670	502	402	335	287	251	223	201	167	143	125	111	100
10	1 047	785	628	523	448	392	349	314	261	224	196	174	157
12	1 507	1 130	904	753	646	565	502	452	376	323	282	251	226
16	2 680	2 010	1 608	1 340	1 148	1 005	893	804	670	574	502	446	402
20	4 188	3 141	2 513	2 094	1 795	1 570	1 396	1 256	1 047	897	785	698	628
25	6 544	4 908	3 926	3 272	2 804	2 454	2 181	1 963	1 636	1 402	1 227	1 090	981
32	–	8 042	6 433	5 361	4 595	4 021	3 574	3 216	2 680	2 297	2 010	1 787	1 608
40	–	–	10 053	8 377	7 180	6 283	5 585	5 026	4 188	3 590	3 141	2 792	2 513

All areas are rounded to the nearest mm² below exact value.

Minimum amounts of reinforcement

Slab thickness h_f (mm)	Self-weight (N/mm²)	High-yield steel $\varrho_1 \not< 0.0013$				Mild steel $\varrho_1 \not< 0.0024$				Over beam flanges near top face of slab $\varrho_1 \not< 0.0015$		
		Assumed effective depth	$A_{s\,req}$	Suggested arrangement	$A_{s\,prov}$	Assumed effective depth	$A_{s\,req}$	Suggested arrangement	$A_{s\,prov}$	$A_{s\,req}$	Suggested arrangement	$A_{s\,prov}$
75	1.77	57	98	6 @ 150	188	57	180	6 @ 150	188	112	6 @ 150	188
100	2.36	82	131	6 @ 200	141	81	240	8 @ 200	251	150	6 @ 150	188
125	2.95	106	163	8 @ 300	167	105	300	10 @ 250	314	187	8 @ 250	201
150	3.54	131	196	8 @ 250	201	130	360	10 @ 200	392	225	8 @ 200	251
175	4.13	156	228	8 @ 225	223	154	420	12 @ 250	452	262	10 @ 250	314
200	4.72	180	261	10 @ 300	261	179	480	12 @ 225	502	300	10 @ 250	314
225	5.31	205	293	10 @ 250	314	204	540	12 @ 200	565	337	10 @ 225	349
250	5.90	230	326	10 @ 225	349	222	600	16 @ 300	670	375	12 @ 300	376
275	6.49	255	358	10 @ 200	392	247	660	16 @ 300	670	412	12 @ 250	452
300	7.08	280	391	10 @ 200	392	272	720	16 @ 275	731	450	12 @ 250	452
325	7.67	304	423	12 @ 250	452	297	780	16 @ 250	804	487	12 @ 225	502
350	8.26	329	456	12 @ 225	502	322	840	16 @ 225	893	525	12 @ 200	565
375	8.85	354	488	12 @ 225	502	347	900	16 @ 200	1 005	562	12 @ 200	565
400	9.44	379	521	12 @ 200	565	372	960	16 @ 200	1 005	600	16 @ 300	670

Maximum bar spacing: if $A_s/bd < 0.003$, clear spacing between bars must not exceed lesser of $3d$ or 750 mm.
$\varrho_1 = A_s/bh$.
Assumed effective depth is based on minimum (15 mm or, with 16 mm bars, 20 mm) cover and on bar size given. With greater cover, maximum bar spacing ($\not> 3d$) must be rechecked.

Maximum effective width of flanged beam

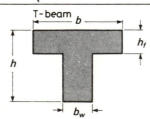

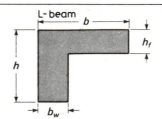

Flange reinforcement must extend across full width b.

$b \not> b_w + \frac{1}{5} l_e$ or actual width.
$b \not> b_w + \frac{1}{10} l_e$ or actual width.
l_e is length of flange in compression.

Simplified detailing requirements

Continuous beam with spans approximately equal

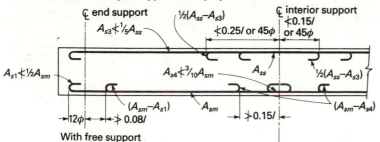

With free support

Continuous slab with spans approximately equal and designed to resist simplified load arrangement specified in Code

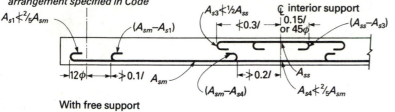

With free support

Freely supported beam or slab

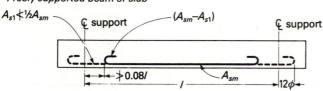

Monolithic with supporting beam or wall

A_{s5} depends on fixity provided:
$\tfrac{1}{2}A_{sm}$ generally sufficient.

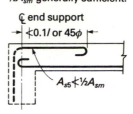

Cantilevered beam or slab

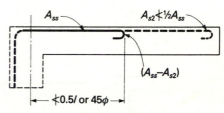

The simplified requirements illustrated are in accordance with clauses 3.12.10.2 and 3.12.10.3 of BS8110 and apply to members supporting substantially uniform loads.

If A_{sm} is the area of tension steel required in the bottom of the member at midspan, and A_{ss} is the area of tension steel required in the top of the member over a support, then A_{s1} to A_{s5} are the areas of tension reinforcement required at various points, in terms of A_{sm} and A_{ss}, and terminating at the positions indicated.

Minimum spacing required for bars or groups of bars

Individual bars or pairs one above the other:

Minimum horizontal spacing required $= h_{agg} + 5\,\text{mm}$
Minimum vertical spacing required $= \tfrac{2}{3}h_{agg}$

Pairs of bars side by side:

Minimum spacing required in each direction $= h_{agg} + 5\,\text{mm}$

Groups of more than two bars:

Minimum spacing required in each direction $= h_{agg} + 15\,\text{mm}$

h_{agg} is maximum size of coarse aggregate.

In all cases horizontal spaces between bars or groups should be vertically in line.

Nominal cover (mm) to all reinforcement

Exposure	Typical conditions	Concrete grade				
		C30*	C35*	C40*	C45*	C50 or more
Mild exposure	Complete protection from weather or from aggressive conditions except briefly during construction	25	20	15†	15†	15†
Moderate exposure	Protected from heavy rain and from freezing while saturated. Concrete continuously under water; in contact with non-aggressive soil; subject to condensation	‡	35	30	25	20
Severe exposure	Exposed to driving rain, to alternate wetting and drying, or to freezing while wet. Liable to heavy condensation	‡	‡	40	30	25
Very severe exposure	Exposed to sea-water spray; severe freezing while wet; de-icing salt; corrosive fumes	‡	‡	50§	40§	30
Extreme exposure	Exposed to abrasion, e.g. sea with shingle, acidic flowing water, machinery, or vehicles	‡	‡	‡	60§	50

* BS8110 permits these grades to be reduced to C25, C30, C35 and C40 respectively if a 'systematic checking regime is established to ensure compliance with limits on free-water/cement ratio and cement content'.
† Increase to 20 mm if maximum aggregate size exceeds 15 mm.
‡ Denotes grade of concrete unsuitable for given conditions.
§ Air-entrained mix must be used if concrete is subjected to freezing while wet.
Code also specifies limits for maximum free-water/cement ratio and minimum cement content corresponding to nominal cover values.

One-way slabs: Pigeaud's coefficients (*see section 9.4*)

Coefficients for loads positioned symmetrically

a_x/l_x	a_y/l_x	a_x	a_y	a_x/l_x	a_y/l_x	a_x	a_y
0	0	—	—	0.6	0	0.164	0.164
	0.5	0.249	0.126		0.5	0.138	0.086
	1.0	0.182	0.070		1.0	0.116	0.051
	1.5	0.143	0.045		1.5	0.096	0.032
	2.0	0.116	0.031		2.0	0.079	0.021
	2.5	0.096	0.023		2.5	0.066	0.016
	3.0	0.083	0.019		3.0	0.056	0.013
0.2	0	0.268	0.268	0.8	0	0.135	0.133
	0.5	0.202	0.113		0.5	0.117	0.074
	1.0	0.157	0.064		1.0	0.098	0.043
	1.5	0.126	0.041		1.5	0.081	0.027
	2.0	0.103	0.028		2.0	0.068	0.018
	2.5	0.086	0.021		2.5	0.058	0.014
	3.0	0.073	0.017		3.0	0.048	0.011
0.4	0	0.205	0.205	1.0	0	0.113	0.110
	0.5	0.167	0.101		0.5	0.097	0.063
	1.0	0.133	0.056		1.0	0.081	0.035
	1.5	0.110	0.036		1.5	0.068	0.022
	2.0	0.092	0.025		2.0	0.057	0.015
	2.5	0.077	0.019		2.5	0.048	0.012
	3.0	0.065	0.015		3.0	0.040	0.009

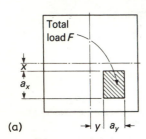

(a)

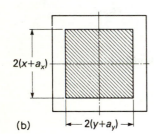

(b)

Poisson's ratio is assumed to be 0.2.

$l_y = \infty$.

In direction of span l_x: $M = a_x F$ per unit width.

At right angles to span l_x: $M = a_y F$ per unit width.

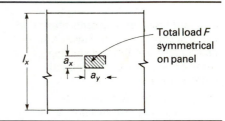

Total load F symmetrical on panel

The following notes also apply to *Data Sheet 45*.

Treatment of loads positioned asymmetrically

To determine bending moments due to concentrated load F shown in diagram (a):

1. Determine coefficients a_x and a_y for symmetrical load spread over area of $2(x + a_x)$ by $2(y + a_y)$, as shown in diagram (b), and multiply these by $(x + a_y)(y + a_x)$.
2. Determine coefficients a_x and a_y for symmetrical load spread over area of $2x$ by $2y$ (diagram (c)) and multiply these by xy.
3. Sum the coefficients for a_x and a_y obtained from steps 1 and 2.
4. Determine coefficients a_x and a_y for symmetrical load spread over area of $2(x + a_x)$ by $2y$ (diagram (d)) and multiply these by $(x + a_x)y$.
5. Determine coefficients a_x and a_y for symmetrical load spread over area of $2x$ by $2(y + a_y)$ (diagram (e)) and multiply these by $x(y + a_y)$.
6. Sum the coefficients for a_x and a_y obtained from steps 4 and 5.
7. Substract the coefficients for a_x and a_y obtained from step 6 from those obtained from step 3 and divide both by $a_x a_y$.

Then $M_x = a_x F$ per unit width and $M_y = a_y F$ per unit width, where a_x and a_y are the coefficients obtained from step 7.

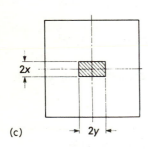

(c)

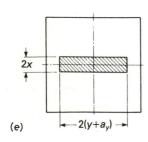

(d)

For values of Poisson's ratio other than 0.2

With a value of Poisson's ratio of k, the modified coefficients a_x' and a_y' may be obtained from the expressions

$$a_x' = [a_x(5 - k) + a_y(5k - 1)]/4.8$$
$$a_y' = [a_y(5 - k) + a_x(5k - 1)]/4.8$$

where a_x and a_y are the coefficients given in the above table or that on *Data Sheet 45*.

(e)

Two-way slabs: BS8110 design coefficients for moments and loads on supporting beams (*see section 10.1*)

β_{sx} = bending-moment coefficients for short span (middle strip)

Ratio of spans l_y/l_k	Corners not restrained — Shorter span	Corners not restrained — Longer span	Corners rest'd — Shorter span	① Pos. BM at midspan	② Neg. BM at support	② Pos. BM at midspan	③ Pos. BM at midspan	④ Neg. BM at support	④ Pos. BM at midspan	⑤ Neg. BM at support	⑤ Pos. BM at midspan	⑥ Neg. BM at support	⑥ Pos. BM at midspan	⑦ Neg. BM at support	⑦ Pos. BM at midspan	⑧ Neg. BM at support	⑧ Pos. BM at midspan
	Freely supported at all edges — α = bending-moment coefficients at midspan																
1.00	0.062	0.062	0.055	0.042	0.057	0.043	0.034	0.046	0.034	0.047	0.036	0.040	0.030	0.039	0.029	0.032	0.024
1.10	0.074	0.061	0.065	0.054	0.065	0.048	0.046	0.050	0.038	0.056	0.042	0.049	0.036	0.044	0.033	0.037	0.028
1.20	0.084	0.059	0.074	0.063	0.071	0.053	0.056	0.054	0.040	0.063	0.047	0.056	0.042	0.048	0.036	0.042	0.032
1.30	0.093	0.055	0.081	0.071	0.076	0.057	0.065	0.057	0.043	0.069	0.051	0.063	0.047	0.052	0.039	0.046	0.035
1.40	0.099	0.051	0.087	0.078	0.081	0.060	0.072	0.060	0.045	0.074	0.055	0.068	0.051	0.055	0.041	0.050	0.037
1.50	0.104	0.046	0.092	0.084	0.084	0.063	0.078	0.062	0.047	0.078	0.059	0.073	0.055	0.058	0.043	0.053	0.040
1.60	0.108	0.042	0.097	0.089	0.088	0.066	0.084	0.064	0.048	0.082	0.061	0.077	0.058	0.060	0.045	0.055	0.042
1.70	0.112	0.039	0.101	0.094	0.091	0.068	0.089	0.066	0.050	0.085	0.064	0.081	0.060	0.062	0.047	0.058	0.043
1.80	0.114	0.035	0.105	0.098	0.094	0.070	0.093	0.068	0.051	0.088	0.066	0.084	0.063	0.064	0.048	0.060	0.045
1.90	0.116	0.032	0.108	0.101	0.096	0.072	0.097	0.069	0.052	0.091	0.068	0.087	0.065	0.066	0.049	0.062	0.046
2.00	0.118	0.029	0.111	0.105	0.098	0.074	0.100	0.070	0.053	0.093	0.070	0.089	0.067	0.067	0.050	0.063	0.048
β_{sy} Neg. BM at edge				0.058				0.045		0.045		0.037		0.037		0.032	
β_{sy} Pos. BM at midspan			0.056	0.044		0.044	0.034		0.034		0.034		0.028		0.028		0.024

β_{vx} = shear-force coefficients for short span: discontinuous and continuous edge

l_x/l_y	① Discont.	② Discont.	② Cont.	③ Discont.	④ Discont.	④ Cont.	⑤ Cont.	⑤ Discont.	⑥ Cont.	⑥ Discont.	⑦ Cont.	⑧ Cont.
1.00	0.33	0.29	0.45	0.30	0.26	0.40	0.40	0.26	0.36	0.24	0.36	0.33
1.10	0.36	0.33	0.48	0.32	0.30	0.43	0.44	0.29	0.40	0.27	0.39	0.36
1.20	0.39	0.36	0.51	0.34	0.33	0.45	0.47	0.31	0.44	0.29	0.42	0.39
1.30	0.41	0.38	0.53	0.35	0.36	0.47	0.50	0.33	0.47	0.31	0.44	0.41
1.40	0.43	0.40	0.55	0.36	0.38	0.48	0.52	0.34	0.49	0.32	0.45	0.43
1.50	0.45	0.42	0.57	0.37	0.40	0.49	0.54	0.35	0.51	0.34	0.47	0.45
1.60	0.46	0.43	0.58	0.38	0.41	0.50	0.55	0.36	0.53	0.35	0.48	0.46
1.70	0.47	0.44	0.59	0.39	0.43	0.51	0.56	0.37	0.54	0.36	0.49	0.47
1.80	0.48	0.46	0.61	0.39	0.44	0.52	0.58	0.38	0.56	0.36	0.50	0.48
1.90	0.49	0.47	0.62	0.40	0.46	0.53	0.59	0.39	0.57	0.37	0.51	0.49
2.00	0.50	0.48	0.63	0.41	0.47	0.54	0.60	0.40	0.59	0.38	0.52	0.50
β_{vy} Discont. edge	0.33	0.30		0.29	0.26			0.26		0.24		
β_{vy} Cont. edge			0.45			0.40	0.40		0.36		0.36	0.33

β_{sy} = bending moment coefficients for long span (middlestrip).
β_{vy} = shear-force coefficients for long span.

Bending moments

For slabs freely supported at all edges:

Bending moment per unit width of middle strip = coefficient × load per unit area × (short span)^2 = $\alpha n l_x^2$, where coefficient for short span is α_{sx} and coefficient for long span is α_{sy}.

For slabs continuous at one or more edges:

Bending moment per unit width of middle strip = coefficient × load per unit area × (short span)^2 = $\beta n l_x^2$, where coefficient for short span is β_{sx} and coefficient for long span is β_{sy}.

Bending moment at discontinuous edge or edge monolithic with support = -0.5 × bending moment on middle strip in adjoining span.

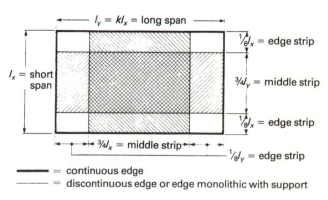

$l_y = k l_x$ = long span
l_x = short span
$\tfrac{1}{8}l_x$ = edge strip
$\tfrac{3}{4}l_y$ = middle strip
$\tfrac{1}{8}l_x$ = edge strip
$\tfrac{3}{4}l_x$ = middle strip
$\tfrac{1}{8}l_y$ = edge strip

—— = continuous edge
‑‑‑‑ = discontinuous edge or edge monolithic with support

Shearing forces and loads on supporting beams

Load per unit length transferred from middle strip of slab to supporting beam = coefficient × load per unit area of slab × short span = $\beta n l_x$, where coefficient for short span is β_{vx} and coefficient for long span is $_{vy}$.

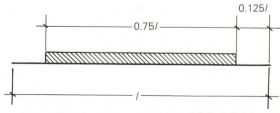

0.125*l*
0.75*l*
l

Assumed distribution of load on supporting beam

Span ratio l_y/l_x

a_x/l_x	a_y/l_y	1 a_x	1 a_y	$1\frac{1}{4}$ a_x	$1\frac{1}{4}$ a_y	$\sqrt{2}$ a_x	$\sqrt{2}$ a_y	$1\frac{2}{3}$ a_x	$1\frac{2}{3}$ a_y	2 a_x	2 a_y
0	0	—	—	—	—	—	—	—	—	—	—
	0.2	0.277	0.219	0.287	0.211	0.288	0.183	0.277	0.161	0.265	0.147
	0.4	0.215	0.157	0.226	0.148	0.218	0.124	0.211	0.104	0.195	0.087
	0.6	0.179	0.125	0.181	0.115	0.178	0.093	0.169	0.076	0.157	0.060
	0.8	0.148	0.101	0.147	0.087	0.147	0.073	0.138	0.059	0.128	0.045
	1.0	0.122	0.081	0.124	0.073	0.121	0.058	0.114	0.045	0.106	0.035
0.2	0	0.219	0.277	0.247	0.270	0.257	0.275	0.262	0.273	0.266	0.267
	0.2	0.192	0.192	0.213	0.177	0.215	0.162	0.215	0.147	0.212	0.134
	0.4	0.166	0.144	0.179	0.128	0.179	0.113	0.175	0.095	0.170	0.081
	0.6	0.141	0.115	0.149	0.099	0.150	0.086	0.145	0.071	0.138	0.055
	0.8	0.119	0.094	0.124	0.080	0.125	0.067	0.121	0.056	0.113	0.041
	1.0	0.098	0.075	0.106	0.065	0.104	0.054	0.101	0.042	0.093	0.032
0.4	0	0.157	0.215	0.180	0.203	0.189	0.211	0.197	0.212	0.202	0.206
	0.2	0.144	0.166	0.164	0.148	0.166	0.140	0.171	0.130	0.171	0.117
	0.4	0.128	0.128	0.144	0.111	0.145	0.101	0.147	0.087	0.144	0.073
	0.6	0.112	0.103	0.124	0.088	0.125	0.077	0.124	0.066	0.120	0.051
	0.8	0.096	0.084	0.104	0.071	0.106	0.061	0.104	0.052	0.099	0.037
	1.0	0.080	0.068	0.088	0.057	0.089	0.049	0.086	0.038	0.080	0.028
0.6	0	0.125	0.179	0.139	0.163	0.151	0.171	0.157	0.175	0.162	0.168
	0.2	0.115	0.141	0.128	0.124	0.137	0.122	0.140	0.114	0.142	0.103
	0.4	0.103	0.112	0.115	0.096	0.122	0.089	0.122	0.075	0.122	0.064
	0.6	0.091	0.091	0.101	0.077	0.107	0.069	0.105	0.059	0.102	0.045
	0.8	0.078	0.074	0.087	0.062	0.091	0.055	0.089	0.046	0.084	0.033
	1.0	0.066	0.061	0.074	0.050	0.074	0.042	0.073	0.034	0.069	0.025
0.8	0	0.101	0.148	0.113	0.133	0.123	0.141	0.128	0.141	0.134	0.136
	0.2	0.094	0.119	0.105	0.105	0.113	0.104	0.115	0.095	0.118	0.086
	0.4	0.084	0.096	0.096	0.082	0.101	0.077	0.101	0.066	0.102	0.055
	0.6	0.074	0.078	0.085	0.065	0.088	0.060	0.088	0.052	0.086	0.038
	0.8	0.064	0.064	0.073	0.053	0.075	0.047	0.074	0.040	0.071	0.029
	1.0	0.055	0.054	0.061	0.042	0.062	0.036	0.061	0.029	0.059	0.021
1.0	0	0.081	0.122	0.095	0.110	0.100	0.116	0.105	0.112	0.109	0.112
	0.2	0.075	0.098	0.089	0.085	0.093	0.087	0.095	0.077	0.100	0.073
	0.4	0.068	0.080	0.080	0.068	0.084	0.065	0.084	0.058	0.086	0.046
	0.6	0.061	0.066	0.070	0.055	0.073	0.050	0.073	0.044	0.073	0.033
	0.8	0.054	0.055	0.060	0.045	0.063	0.039	0.062	0.033	0.062	0.025
	1.0	0.048	0.048	0.050	0.036	0.052	0.031	0.052	0.025	0.051	0.019

Poisson's ratio assumed to be 0.2.

See also notes on *Data Sheet 43*.

Bending moments:
  On short span (l_x): $M = k_x a_x F$ per unit width
  On long span (l_y): $M = k_y a_y F$ per unit width

Total load F (symmetrical on panel)

Adjustment factors to allow for continuity

Ratio of spans l_y/l_x	Pos. BM at mid-span	Pos. BM at mid-span	Neg. BM at cont. edge	Pos. BM at mid-span	Pos. BM at mid-span	Neg. BM at cont. edge	Pos. BM at mid-span	Neg. BM at cont. edge	Pos. BM at mid-span	Neg. BM at cont. edge	Pos. BM at mid-span	Neg. BM at cont. edge	Pos. BM at mid-span	Neg. BM at cont. edge	Pos. BM at mid-span
k_x for short span — 1	1.0	0.9 to 0.8	1.15 to 1.05	0.9 to 0.8	0.8 to 0.65	1.1 to 0.8	0.8 to 0.65	1.1 to 0.85	0.9 to 0.65	1.05 to 0.7	0.85 to 0.5	1.05 to 0.65	0.85 to 0.5	1.0 to 0.6	0.8 to 0.45
$1\frac{1}{4}$	1.0	0.9 to 0.85	1.15 to 1.0	0.9 to 0.75	0.95 to 0.75	1.05 to 0.75	0.8 to 0.55	1.1 to 0.85	0.9 to 0.65	1.05 to 0.75	0.85 to 0.55	1.05 to 0.65	0.85 to 0.5	1.0 to 0.6	0.8 to 0.45
$\sqrt{2}$	1.0	0.9 to 0.85	1.15 to 0.95	0.9 to 0.75	0.95 to 0.75	1.05 to 0.7	0.75 to 0.55	1.1 to 0.85	0.9 to 0.65	1.05 to 0.75	0.85 to 0.55	1.05 to 0.65	0.85 to 0.5	1.0 to 0.6	0.8 to 0.45
$1\frac{2}{3}$	1.0	0.9	1.15 to 0.95	0.9 to 0.7	0.95 to 0.8	1.05 to 0.7	0.75 to 0.5	1.1 to 0.85	0.9 to 0.65	1.05 to 0.8	0.85 to 0.6	1.05 to 0.65	0.85 to 0.5	1.0 to 0.6	0.8 to 0.45
2	1.0	0.9	1.2 to 0.95	0.9 to 0.7	0.95 to 0.85	1.05 to 0.65	0.75 to 0.5	1.1 to 0.85	0.9 to 0.65	1.05 to 0.8	0.85 to 0.6	1.05 to 0.65	0.85 to 0.5	1.0 to 0.6	0.8 to 0.45
k_y for long span	1.0	0.9 to 0.8	1.15 to 1.05	0.9 to 0.8	mid-span 0.8 to 0.65 edges 1.1 to 0.8	—	0.8 to 0.65	1.1 to 0.85	0.9 to 0.65	1.05 to 0.7	0.85 to 0.5	1.05 to 0.65	0.85 to 0.5	1.0 to 0.6	0.8 to 0.45

Where two coefficients are given: upper coefficient applies for load on small area; lower coefficient applies for load on entire area. Interpolate approximately for intermediate conditions.

Flat slabs: simplified method (*see section 11.1*)

Values of bending-moment coefficients k

	Column strip	Middle strip
Negative bending moment at interior support	−0.041 25	−0.013 75
Positive bending moment in interior span	+0.039 05	+0.031 95
Negative bending moment at penultimate support	−0.047 25	−0.015 75
Positive moment in end span	+0.045 65	+0.037 35
Negative moment at *column* end support	−0.030 00*	−0.010 00
Negative moment at *wall* end support	−0.015 00	−0.005 00

* If the bending moments calculated using this coefficient exceed those which may be transferred from the column strip to the columns at the edge of the slab, as calculated below, the edge moments must be reduced to meet these requirements and the positive moments in the end span increased accordingly.

Total ultimate bending moment on strip of width $0.5l_2$:
$M_{ds} = knl_1 l_2(l_1 - 2h_c/3)$, where

n: total ultimate load per unit area (i.e. $1.4g_k + 1.6q_k$)
l_1: distance between column centres in direction of span
l_2: distance between column centres at right angles to l_1
h_c: effective diameter of column head, i.e. diameter at 40 mm below the underside of the slab or drop panel

h_c must not exceed one-quarter of shorter span. If column head is not circular, h_c should be taken as the diameter of a circle having an equivalent area.

Column strip and middle strip of unequal width

Bending-moment factors are based on strips of equal width. If this is not so, first calculate the moments for strips of equal width. Then reduce the moment in the column strip by multiplying this value by the ratio of the actual strip width to that originally assumed. Finally, increase the moment in the middle strip by the same amount that the moment in the column strip has been reduced.

Transfer of moment from slab to edge and corner columns

The maximum moment that can be transferred to a column via the column strip is $0.15b_e d^2 f_{cu}$, where b_e is the effective breadth of the column strip and d is the effective depth of the top steel in the strip. Where the column is near an edge, $b_e = x + y$, where x is the contact length between the innermost column face and the slab, and y is one-half of the sum of the column faces at right angles, also in contact with the slab. Where a column is inset from the slab edge, y is the distance from the edge to face x. Width b_e must not exceed the column strip width for an internal panel.

Other variants are based on these two limiting cases.

Nomogram for axially loaded columns
(see section 13.3)

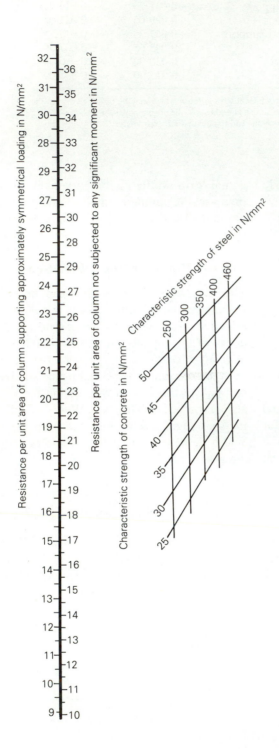

Fixed-end moments for beams and coefficients for slender columns (*see section 14.1*)

Fixed-end moments

Load	C both supports	Load	C support L	C support R
	$\frac{1}{12}$		$\frac{a}{12}(3a^2-8a+6)$	$\frac{a^2}{12}(4-3a)$
	$\frac{5}{48}$		$\beta(1-\beta)^2-\frac{a^2}{12}(2-3\beta)$	$\beta^2(1-\beta)-\frac{a^2}{12}(3\beta-1)$
	$\frac{1}{8}$		$a(1-a)^2$	$a^2(1-a)$
	$\frac{j+2}{12(j+1)}$		$\frac{1}{10}$	$\frac{1}{15}$

j	Factor
1	0.125
2	0.111
3	0.104
4	0.100
5	0.097
∞	0.083

Any number j of equal loads equally spaced

Fixed-end moment $FEM = CFl_b$
where C is coefficient from this table
and F is total load on span l_b

Slender-column coefficients

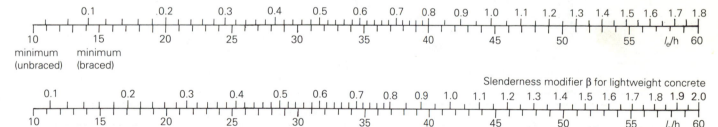

Slenderness modifier β for normal-weight concrete

Slenderness modifier β for lightweight concrete

Total moment to be resisted $M_t = M_i + \beta Nh$, where

M_i: maximum moment due to normal ultimate load, calculated by simple elastic analysis: must not be taken as less than $Nh/20$

h: overall depth of cross-section in plane of bending

a: lesser of two cross-sectional dimensions

l_e: greater of two effective heights of column (i.e. parallel to and at right angles to plane of bending)

β: modifying factor read from above scales

Basic design procedure for slender columns

1. Calculate M_t.
2. Design column section using charts on *Data Sheets 50 to 54* and read off appropriate value of K.
3. Recalculate $M_t = M_i + K\beta Nh$.
4. Repeat steps 2 and 3 until adjustment becomes insignificant.

Since this cycling procedure always tends to decrease A_s, it may be safely terminated at any time, although clearly, if this is done, the resulting design may not be the most economic that it is possible to achieve.

Notation

l_b span of beam (or l_{b1} or l_{b2})
l_u length of upper column
l_l length of lower column
I_b moment of inertia of beam
I_u moment of inertia of upper column
I_l moment of inertia of lower column
FEM   fixed-end moment at end of beam monolithic with exterior column
F'_{b1b2} difference between fixed-end moments at adjacent ends of two beams l_{b1} and l_{b2} monolithic with interior column, if one beam carries maximum load and other beam carries minimum load

Stiffness coefficients

Exterior column:

$$\overline{K}_u = k\frac{I_u l_b}{I_b l_u} \;; \qquad \overline{K}_l = k\frac{I_l l_b}{I_b l_l} \;; \qquad \overline{K}_{b1} = k\frac{I_b l_b}{I_b l_b} = k$$

Interior column:

$$\overline{K}_u = k\frac{I_u l_{b1}}{I_{b1} l_u} \;; \quad \overline{K}_l = k\frac{I_l l_{b1}}{I_{b1} l_l} \;; \quad \overline{K}_{b1} = k \;; \quad \overline{K}_{b2} = k\frac{I_{b2} l_{b1}}{I_{b1} l_{b2}}$$

Coefficients for end conditions

End conditions		k
Beam	**Column**	
Fixed	Fixed	1.00
Hinged	Fixed	1.33
Fixed	Hinged	0.75
Hinged	Hinged	1.00

Moment of inertia (second moment of area) of gross concrete sections (neglecting reinforcement)

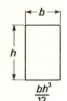
$$\frac{bh^3}{12}$$

$$\frac{h^4}{12}$$

$$0.055h^4$$

$$\frac{\pi}{64}h^4 = 0.049h^4$$

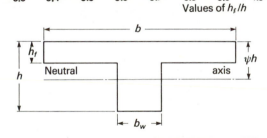

Moment of inertia of gross concrete section $= Kbh^3$

Depth to neutral axis $= \psi h$

(Graph: Values of K (vertical axis, 0 to 0.08) versus Values of h_f/h (horizontal axis, 0.1 to 1.0); curves labelled $b_w/b = 0.9, 0.8, 0.7, 0.6, 0.5, 0.4, 0.3, 0.2, 0.1, b_w/b = 0$; dashed curves labelled $\psi = 0.45, 0.40, 0.35, 0.30, 0.25, 0.20, 0.15, \psi = 0.10$)

Bending moments on columns

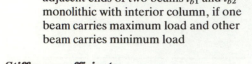

		Member		
Exterior column	$M_b = M_l$ l_b $M_b = M_u + M_l$ M_u M_l	**Top storey** ($M_u = 0$)	BM at head of column and end of beam $M_l = M_b$	$\dfrac{\overline{K}_l}{\overline{K}_l + \frac{1}{2}}FEM$
		Intermediate storeys	BM at head of lower column M_l	$\dfrac{\overline{K}_l}{\overline{K}_l + \overline{K}_u + \frac{1}{2}}FEM$
			BM at foot of upper column M_u	$\dfrac{\overline{K}_u}{\overline{K}_l + \overline{K}_u + \frac{1}{2}}FEM$
			BM at end of beam M_b	$\dfrac{\overline{K}_l + \overline{K}_u}{\overline{K}_l + \overline{K}_u + \frac{1}{2}}FEM$
Interior column	M_l M_u l_{b1} l_{b2} Loaded with $1.4G_k + 1.6Q_k$ Loaded with $1.0G_k$	**Top storey** ($M_u = 0$)	BM at head of column M_l	$\dfrac{\overline{K}_l}{\overline{K}_l + \frac{1}{2} + \frac{1}{2}\overline{K}_{b2}}F_{b1b2}{}'$
		Intermediate storeys	BM at head of lower column M_l	$\dfrac{\overline{K}_l}{\overline{K}_l + \overline{K}_u + \frac{1}{2} + \frac{1}{2}\overline{K}_{b2}}F_{b1b2}{}'$
			BM at foot of upper column M_u	$\dfrac{\overline{K}_u}{\overline{K}_l + \overline{K}_u + \frac{1}{2} + \frac{1}{2}\overline{K}_{b2}}F_{b1b2}{}'$

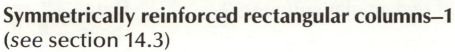

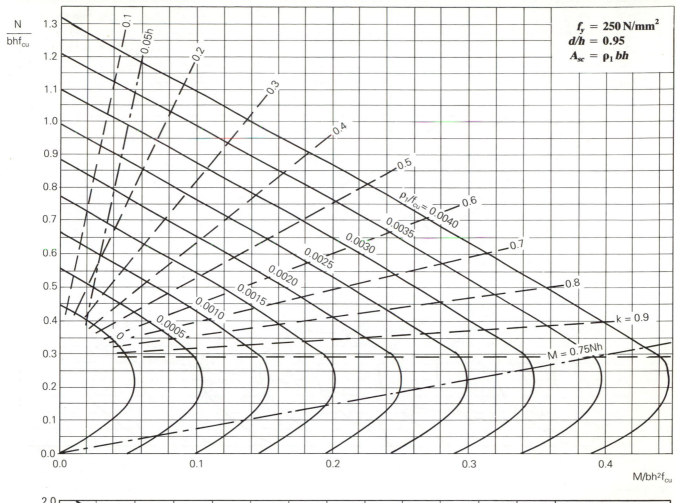

$$\frac{N}{bhf_{cu}}$$

$f_y = 250\,\text{N/mm}^2$
$d/h = 0.95$
$A_{sc} = \rho_1\,bh$

0.1
0.05h
0.2
0.3
0.4
0.5
$\rho_1/f_{cu} = 0.0040$
0.6
0.0035
0.0030
0.0025
0.7
0.0020
0.0015
0.8
0.0010
0.0005
k = 0.9
0
M = 0.75Nh

$$M/bh^2f_{cu}$$

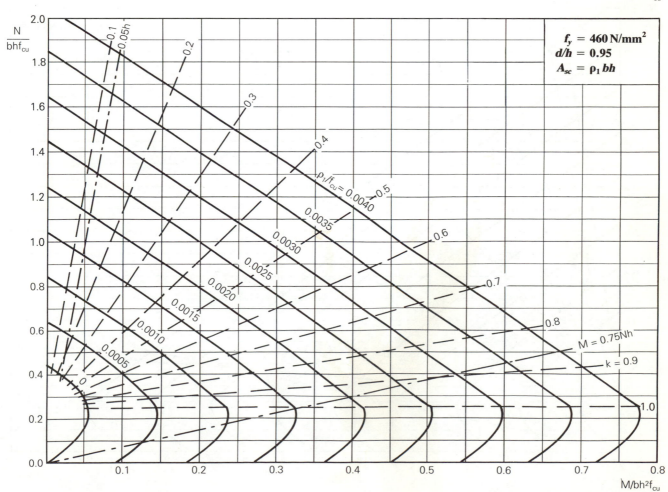

$$\frac{N}{bhf_{cu}}$$

$f_y = 460\,\text{N/mm}^2$
$d/h = 0.95$
$A_{sc} = \rho_1\,bh$

0.1
0.05h
0.2
0.3
0.4
$\rho_1/f_{cu} = 0.0040$
0.5
0.0035
0.0030
0.6
0.0025
0.0020
0.7
0.0015
0.0010
0.8
0.0005
M = 0.75Nh
0
k = 0.9
1.0

$$M/bh^2f_{cu}$$

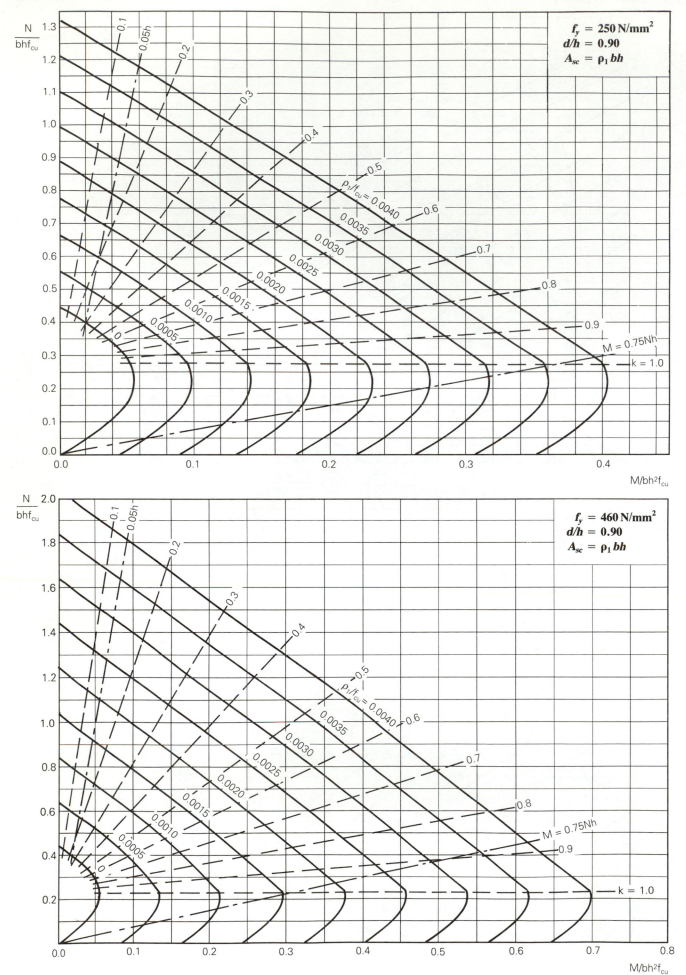

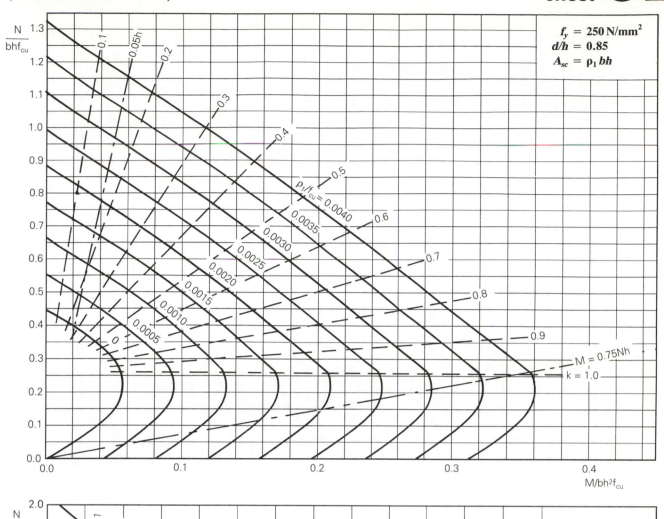

$f_y = 250 \text{ N/mm}^2$
$d/h = 0.85$
$A_{sc} = \rho_1 bh$

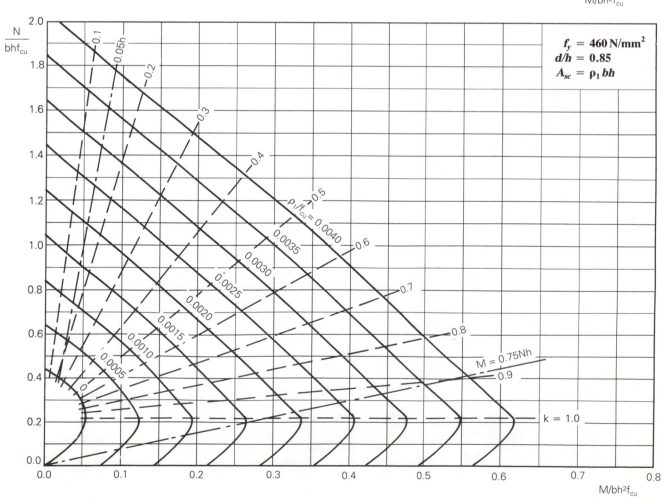

$f_y = 460 \text{ N/mm}^2$
$d/h = 0.85$
$A_{sc} = \rho_1 bh$

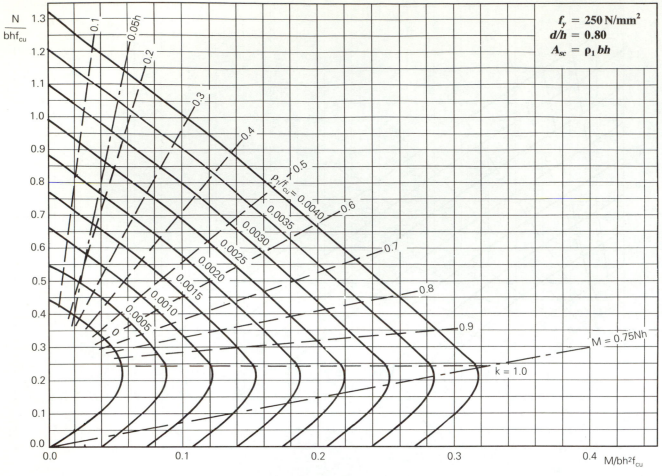

$f_y = 250 \, \text{N/mm}^2$
$d/h = 0.80$
$A_{sc} = \rho_1 \, bh$

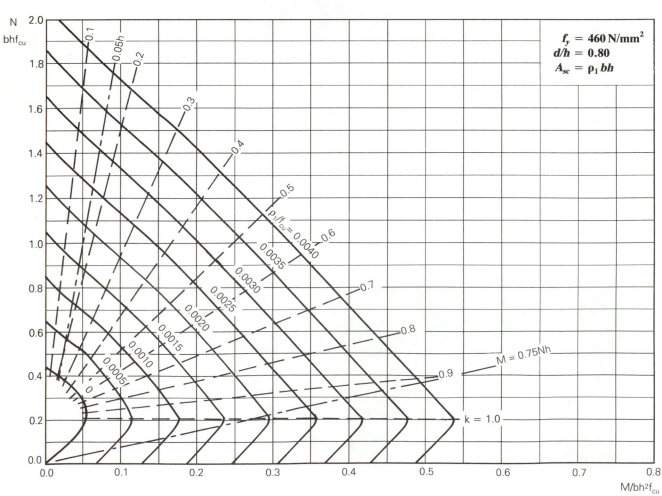

$f_y = 460 \, \text{N/mm}^2$
$d/h = 0.80$
$A_{sc} = \rho_1 \, bh$

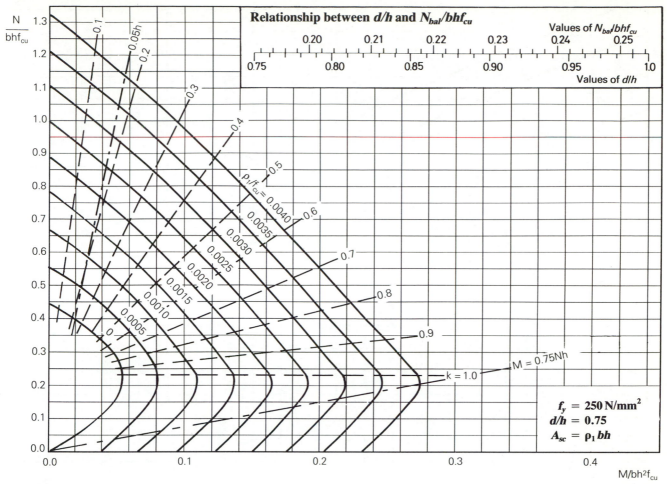

Relationship between d/h and N_{bal}/bhf_{cu}

Values of N_{bal}/bhf_{cu}

$\frac{N}{bhf_{cu}}$

$p_1/f_{cu} = 0.0040$

$M = 0.75Nh$

$k = 1.0$

$$f_y = 250 \text{ N/mm}^2$$
$$d/h = 0.75$$
$$A_{sc} = \rho_1 bh$$

M/bh^2f_{cu}

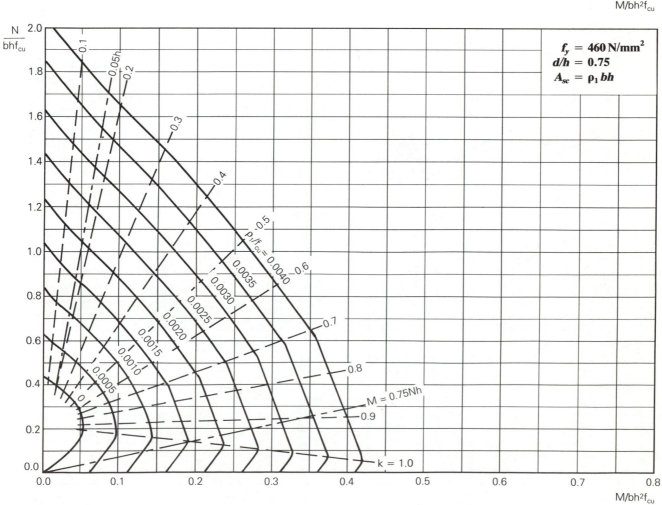

$$f_y = 460 \text{ N/mm}^2$$
$$d/h = 0.75$$
$$A_{sc} = \rho_1 bh$$

$\frac{N}{bhf_{cu}}$

$p_1/f_{cu} = 0.0040$

$M = 0.75Nh$

$k = 1.0$

M/bh^2f_{cu}

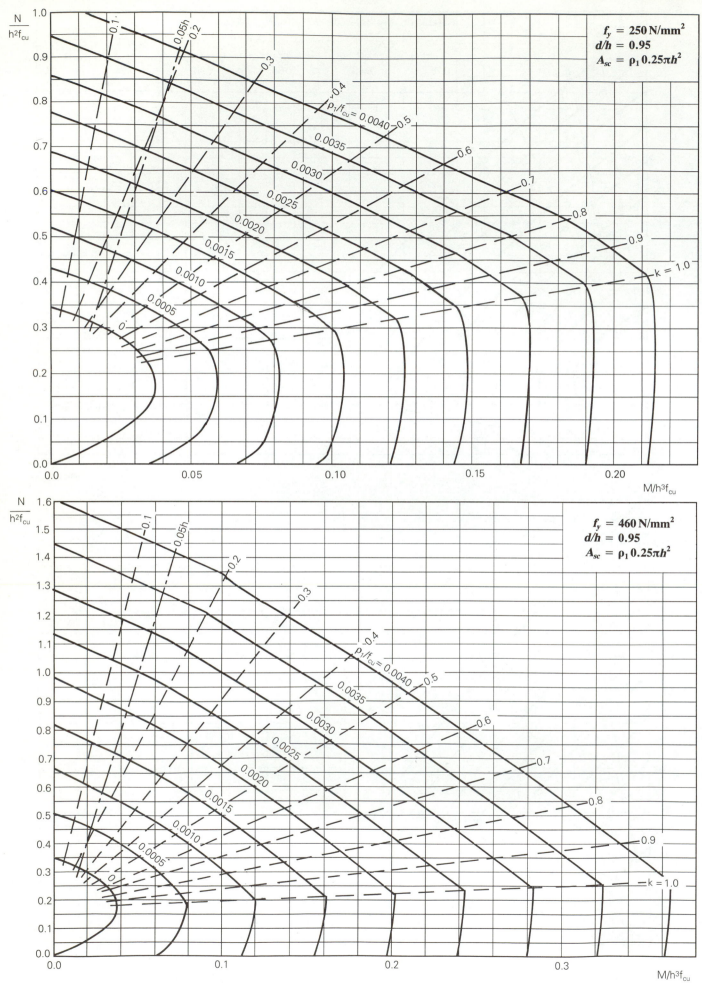

$f_y = 250\,\text{N/mm}^2$
$d/h = 0.95$
$A_{sc} = \rho_1\,0.25\pi h^2$

$f_y = 460\,\text{N/mm}^2$
$d/h = 0.95$
$A_{sc} = \rho_1\,0.25\pi h^2$

Circular columns–2 (*see section 14.5*)

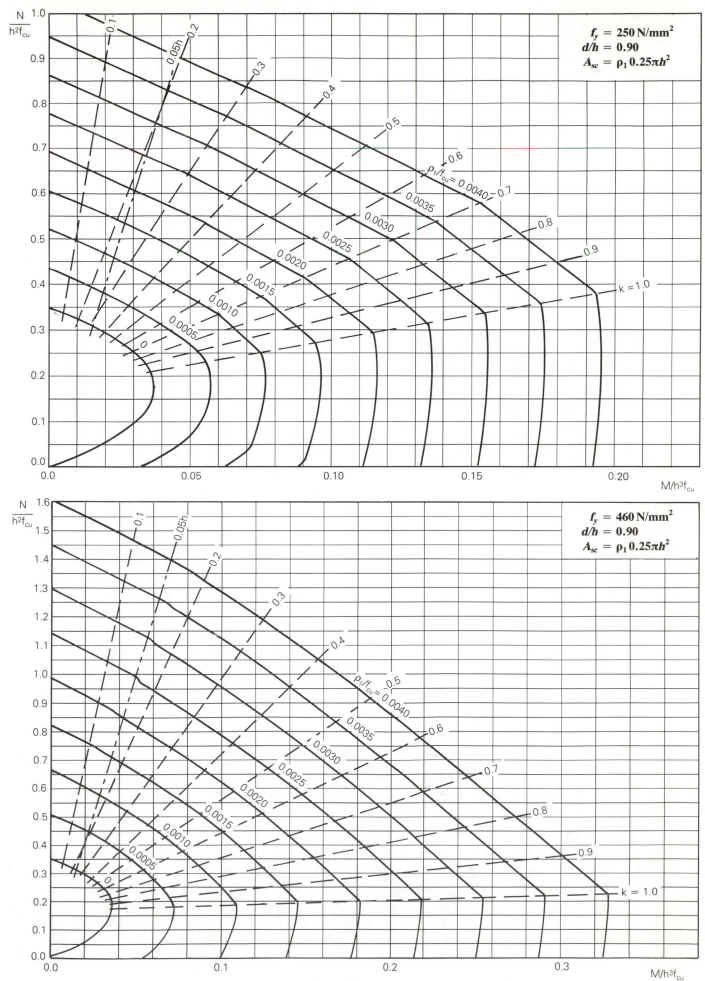

$f_y = 250 \, \text{N/mm}^2$
$d/h = 0.90$
$A_{sc} = \rho_1 \, 0.25\pi h^2$

$f_y = 460 \, \text{N/mm}^2$
$d/h = 0.90$
$A_{sc} = \rho_1 \, 0.25\pi h^2$

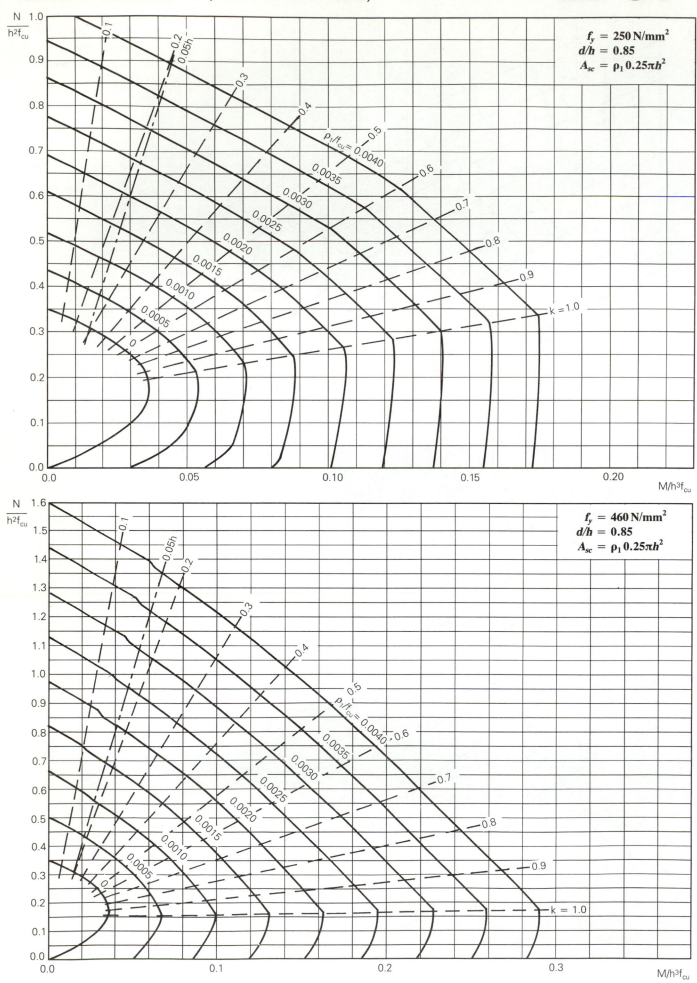

$f_y = 250\,\text{N/mm}^2$
$d/h = 0.85$
$A_{sc} = \rho_1\, 0.25\pi h^2$

$f_y = 460\,\text{N/mm}^2$
$d/h = 0.85$
$A_{sc} = \rho_1\, 0.25\pi h^2$

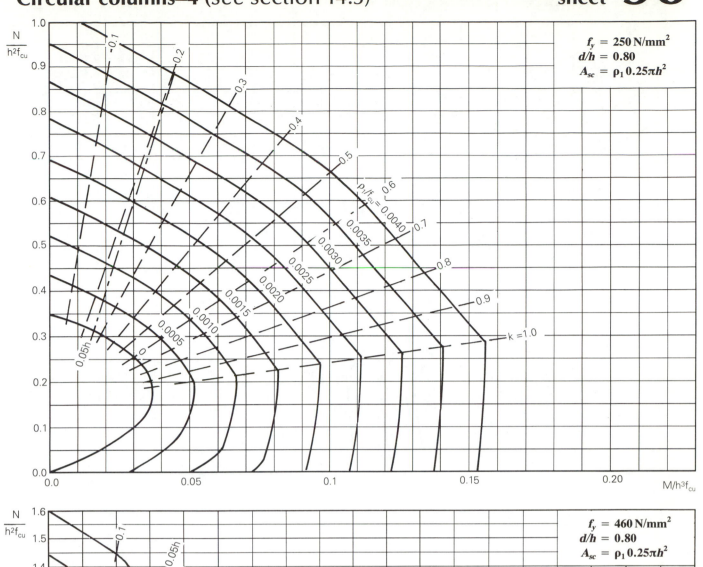

$f_y = 250 \, \text{N/mm}^2$
$d/h = 0.80$
$A_{sc} = \rho_1 \, 0.25 \pi h^2$

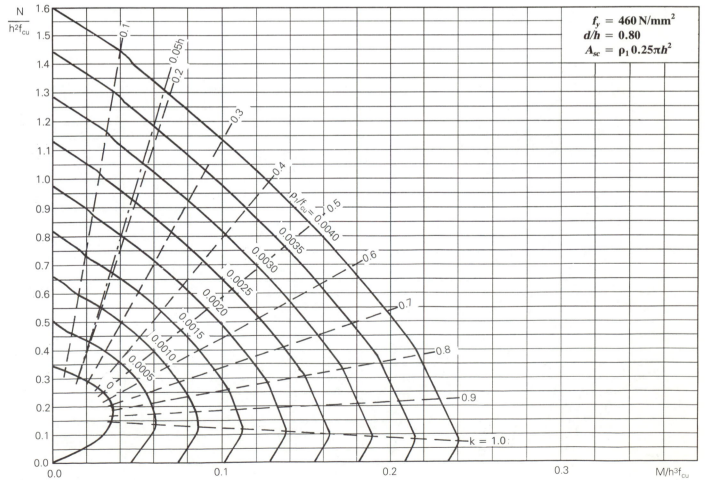

$f_y = 460 \, \text{N/mm}^2$
$d/h = 0.80$
$A_{sc} = \rho_1 \, 0.25 \pi h^2$

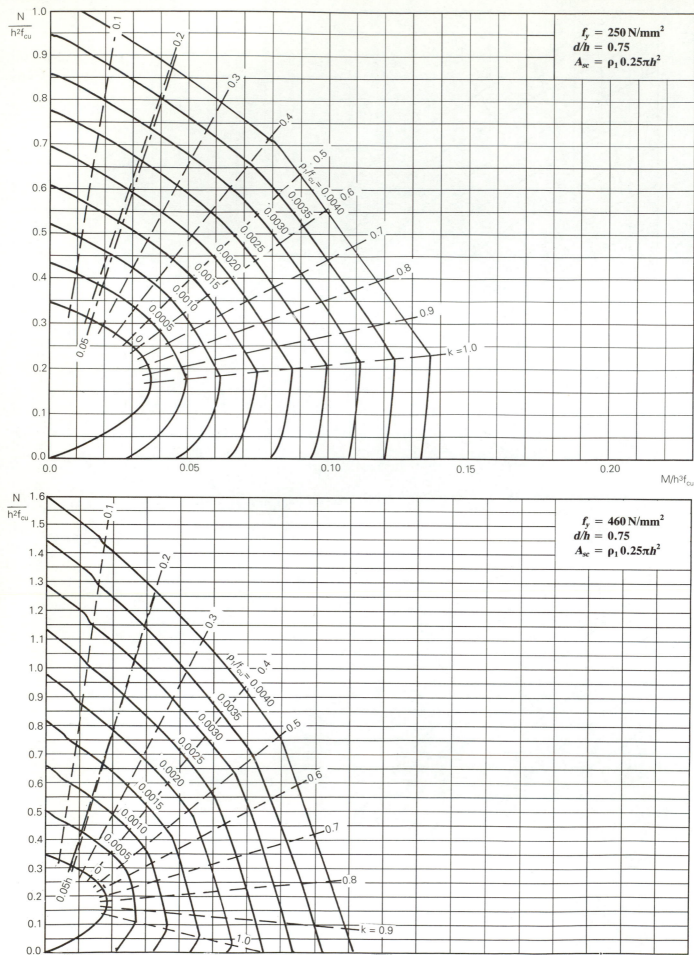

$f_y = 250 \, \text{N/mm}^2$
$d/h = 0.75$
$A_{sc} = \rho_1 \, 0.25 \pi h^2$

$f_y = 460 \, \text{N/mm}^2$
$d/h = 0.75$
$A_{sc} = \rho_1 \, 0.25 \pi h^2$

Biaxial bending data

Relationship between N/N$_{uz}$, M$_y$/M$_{uy}$ and M$_x$/M$_{ux}$

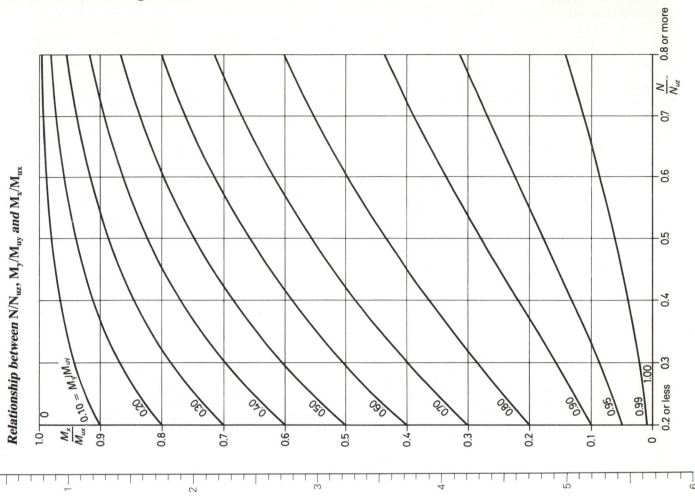

Percentage of reinforcement

Resistance of column to pure axial load

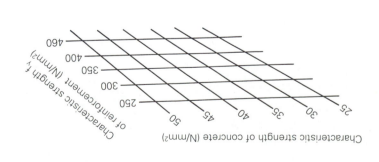

Resistance per unit area to pure axial load (N/mm²)

Load-bearing walls (*see section 15.5*)

Reinforcement in load-bearing walls

Specified proportion of steel	100	125	150	175	200	225	250	275	300
0.0025bh Near *each* face	250 mm²	313 mm²	375 mm²	438 mm²	500 mm²	563 mm²	625 mm²	688 mm²	750 mm²
	6 @ 225	6 @ 180	6 @ 150	6 @ 125	6 @ 110	6 @ 100	6 @ 90	6 @ 80	6 @ 75
	8 @ 300	8 @ 300	8 @ 250	8 @ 225	8 @ 200	8 @ 175	8 @ 160	8 @ 145	8 @ 130
	10 @ 300	10 @ 300	10 @ 300	10 @ 300	10 @ 300	10 @ 275	10 @ 250	10 @ 225	10 @ 200
	12 @ 300	12 @ 300	12 @ 300	12 @ 300	12 @ 300	12 @ 300	12 @ 300	12 @ 300	12 @ 300
0.003bh Near *each* face	300 mm²	375 mm²	450 mm²	525 mm²	600 mm²	675 mm²	750 mm²	825 mm²	900 mm²
	6 @ 185	6 @ 150	6 @ 125	6 @ 105	6 @ 90	6 @ 80	6 @ 75	—	
	8 @ 300	8 @ 250	8 @ 200	8 @ 190	8 @ 165	8 @ 145	8 @ 130	8 @ 120	8 @ 110
	10 @ 300	10 @ 300	10 @ 300	10 @ 275	10 @ 250	10 @ 225	10 @ 200	10 @ 190	10 @ 170
	12 @ 300	12 @ 300	12 @ 300	12 @ 300	12 @ 300	12 @ 300	12 @ 300	12 @ 250	12 @ 250
	16 @ 300	16 @ 300	16 @ 300	16 @ 300	16 @ 300	16 @ 300	16 @ 300	16 @ 300	16 @ 300
0.004bh Near *each* face	400 mm²	500 mm²	600 mm²	700 mm²	800 mm²	900 mm²	1 000 mm²	1 100 mm²	1 200 mm²
	6 @ 140	6 @ 110	6 @ 90	6 @ 80					
	8 @ 250	8 @ 200	8 @ 165	8 @ 140	8 @ 125	8 @ 110	8 @ 100	8 @ 90	8 @ 80
	10 @ 300	10 @ 300	10 @ 250	10 @ 200	10 @ 195	10 @ 170	10 @ 155	10 @ 140	10 @ 130
	12 @ 300	12 @ 300	12 @ 300	12 @ 300	12 @ 275	12 @ 250	12 @ 225	12 @ 200	12 @ 185
	16 @ 300	16 @ 300	16 @ 300	16 @ 300	16 @ 300	16 @ 300	16 @ 300	16 @ 300	16 @ 300
0.01bh In single layer	1 000 mm²	1 250 mm²	1 500 mm²	1 750 mm²	2 000 mm²	2 250 mm²	2 500 mm²	2 750 mm²	3 000 mm²
	10 @ 75	—	—						
	12 @ 110	12 @ 90	12 @ 75	—	—				
	16 @ 200	16 @ 160	16 @ 130	16 @ 110	16 @ 100	16 @ 85	16 @ 80	—	—
	20 @ 300	20 @ 250	20 @ 200	20 @ 175	20 @ 155	20 @ 135	20 @ 125	20 @ 110	20 @ 100
	25 @ 300	25 @ 300	25 @ 300	25 @ 275	25 @ 225	25 @ 200	25 @ 195	25 @ 175	25 @ 160
	32 @ 300	32 @ 300	32 @ 300	32 @ 300	32 @ 300	32 @ 300	32 @ 300	32 @ 275	32 @ 250
0.02bh In single layer	2 000 mm²	2 500 mm²	3 000 mm²	3 500 mm²	4 000 mm²	4 500 mm²	5 000 mm²	5 500 mm²	6 000 mm²
	16 @ 100	16 @ 80	—						
	20 @ 155	20 @ 125	20 @ 100	20 @ 85	20 @ 75	—	—		
	25 @ 225	25 @ 195	25 @ 160	25 @ 140	25 @ 120	25 @ 105	25 @ 95	25 @ 85	25 @ 80
	32 @ 300	32 @ 300	32 @ 250	32 @ 225	32 @ 200	32 @ 175	32 @ 160	32 @ 145	32 @ 130
0.04bh In single layer	4 000 mm²	5 000 mm²	6 000 mm²	7 000 mm²	8 000 mm²	9 000 mm²	10 000 mm²	11 000 mm²	12 000 mm²
	20 @ 75	—	—						
	25 @ 120	25 @ 95	25 @ 80	—	—	—	—	—	
	32 @ 200	32 @ 160	32 @ 130	32 @ 110	32 @ 100	32 @ 85	32 @ 80	—	—

Wall thickness (mm)

Area of reinforcement given is proportion of cross-sectional area *bh* of wall. For each combination of wall thickness and specified proportion of reinforcement, top line is total cross-sectional area of steel needed in mm² per metre length or height, taking account of bars necessary near both faces of wall. Remaining values are bar combinations meeting these requirements; for example, '6 @ 225' denotes that a suitable arrangement would be 6 mm bars at 225 mm centres. For steel percentages of less than 1%, the indicated bar arrangement must be provided near *each* face to meet the requirement specified in BS 8110. For greater amounts of steel the bar arrangement involves a single layer only.

Limiting bar spacings adopted in table are 75 mm (minimum) and 300 mm (maximum). For specified criteria in various documents, see section 15.4.

Applicability

0.0025bh	Minimum amount of horizontal high-yield steel in reinforced concrete wall. Minimum amount of anti-crack high-yield steel in plain concrete wall (in each direction).
0.003bh	Minimum amount of horizontal mild steel in reinforced concrete wall. Minimum amount of anti-crack mild steel in plain concrete wall (in each direction).
0.004bh	Minimum amount of vertical reinforcement in reinforced concrete wall. For greater amount of vertical reinforcement, fire-resistance criteria change.
0.01bh	For greater amount of vertical reinforcement, fire-resistance criteria change.
0.02bh	For greater areas of vertical reinforcement, if bars are provided near both faces, links must be provided through the wall to tie such vertical bars.
0.04bh	Maximum area of vertical reinforcement permitted in a reinforced concrete wall.

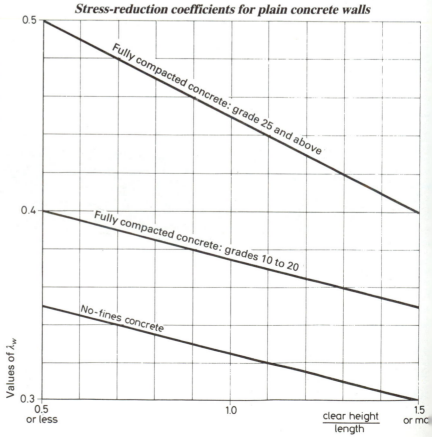

Stress-reduction coefficients for plain concrete walls

Fully compacted concrete: grade 25 and above

Fully compacted concrete: grades 10 to 20

No-fines concrete

Values of λ_w

$\dfrac{\text{clear height}}{\text{length}}$

0.5 or less — 1.0 — 1.5 or more

Fire resistance (*see* Chapter 19)

Floor construction

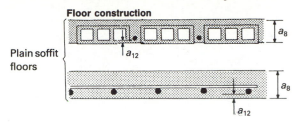

Plain soffit floors

Ribbed soffit floor

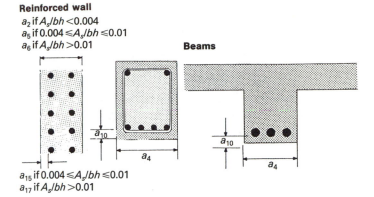

Reinforced wall

a_2 if $A_s/bh < 0.004$
a_5 if $0.004 \leq A_s/bh \leq 0.01$
a_6 if $A_s/bh > 0.01$

Beams

Columns

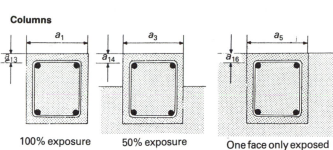

100% exposure   50% exposure   One face only exposed

a_{15} if $0.004 \leq A_s/bh \leq 0.01$
a_{17} if $A_s/bh > 0.01$

Minimum dimensions (mm)

Type of concrete	Normal-weight aggregate						Lightweight aggregate						Notes
Fire period (hours)	0.5	1	1.5	2	3	4	0.5	1	1.5	2	3	4	
a_1	150	200	250	300	400	450	150	160	200	240	320	360	
a_2	150	150	175	—	—	—	—	—	—	—	—	—	
a_3	125	160	200	200	300	350	125	130	160	185	250	275	
a_{4F}	80	120	150	200	240	280	80	100	130	160	200	250	
a_{4C}	80	80	120	150	200	240	60	80	90	110	150	200	
a_5	100	120	140	160	200	240	100	100	115	130	160	190	3
a_6	75	75	100	100	150	180	—	—	—	—	—	—	
a_{7F}	75	90	110	125	150	175	60	75	85	100	125	150	
a_{7C}	75	80	90	110	125	150	70	75	80	90	100	125	
a_{8FC}	75	95	110	125	150	170	70	90	105	115	135	150	
a_{9FC}	70	90	105	115	135	150	70	85	95	100	115	130	
a_{10F}	20	30	40	50	70	80	15	20	35	45	55	65	1, 2
a_{10C}	20	20	35	50	60	70	15	20	25	35	45	55	1, 2
a_{11F}	15	25	35	45	55	65	15	25	30	35	45	55	1, 2
a_{11C}	15	20	25	35	45	55	15	20	25	30	35	45	1, 2
a_{12F}	15	20	25	35	45	55	15	15	20	25	35	45	1, 2
a_{12C}	15	20	20	25	35	45	15	15	20	20	25	35	1
a_{13}	20	25	30	35	35	35	20	20	25	35	35	35	1
a_{14}	20	25	25	25	30	35	20	20	25	25	30	30	1
a_{15}	25	25	25	25	25	25	10	20	20	25	25	25	1
a_{16}	20	25	25	25	25	25	10	20	20	25	25	25	1
a_{17}	15	15	25	25	25	25	—	—	—	—	—	—	1

Subscripts F and C denote values for freely supported and continuous spans respectively.

Notes

1. 'Average cover' dimensions illustrated are nominal values to main reinforcement, evaluated by multiplying area of each bar by its distance from the nearest exposed face and then dividing the resulting summation by the total steel area. Compliance with 'minimum cover' requirements related to durability and concrete grade etc. must also be observed.
2. Where actual cover to outermost steel exceeds 40 mm for normal-weight concrete or 50 mm for lightweight concrete, spalling may occur and special measures must be taken to prevent this in those cases where the member would be endangered if this should happen.
3. Lightweight values are for a concrete density of 1.2 tonnes per cubic metre. For concretes having densities of between this value and that of normal-weight concrete (i.e. 2.4 t/m^3) the required minimum thickness may be interpolated.
4. Additional protection can be achieved by applying a finish of mortar, gypsum, plaster, vermiculite etc. (see BS8110, Part 2, clause 4.2.4).

5. Values of average cover tabulated relate specifically to members having the minimum widths set out above. For wider members the average cover dimensions may be decreased as follows:

Minimum increase in width (mm)	Decrease in cover (mm)	
	Normal-weight concrete	Lightweight concrete
25	5	5
50	10	10
100	15	15
150	15	20

Metric/imperial equivalents for common units

Basic conversion factors

The following equivalents of SI units are given in imperial and, where applicable, metric technical units.

1 mm = 0.039 37 in	1 in = 25.4 mm	1 m^2 = 1.196 yd^2	1 yd^2 = 0.8361 m^2
1 m = 3.281 ft	1 ft = 0.3048 m	1 hectare = 2.471 acres	1 acre = 0.4047 hectares
= 1.094 yd	1 yd = 0.9144 m	1 mm^3 = 0.000 061 02 in^3	1 in^3 = 16 390 mm^3
1 km = 0.6214 mile	1 mile = 1.609 km	1 m^3 = 35.31 ft^3	1 ft^3 = 0.028 32 m^3
1 mm^2 = 0.001 55 in^2	1 in^2 = 645.2 mm^2	= 1.308 yd^3	1 yd^3 = 0.7646 m^3
1 m^2 = 10.76 ft^2	1 ft^2 = 0.0929 m^2	1 mm^4 (M of I) = 0.000 002 403 in^4	1 in^4 = 416 200 mm^4

Force

1 N = 0.2248 lbf = 0.1020 kgf	1 kN = 0.1004 tonf = 102.0 kgf = 0.1020 tonne f
4.448 N = 1 lbf = 0.4536 kgf	9.964 kN = 1 tonf = 1016 kgf = 1.016 tonne f
9.807 N = 2.205 lbf = 1 kgf	9.807 kN = 0.9842 tonf = 1000 kgf = 1 tonne f

Force per unit length

1 N/m = 0.068 52 lbf/ft = 0.1020 kgf/m	1 kN/m = 0.0306 tonf/ft = 0.1020 tonne f/m
14.59 N/m = 1 lbf/ft = 1.488 kgf/m	32.69 kN/m = 1 tonf/ft = 3.333 tonne f/m
9.807 N/m = 0.672 lbf/ft = 1 kgf/m	9.807 kN/m = 0.3000 tonf/ft = 1 tonnef/m

Force per unit area

1 N/mm^2 = 145.0 lbf/in^2 = 10.20 kgf/cm^2	1 N/mm^2 = 0.064 75 tonf/in^2 = 10.20 kgf/cm^2
0.006 895 N/mm^2 = 1 lbf/in^2 = 0.0703 kgf/cm^2	15.44 N/mm^2 = 1 tonf/in^2 = 157.5 kgf/cm^2
0.098 07 N/mm^2 = 14.22 lbf/in^2 = 1 kgf/cm^2	0.098 07 N/mm^2 = 0.006 350 tonf/in^2 = 1 kgf/cm^2
1 N/m^2 = 0.020 89 lbf/ft^2 = 0.102 kgf/m^2	1 N/mm^2 = 9.324 tonf/ft^2 = 10.20 kgf/cm^2
47.88 N/m^2 = 1 lbf/ft^2 = 4.882 kgf/m^2	0.1073 N/mm^2 = 1 tonf/ft^2 = 1.094 kgf/cm^2
9.807 N/m^2 = 0.2048 lbf/ft^2 = 1 kgf/m^2	0.098 07 N/mm^2 = 0.9144 tonf/ft^2 = 1 kgf/cm^2

Force per unit volume

1 N/m^3 = 0.006 366 lbf/ft^3 = 0.102 kgf/m^3	1 kN/m^3 = 0.002 842 tonf/ft^3 = 0.1020 tonne f/m^3
157.1 N/m^3 = 1 lbf/ft^3 = 16.02 kgf/m^3	351.9 kN m^3 = 1 ton/ft^3 = 35.88 tonne f/m^3
9.807 N/m^3 = 0.0624 lbf/ft^3 = 1 kgf/m^3	9.807 kN/m^3 = 0.027 87 tonf/ft^3 = 1 tonne f/m^3
1 kN/m^3 = 0.003 684 lbf/in^3 = 0.1020 tonne f/m^3	
271.4 kN/m^3 = 1 lbf/in^3 = 27.68 tonne f/m^3	
9.807 kN/m^3 = 0.036 13 lbf/in^3 = 1 tonne f/m^3	

Moment

1 N m	= 8.851 lbf in	= 0.7376 lbf ft	= 0.1020 kgf m
0.1130 N m	= 1 lbf in	= 0.083 33 lbf ft	= 0.011 52 kgf m
1.356 N m	= 12 lbf in	= 1 lbf ft	= 0.1383 kgf m
9.807 N m	= 86.80 lbf in	= 7.233 lbf ft	= 1 kgf m

Fluid capacity

1 litre	= 0.22 imperial gallons	= 0.2642 US gallons
4.546 litres	= 1 imperial gallon	= 1.201 US gallons
3.785 litres	= 0.8327 imperial gallons	= 1 US gallon

Useful data

1000 kg/m^3 = 62.4 lb/ft^3 (density of water)

23.6 kN/m^3 = 2400 kg/m^3 = 150 lb/ft^3 (nominal weight of reinforced concrete)

14 kN/mm^2 (approx.) = 140×10^3 kg/cm^2 (approx.) = 2×10^6 lb/in^2 (nominal elastic modulus of concrete)

10×10^{-6} per °C = 5.5×10^{-6} per °F (nominal coefficient of linear expansion of concrete)

References

1. Manning, G. P. (1966) *Reinforced Concrete Design*. London, Longmans, Green, third edition, p. 413
2. Institution of Structural Engineers and Concrete Society (1989) *Standard Method of Detailing Structural Concrete*. London, Institution of Structural Engineers, p. 138
3. Institution of Structural Engineers (1991) *Recommendations for the Permissible-Stress Design of Reinforced Concrete Building Structures*. London, Institution of Structural Engineers, p. 127
4. Somerville, G. (1989) The writing of Eurocode 2. *The Structural Engineer*, 67(13), 216–18
5. Bate, S. C. C. (1968) Why limit state design? *Concrete* 2(3), March, pp. 103–8
6. Holdaway, A. E. (1958) A precise method of moment distribution. *Concrete and Constructional Engineering* 53(2), February, pp. 73–8
7. Steedman, J. C. (1962) Charts for the determination of moment-distribution factors. *Concrete and Constructional Engineering* 57(9), September, pp. 348–53, and 57(10), October, pp. 395–6
8. Rygol, J. (1968) *Structural Analysis by Direct Moment Distribution*. London, Crosby Lockwood, p. 407
9. Allen, A. H. (1973) *Safe-Load Tables for Solid Slabs*. Wexham Springs, Cement and Concrete Association, p. 43
10. Allen, A. H. (1987) *Design Data for Rectangular Beams and Slabs to BS8110:1985*, E & FN Spon, London
11. Steffens, R. J. (1966) Some aspects of structural vibration. *Proceedings of a Symposium on Vibration in Civil Engineering*. London, Imperial College of Science and Technology, April 1965, Butterworth, pp. 1–30, session 1. See also Building Research Station Current Paper (Engineering Series) 37, p. 30
12. Beeby, A. W. (1971) *Modified Proposals for Controlling Deflections by means of Ratios of Span to Effective Depth*. Wexham Springs, Cement and Concrete Association, publication 42.456, p. 19
13. Comité Européen du Béton–Fédération Internationale Précontrainte (1970) *International Recommendations for the Design and Construction of Concrete Structures*. English edition, Wexham Springs, Cement and Concrete Association, principles and recommendations p. 80, appendices p. 47
14. Newmark, M. N. (1942) Numerical procedure for computing deflections, moments and buckling loads. *Proceedings of the American Society of Civil Engineers*. May
15. Johansen, K. W. (1962) *Yield-Line Theory*. English edition, Wexham Springs, Cement and Concrete Association, p. 181. *Yield-line Formulae for Slabs*. English edition, London, E & FN Spon, 1972, p. 106
16. Hillerborg, A. (1974) *Strip Method of Design*. English edition, London, E & FN Spon, p. 256
17. Concrete Society (1986) *Concrete Detail Design*. London, Architectural Press
18. Whittle, R. T. (1985) *Design of Reinforced Concrete Flat Slabs to BS8110*. CIRIA report 110, London, CIRIA, p. 49
19. Hulse, R. and Mosley, W. H. (1986) *Reinforced Concrete Design by Computer*. London, Macmillan, p. 288
20. Pinfold, G. M. (1984) *Reinforced Concrete Chimneys and Towers*. London, E & FN Spon, second edition, p. 186
21. Cranston, W. B. (1972) *Analysis and Design of Reinforced Concrete Columns*. Wexham Springs, Cement and Concrete Association, publication 41.020, p. 28
22. Terzaghi, K. (1934) Large retaining wall tests, Parts 1–5. *Engineering Record*, 112, 136–40, 259–62, 316–8, 403–6, 503–8. (Also see Lee, I. K., White and Ingles, O. G. (1983) *Geotechnical Engineering*, Pitman, p. 252)
23. CIRIA (1978) *Guide to the Design of Waterproof Basements*. CIRIA Guide 5. London, CIRIA, p. 38
24. Ove Arup & Partners. (1987) 'The design of deep beams in reinforced concrete'. CIRIA Guide 2. London, CIRIA. January
25. Cusens, A. R., Wang, Y. B. and Wong, K. K. 'Bearing capacity of plain and reinforced concrete blocks'. *The Structural Engineer*. 67(23), p. 407–14
26. Cusens, A. R. and Santathadaporn, Sakda (1966) *Design Charts for Helical Stairs with Fixed Supports*. London, Concrete Publications, p. 35
27. Cusens, A. R. and Kuang, Jing-Gwo (1965) A simplified method of analysing free-standing stairs. *Concrete and Constructional Engineering* 60(5), May, pp. 167–72 and 194
28. Cusens, A. R. (1966) Analysis of slabless stairs. *Concrete and Constructional Engineering* 61(10), October, pp. 359–64
29. Tomlinson, M. J. (1987) *Pile Design and Construction Practice*. London, E & FN Spon, 3rd edition, p. 310
30. Concrete Society (1981) Model Procedures for the Presentation of Calculations, 2nd edn, The Concrete Society, Wexham, Slough, pp. 20
31. Higgins, J. B. and Rogers, B. R. (1986) *Designed and Detailed (BS8110:1985)*, Cement and Concrete Association. publication 43.501, Wexham Springs, pp. 28
32. McKelvey, K. K. (1974) *Drawing for the Structural Concrete Engineer*, E & FN Spon, London, pp. 266

Index